D0170284

WHY WE LOST

WHY WE LOST

A GENERAL'S INSIDE ACCOUNT OF THE IRAQ AND AFGHANISTAN WARS

DANIEL BOLGER

AN EAMON DOLAN BOOK
MARINER BOOKS
HOUGHTON MIFFLIN HARCOURT
BOSTON NEW YORK

First Mariner Books edition 2015
Copyright © 2014 by Daniel Bolger

All rights reserved

For information about permission to reproduce selections from this book, write to trade.permissions@hmhco.com or to Permissions, Houghton Mifflin Harcourt Publishing Company, 3 Park Avenue, 19th Floor, New York, New York 10016.

www.hmhco.com

Library of Congress Cataloging-in-Publication Data

Bolger, Daniel P., date.

Why we lost : a general's inside account of the Iraq and Afghanistan Wars / Daniel Bolger.

pages cm

"An Eamon Dolan Book."

ISBN 978-0-544-37048-7 ISBN 978-0-544-57041-2 (pbk.)

1. Iraq War, 2003–2011 — Campaigns. 2. Afghan War, 2001– — Campaigns. 3. Iraq War, 2003–2011 — Personal narratives, American. 4. Afghan War, 2001– — Personal narratives, American 5. War on Terrorism, 2001–2009 — Personal narratives, American. 6. Civil-military relations — United States — History — 21st century. 7. Strategic culture — United States. 8. Leadership — United States. I. Title. II. Title: General's inside account of the Iraq and Afghanistan Wars.

DS79.76.B653 2014

956.7044'3 — dc23 2014026908

Book design by Brian Moore

Maps by Don Larson / Mapping Specialists

Printed in the United States of America

DOC 10 9 8 7 6 5 4 3 2 1

The views expressed herein are those of the author and do not reflect the official positions of the Department of the Army, the Department of Defense, and the United States government.

For Joy, who knew
For Philip, who fought
For Curolyn, who learned
For our honored dead, lest we forget

CONTENTS

PART III: NEMESIS: THE AFGHAN CAMPAIGN, APRIL 2003 TO DECEMBER 2014

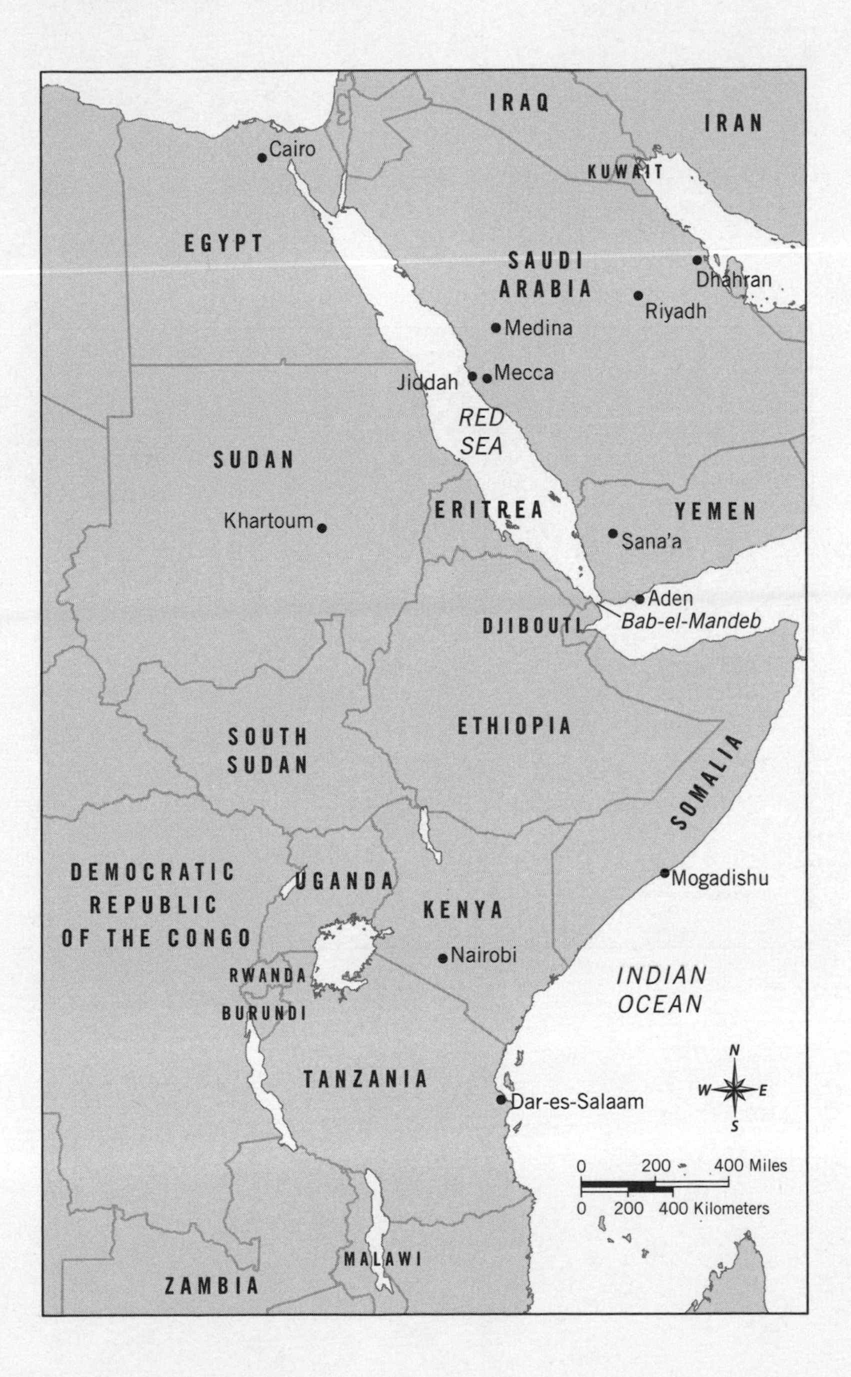
IRAQ
IRAN
Cairo
KUWAIT
EGYPT
SAUDI ARABIA
Dhahran
Riyadh
Medina
Jiddah
Mecca
RED SEA
SUDAN
ERITREA
YEMEN
Khartoum
Sana'a
Aden
Bab-el-Mandeb
DJIBOUTI
ETHIOPIA
SOUTH SUDAN
SOMALIA
DEMOCRATIC REPUBLIC OF THE CONGO
UGANDA
KENYA
Mogadishu
Nairobi
RWANDA
BURUNDI
INDIAN OCEAN
TANZANIA
Dar-es-Salaam
N
W
E
S
0 200 400 Miles
0 200 400 Kilometers
MALAWI
ZAMBIA

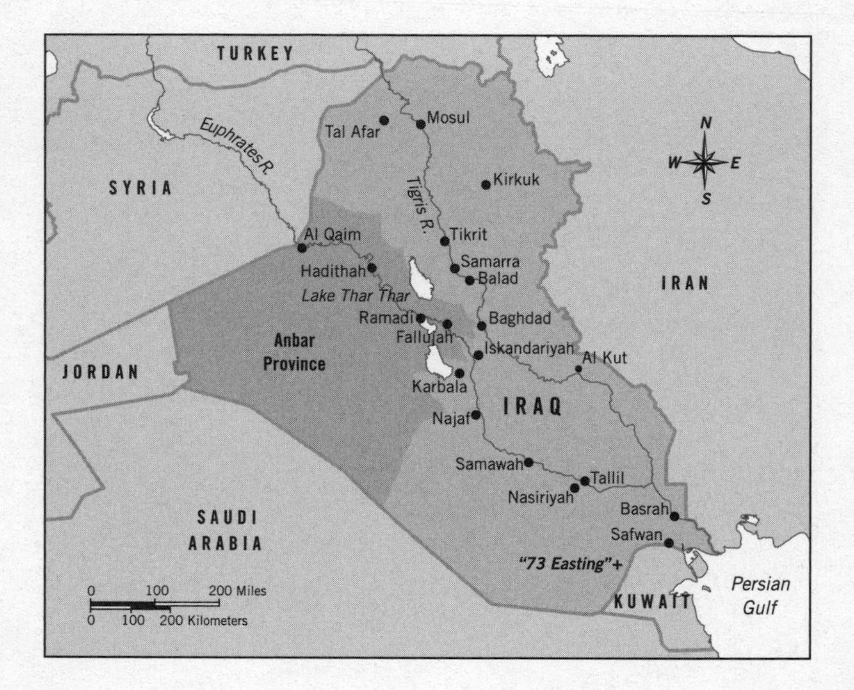
TURKEY
SYRIA
IRAN
JORDAN
SAUDI ARABIA
KUWAIT
IRAQ
Euphrates R.
Tigris R.
Tal Afar
Mosul
Kirkuk
Al Qaim
Tikrit
Samarra
Balad
Hadithah
Lake Thar Thar
Ramadi
Fallujah
Baghdad
Anbar Province
Iskandariyah
Al Kut
Karbala
Najaf
Samawah
Tallil
Nasiriyah
Basrah
Safwan
"73 Easting"+
Persian Gulf
N
W
E
S
0 100 200 Miles
0 100 200 Kilometers

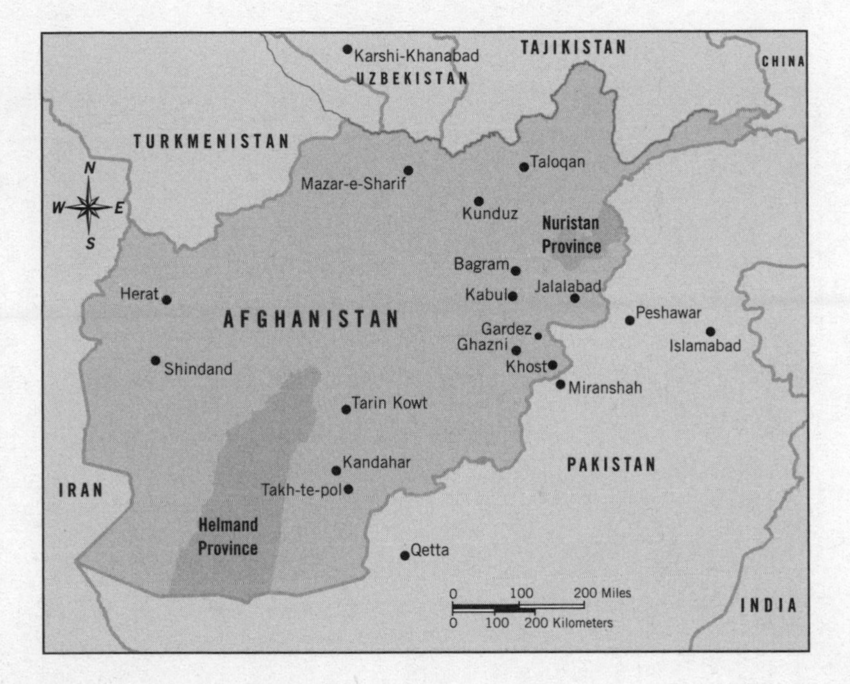
Karshi-Khanabad
UZBEKISTAN
TAJIKISTAN
CHINA
TURKMENISTAN
N
W
E
S
Mazar-e-Sharif
Taloqan
Kunduz
Nuristan Province
Bagram
Kabul
Jalalabad
Herat
AFGHANISTAN
Peshawar
Islamabad
Gardez
Ghazni
Khost
Miranshah
Shindand
Tarin Kowt
Kandahar
Takh-te-pol
PAKISTAN
IRAN
Helmand Province
Qetta
0
100
200 Miles
0
100
200 Kilometers
INDIA

AUTHOR'S NOTE

I am a United States Army general, and I lost the Global War on Terrorism. It's like Alcoholics Anonymous; step one is admitting you have a problem. Well, I have a problem. So do my peers. And thanks to our problem, now all of America has a problem, to wit: two lost campaigns and a war gone awry.

We should have known this one was going to go bad when we couldn't even settle on a name. In the wake of the horrific al-Qaeda attacks on September 11, 2001, we tried out various labels. The guys in the Pentagon basement at first offered Operation Infinite Justice, which sounded fine, both almighty and righteous. Then various handwringers noted that it might upset the Muslims. These were presumably different kinds of followers of Islam than the nineteen zealots who had just slaughtered thousands of our fellow citizens. Well, better incoherent than insensitive, I guess.

So we settled on Operation Enduring Freedom (OEF). Our efforts in Afghanistan certainly lived up to the "enduring" part, dragging out longer than the ten-year Trojan War as we desperately tried to impose "freedom" on surly Pashtuns. The big guys in Washington then riffed off that OEF theme. We embarked on OEF-CCA (Caribbean and Central America), OEF-HOA (Horn of Africa, run out of Djibouti), OEF-K (Kyrgyzstan), OEF-P (the Philippines), OEF-PG (Pankisi Gorge, in the Caucasus republic of Georgia), and OEF-TS (Trans-Sahara, in northern Africa), among others not publicly acknowledged. We even found time, and nomenclature, for loosely related campaigns. One was the 2011 imbroglio in Libya known at the outset as Operation Odyssey Dawn, a good name for a Las Vegas pole dancer but a bit exotic for a military campaign. Cooler heads at NATO headquarters quickly substituted the boring but less provocative Unified Protector. We are a long way from Korean War operations like Killer and Ripper. You didn't have to guess what those names meant.

Still, that Enduring Freedom idea reflected the preferred brand. Few could have been much surprised when, in 2003, the next major campaign in the ill-named war drew the title Operation Iraqi Freedom. As in World War II, the Iraq intervention was seen, rightly, as yet another theater in what the military formally called the Global War on Terrorism (GWOT). Like many veterans, I earned campaign ribbons with that designation.

Other names were tried for the parts or the whole of the conflict. These included the Long War (true enough), the Afghan War, the Iraq War, the 9/11 War, the War on Terrorism (minus the *Global*), the War on Terror (minus the *-ism*), and — a real bureaucratic gem — the Overseas Contingency Operation. That inane euphemism arose as a result of a related phrase, the "man-caused disaster" on 9/11. Not one title for this war identified the enemy: anti-Western Islamists and the ramshackle, quasi-fascist Middle East states that enabled them.

So instead we waged a Global War on Terrorism against enemies referred to vaguely as terrorists, cowards, evildoers, and extremists. Although those descriptions were rather generic, somehow we always ended up going after the same old bunch of Islamists and their ilk. Our opponents had no illusions about who our targets were, even if some of us did. This GWOT sputtered along for years, with me in it, along with many others much more capable, brave, and distinguished.

I was never the overall commander in either Afghanistan or Iraq. You'd find me lower down on the food chain, but high enough. I commanded a one-star advisory team in Iraq in 2005–06, an Army division (about twenty thousand soldiers) in Baghdad in 2009–10, and a three-star advisory organization in Afghanistan in 2011–13. I was present when key decisions were made, delayed, or avoided. I made, delayed, or avoided a few myself. I got out on the ground a lot with small units as we patrolled and raided. Sometimes, I communed with the strategic-headquarters types in the morning and at sunset grubbed through a village with a rifle platoon. Now and then, Iraqi and Afghan insurgents tried to kill me. By the enemy's hand, abetted by my ignorance, my arrogance, and the inexorable fortunes of war, I lost eighty men and women under my charge; more than three times that number were wounded. Those sad losses are, to borrow the words of Robert E. Lee on that awful third day at Gettysburg, all my fault.

This history does not purport to tell the whole story of the war. At best, it's a start. A century hence, if our society still exists and still cares, much better chroniclers than I will still be laboring to render a complete account. All I can offer at this point is a study in what went right and what went wrong, colored — perhaps too much — by experience.

What went right involved the men and women who fought. Most of them were not Americans, as these wars among the people always feature a lot of locals helping the cause. Like our enemies, our regional comrades were almost all Muslims. In addition, both the Afghan and Iraq campaigns included partner countries, from old allies like Australia, Britain, and Canada to newer teammates like El Salvador, Korea, and Mongolia. All of that help meant a lot, and although we didn't always appreciate it enough at the time, we can never forget those others and their sacrifices. But in the end, the outcome rode on the United States.

This look at the war focuses on us, the Americans. We didn't start it, but once it began, we drove the pace and course of the conflict. At the tactical level — Army-speak for the realm of vicious firefights and night raids — the courage, discipline, and lethality of our Americans in uniform stand with anything accomplished in the Civil War, both world wars, Korea, or Vietnam. That all went very right.

What went wrong squandered the bravery, sweat, and blood of these fine Americans. Our primary failing in the war involved generalship. If you prefer the war-college lexicon, we — guys like me — demonstrated poor strategic and operational leadership. For soldiers, *strategy* and *operational art* translate to "the big picture" (your goal) and "the plan" (how you get there). We got both wrong, the latter more than the former. Some might blame the elected and appointed civilian leaders. There's enough fault to go around, and in this telling, the suits will get their share. But I know better, and so do the rest of the generals. We have been trained and educated all our lives on how to fight and win. This was our war to lose, and we did.

We should have known better. In the military schools, like West Point, Fort Leavenworth, Quantico, and Carlisle Barracks, soldiers study the work of the great thinkers who have wrestled with winning wars across the ages. Along with Thucydides, Julius Caesar, and Carl von Clausewitz, the instructors introduce the ancient wisdom of Sun Tzu, the Chinese general and theorist who penned his poetic, elliptical, sometimes cryptic *Art of War* some twenty-three centuries ago. Master Sun put it simply: "Know the enemy and know yourself; in a hundred battles you will never be in peril." We failed on both counts. I know I sure did. As generals, we did not know our enemy — never pinned him down, never focused our efforts, and got all too good at making new opponents before we'd handled the old ones.

We then added to our troubles by misusing the U.S. Armed Forces, which are designed, manned, and equipped for short, decisive, conventional conflict. Instead, certain of our tremendously able, disciplined troops, buoyed by dazzling early victories, we backed into not one but two long, indecisive counterinsurgent struggles ill suited to the nature of our forces. Time after time, despite the fact that I and my fellow generals saw it wasn't working, we failed to reconsider our basic assumptions. We failed to question our flawed understanding of our foe or ourselves. We simply asked for more time. Given enough months, then years, then decades — always just a few more, please — we trusted that our great men and women would pull it out. In the end, all the courage and skill in the world could not overcome ignorance and arrogance. As a general, I got it wrong. And I did so in the company of my peers.

Just as the soldiers, sailors, airmen, Marines, and coastguardsmen of this conflict very much resembled Johnny Reb, Billy Yank, the frontier regulars, the doughboys, GI Joe, and the grunts of Southeast Asia, our generals also often ran true to types and archetypes. The American character has not changed all that much in two centuries and a few decades, and so we see more than a few echoes of our military heritage. Certainly in David Petraeus there is something of the innovative yet overly ambitious Douglas MacArthur. Tough Marine Jim Mattis filled the role of a latter-day George Patton or, if you prefer, Mattis's fellow Marine Lewis B. "Chesty" Puller. Stan McChrystal definitely evokes hard-bitten Matthew Ridgway, come to energize a floundering war effort. George Casey conjures up thoughts of the stolid U. S. Grant, and John Allen's overriding regard for the alliance reminds one of the collegial Dwight D. Eisenhower. Ray Odierno mirrors Omar Bradley, schooled in this hard war he rose to run. And there are others — the strong, the middling, the overmatched, the unappreciated — good people, hard-working, tough-minded, fair, and mostly honest and decent. Yet in the end, collectively, we all proved unequal to the moment.

As for myself, I make no excuse. I'm just a soldier who tried, got a few things right, but, in the end, failed. If I remind you of anyone at all, maybe it's Joe Stilwell, "Vinegar Joe," of the China-Burma-India theater in World War II. He told it like it was, eventually got sent home for it, and deserved a better war.

PROLOGUE

APOCALYPSE THEN

First we're going to cut it off, and then we're going to kill it.

— GENERAL COLIN POWELL, JANUARY 23, 1991

"CONTACT, FIVE ARMORED vehicles direct front. Three more off to the left."

Captain Herbert R. "H.R." McMaster and his troopers had been looking for the enemy. And now, here they were, just past this shrug of a rise on the hardpan desert floor. You could barely make them out: ragged lines of fuzzy black spots, swimming up like sharks out of the gray murk of blowing sand and fine, spitting rain.

There were a lot of them. It was a spectacle no American had seen since World War II: a row of eight enemy tanks arrayed for action, stationary, their long cannon slowly turning toward the advancing Americans. The Iraqis were lining up to shoot. McMaster and his men could see only a fraction, but they knew more were out there. The entire Iraqi Republican Guard Tawakalna Division waited, hidden in that shadowy, gritty mist. You could feel it.

Now it would come down to training, discipline, and timing. It always did. In contact, the side that acted first gained the advantage. Many of the Iraqis had fought in the grinding eight-year war with Iran. McMaster's tank crew, like most on his side, had never been under fire. Only a few U.S. senior officers and sergeants had served in Vietnam. None of the Americans had ever been in a tank battle. The Iraqis had fought in plenty. But the Americans knew what to do. The Iraqis did not.

As McMaster sounded off with his report, Sergeant Craig Koch, his tank gunner, pressed a button. An invisible laser lanced out through the gloom, and the number came up: 1,420 meters, almost a mile away. He stated, as he had on dozens of gunnery exercises on U.S. Army tank ranges, "Identified." But this was not a range, and these were not plywood targets.

The loader, Specialist Jeffrey Taylor, checked to ensure a 120 mm tank round was seated in the big bore. It was. Ideally, to blow holes in hostile tanks, you'd prefer to use a sabot round, a vicious, super-hardened heavy-metal spear that can rip through almost anything. But aboard McMaster's tank that late afternoon in the soup, the men had already preloaded a HEAT (high-explosive antitank) round, which used a molten-metal chemical mix to burn through the foe's armor. It wasn't as sure as a sabot slug; HEAT was better for trucks and lightly plated vehicles, but it would have to do. Taylor said: "Up."

McMaster commanded, "Fire." Koch did.

The HEAT round proved more than good enough, ripping through the Iraqi T-72 tank. Shards of metal sparked and spiraled out of the smoking hull. A shape, a man, maybe, rolled out and over the front slope of the stricken tank. But McMaster, Koch, and Taylor had already moved on.

The second shot — a deadly sabot penetrator — hit the second T-72. The turret and its encased 125 mm Russian-made cannon popped off like a hot frying pan knocked from a stove. McMaster's crew fired again, blowing open another Iraqi tank. Flames burned bright and hot. It had been ten seconds since the first sighting.

The Republican Guard tried to fire back. Their wildly mis-aimed 125 mm shots churned dirt on either side of McMaster's tank. Iraqi machine-gun fire zipped overhead, uselessly high. Meanwhile, the other American crews in their M1A1 Abrams tanks also went to work. They rolled on, firing on the move with computer-stabilized main guns, blowing right through what the Iraqis styled as a defensive position. For the tanks, it was barely a speed bump.

In twenty-three minutes, McMaster and fewer than a hundred men of Troop E, Second Squadron, Second Armored Cavalry Regiment knocked out twenty-eight T-72 tanks, sixteen armored infantry fighting vehicles, and thirty-nine trucks of the Eighteenth Brigade, Third Tawakalna Division. A reinforced Iraqi battalion of more than nine hundred men could muster only forty after the clash. Those forty stood near their confused commander, grouped in threes and fours, uncertain, dazed, helmetless, with hands up. The Iraqis never even hit a U.S. vehicle.

The Americans were veterans now. They kept going.

In later years, it became fashionable to denigrate the inept Iraqi performance in the 1990–91 war. The old cliché holds that hindsight is 20/20. It might be as good as 20/10—Ted Williams eyeing a curve ball—inside the Washington, DC, establishment press. All the smart guys always knew (didn't they) that the Iraqis were lightweights, sad sacks, and comic-opera extras. They were a bunch of superstitious, ignorant misfits led, when they were led at all, by buffoonish, cowardly officers. Their Russian gear was no damn good. As for their once celebrated combat experience against the even more incompetent Iranians . . . well, don't even talk about that. The overwhelming success of Operation Desert Storm in 1991, the devastating air campaign followed by the hundred-hour blitzkrieg through Kuwait and southern Iraq, ruined the Iraqi reputation as the battle-hardened fourth-largest army in the world. No wonder the Israelis routinely mopped the floor with opponents like this. Yes, the clever set found it all very amusing once it ended like it did.

Things did not seem quite so clear-cut at the outset. On August 2, 1990, as Iraq seized Kuwait between dawn and sunset, America's Saudi Arabian friends were not making any jokes. Iraq's bombastic dictator Saddam Hussein looked and acted like Joseph Stalin of Soviet Russia, and he consciously emulated Adolf Hitler of Nazi Germany in his pitiless slaughters of his own and his steady aggression on his borders. Between 1980 and 1988, Saddam and his substantial forces waged a bitter attrition bloodbath against revolutionary Persian Iran to the east. Now this vile and violent maximum leader turned his attention to the south. Kuwait ceased to exist, absorbed as the nineteenth province of Saddam's Iraq. Would Saudi Arabia be next?

Saddam's Republican Guard T-72 tank battalions needed only to refuel to begin a strong lunge down the Saudi coast. That would do it for much of the developed world's oil supply. The Saudis had no illusions about their own military. Fresh from their murderous struggle with Iran, the Iraqis would not be delayed long by untested, unsteady Saudi units. King Fahd and his brother princes pleaded for immediate U.S. help, the more the better.

The U.S. had a plan to defend Saudi Arabia from a Soviet Russian invasion. It would work well enough against Russian-equipped Iraqis. The U.S. Central Command (USCENTCOM) oversaw the entire Middle East and so prepared and maintained war plans to address potential threats affecting dozens of countries, Saudi Arabia among them. Our man in Saudi Arabia was thus the USCENTCOM commander at the time, General H. Norman Schwarzkopf Jr., U.S. Army. This large, forceful Vietnam veteran, a West Point graduate like his father, who was a World War II–era general, had lived in Iran from 1946 to 1947 while his father reorganized the shah's police. The shah was long gone, and the Iranians anything but friendly, but the experience made Schwarzkopf unique among senior American officers. He had lived in a Muslim country. True, it wasn't an Arab country, but it still counted for something. Schwarzkopf knew the region, knew the Saudis, and knew the plan. According to journalists hungry for color, the troops supposedly called him "Stormin' Norman" and "the Bear." Most military men and women who later served under Schwarzkopf had never heard of those nicknames or of him either.

As for the soldier who was backing Schwarzkopf from Washington, DC, almost every American in uniform, and most citizens, knew his name quite well. General Colin L. Powell was the chairman of the Joint Chiefs of Staff, the senior military officer in the U.S. Armed Forces, and the highest-ranking African American officer in the country's history. Powell's station reflected his talent and experience. It especially demonstrated his patience and self-discipline; in 1958 in Fort Benning, Georgia, Lieutenant Powell, wearing his country's uniform, suffered the humiliation of being shunted to the dirty back rooms of segregated diners and hotels. Although clearly not the first important black general officer in the U.S. Army, Powell certainly became the most widely known and influential one of his time, admired across the country as an exceptionally able manifestation of how far the country had progressed in race relations. If you asked an American in 1990 to name a serving general, he or she would say, "Powell." He stood out.

Yet in many ways, he was very much like Schwarzkopf. Powell shared Schwarzkopf's background as a lifelong infantryman and Vietnam veteran. Powell had also done a memorable stint as national security adviser for President Ronald W. Reagan from 1987 to 1989, and so he was seen by some as a political general. That was unfair to Powell, who had certainly done all the hard jobs at battalion, brigade, and division levels coming up through the ranks. That recent high-profile detour to the inner circle of the White House was unusual. It suited Powell well, however, to play Mr. Inside in Washington to Schwarzkopf's very much Mr. Outside in Southwest Asia.

There were other key common threads. Both men had served as small-unit advisers to the doomed South Vietnamese. Both had done their second assignments in Vietnam with the ill-starred Twenty-Third Americal Infantry Division. Major Powell held things together as the division G-3 (the G-3 is the operations chief, usually a lieutenant colonel's job) in the wake of the awful My Lai massacre of Vietnamese villagers by vengeful Americal soldiers. He went through a wringer of stringent investigations that cleared him completely of any wrongdoing but nonetheless put his judgment under a microscope and, at times, a proctoscope. About that same time, Lieutenant Colonel Schwarzkopf commanded an infantry battalion during a tough stretch of combat operations cursed by an ugly accidental-firing incident dissected in C.D.B. Bryan's haunting book *Friendly Fire*. In a 1979 film of that book, the character based on Schwarzkopf, Lieutenant Colonel Byron Schindler, was played by a journeyman character actor opposite the popular beloved Carol Burnett as the dead soldier's grieving mother. Audience sympathies aligned about as you'd expect. Powell, who gritted his teeth through the excruciating Iran-Contra scandal in the latter days of the Reagan administration and bore the coincident rehash of what he'd done and had not done regarding the My Lai war crime, understood what it meant to have one's failings held up to the hot glare of public scrutiny. Such episodes injected a degree of useful humility into both generals.

Yet, whatever else they did or did not share, the two men, like their lesser-known contemporaries, were bound as brothers by one thing:

they were sons of the long, failed Vietnam War. They did not accept the breezy assurances of figures like their former theater commander General William C. Westmoreland, who wrote, "Despite the final failure of the South Vietnamese, the record of the American military services of never having lost a war is still intact." Maybe he thought so. Schwarzkopf and Powell knew they had lost the war. They did not intend to lose another.

Within days of the Iraqi invasion of Kuwait, as Powell set the strategic framework in Washington, Schwarzkopf joined an American delegation flying to the Saudi summer capital in Jiddah to meet the king and his key princes. Secretary of Defense Richard B. Cheney headed the group, accompanied by deputy national security adviser Robert M. Gates. Both men knew the Saudis, and both would have a lot more to do with the region in future years. This amounted to an opening round in a very, very long fight. Nobody could see that yet. It was just as well. The immediate challenge of Iraqis massing on the Saudi border more than sufficed to hold their attention.

On August 6, 1990, the senior American officials arrived in Jiddah on the Red Sea coast. They carried a clear message, the same one stated forcefully by President George H. W. Bush in public the day before: "This will not stand." The Americans intended to defend Saudi Arabia and, in time, liberate Kuwait.

King Fahd certainly wanted defenders. Whether or not Kuwait would be reclaimed remained to be seen. More immediate matters had to be settled. How many Americans would come? More than a hundred thousand to start, Cheney answered, followed by as many as were needed to win the war. The actual number, counting Coalition elements on land, sea, and air, eventually reached nearly ten times that first figure.

King Fahd liked what he heard. The U.S. was committed. This meant the Iraqis could not succeed. The details of how quickly the Americans arrived, where they staged, and how they operated all needed to be settled. Unlike Emir Jaber of Kuwait, holed up with his retinue in Dhahran on the Saudi shore of the Persian Gulf, Fahd and his brothers no longer needed to fear the unpleasant fate of hiding out in a foreign hotel room hoping somebody else would get their country back. Saudi Arabia would endure.

But there was a problem. Like his fellow members and his predecessors in the House of Saud, Fahd played a double (or triple, or quadruple) game, doing business with the U.S. and the rest of the Western powers — and sometimes the Communist Russians and Chinese — while simultaneously paying off and playing off fundamental Islamists throughout the country and region. The severe, uncompromising Wahhabi sect received favored treatment. The king's mouthpieces decried Israel and warned of the perils of godless Western evils even as the king and his brothers, sons, nephews, and cousins enjoyed Mercedes sedans, Hollywood films, and Jack Daniel's whiskey courtesy of boundless oil revenues. Apparently, Allah could not see into modern garages and houses very well.

The House of Saud claimed the mantle of the protector of the holy places of Mecca and Medina. Muslims from Morocco to Indonesia, Kazakhstan to Yemen, longed to make the haj to Mecca. Thousands did every year. Five times a day, all practicing Muslims faced the city in Saudi Arabia as they prayed. King Fahd did more than run a country. He ruled the central physical manifestation of a vibrant, proselytizing world religion. Fahd's tightrope scimitar dance between infidels and believers was about to get very, very dicey as thousands of Americans and other nonbelievers poured into the Arabian Peninsula. An obscure former jihadist who fought the Soviets in Afghanistan, a wealthy young fellow from Jiddah named Osama bin Laden, watched all of this with great interest. He was not alone.

So, to appease his Arab and other Muslim supporters and to placate his Islamist clients, King Fahd made three decrees. First, General (and Prince) Khalid bin Sultan, his kinsman, would be co-commander with Schwarzkopf. Khalid had attended the British Royal Military Academy at Sandhurst, Auburn University in Alabama, the U.S. Army Command and General Staff College at Fort Leavenworth in Kansas, and the U.S. Air War College at Maxwell Air Force Base in Alabama. Khalid understood his new comrades pretty well.

By agreement, Schwarzkopf directed the U.S. forces and most of the other allies while Khalid led the Saudi units as well as the Arab and Muslim contingents and, for their own perverse reasons, the French. Of course, in the real world, Schwarzkopf ran the preparation, the defense scheme, and, eventually, the offensive. In the never-never land of Saudi Arabia's princely courts, out in the souks (markets) that made up the so-called Arab Street across North Africa and Southwest Asia, and in the anxious capitals in much of the Third World, Khalid held equal if not greater authority than Schwarzkopf. Nobody believed it. Everybody believed it. This is the sort of thing that passes for cooperation between infidels and believers.

Along with appointing Khalid to co-command, King Fahd insisted on keeping the Americans far out in the desert. It would not do for his faithful subjects to see female Marine truck drivers and Air Force C-130 transport pilots boldly unhidden by full-body black chadors, lacking even hijab headscarfs. The Americans and their European partners probably carried Bibles and rosary beads, might well gawk at Saudi women, and seemed unlikely to heed calls to prayer from the local minarets. No, these outlanders had to go way out into the wilderness. That accorded well with Schwarzkopf's plan anyway. Saudi Arabia had to be defended along its desolate border with Iraq and Kuwait. When the time came to attack by air and ground, that, too, would have to originate far out in the distant hardscrabble or far off the coast at sea.

Finally, Fahd insisted on, and the Americans agreed to, a very strict code of conduct. As a former adviser to the Vietnamese airborne troops, Schwarzkopf remembered only too well that in Southeast Asia, the ugly American was more than just a clever title for a novel. Incidents involving drunken, violent off-duty GIs poisoned many relationships with the already suspicious Vietnamese villagers. When the strict Saudis said no alcohol, Schwarzkopf agreed, a major departure from the practice in other American wars. Long-standing and extensive drug testing made Vietnam-era marijuana and hard-drug use very rare in the American ranks. Some raised eyebrows when Schwarzkopf and his staff curbed the chaplains' display of Christian and Jewish insignia and carefully concealed religious services, all at the request of the Saudi "hosts."

The Saudis did not see themselves as hosts. In their own eyes, they were customers, buying Americans and other defenders in much the same way they hired hundreds of thousands of Filipino, Bangladeshi, and Pakistani contract laborers to build their homes, run their oil wells, and clean their streets. Islamic practice allowed and even encouraged such relationships as part of the proper order of life. Those outsiders who were not Muslims lived as dhimmis, second-class folk who paid due tribute to local Islamic traditions. Schwarzkopf agreed to the king's directives, and the Americans accepted substantial Saudi financing and, later, material support like tentage, bottled water, trucks, and fuel, so the transactional relationship looked very clear to the Arab authorities. From this viewpoint, the House of Saud, in the happy spirit of the famous Arab traders of history, had rented the best armed forces in the world.

King Fahd got his money's worth. The Coalition forces arrayed for what was briefly Operation Peninsula Shield, then Desert Shield (defense), and finally Desert Storm (attack) proved to be both vast and powerful, totaling nearly a million armed combatants from thirty participating countries. The Americans provided 73 percent (almost 700,000) of the air, sea, and land contingents. The Saudis fielded about 100,000. The British sent 45,400. Egypt provided 33,600. France and Syria each sent about 14,500; in the end, the French agreed to place their forces under U.S. orders for the actual fighting. Morocco deployed almost 13,000. Some 10,000 Kuwaitis, reequipped and zealous, participated in the liberation of their country. Pakistan put nearly 5,000 in theater. Even the Afghan jihadist fronts, not yet in charge of their fractured state, cooperated enough to send about 300 riflemen. Other contingents came from Argentina, Australia, Bahrain, Bangladesh, Belgium, Canada, Denmark, Greece, Hungary, Italy, Korea, the Netherlands, Niger, Norway, Oman, Poland, Qatar, Senegal, Spain, and the United Arab Emirates. The Coalition employed almost 4,600 fixed-wing aircraft and helicopters, about 100 major warships, and nearly 3,400 tanks. Many other countries furnished money, overflight rights, port rights, and diplomatic backing. It was indeed a large and impressive coalition.

Yet the existence of Prince Khalid as co-commander and the odd relationship of the Saudis' guests to their hosts exemplified the major fissure in the alliance. The Coalition included the able and the not so able, the willing and the not so willing, those inclined to fight hard and those hardly inclined to fight. The British and French certainly took it all seriously and drew on their strong military traditions. But even they depended wholly on the will and skill of the U.S. forces. As went Schwarzkopf's Americans, so would go the war. Many of the rest were along for the ride.

The ride's course was not preordained. What of the other side? As the Coalition built up strength and spread out to defend Saudi Arabia, various intelligence entities studied the Iraqis. They, too, had impressive roll-ups: eleven hundred warplanes (including helicopters), fifty fighting ships (admittedly, small missile boats and minelayers), and fifty-eight hundred tanks, all parts of an organization that fielded more than a million men. Their weapons came from Russian factories and were pretty much what American battalions might have faced if the Cold War had ever gotten hot. In addition, these guys possessed and used chemical weapons inside Iraq to suppress the Kurds and in barrages inflicted on the hapless Iranians during the 1980–88 war. Iraq had thousands of howitzers and rockets, ranging from little frontline models to big, long-range surface-to-surface missiles that NATO referred to as Scuds. All could deliver nerve gas. The daunting Iraqi numbers corroborated with the usual summary: fourth-largest army in the world, backed by a big air force and a coastal defense navy. *Plus chemicals,* whispered the analysts. *Don't forget those.*

This massive tabulation seemed reasonable to those who received it in Washington and at USCENTCOM headquarters in Riyadh, Saudi Arabia. It accorded well with common U.S. military practice. In the West, military intelligence (MI) analysts have long followed a simple premise: Assess enemy capabilities, not intentions. Put another way, look at what an opponent *can* do rather than what you think he *will* do. American intelligence experts preferred to stick to hard numbers that could be verified by photographs, signal intercepts, and visual sightings. Non-intelligence types called this bean counting and implored the MI people to go beyond tallying and guess what the enemy would do.

Like economists and weathermen, intel guys rendered forecasts. They began with the most likely enemy course of action. But they also added in the most dangerous enemy course of action. It led to comments like this: "He will probably go right . . . but he might go left, which would be, of course, much worse . . ." Then the staff experts would throw in a few more warnings: there might be counterattacks, land mines, sea mines, chemical strikes, biowarfare, nuclear radiation, and so on; almost everything except locusts and fiery hail. You learned to expect the intelligence community to predict disaster ten out of every three times.

Infantrymen like Norman Schwarzkopf and Colin Powell might well object, but the MI officers had sound reasons to stick to counting real beans and only real beans. In the living memory of many in 1990, including World War II veteran President George H. W. Bush (himself a former CIA director), some of the intel folks had strayed too far from hardware metrics to software opinions. In 1941, experts believed the Japanese could not fly modern fighter-bombers hundreds of miles from their strange-looking aircraft carriers. The men from Nippon, it was believed, surely suffered from poor night vision, allegedly caused by too much rice in the diet and not enough vegetables. These prejudices were overturned in blood and fire, beginning at Pearl Harbor and then in many battles to follow. The Imperial Japanese Navy proved quite capable, by day and by night, on the sea, in the air, and beneath the waves.

Even chastened by Pearl Harbor, Bataan, and Savo Island, American MI elements fumbled again and again when they went beyond the numbers game. They insisted that the peasant North Koreans could not be so tough; they were. They were sure that the Chinese Communist regiments could never cross the Yalu River undetected; they did. They believed the North Vietnamese must fold under U.S. bombing; they did not. Any excursions into the cultural stuff tended to reinforce biases. No good intelligence professional dared to chance such errors, nor could America.

By comparison, the bean counting that characterized the Cold War with the Soviet Union was touted as a great success. Year after year, using aircraft, agents, and "national technical means," the U.S. and its allies built a very reliable picture of the Union of Soviet Socialist Republics and its armed forces. Of course, that country's rapid collapse in 1991 caught us by surprise. But we sure knew how many missiles, bombers, and tanks it had, and where to find them, even while the whole lash-up went under.

In addition to counting the beans and checking the bean counts, the MI units learned a great deal about each kind of bean. Our collectors vacuumed up a lot of technical intelligence. H. R. McMaster and his cavalry troopers at 73 Easting knew exactly what a T-72 tank looked like, how fast it drove, and how far its 125 mm cannon shot. That kind of information had been incorporated in training for all U.S. services, with specially schooled units actually playing surrogate Russian air, land, and sea threats. As the Iraqis used almost the same equipment, so much the better. War games such as the Air Force Red Flag series, the Navy's Top

Gun courses, and the Army's National Training Center unit rotations taught hard lessons about how to battle the Russians and thus the Iraqis. McMaster, Koch, and Taylor had been running practice plays against their enemies every time they'd gone to the gunnery range or played full-up laser-tag tactical exercises. The real clash matched the U.S. drills, and that did great credit to years of hard, drudging intelligence analysis.

Yet in this mass of numbers and weapons facts and figures, one looks for a glimpse of the Iraqis as people and soldiers. Would the Republican Guardsmen fight? How about the regular line infantry? What of the Iraqi air force? The Americans and their Coalition friends just stuck to the numbers and calculated that tanks were tanks were tanks. They ran computer models of the campaign and came back with predictions of U.S. casualties ranging as high as ten thousand, with fifteen hundred dead. In all of those crunched numbers, the programmers assumed the Iraqis would fight.

Some said otherwise. The Israelis, naturally, thought little of Iraq's military and said so to anyone who would listen. British officers with extensive Middle Eastern time made the same observations. A look at the Coalition partners from the region told the tale. Could the Iraqis really be all that different, or better, than the Egyptians, Syrians, or Saudis? Schwarzkopf had been careful to assign them rather minor roles as they struggled with logistics, combined-arms tactics, close air support, and small-unit discipline. No, the Iraqis appeared unlikely to operate up to Western standards. Despite the Iraqis' outward appearances and bold aspirations, it just wasn't in their culture. People in USCENTCOM sensed it, but the older ones remembered Vietnam and what those allegedly pissant hostile riflemen had accomplished.

American military leaders with long service in the region, such as Colonel Norvell B. De Atkine, knew what to expect: "over-centralization, discouraging initiative, lack of flexibility, manipulation of information, and discouragement of leadership at the junior level." In essence, the Iraqis had been forced to play by foreign rules and Western norms that didn't align with their upbringing and experience. The Arab world, and indeed the larger Islamic realm, at times resembles what we think we know of life in prescientific Europe during the Middle Ages. Uneducated Iraqis tended to be overly accepting of natural and manmade events, deferential to distant authorities, subservient to family and clan elders, suspicious of outsiders, and strongly influenced by their pervasive religious faith. It all made for an interesting and rich civilization. But it didn't teach anyone how to repair a tank, trust a laser-range finder, or practice marksmanship until you got it right. *Inshallah,* God willing, if the bullet was supposed to hit the target, it would hit. If not, move on.

The Iraqis did have a preferred fighting style, but it did not involve attack helicopters or main battle tanks, let alone uniforms and sergeants barking orders. Eccentric British guerrilla leader Colonel T. E. Lawrence made much of this preferred Arab way of combat while fighting the Germans and Turks during World War I. In a series of actions not far north and west of the main ground battles of Desert Storm, Lawrence of Arabia, lionized in print and on film, lived with the tribal warriors and spoke their language. Schooled in history and archaeology, he found that the locals preferred raids and ambushes, short skirmishes, and quick hit-and-run escapades to Western maneuvers. They disdained uniforms, discipline, and stand-and-fight tactics. But they could make entire desert districts wholly untenable for conventional adversaries, demolishing rail lines, blowing out bridges, sniping, stealing, and slowly bleeding the big regiments to death. Lawrence called it "winning wars without battles." That method did not apply in the 1990–91 campaign. In one of his dumber decisions, Saddam Hussein chose to make war in the Western style. It destroyed his military and damn near finished him.

That the Iraqis were playing the wrong game in the wrong league occurred to some in the Coalition, especially after Coalition members had spent a few months dealing with the Iraqis' cousins on their own side. Yet maybe, just maybe, all those tanks and guns on the Kuwaiti border constituted the real deal. Burned badly in Vietnam, the Americans stuck to the bean counting, the worst-case scenarios. General Colin Powell in particular emphasized the use of overwhelming force and numbers, hence the massive buildup. In football terms, the Americans were not playing to win by a field goal; if the Iraqis cracked — as a few thought they might — Powell and Schwarzkopf intended to run up the score.

The Iraqis cracked, all right. They folded over like cardboard in the rain. Their vaunted air force mounted a pathetic 910 sorties (individual aircraft flights), compared to more than 69,000 by the Coalition. Iraqi pilots lost every single air-to-air encounter, and 132 planes roared off to unfriendly Iran rather than staying and getting blown off their runways. The tiny Iraqi navy swung at anchor and waited to get sunk, which occurred in due course. As for the Iraqi army and Republican Guard, the best summary comes from the end of the 1933 version of the movie *King Kong:* "Well, Denham, the airplanes got him."

That is not to ignore the ground campaign that swiftly liberated Kuwait. Actions like the clash at 73 Easting ripped apart Saddam Hussein's divisions. Later accounts made much of the great western flanking maneuver that Schwarzkopf labeled his Hail Mary pass, referring to a long throw toward the end zone in football that's often made in desperation. The Coalition effort was anything but desperate. Later books and documentaries described battles and firefights, aerial dogfights and naval encounters, all real enough, yet all largely one-sided. The Americans and their partners maneuvered and closed. The Iraqis shot back a little, then died or ran away.

Conservative estimates of Iraqi casualties ran about 20,000 dead and up to 75,000 wounded; nobody on the Coalition side really knew. Prisoners were counted precisely as 86,743, with 63,948 taken by U.S. units. The Iraqis tried to shoot back; they killed 148 Americans, wounded 458, and captured 21, all of whom were repatriated. The number of Iraqi tanks, artillery, armored personnel carriers, and trucks destroyed ran into the thousands. Today, their torn, blackened, abandoned hulks still dot the Kuwaiti and southern Iraq deserts.

When the cease-fire came at Powell's urging after forty-two days of conflict, Schwarzkopf moved immediately to meet with the Iraqi commander at the crossroads of Safwan, just inside Iraq, north of the Kuwaiti border. In truth, Schwarzkopf's counterpart was Saddam Hussein himself. Like his idol Hitler in the Wolfschanze (Wolf's Lair) bunker in mid-1944, sticking pins in the map and issuing detailed orders to individual battalions as the Allied forces closed the ring, Saddam liked to keep his hands on all the controls and buttons even as the plane of state spiraled into a death dive. He alone directed the movements of Iraqi units great and small.

Yet as an ultimate survivor, Saddam knew better than to trust Schwarzkopf's flag of truce. The Iraqi dictator did not wish to get snagged in Safwan, placed in handcuffs and an orange jumpsuit, and, like former Panamanian strongman Manuel Noriega, hauled to the docket of a U.S.

courtroom. Already beset with rebellions in the Kurdish north and Shiite Arab south, the Iraqi president squatted safely in his Baghdad enclave. He dispatched two senior representatives: Lieutenant General Sultan Hashim Ahmad, the deputy chief of the general staff, and Lieutenant General Salah Abud Mahmud, the commander of Third Corps, or what remained of it. It wasn't quite the Japanese motoring out to the battleship USS *Missouri* to meet Douglas MacArthur on September 2, 1945, but it would do.

At half past eleven on the morning of March 3, the two Iraqis arrived to meet Schwarzkopf and General Khalid, the Saudi co-commander. American soldiers escorted the Iraqis. They walked past U.S. tanks and tracked infantry carriers, the slab-sided armored vehicles fronted by lines of troops. Apache attack helicopters clattered overhead. The Iraqis looked small, old, and nervous. One trembled a bit, but maybe it was just due to the desert breeze.

The Iraqis moved into the tent and sat down. Behind the Iraqis sat a few of their subordinate officers, taking notes. Schwarzkopf and Khalid entered and took their seats. They also had their people behind them.

After some photographs for posterity, the big American spoke. It was 11:34 a.m. "The purpose of this meeting," said Schwarzkopf, "is to discuss and resolve conditions that we feel are necessary to ensure that we continue the suspension of offensive operations on the part of the Coalition." The agenda adhered to military matters. Someday, the diplomats hoped to hammer out a true peace treaty, or a pact, or a convention, or whatever. It never happened.

"We are authorized," replied Ahmad, "to make this meeting a successful one in the spirit of cooperation." He spoke deliberately, careful to make eye contact.

Schwarzkopf nodded and pressed on. He talked about the cease-fire boundary, referring to a map. The Iraqis leaned forward as the American confirmed their fears. The Coalition held the southern fifth of Iraq. Schwarzkopf made it clear that the U.S. had no permanent territorial designs provided the Iraqis met their obligations regarding withdrawal of the surviving Iraqi forces, return of prisoners, transfer of the dead, and marking of minefields. "But until that time we intend to remain where we are," noted the U.S. general.

Discussion continued for a while on details of unit positions and movements, then broadened to the other agenda items. The American went down the list. Those present later remembered that the USCENT-COM commander did almost all of the talking. The Iraqis listened. Sub-ordinates took notes. The tent heated up. The air grew stuffy. Finally, Schwarzkopf finished.

Ahmad spoke up. He pressed on the Iraqi prisoners. How many?

"We have, as of last night, sixty thousand," replied the American. "Sixty thousand plus." Ahmad looked stunned.

His comrade Mahmud, who had watched his units shredded by the American Marines and soldiers, offered: "It's possible. I don't know."

An awkward silence ensued.

Schwarzkopf broke it, trying to wrap up. "Are there any other matters the general would like to discuss?"

There were. Ahmad said, "We have a point, one point."

Schwarzkopf waited.

"You might very well know," Ahmad continued, "the situation of our roads and bridges and communications." The USCENTCOM commander definitely knew. On his orders, his airmen had severed most of those links.

Ahmad went on. "We would like to agree," he offered, "that helicopter flights sometimes are needed to carry officials from one place to the other because the roads and bridges are out." That seemed reasonable enough, though it was anything but a casual request.

Thus far, the Safwan conference had been all about sticking it to the Iraqis in a most public way. It featured Vietnam veteran H. Norman Schwarzkopf making sure that this time, this war, ended in the old style, with the beaten foe hangdog and helpless, at the mercy of the victor. America had suffered through the humiliation of North Korean and Chinese propaganda ploys at the truce talks in Panmunjom, Korea, from 1951 to 1953 and the North Vietnamese bluster and circumlocutions in Paris from 1968 to 1973. Working from battlefield parity or worse, the Americans in those conflicts got bamboozled and hoodwinked, strung along, and embarrassed over and over; they were played for fools and suckers by the much more savvy enemy negotiators, who had a simple premise: What was theirs was theirs. What was America's was negotiable. Well, in Desert Storm, America crushed Iraq. This time, nobody would play Norm Schwarzkopf for a fool or sucker.

The wily Arab trader, however, has many means to address his infidel mark. Ahmad and Mahmud, well briefed in Baghdad, figured on a bit of magnanimity from Schwarzkopf as long as they asked for only one thing. After all, in the iconic surrender at Appomattox, Virginia, in April of 1865, hadn't Union general Ulysses S. Grant agreed to Confederate general Robert E. Lee's request that his men be allowed to keep their horses so they could do their spring plowing? The Iraqis wanted to keep their modern horses, the helicopters. But they wouldn't be used for agriculture.

Schwarzkopf walked right into the snare. "As long as the flights are not over the part we are in, that is absolutely no problem." Ahmad pushed a bit: "So you mean even armed helicopters can fly in Iraqi skies?" Schwarzkopf agreed. "You have my word," said the American.

As the meeting concluded, Ahmad saluted and offered his hand. Schwarzkopf returned the salute and shook the Iraqi general's hand. "As an Arab, I hold no hate in my heart," Ahmad said.

He did hold those helicopters, though.

The Iraqi authorities needed them, every airframe, in order to keep their heads connected to their bodies. The huge, diverse Desert Storm Coalition wasn't who scared them. That group had agreed on defending Saudi Arabia, mostly concurred with liberating Kuwait, and split completely on ousting Saddam Hussein. Wiser heads like Colin Powell's kept the strategy limited and focused on clearing Kuwait. Although some plans were made, few displayed interest in marching on to Baghdad. Enough was enough.

Some strong rhetoric, though, encouraged the Iraqis to take advantage of Desert Storm, wash their dirty laundry, and toss out Saddam with the resultant wastewater. In a February 15, 1991, Voice of America broadcast, President Bush asked "the Iraqi military and Iraqi people to take matters into their own hands." Plenty of them did so. The ethnic minority Kurds and downtrodden Shiite Arabs needed little encouragement to rise against a man who'd killed their families in droves. In the north and in the south, they gave it a shot.

The helicopter ploy alone did not save Saddam Hussein's regime from the twin revolts. Combined with the rapid U.S. departure from Iraq, though, the use of attack helicopters gave hope to Saddam's faithful. Though they'd fumbled and stumbled in engagements with determined armed men like H. R. McMaster's Troop E, the battered remnants of the Republican Guard brigades confidently gunned down Kurd and, especially, Shiite rebels. AK-47s and hand grenades availed the rebels little against tanks and heavy artillery. The helicopters reinforced the government's morale and momentum. Altogether, it did the job.

In the north, under the rubric Operation Provide Comfort, U.S. and NATO airpower staged out of Turkey to keep Saddam's armored elements and helicopters out of the mountainous Kurdish homeland. Above the Kurds, even Saddam's rotary-wing fleet dared not fly. Some NATO units went in on the ground. The intervention offered significant succor to Kurdish refugees and helped stabilize a place often battered by Saddam Hussein's vengeful military. Participants included American paratroopers commanded by Lieutenant Colonel John Abizaid, an Arabic-speaking battalion commander of Lebanese ancestry. Colonel James L. Jones led the Marine expeditionary unit. Like Cheney, Powell, and Gates, Abizaid and Jones would have future business in the area. Abizaid's airborne battalion, Jones's Marines, and the other NATO ground units stayed only three months, but the air cap remained up until the second U.S. war with Iraq, in 2003. It allowed relative security in a semiautonomous Kurdistan, one persistent rollback of Saddam in the aftermath of Desert Storm. The Kurds enjoyed only a partial respite as long as Saddam remained in power and mounted some incursions. But thanks to Provide Comfort, the Kurds developed a degree of trust with the U.S. and its allies. Kurdistan offered a potential springboard for future operations, as did newly restored Kuwait in the south.

Shiite southern Iraq, however, went under in a welter of blood and fire. While U.S. units stood by, Iraqi battalions moved methodically up the Euphrates and Tigris River Valleys, shooting as they went. Tanks bulldozed houses. Artillery barrages smashed villages. Street fighting erupted in the city of Basrah, but Iraqi military machine guns and cannon put paid to that. Above it all, those helicopters buzzed and darted, dropping down to disgorge raiding teams or pump rockets into burning houses. By the end of March 1991, the major spasm was over. Follow-up operations continued into the summer of 1992, when President Bush finally extended a Kurdish-type no-fly zone over the Shiite south. By then, however, the allied aircraft flew cover for gravesites. Some ten thousand Shiites died in Saddam's brutal riposte. Whereas the Kurds saw reason to work with Americans after Desert Storm, the Shiites of the south found the U.S. unreliable, a suspicious view gleefully encouraged by their Persian Shiite benefactors in Iran.

The two upheavals, both contained, left Saddam Hussein in charge in Baghdad. That became evident by the summer of 1991, not long after the welcome-home banners came down and the yellow ribbons faded all across America. In the usual late-twentieth-century American way, Bush's great victory of the winter begat a summer of reassessment and, soon enough, recriminations.

Some of the inside baseball truly got petty. The services endured a seemingly endless debate over how many tanks the Air Force had killed versus the number taken out by the Army. The Navy felt unloved and unappreciated for its long vigil out in the Persian Gulf. The Marines argued that they had won the war by freeing Kuwait. Prince Khalid emphasized his clearly critical and obviously equal command role and received a promotion to field marshal, a validation of his claims. Various random colonels and generals complained that Schwarzkopf "stormed" too much at his fellow generals in Riyadh and across the theater. All of this stirred some interest but in the long run meant little.

A more serious debate began regarding the end of the war. The abortive Kurdish and Shiite revolts emphasized the incomplete nature of the victory. Saddam Hussein remained in power, and although most U.S. troops departed in 1991, a strong air component stayed behind to patrol the no-fly zone over the Kurdish north and eventually the Shiite south. Some former senior officers, such as Lieutenant General Jack Cushman of the Army and Lieutenant General Bernard Trainor of the Marines, pulled out their maps, reviewed reports, questioned participants, and figured out what the Iraqis had known as early as the Safwan armistice meeting: a large portion of the Republican Guard — enough to menace the Kurds and shred the Shiite rebels — escaped the firepower of Desert Storm. As a result, oppressed groups inside Iraq and, soon enough, Kuwait, Saudi Arabia, and the U.S. faced an Iraqi threat that appeared to regain strength during the 1990s. So the war was over, but it wasn't over. Not that it mattered in hometowns across America. A win was a win. This one looked, smelled, and played out like a win.

That view also prevailed at the Pentagon, where little time was spent agonizing over how many Republican Guard tanks had or had not escaped across the Basrah causeway. Operation Desert Storm freed Kuwait, the mission's goal. Doing more inside Iraq, hanging around in a very rough neighborhood, appeared to be a bad bet for American forces. For leaders such as Powell and Schwarzkopf, men scarred by Vietnam, the boiling blood stew of quarrelsome Kurdish factions and squabbling Shiite Arab religious elements looked like a bottomless pit of pain, as indeed it was, and they had no desire to plunge into it. Leave some airpower, leave some equipment and a few support troops in Kuwait, but mostly just leave. If Iraq demanded another beating, we knew just how to do it.

Indeed, the scale and speed of the U.S. victory brought a degree of certainty to American military leaders confronted with defining their role in a post-Soviet world. The same guys who wanted to count every Iraqi T-72 in January of 1991 now spoke of "Desert Storm equivalents" in war planning, as if one could just package up the thing and drop it on the miscreant du jour in North Korea, Iran, or wherever — even in Saddam's Iraq, should he try some new mischief. The U.S. Department of Defense leadership began talking about maintaining a capacity to fight two such operations at once, or nearly at once. This construct endured throughout the 1990s even as the American forces drew down almost 40 percent. By this arithmetic, 60 percent of a Desert Storm should do the trick. After all, hadn't we learned that these Third World opponents really weren't all that good? Somewhere in the afterworld, former members of the Imperial Japanese Navy, the North Korean People's Army, and the North Vietnamese Army must have been laughing ruefully.

Three ideas emerged from the blitzkrieg in the desert: the value of a volunteer military, the key role of information technology, and the utility of joint decisive operations. The U.S. Armed Forces expected to get smaller as the Soviet Union went away. Operation Desert Storm showed how to do that wisely. Investing in these three areas meant the country would have the capacity to replicate the Gulf War when any threat arose.

The volunteer armed forces avoided the mess of draft calls and pleas for deferments and exemptions. Volunteers *wanted* to join. The military could get by with fewer recruits because more in the ranks reenlisted. The quality of the volunteers turned out to be good, because the services insisted on drug-free high-school graduates with clean criminal records, criteria that ruled out 70 percent of American youth. (There is an unfortunate message in that statistic.) Smarter, tougher, and willing, volunteers trained and worked to their limits. The volunteer Marine Corps, U.S. Navy submarine and aviation arms, and U.S. Army Airborne of 1941 to 1945 proved to be the elite during World War II. Now the entire armed forces followed that model.

The volunteer approach had only two drawbacks. As it outsourced defense to the willing, going to war no longer involved a large portion of the entire population of U.S. citizens. In addition, if you needed to expand the force, it took a long time. You could not mass-produce highly trained, well-led, technically educated modern units, especially the ones you really needed, like bomb-disposal detachments, helicopter companies, and Special Forces teams.

The volunteers made up an armed force that increasingly adopted civilian information technologies to make unit movements more certain, weapons more accurate, and resupply more effective. The U.S. Air Force and Navy had pioneered these ideas as early as the 1940s with radar and early large-capacity calculating machines. By the Gulf War, the entire architecture was coming together nicely. Primitive e-mail disseminated orders. Global positioning systems facilitated navigation that had previously depended all too much on the fabled well-meaning second lieutenant with a map and a compass. Captured Iraqis marveled that the Americans came at them out of the open desert, unbound to roads. Early multiservice tracking screens showed the blue (friendly) units, although the red (enemy) kind stubbornly refused to play along except as placed by intelligence staffers. Computers integrated IT into weapons, vehicles, planes, ships, and many missiles, bombs, and artillery shells. The American troops called the devices smart, in that the things recognized where they were and went where they were pointed. The U.S. military knew their location, knew the enemy location (almost), and delivered weapons that went exactly where they were sent. It greatly multiplied U.S. firepower.

An air example illustrates the advantage. In 1944, B-17 bomber formations dropped 9,070 bombs in order to hit one German building. In 1967, F-105 jet fighter-bombers used 176 munitions to knock out a single North Vietnamese building. By 1991, a smart F-16 fighter-bomber could do the job with thirty bombs, or just one, if the bomb was smart too.

Similar improvements applied to sea and ground armament enhanced with IT. It meshed smoothly when used by tech-savvy, educated volunteer soldiers, sailors, airmen, and Marines.

Composed of quality men and women, equipped with the fruits of U.S. information technology, the U.S. Armed Forces developed joint (multiservice) decisive operations to win quickly against most enemies. General Colin Powell sometimes got tagged as the author of this doctrine, but he merely expressed it better than most. Working as a team, America's armed forces would take the strategic objective given by the U.S. president (liberate Kuwait), fashion an operation (Desert Storm), and then execute (73 Easting). Powell and his fellow generals and admirals expressed no inclination to have the U.S. military diddle with long, tit-for-tat attrition contests or unending "we're here because we're here because we're here" peacekeeping missions. Those might happen — things you didn't expect always happened, like the Kurdish business — but they would be lesser included cases, not the real thing. Those "operations other than war" did not form the basis for force sizing or war planning. No, you didn't waste time monkeying around in the bush leagues. You opened the cage, the pit bull came out, ripped the head off the snake, then went back in its cage to get ready for the next time.

Operation Desert Storm represented the favored American way of war, all right. From now on, that was the way to go. As President Bush put it, "By God, we've licked the Vietnam syndrome once and for all." So we hoped. But as the U.S. Army chief of staff at that time, General Gordon R. Sullivan, liked to remind his soldiers, hope is not a method.

Hope had a short half-life in Baghdad too. Saddam Hussein and his Baathist Party inner elite, having just barely survived "the Mother of All Battles," took stock. Objectively, Iraq lost the war in a most humiliating fashion. In the rubble, Saddam, like the Americans, discerned the future. He identified three key lessons.

First, at the strategic level, Saddam Hussein realized that victory came from persistence, not a bean count or a flashy road race. The Americans just did not stick with it; the energetic but quickly-played-out hares were eventually overtaken by the plodding Iraqi tortoises (with apologies to Aesop). Saddam chose to define his regime's continued existence as a victory over the vast American-led Coalition. As the wags put it when

President Bush lost the 1992 election, "Saddam still has his job. How about you?" The Arab Street bought it, as did many in the U.S. tired of President Bush and his foreign adventures. By the mid-1990s, the containment of Iraq got harder and harder as Saddam claimed to be the victim of the infidels. By the late 1990s, with the Kurds boxed, the Shiites blasted, and Saddam rearming, Desert Storm looked like it had been a lot less decisive.

Second, at the operational level, Saddam figured out that propaganda trumped reality. The Vietnam War featured the first comprehensive television coverage of an ongoing conflict. The Gulf War expanded and deepened that trend. Intense media interest focused on the Gulf War. CNN introduced a twenty-four-hour news cycle for viewers hungry for pictures, interviews, and on-the-spot reporting, courtesy of miniaturized cameras and satellite uplinks based on the same IT that made the smart bombs so smart. Soon to be joined by Fox, MSNBC (formed by Microsoft and NBC), a revitalized BBC, Al Jazeera (out of Qatar on the Persian Gulf), and others, CNN blazed the trail in 1990. The infant Internet promised to expand that media aperture even more, but those days had not yet arrived. In the meantime, CNN had to fill a whole day. Old soldiers like Schwarzkopf kept their intentions to themselves — "loose lips sink ships" and all of that — adhering to traditional operational security. Why give the enemy free intelligence?

Saddam and his Baathist cronies didn't see it like Schwarzkopf — quite the opposite, in fact. Iraq's biggest successes of the war involved actions of minimal military utility that created maximum disruption in U.S. and world public opinion. Early on, his men seized Western civilians and used them as hostages and human shields in Kuwait. Saddam foolishly released them without gaining anything but time. He did a lot better in subsequent rounds.

As the airstrikes started on January 17, 1991, Saddam found his surface-to-air missiles wanting, his fighter jets incapable, and his airspace porous. Then he discovered a more useful countermeasure in CNN reporter Peter Arnett. Saddam's subordinates carted Arnett and other journalists to the site of purported American aerial atrocities: bunkers full of civilians immolated, baby-formula factories ravaged, hospitals hit, and the like. Some events really had happened due to U.S. mistakes or stray munitions. Most had not. A good number actually resulted from the large, burned-out carcasses of Iraqi air-defense missiles falling to the ground. But like Hollywood moguls determined to milk one good first weekend out of a turkey movie, Saddam didn't much care if the stories were truth or lies, as long as they opened well. At one point, as a result of this Iraqi media stream, President Bush severely limited air attacks in Baghdad, thereby achieving something for Saddam that Iraq's air force could not. In a similar way, the final cease-fire on February 28, 1991, followed concerns in Washington about relentless CNN images depicting the destruction wrought along the "Highway of Death" leading north out of Kuwait City. Rarely have rapists and pillagers garnered such thoughtful consideration. On the Arab Street, they really ate it up.

A strategy of hanging in there and an operational approach of highlighting propaganda matched well with the favored local fighting style of winning wars without battles. Hit and run, opportunism, and wearing out the opponent trumped trying to stand and fight like Western battalions. Terror tactics paid off when Saddam's rocket men responded to the Coalition aerial onslaught by launching eighty-eight Scud missiles, split about evenly between Israel and Saudi Arabia. The ungainly, inaccurate Scuds mostly turned dirt, killing one Israeli and one Saudi. But the rockets nearly brought Israel into the war, and that guaranteed trouble with the Coalition's Muslim members. The U.S. sent air-defense assistance to Israel and diverted thousands of air sorties and a major special operations force to hunt the Iraqi rocket launchers hopscotching around the western desert.

Indicative of why all this got Schwarzkopf and his people excited, a single Scud struck a warehouse in Dhahran, killing twenty-nine and wounding ninety-nine Army reservists from Pennsylvania. The things mostly missed population centers, but if they hit, casualties followed. Had any of the Scuds carried poison gas — they didn't — valid fears might have transformed into outright hysteria. By any measure, attacking civilians advanced Saddam's goals of protracting the war and harming Coalition public support. If some Coalition soldiers got it too, that helped the Baathist cause. Terrorism worked. Saddam and his key people kept all of this in mind going forward.

Another interested observer drew conclusions of his own. Osama bin Laden, "the Contractor" to his clandestine al-Qaeda network, objected most strongly to the huge number of infidels that entered Saudi Arabia and thereby polluted the holy land of the Prophet Muhammad and the venerated cities of Mecca and Medina. Bin Laden thought little of the secular Saddam Hussein, whose wartime embrace of Islam struck many as a cynical subterfuge to gain popular support in Iraq. But at least the Iraqi president actually fought the West. Saddam's thoughts regarding persistence, propaganda, and terror tactics against civilians made a lot of sense to the al-Qaeda leader.

Saddam showed what did not work: tank battles, jet dogfights, and the like. Bin Laden knew from his time in Afghanistan that pious Muslims, especially Arabs, worked better in small groups, wearing civilian clothes, mixed into the wider populace of believers. The Americans brought impressive weapons and large numbers, but they lacked faith and staying power. That's what Osama bin Laden learned from the 1990–91 Gulf War.

PART I

TRIUMPH

THE GLOBAL WAR ON TERRORISM

SEPTEMBER 2001 TO APRIL 2003

Americans should not expect one battle, but a lengthy campaign unlike any other we have ever seen.

— PRESIDENT GEORGE W. BUSH, SEPTEMBER 20, 2001

1

HARBINGERS

A destroyer: even the brave fear its might.
It inspires horror in the harbor and in the open sea.
She sails into the waves
Flanked by arrogance, haughtiness and false power.
To her doom she moves slowly.
A dinghy awaits her, riding the waves.

— OSAMA BIN LADEN, JANUARY 2001

AMERICAN SAILORS HATED to stop in Aden. They got no liberty, didn't even tie up at a pier — going to the great harbor of backward, hostile Yemen amounted to a very unpleasant way to get some fuel. The craggy peaks of a long-dead volcano marked the eastern promontory that defined the anchorage. The great pinnacles loomed above, a reminder of why soldiers always want to hold the high ground. The hills had eyes, or worse.

Still, one of the duties of the U.S. Navy, going all the way back to the early 1800s, the days of the Barbary pirates of North Africa, involves showing the flag. Safe passage of Navy ships ensures unmolested transit of merchant shipping, always the main conduit of all overseas trade whether in 1800 or 2000. Port calls projected U.S. influence ashore and kept markets open. Freedom of the seas, like all freedoms, must be exercised or it will atrophy.

A glance at a map showed why naval officers valued Aden. It dominated the Bab-el-Mandeb, Arabic for "Gate of Grief," which divided the Arabian Peninsula from Africa. Rarely has a geographic location been better named. Grief flowed in abundance on all sides and had done so for centuries. The Bab-el-Mandeb marked the entrance from the Indian Ocean to the southern end of the Red Sea. Those troubled waters separated Somalia, Djibouti, and Eritrea from Yemen. All four countries suffered from terrorist activities and internal turmoil. As it widened, the Red Sea divided Egypt from Saudi Arabia, both countries that balanced the usual requirement for help from the U.S. infidels with strong anti-Western sentiments in their local souk, the ever-aggrieved Arab Street. At its northern end, that same Red Sea fed into the Suez Canal, central to many Egyptian-Israeli conflicts, although quiescent since the Camp David Accords of 1978. Beyond lay the Mediterranean Sea, with all of its travails.

The Red Sea and its approaches have been contested since the time of Moses and the pharaohs. Greeks, Romans, Arabs, Crusaders, Turks, Mamelukes, Napoleon's French, British regulars, the Nazi German Afrika Korps, Egyptian armored brigades, and Israeli paratroopers came and went, leaving their bones in the desert. The U.S. Navy stationed missile-firing ships there during the 1990–91 war with Iraq. The advent of the Suez Canal made the Red Sea a critical waterway feeding maritime trade, notably oil, to Europe. Despite all the poverty and violence on its shores, some 3.4 million barrels of the world's oil, about 10 percent of all petroleum that went by sea, moved each day through the Bab-el-Mandeb. That's why the U.S. Navy stopped in Aden.

Like Singapore on the busy Strait of Malacca and Gibraltar in Spain, Aden attracted the British long before the Yankees got there. Dependent on sea commerce, Great Britain's Royal Navy identified such key chokepoints and took them. The British East India Company secured the port of Aden in 1832. Redcoats followed, and with Aden in hand, the British held the surrounding territory as the Aden Protectorate until 1967. The British ignored the wild, poor, mountainous north, leaving that to the ostensible rule of the far-off, feeble Ottoman Turks. For London, only the strategic Bab-el-Mandeb really mattered, and it mattered even more once the Suez Canal to the north opened for traffic in 1869.

When the British left under fire in 1967, a pro-Soviet front took over in the newly proclaimed People's Democratic Republic of Yemen. The

Russians knew all about the Bab-el-Mandeb and wanted a boot on that Western windpipe. North Yemen, always quasi-independent, split with the leftists in Aden city and sought Saudi and Western help. The two states and their various subfactions skirmished throughout the rest of the Cold War. In a reversal of its familiar pattern in both Korea and Vietnam, the U.S. backed the north.

After a surfeit of blood and a lot of haggling guided by the Arab League, mainly the Saudis, the two countries took advantage of the end of the Cold War. In 1990, Yemen unified under President Ali Abdullah Salah, an authoritarian autocrat in the style of Hosni Mubarak of Egypt. In the usual Middle Eastern fan dance of dealing with necessary outsiders while placating militant Muslims at home, Salah said all the right things to America and the West. Meanwhile, inside his own country, official and unofficial spokesmen impugned the West daily, reserving special venom for Israel and its superpower ally, the United States. Yemen's weak security forces had hit-or-miss success in trying to quash a rat's nest of local terror networks, insurgents, and general bad actors.

Into this hothouse sailed the guided-missile destroyer USS *Cole* (DDG-67). The USS *Cole* stretched 505 feet long and 66 feet wide, and it displaced 9,000 tons, making it about the size of a World War II light cruiser. The *Cole* bristled with firepower: ninety launchers for various long-range missiles, an automated 127 mm (5-inch) gun, six torpedo tubes, and an armed helicopter, the Navy version of the Army Black Hawk. Its radar allowed the ship to track and destroy attacking aircraft hundreds of miles away; its capable sonar and affiliated weapons ensured the detection and engagement of lurking enemy submarines. The *Cole* could plaster up to ninety different targets with smart Tomahawk cruise missiles, and each of those big sluggers could range out some fifteen hundred miles to deliver a thousand-pound-bomb-equivalent warhead. Whereas a 1945 light cruiser carried nearly 900 sailors, thanks to modern IT, only 281 were needed to run the *Cole.*

One thing that hadn't changed since the olden days was the risk to the crew. In the armed forces, those who fight on the ground generally see those on ships as much better off. The Marines live in both worlds, and they have strong views. Major General Julian C. Smith put it well on the eve of the bloody 1943 Tarawa landing: "Even though you Navy officers do come in to about a thousand yards, I remind you that you have a little armor. I want you to know that Marines are crossing that beach with bayonets, and the only armor they will have is a khaki shirt." As an admiral who had risen from the ranks once told an Army infantryman, the worst wardroom always trumps the best foxhole.

Yet that same admiral went on to explain that the life of the sea is inherently exhausting and dangerous. To the unforgiving challenges of wind and waves, familiar to Ulysses or John Paul Jones, one must add the back-straining, knuckle-scraping daily rhythms of checking and fixing the seagoing industrial plants that are modern ships. Each day sees struggles to move on icy decks or wrestle outsize, greasy equipment through narrow passageways or up slippery ladders. Engine rooms run hot as furnaces; you can't drink enough water. Things break down in odd, oily corners. Somebody must wriggle in to make repairs. To this, append the pleasures of living cheek by jowl with hundreds of other people. Often, there's another sailor six inches above your head when you go to sleep for a few hours. And the whole time, whether you're tossing in your narrow bunk or yanking on a flailing cable in the howling wind, you're one massive rogue wave or hull breach from going right to the bottom, full *Titanic,* in the time it would take someone to read this page. "For a sailor," the admiral concluded, "there's only about 5% difference, where you aim the Tomahawks or how much you shoot the 5-incher, between being in combat and just doing your job at sea. Unless something bad happens . . ."

Unless something bad happens. On October 12, 2000, it did.

Whatever the *Cole* and her crew did or didn't do, Commander Kirk S. Lippold owned it. A U.S. Navy captain — and the ship's commander always holds that title, regardless of his or her official rank — enjoys near-absolute authority. But as both Voltaire and World War II veteran Stan Lee wrote (Lee in reference to his superhero Spider-Man), with great power comes great responsibility. That, too, had not changed since the days of John Paul Jones.

An Annapolis graduate with a wealth of time at sea, Lippold knew his crew, his ship, and the dangerous waters they sailed. On the landing ship USS *Fairfax County,* he served off Lebanon in 1983 during the ill-fated Marine expedition to that troubled country. Aboard the cruiser USS *Yorktown,* Lippold participated in the fighting against Libya in 1985–86.

Not only had he heard of terrorist threats, he had been in operations that dealt with them, poorly in Lebanon, but well enough off Libya. Now he commanded the relatively new and very powerful USS *Cole,* transiting from the Mediterranean Sea to join the Fifth Fleet in the Persian Gulf.

Although today's U.S. Navy submarines and aircraft carriers use nuclear power and hence can sail for years on a single reactor core, a destroyer like the *Cole* runs on diesel fuel — and lots of it. Lippold and his sailors calculated that to get to the Persian Gulf with a proper reserve of fuel aboard, they needed to fill up with some 220,000 gallons of F-76 naval distillate diesel at about the halfway point. That meant a stop at either French-influenced Djibouti or Aden, Yemen. Djibouti was poor but pretty safe for this part of the world.

From the perspective of the U.S. embassy up in the mountain capital of San'a — that old Cold War North Yemen connection — refueling in Aden reinforced the U.S. commitment to the fractious country and its client in Yemen, President Salah. In 1999 and 2000, twenty-six U.S. Navy vessels had refueled before *Cole.* The Navy categorized it as a "brief stop for fuel," a few hours the destroyer would spend tied up to a narrow concrete manmade island with a diesel pipe stuck into it. The fuel dolphin was way out in the middle of the anchorage, well off the Aden waterfront. As long as the *Cole* stayed in the open harbor, what could go wrong?

So Commander Lippold and his crew made their way toward Aden. The *Cole* played by the rules, arriving the night before and steaming slowly back and forth outside the twelve-mile limit of Yemen's territorial waters. The *Cole* awaited clearance to enter, arranged for 7:30 a.m. on Thursday, October 12. The Americans, of course, were ready to get in and get going early. Lippold set his sea and anchor details at 5:46 a.m., just as the sun rose. His communications officer got on the radio.

"Aden Port Control, this is U.S. Navy warship USS *Cole,* channel one-six, over."

Nothing came back. The American tried again; no answer. This went on for about twenty minutes. Then the radio crackled.

"This is Aden Port Control. What time are you scheduled to come in?"

Kirk Lippold was about to find out why Indian-born British poet

Rudyard Kipling disdained the "fool . . . who tried to hustle the East." In Aden's harbor, adhering to a bureaucratic fashion that would please their long-gone Soviet mentors, the Yemenis treated the schedule as if it were Holy Writ, just below the Koran. If that accorded with a casual arrival to start the day, so much the better. The schedule offered a way to direct the pushy infidels. That suited the locals just fine. When Lippold asked to enter early, the predictable response came back: "You are going to have to wait."

The Americans waited. The day grew warm. Finally, at 7:46 a.m., close enough by Yemeni clocks, the harbor pilot arrived to guide the *Cole.* The Yemenis wanted to do it the easy way: push straight in and moor the *Cole* with its bow facing the shore and its left (port) side tied to the fuel dolphin. Since the cement structure stretched 350 feet, the U.S. destroyer would stick out 75 feet on either end and have a tough time backing out. Lippold knew there was a better way.

After Lippold plied the pilot with some tea and talked him slowly through the benefits of turning the ship's bow to the open sea, the Yemeni consented to Lippold's request. That did not speed up the Yemeni tugboats. Instead, confused by the change in direction, the two Aden harbor tugs took more than an hour to get the *Cole* next to the fuel dolphin. Then mooring lines had to be secured, checklists checked, and fuel pipes connected and tested. It all took until 10:31 a.m. By Lippold's reckoning, passing diesel looked to take up to eight more hours. The American skipper definitely wanted to get out of Aden in daylight. It would be tight.

Though the captain of a warship is, indeed, the sole accountable officer, he runs his crew through his key subordinates. Aboard USS *Cole,* Lieutenant Commander John Christopher Peterschmidt served as the executive officer (XO). As second in command, he was backed by a brace of lieutenants heading the various departments, such as engineering, weapons, communications, and supply. Command Master Chief James Parlier, the senior enlisted sailor, enforced standards, represented the interests of the crew, and supervised a group of tough, experienced chiefs who served as de facto foremen in each shop of this seagoing factory. Together, Peterschmidt and Parlier acted as Lippold's right and left arms, making things happen aboard the destroyer.

With the diesel flowing, *Cole* settled into a well-rehearsed routine.

Because the ship was not under way, nobody manned the bridge. On the deck, a dozen armed lookouts ensured that no unauthorized swimmers or craft approached the ship. Lippold considered putting out a rigid-hull inflatable boat with armed sailors, and he even got one ready to go. It didn't happen. The boat had to drop off the right (starboard) side, but that now aligned with the cement island. So the security boat stayed stowed. The armed sentinels proved more than capable of waving off one overly curious Yemeni small craft.

With the ship stationary and his crew carrying out their tasks, Lippold went to his cabin to catch up on paperwork. While he worked, Command Master Chief Parlier was near the helicopter deck at the ship's stern. The XO, Peterschmidt, fulfilled his role and rode herd on the refueling and the deck security.

About twenty minutes after Lippold went to his cabin, Peterschmidt came in. He and the ship's supply officer wanted to get rid of garbage. Enterprising Yemenis had offered to do it with three small boats. At first, Lippold said no. The *Cole* would be in the Persian Gulf in a few days and could dump trash easily at the big U.S. Navy base at Bahrain. Peterschmidt insisted. The Yemenis wanted only $150, and it would clear out a lot of the useless clutter that had accumulated since the ship left Norfolk, Virginia, almost two months earlier. The XO made a good point.

"Okay, fine," Lippold conceded. "Go ahead and bring the boats out to us. Let's pass the word and get everything off the ship." Anticipating approval, two trash boats had already snuggled alongside. Up on the deck, sailors passed bagged garbage down to the Yemenis.

Peterschmidt had more for his captain. The diesel flow ran fast that morning. He estimated they might be finished as early as 12:30 p.m., a welcome break. He recommended opening the galley early for lunch. Getting the crew fed quickly should help get the *Cole* on its way as soon as possible. Lippold agreed.

The XO announced the change in mealtime over the ship's intercom. Hungry sailors lined up gratefully; the *Cole* was known as a good feeder, and this day had started way too early. With all the activities humming and the lieutenants and chiefs in charge, Peterschmidt stayed below. He went to a compartment near the ship's stern. There he held a meeting of the destroyer's morale and welfare committee.

Topside, the two Yemeni garbage boats finished at about 11:15. Fireman Raymond Mooney watched them begin to pull away. A third boat approached from the shore, swinging in a nice wide arc off the port bow. White with red trim, thirty-five feet long, with two thin Yemeni men aboard, the boat slowed to approach the *Cole*. Mooney noticed that both of the locals smiled and waved. Well, they would soon get rid of the rest of the trash. The dinghy touched the *Cole*'s great gray flank, then it detonated with a brilliant flash and a thunderous roar.

The *Cole* heeled up and away from the explosion, then rolled down and in, its steel hull whipsawed by the burst of the equivalent of a five-hundred-pound bomb. The blast tore open a massive hole, forty feet across, like ripping an entire railroad boxcar out of the destroyer's port side. A third of the damage extended well below the waterline. Water poured into the ragged gap. The *Cole* began to list to port. If the breach went unaddressed, the ship would roll over and sink.

Only fast, purposeful action by the crew could keep their destroyer alive. A modern U.S. warship has numerous watertight compartments. Even in this massive explosion, only a few spaces had been "opened to free communication with the sea," to use the cold language of naval engineers. If the right hatches were closed fast enough, and if the electricity got going, and if the pumps worked, then the *Cole* would make it. No computer or machine held the answer. Saving the ship depended on people, trained sailors working with a will.

Such people were aboard, but not all could contribute. Fatefully, the device had blown right next to the mess deck, which was filled with sailors getting early chow. A quarter of the *Cole*'s sailors were out of the fray: four killed, forty-two wounded, and thirteen missing. Many of those lost possessed exactly the skills and experience needed most to stabilize the stricken warship. The bomb had knocked out all but one of the four engines. Flying debris severed communications, slashed electrical wires, and wrecked pumps.

The U.S. Navy trains its sailors, starting in boot camp, on how to keep their ships from sinking. Damage-control drills are run over and over until sailors can apply the measures in the dark, without electricity, with water sloshing around their ankles and the wounded moaning. Aboard *Cole*, that training kicked in. Small groups of sailors, some the designated damage-control teams, some in partially organized groups, some pickup gatherings, went to work. Slowly, very slowly, things got sorted out, not enough to restore the destroyer but enough to keep it above the waves.

Below decks, Command Master Chief Parlier, a medical specialist by rating, took charge of the wounded. Up in the sun, Lippold and Peterschmidt reestablished their top-deck watch. One attack might well lead to a follow-up. After the explosion, any boats that approached the *Cole* received warning to anticipate a very hot reception. Then the battle turned to containing the influx of water and restoring electrical power. Casualty treatment and evacuation continued. Through heroic actions, uncompromising on-scene leadership, and some plain old grit among a lot of sailors, the *Cole* stayed afloat. The destroyer departed Aden seventeen days later, bloodied, battered, but game, cradled on the deck of a massive commercial salvage ship.

The Yemen government initially blamed the event on an internal explosion on the *Cole.* They'd never heard of terrorists. Well, maybe they'd heard of some. But not in Aden. And not this time. Weeks later, President Salah's people coughed up some suspects.

The U.S. reaction improved only marginally on that of Yemen. Like all its predecessors, the current presidential administration could not determine how to address terrorism. Was it a crime, to be handled by law enforcement? Was it an act of war, to be met by retaliation? As Presidents Richard Nixon, Gerald Ford, Ronald Reagan, and George H. W. Bush had before him, President William J. Clinton chose both of the above.

In Clinton's time, the military option had been tried on multiple occasions in Iraq and even once against Osama bin Laden's al-Qaeda. Bitterly condemned for launching Tomahawk cruise missiles into Sudan and Afghanistan on August 20, 1998, in response to al-Qaeda bombings of the U.S. embassies in Tanzania and Kenya, Clinton endured a welter of carping from all segments of the political spectrum; he was criticized for the obviously ineffective targeting, for swatting flies with high-tech hammers, and for blowing over dingy tents with millions of dollars of smart weapons. Some pundits and Clinton political adversaries argued that the whole episode amounted to a calculated distraction from an ongoing political impasse involving a White House sex scandal. It all smacked of a contemporary Hollywood satirical film, *Wag the Dog*, dramatizing such an occurrence.

Now, with the *Cole* listing and smoking in Aden harbor less than a month before the hotly contested 2000 presidential election, Clinton leaned toward the deliberate legal approach, treating terrorism as a felony. Dispatched in the wake of the attack, investigators from the FBI combed the *Cole*'s deck and the fuel dolphin for physical evidence, took statements from survivors, and tried to assemble a case. The forensics impressed all involved, except the Yemenis, who didn't want to hear any of it. The evidence pointed clearly to al-Qaeda, of course. But it took time to be sure. Nobody dared raise another *Wag the Dog* episode.

In Washington, all the right things were said about bringing the attackers to justice. Nothing much happened beyond handwringing. Then came the 2000 election. That dragged on for five weeks, into the Florida recount, replete with hanging chads, partisan charges and countercharges, and finally ended by a highly controversial Supreme Court ruling. The incident in Aden got moved to the back burner, then off the stove, and then it drifted out of the kitchen altogether. Meanwhile, the *Cole*'s dead came home and were buried. The *Cole*'s crew flew home and were honored. The *Cole* itself was carried home and put into dry dock.

The institutional U.S. Navy, admirals and civilian officials alike, tried to look out for Kirk Lippold, the man who had held his crew together and led the highly successful effort to save his ship. A strong push aimed to get him promoted; indeed, the selection board recommended he be promoted to captain. The Senate Armed Services Committee, led by former enlisted sailor, Marine officer, and secretary of the Navy John Warner, refused to consider it. Like the Arab Street, the Navy has its grapevine. Just enough questions had been raised about Lippold's judgment. Yes, the opposition had been clever. True, the suicide-boat strike had been tough to predict. But the whispers went around: there'd been no U.S. small craft circling the ship; the bridge was unmanned; garbage dinghies were allowed to come alongside when a trash offload wasn't really necessary; sailors had been sent to lunch while the ship was fueling in a dangerous foreign harbor; and the captain had been sitting in his cabin in an armchair, doing paperwork like a clerk, while the entire awful event transpired. Throughout the fleet, the officers and chiefs nodded knowingly. The captain bore full responsibility. Unfair or not, in the end, that settled it. Lippold never led sailors again.

The *Cole*, however, did sail and fight once more. It took much time and massive effort and had its own heart-wrenching aspects that went beyond the hard work. When rebuilding started at the Ingalls shipyard in Pascagoula, Mississippi, and cutters took apart the crumpled hull and bulkheads, sad reminders turned up. While ship fitters labored, dedicated teams recovered human remains. For almost a year, skilled technicians found and honorably returned the pieces of the dead sailors to their grieving families. By then, the U.S. was at war with the people who had hit the *Cole*.

The explosion abeam the USS *Cole* gladdened the heart of Osama bin Laden. "The pieces of the bodies of infidels were flying like dust particles," the tall Arab told his followers. "If you would have seen it with your own eyes, you would have been very pleased, and your heart would have been filled with joy."

Osama bin Laden saw visions, all right. His designated cameraman never made it into position to record the strike on the *Cole*. Plenty of public images depicted the aftermath but not the actual impact. Nonetheless, the charismatic terrorist chief saw it all in his active mind. He dreamed of it, described it, and even wrote a poem about the event.

Who was this ascetic visionary who had taken on the mighty United States Navy?

Born in 1957 in Jiddah, Saudi Arabia, on the Red Sea coast, Osama bin Laden came from a wealthy family with roots in Yemen. His billionaire father, Mohammed, headed a huge construction firm that employed up to forty thousand people at times and did business throughout the Middle East. A favorite of the Saudi royal family, the firm even did renovation work on the grand mosque at Mecca, the destination of every Muslim pilgrim who made the haj. Young bin Laden's money and extensive regional connections earned him the nickname "the Contractor" among his followers. Bin Laden grew up working with heavy equipment and helping in and around building projects. He learned how to drive a bulldozer and handle an excavator, and evidently he liked it. Educated well, if incompletely, at King Abdulaziz University in Jiddah, bin Laden left school in 1979. He traveled to Afghanistan to join the jihad, the holy war against the Soviet Fortieth Army occupiers. Already a devout Wahhabi Muslim, bin Laden saw it as his religious duty to defeat the atheistic Russian Communists. God had called him.

In Afghanistan and even more in the jihadi sanctuaries in neighboring Pakistan, bin Laden played to his strengths. Although tall and athletic, he proved an indifferent rifleman and guerrilla leader, and he may have seen action in a firefight or two. The tough Pashtuns and Tajiks thought him too rich and too Arab for their rough tastes. Instead of leading raids and ambushes, bin Laden focused on building training sites, boring tunnels in mountainsides, raising money, buying arms, and recruiting non-Afghan volunteer fighters. The Contractor proved quite adept at these administrative and logistical tasks. Given that most of the Afghan jihadis could barely read or write, an organizer like bin Laden was a handy sort to have on the team. As bin Laden's emissaries canvassed mosques and madrasas (religious schools) in Saudi Arabia and the smaller Gulf states, dunning believers for donations to support the jihad, he built a funding network that proved resilient and useful for follow-on endeavors. The Contractor called his network of string pullers al-Qaeda, meaning "the Base." It proved to be a firm foundation for expanded jihads.

The Soviet withdrawal from Afghanistan in 1989 convinced Osama bin Laden that Allah favored a vigorous jihadi campaign. The collapse of the powerful Communist state in late 1991 seemed to validate his assessment. Prayer, money, fighting, and patience had bested Soviet tanks, jets, and firepower. Now one of the two great superpowers was gone. Bin Laden began to think about how to eliminate the other one, the secular, arrogant United States.

Bin Laden believed the time had come to re-create the ideal Islamic world as it had once existed, at least in his mind. Strongly influenced by the Islamist ravings of Sayyid Qutb of Egypt's Muslim Brotherhood, bin Laden saw the world as divided between the believers in the Dar-es-Salaam (the "House of Peace," places under strict Islamic shari'a law) and the infidels in the Dar-al-Harb (the "House of War," those places not under shari'a). Bin Laden saw America as the far enemy and accommodators like King Fahd of Saudi Arabia, Hosni Mubarak of Egypt, and Ali Abdullah Salah of Yemen as the near enemy. Both had to be rolled back and then destroyed. Once these interlopers departed, then a great Islamic caliphate would rise. As the Prophet Muhammad's true successor, the caliph would combine completely all matters of church and state, as supposedly had been done in the seventh century. An outer world would still hobble on, as backward Europe had struggled through its Dark Ages during the great epoch of Arab-led Islamic conquest. Whatever might happen next out there in the shrunken, ravaged, infidel House of War mattered little to bin Laden. He focused on establishing the caliphate.

Hearkening back to a fabled golden age and designating enemies inside and outside the Islamic world appealed greatly to the poor, uneducated Middle Eastern, North African, and South Asian people longing for something better. To those with more circumspection and longer memories, it all smacked way too much of Adolf Hitler's National Socialist movement with its bloody-minded appeal to a supposedly glorious past and designation of nefarious foes. Of course, Osama bin Laden modestly assumed that he'd make a fine caliph. People always require a führer.

Bin Laden's other nickname was the Sheikh (meaning "Chief"), a title normally awarded to a tribal elder in a nomadic Bedouin family. Inside al-Qaeda, bin Laden acted as the contractor, forever taking pains with details of planning, preparation, and execution. To the wider world, though, he relished the role of the sheikh, garbing himself in traditional flowing robes and, in later days, accessorizing with a camouflage jacket and a clean little AK-74 Kalikov (a pintsize Kalashnikov), an automatic carbine taken from a dead Russian paratrooper. The term *sheikh* also denoted religious learning well above the norm. Bin Laden liked to style himself as a scholar of Islam and an explainer of Islamist Muslim Brotherhood theory. But bin Laden was better at contracting than religious thinking.

Bin Laden needed a more accomplished intellectual to do the theological heavy lifting. That man, bin Laden's number two, was the Egyptian surgeon Ayman al Zawahiri, a Muslim Brotherhood type who'd proved too devious and violent even for that vicious group. Zawahiri formed Egyptian Islamic Jihad, which strongly supported al-Qaeda during the Soviet-Afghan War. Much brighter than bin Laden, Zawahiri became the Sheikh's physician, spiritual adviser, and source of Islamic learning. When bin Laden spoke or wrote in his guise of Sheikh, the words often sprang from the stern, doctrinaire, but much more prolific Zawahiri.

In 1990, bin Laden objected strongly when King Fahd of Saudi Arabia invited the Americans into that country. The idea of the far enemy consorting with the near enemy incensed bin Laden, especially when it polluted the land of Mecca, Medina, and the Prophet Muhammad himself. When the Sheikh got too vocal, Fahd's security services pulled bin Laden's passport. Arrest seemed certain to follow, but the Contractor did what he did best. He made arrangements and headed to lawless Sudan, with some assistance from his old comrades from the anti-Soviet campaign. Once the horse had left the barn, the Saudis froze his financial assets and revoked his citizenship. For bin Laden, these became credentials, not impediments. He always found ways to get his money.

In Sudan, unimpeded by the weak government, bin Laden strengthened the al-Qaeda network. He built a road from Khartoum to the Red Sea coast and set up charitable offices in Muslim areas near and far. Along with its established Arab-country connections, al-Qaeda gathered money and recruited operatives in Azerbaijan, Bosnia, Burma, Chad, Chechnya, Cyprus, Indonesia, Kenya, Malaysia, Mali, Niger, Nigeria, the Philippines, Thailand, and Uganda. The group also developed ties in European cities such as Budapest and Vienna, as well as in American cities such as Atlanta, Boston, and New York. Ever the Contractor, bin Laden built his al-Qaeda, his foundation, first. When bin Laden acted, he would do so with force and imagination.

Bin Laden learned early, while battling the Russians, that spectacular events brought in more money and recruits. If the infidels retaliated, that worked even better. Western bombs and missiles, even the smart ones, left a mess and often killed bystanders — or at least those who could be portrayed as such. Always, bin Laden aimed for a big splash, as much free publicity as possible. It all attracted money, arms, and recruits to his cause.

Throughout the 1990s in Sudan, bin Laden made contacts and experimented. He and his associates got involved in the 1993 World Trade Center attack; the 1993 carnage in Mogadishu, Somalia; and the 1996 Khobar Towers bombing in Saudi Arabia. They did not lead. They facilitated, networked, and learned.

In pursuit of a longer contact list, bin Laden thought broadly. He reached out to Saddam Hussein's embattled Iraq. The dictator had his hands full with American-led no-fly zones squatting on his restive country. Saddam had no interest in bin Laden's caliphate dream, but help was help. The Iraqi leader let bin Laden set up some cells in Kurdistan. If the terrorists eventually killed rebellious Kurds, that suited the Baathists in Baghdad just fine. More important, the Iraqis quit working against al-Qaeda. This nonaggression arrangement aided bin Laden's operations throughout the Middle East.

In an even more difficult initiative, the Sheikh also tried to mend fences with the Shiite Persians of Iran. Although suitably anti-American and vigorously Muslim, Iran's ayatollah clerical clique and their flocks followed Shia, a sect that believed that only a blood descendant of Muhammad could rule the Islamic world. Most Muslims adhered to the Sunni teaching that God designated successors to the Prophet Muhammad without regard to family ties. Shiites constituted a minority of the populace in most Muslim countries, but in a few states, like Iran and Iraq, the Shia formed the majority. As a Wahhabi Sunni, bin Laden formally considered the Shiites apostates, worse than nonbelievers. They were wrong believers and deserved their fate. Still, an enemy of one's enemy can be one's friend. And the Persians of Iran boldly faced the far enemy, America. So bin Laden cut some deals.

All of this activity did not go unnoticed by enemies near and far. The United States and its allies discovered Osama bin Laden and al-Qaeda, although nobody really knew what to make of either the man or his group; they were just another set of troublemakers in a region teeming with the same. Only a few intelligence analysts in America knew the name Osama bin Laden. His colleague Zawahiri had a higher profile. Egypt's President Hosni Mubarak in particular wanted a piece of Zawahiri and his Egyptian Islamic Jihad faction, doubly so once a hit squad took a shot at him in June of 1995 during a trip in nearby Ethiopia.

Frustrated with Sudan's unwillingness to crack down on its tenant terror organizations, al-Qaeda among them, America pressured the Islamist Sudanese government. Washington arranged a series of increasingly harsh economic sanctions to squeeze the poor desert state. The squeeze worked. Early in 1996, Sudanese security services asked the Saudis to take back their wayward son. Saudi Arabia refused. Weak Sudanese and duplicitous Saudis looked the other way when bin Laden and his al-Qaeda leadership flew to Jalalabad, Afghanistan, on May 19, 1996. The United Arab Emirates graciously allowed the charter jet to refuel. The Gulf Arabs relaxed. Let the Sheikh see how many contracts he could arrange among those ignorant Pashtun highlanders.

Bin Laden and Afghanistan made a good combination in mid-1996. After a brutal civil war in the wake of the Soviet withdrawal of 1989, the strongest faction stood ready to take over in the crumbling capital of Kabul. Mullah (meaning "preacher") Mohammed Omar and his Taliban (meaning "student") movement shared bin Laden's vision of a caliphate under unyielding Islamic law. Omar thought that Afghanistan provided a great place to begin. That worked for bin Laden.

Ethnic Pashtuns dominated the Taliban, whose strength came from the south and east along the Pakistan border. As 40 percent of the population, the Pashto speakers made up a plurality that traditionally ruled the mountainous desert country, at least to the extent that the aggressive, illiterate Afghans consented to be ruled at all. Other groups, such as the Dari-speaking Tajiks (Dari is a variant of Persian Farsi) in the north (27 percent), the Shiite Hazara in the central highlands (10 percent), the Uzbeks (10 percent), the Turkmen (10 percent), and smaller groups rounded out the picture. These disparate peoples had once opposed the Communist Russians. When the Soviets left, they turned on one another. The Taliban emerged from the scrum as the best-led, best-organized, and best-funded group. Bin Laden was now their guest while Mullah Omar's Islamist bands prepared their victorious final march on Kabul. In September, the Taliban seized the capital and proclaimed their Islamist republic. The caliphate had its first member state.

Two moves in five years depleted bin Laden's financial resources. Saudi attempts to block his access to his money did not help. Despite his high aspirations and help from Mullah Omar, bin Laden understood that mounting a major operation, the kind he desired, took time. The Contractor's team faced a few rebuilding seasons. He used the time to get organized for the next phase.

Strategy came first. Back in Afghanistan, scene of his glory days, bin Laden received a revelation that he attributed to Allah. The Communist Russian superpower had failed, pulled out, and then imploded, all thanks to holy warriors in Afghanistan. If al-Qaeda could lure the United States to the same killing ground, that might well cripple the far enemy, run them out of the entire region, and clear the way for the wider caliphate. Bin Laden saw a winning formula.

Getting America to come into Afghanistan looked tough. The old Soviet Union had shared a border with Afghanistan and had a long history of meddling in the country, going back to the empire of the Romanov czars. But the U.S., a strong sea power, seemed unlikely to intervene in land-locked Afghanistan. Even American opposition to the Soviet invasion had been handled indirectly through Pakistan. The Sheikh determined to do something so large that the consequent American retaliation had to result in large numbers of people on the ground in the red dust of the Registan Desert and the jagged heights of the Hindu Kush Mountains. Then the hit-and-run tactics, the bomb throwing, the war of a thousand cuts, promised to take its toll. The Americans would run out of patience before bin Laden ran out of motivated terrorists. So that was the plan: a baited ambush to lure, snare, and then shred the American eagle.

His course chosen, bin Laden decided to put himself on the map to keep his side stirred up and interested. He chose to announce his views to the world. On August 23, 1996, as the Taliban staged to take Kabul, the Sheikh issued a fatwa, a religious order, to his al-Qaeda faithful and all like-minded souls. The title aimed high: "Declaration of War Against the Americans Occupying the Land of the Two Holy Places."

Not an easy read, the lengthy, closely spaced sermon combined exhortations to his men, appeals to the wider Islamic community, and admonitions to the hated infidels, most notoriously the United States. It included quotes from the Koran, threats against the collaborating Saudi Arabian royals, and even some snatches of poetry. A few key segments convey bin Laden's strategy, no doubt steered by the grim Zawahiri.

"It should not be hidden from you that the people of Islam had suffered from aggression, iniquity, and injustice imposed on them by the Zionist-Crusaders alliance and their collaborators," bin Laden stated. Here, he identified the enemy and reminded all of what had happened. In his fatwa, bin Laden mixed references to the medieval Crusades with comments about recent events, like the fighting in Mogadishu, Somalia, in 1993 and the attack on the Khobar Towers in Saudi Arabia two months earlier. The infidels had thwarted the believers all too often in the past, and now they would pay for it.

"Nevertheless," the al-Qaeda chief continued, "it must be obvious to you that, due to the imbalance of power between our armed forces and the enemy forces, a suitable means of fighting must be adopted, i.e., using fast moving light forces that work under complete secrecy." In other words, he and his followers had no intention of playing the staked goats in a repeat of Desert Storm. The Sheikh knew his people and their preferred manner of warfare. He freed them to follow their preferences.

Bin Laden explained whom he had in mind for such tasks. "Those youths," he wrote, "will not ask you for explanations, they will tell you singing there is nothing between us need to be explained, there is only killing and neck smiting." Along with a threat, the fatwa was also a recruiting call. By taking on the great, deadly American far enemy, bin Laden sought to inspire many young Muslims to join the historic campaign to come. Certain eighteen-year-old males like to inflict violence and pain. Bin Laden offered them plenty of it.

The Sheikh concluded with a warning to America. "Terrorizing you," he proclaimed, "while you are carrying arms on our land, is a legitimate and morally demanded duty." The al-Qaeda leader had made his case. In the Arab Street, and in the even more wild-eyed Pashtun Street in southern Afghanistan and nearby Pakistan, the fatwa generated excitement, donations, and volunteers. Bin Laden gave a voice to centuries of frustration, and he had poked his sharp spear into the hide of the American beast. A tiny number of Muslims joined him. A much wider swath nodded in approval. An even bigger bunch proved indifferent. It wasn't their fight.

In the United States, the vitriolic fatwa met a collective yawn. Political attention focused on the upcoming presidential election. The rantings of an obscure Islamist out in illiterate, desolate Afghanistan went nowhere in the land of the far enemy. Only a small number of U.S. intelligence analysts noticed. The vast American public did not. Osama bin Laden decided to get their attention.

Along with being an accomplished construction contractor, bin Laden displayed a creative streak, a definite architectural bent. His operations showed this vibrant tendency. The three big ones—the 1998 African embassy explosions, the *Cole* suicide strike, and the 9/11 U.S. airliner plot—all demonstrated his ingenuity, maximized news coverage, and showed his great flair for the dramatic. Had he gone into cinema instead of international terrorism, the Contractor might well have produced large-scale spectacles that rivaled the best work of Cecil B. De Mille. Doing things like this, orchestrated from the deserts and mountains of Afghanistan and out of the reach of watchful U.S. intelligence entities, took time.

For two years after the declaration of war, bin Laden and his followers schemed and readied themselves. After 9/11, American television viewers saw endless loops of al-Qaeda training videos. The propaganda highlighted men going through rudimentary obstacle courses and firing various weapons. Recruits swinging arm over arm along overhead ladders, scaling walls, jumping ditches, and shooting a lot of AK bullets typified the short film snippets. It looked exciting and menacing at the same time, terrorist regiments getting ready to go head-to-head against the products of Fort Benning and Parris Island.

In truth, though, that brand of rigorous, repetitive conventional training flew in the face of the hit-and-run style bin Laden chose. With the exception of attending religious-indoctrination classes, these people didn't train in any coherent or recognizable Western way. They did not spend hours honing their marksmanship or practicing fire and maneuver tactics. Some courses began with a degree of hazing or deprivation more suited to a low-end college fraternity than an organized military program. Physical development, such as weightlifting, running, and marching, was up to each individual and typically honored in the breach. Prospective terrorists brought their talents and experience. Getting stronger, shooting better, and learning skills such as navigating, providing first aid, and maintaining weapons happened haphazardly, if at all. The camps did not teach young men how to be parts of a conventional, battle-drilled tactical unit.

Rah-rah films aside, the real al-Qaeda preparatory work involved master-and-apprentice, do-as-I-do education in, for example, how to make various bombs, how to penetrate government checkpoints, and how to create false documents. The camps taught volunteers how to be furtive criminals, which is to say how to get around without being noticed. Much of that practical work seemed like gang-initiation rituals and happened in the local souk or village or, sometimes, even in prospective target countries. This was why lobbing Tomahawk cruise missiles into a terrorist center rarely did much. Many men passed through these centers, but few stayed long. Hardly anyone was training in them, at least not in a Western sense.

Bin Laden's people did, however, spend many hours observing potential targets. As British soldiers say, time spent on reconnaissance is seldom wasted. Thorough target assessment became a hallmark of al-Qaeda preparations. They selected possible sites, tested security, adjusted, and picked anew. For each proposed major operation, al-Qaeda agents checked and rechecked the routes and the chosen targets. Many plans were rejected because the reconnaissance did not pan out.

Timing also meant a lot. Al-Qaeda operatives brilliantly used the Internet, cell phones, handwritten messages, visual signals, and careful dress rehearsals to coordinate far-flung strikes across continents and oceans. Even highly effective terrorist organizations like the Iranian-backed Hezbollah (Party of God) could not match this aspect of al-Qaeda's planning and execution. In this regard, bin Laden stepped well outside the usual desert-nomad fighting style and adopted a method more typical of Western special operations teams.

These check-ride test runs had to be balanced with the organization's fervent devotion to secrecy. Cell members knew only what happened within their own cell. One cell leader handled contact with higher-ups, which might in fact be a lateral tie to another cutout element. Some cells served as backups, duplicates, or reserves. Zawahiri's Egyptian Islamic Jihad people referred to this as the "bunch of grapes" approach. Each cell did one task but could be removed without compromising the entire plant. Almost all terrorists used this method, but few applied it as rigorously as al-Qaeda.

In the Contractor's network, people followed orders. A watch team observed a designated street. A bomb-construction cell wired explosives to a truck. If one cell got captured or killed, only that element failed. The rest moved on. Despite emerging from a culture that did not carefully case targets, did not value split-second timing, did not rehearse much, and often leaked like a sieve, bin Laden and his lieutenants proved hard taskmasters. It took many months to move all of these pieces onto the board. They put people in place, assessed, adjusted, refined, and re-thought. Sometimes, the complex schemes miscarried or just stopped. Once all stood ready, however, al-Qaeda executed.

On August 7, 1998, just after 10:30 a.m., truck bombs detonated almost simultaneously at the United States embassies in Nairobi, Kenya, and Dar-es-Salaam, Tanzania. The Kenya explosion killed 212, including a dozen Americans, and wounded about 4,000. The smaller Tanzania strike killed 11 and wounded 85. The timing had been chosen to coincide with the eighth anniversary of the arrival of U.S. forces in Saudi Arabia for Desert Shield.

As the name of the Tanzanian capital indicates, many of the victims were Muslims. No matter to bin Laden; they must have been collaborators, near enemies, stooges foolishly hanging around the American legations. Their deaths ran up the jihadis' score and served the greater cause. This butchery rang the bell in Washington. Within days, the U.S. Central Intelligence Agency fingered al-Qaeda, as bin Laden had expected they would.

Destruction done, the Sheikh waited for the U.S. response. And he waited. He moved between safe houses in and around Kandahar, Afghanistan. Some sources placed him in Khost, near the Pakistan border. Finally, on the evening of August 20, 1998, U.S. Navy warships fired sixty-six Tomahawk cruise missiles into Afghanistan, aiming at the Farouk and Jihad Wal terrorist training camps outside the city of Khost. Another thirteen cruise missiles hit a suspected nerve-gas site related to al-Qaeda, the al-Shifa pharmaceutical factory near Khartoum, Sudan. The Americans code-named the retaliatory strike Operation Infinite Reach.

They didn't get bin Laden, although the CIA estimated they came close. Some reports claimed they missed him by a few hours. Other accounts disagreed, saying he'd never been in the targeted camps. Chasing rumors with Tomahawk cruise missiles seemed like a poor use of sophisticated smart weapons. Estimates of casualties ranged from six to thirty-four al-Qaeda trainees, hardly an impressive bag. At the al-Shifa plant, one man died and ten were wounded. The Sudanese claimed the factory held no nerve gas. The CIA argued that it did. The entire attempt highlighted the poor resolution available to U.S. intelligence. It all amounted to a gesture, not a serious counterstrike.

Frustrated by the limited U.S. response, searching for a way to attract a large troop deployment into the Afghan rat's nest, bin Laden reviewed other plans in train. He elected to try something bigger, hitting an actual U.S. warship, the same kind that had launched those Tomahawks. The Aden operation followed.

The first attempt, on January 3, 2000, went after the guided missile destroyer USS *The Sullivans* (DDG-68), sister ship of the *Cole*. The mission failed when the overloaded explosive boat sank. The al-Qaeda men erred, but they learned. They watched other U.S. Navy ships come and go. They rehearsed. Then they recocked and tried again on October 12, 2000, against *Cole*. This time it worked.

Osama bin Laden believed this attack on a U.S. warship would bring a massive American reprisal. Again, the Sheikh and his key leaders moved around to different secure locations: Kabul, Kandahar, Khost, and Jalalabad. Zawahiri split off from bin Laden to ensure one of them survived. Bin Laden awaited the rain of cruise missiles, or maybe, just maybe, U.S. soldiers or Marines. This time, nothing at all happened. The far enemy appeared to be wholly inert.

Hitting two embassies hadn't done it. Almost sinking a U.S. Navy destroyer failed to work. At an acrimonious White House meeting, frustrated with the unwillingness of the military to get entangled in a manhunt in landlocked Afghanistan, State Department counterterrorism coordinator (and former Army infantry officer) Michael A. Sheehan pointed at those in uniform and said: "Does *Al Qaeda* have to hit the Pentagon to get [your] attention?" The sarcasm got a few chuckles around the table. On the far side of the world, Osama bin Laden wasn't laughing. He knew what he had to do next.

2

9/11

Death is art.

— HASAN AL-BANNA

TUESDAY, SEPTEMBER 11, 2001, was a lovely, sunny day up and down the American East Coast. That brilliant morning was like so many others. Mothers got their children ready for school. Men and women headed off to work: to factories, to offices, to hospitals. Thousands filed into the massive cement Pentagon in Washington, the main command center for the entire U.S. Armed Forces. Thousands more entered the gleaming Twin Towers of the World Trade Center in New York, hub of commerce. Morning commuters arrived by bus and train; many others crowded the highways.

Some went to the airports, like Boston's Logan, New Jersey's Newark, and Washington's Dulles. American 11, United 175, American 77, and United 93, four flights among some forty-five hundred scheduled on that warm September morning, would never reach their announced destinations.

"We have some planes."

The radio crackled at 8:24 a.m. The air traffic controller at Boston Air Route Traffic Control Center — referred to as Boston Center — was unsure he'd heard correctly. The transmission originated from American

11, en route from Boston to Los Angeles with ninety-two aboard, including the crew.

"We are returning to the airport," said a man in thickly accented English, definitely not the pilot. "Nobody move," the foreign male continued, broadcasting over the radio, presumably mistaking it for the airliner's intercom. "Everything will be okay," the man said. "If you try to make any moves, you'll endanger yourselves and the airplane. Just stay quiet."

Boston Center flight controllers began to notify other authorities of a likely hijacking. Within ten minutes, calls went out to the military's Northeast Air Defense Sector in Rome, New York. "Is this real-world or exercise?" the Air Force duty officer asked. The Boston controller responded, echoing words said at Pearl Harbor on December 7, 1941: "No, this is not an exercise, not a test."

The first call said it all: "We have some planes." The "we," of course, was al-Qaeda, busily carrying out what had become known inside Osama bin Laden's circle as "the planes operation." The "have" resonated as teams of terrorists took over four aircraft. "Some" kept the infidels guessing; was there more than one taken? How many? It caused the Americans to scramble all over looking for others, unsure, for several vital hours, just how many airliners had been captured. As for "planes," well, a modern jet airliner filled with volatile fuel, a dead crew, screaming passengers, and determined hijackers in the cockpit did not bode well. In all past hijackings, the bad guys eventually landed and then, with passengers held hostage, made demands.

This one went a lot differently. By 8:42 a.m., United 175, another Boeing 767 out of Boston headed for Los Angeles with sixty-five aboard, had been taken too. At 8:51 A.M., American 77, a Boeing 757 going from Washington to Los Angeles with sixty-three passengers and crew, also came under terrorist control. It soon became obvious that these hijackers did not intend to land and negotiate.

By the time the trouble began, American 77's transponder no longer squawked its position to the air traffic controllers. U.S. air-defense radar receivers were pointed out to sea or toward Canada and Mexico, still waiting for the long-gone Soviets. Without a transponder, American 11 disappeared, a stealth airliner crossing eastern America. But some information trickled out. Two cool-headed flight attendants, Betty Ong and

Madeline "Amy" Sweeney, used onboard telephones to call the American Airlines operations center. They reported both pilots and two flight attendants down, as well as a passenger, all stabbed and cut. With five terrorists in charge and at least one at the controls, the big jet made a sharp turn to the east. Ong said the airliner was "flying erratically."

"Something is wrong," Sweeney said in her next call. "We are in a rapid descent." The 767 swung low toward the south end of Manhattan, engines roaring, pushing toward the spires of the New York City skyscrapers below. Sweeney got in one more sentence.

"Oh my God, we are way too low."

At 8:46 a.m., American 11 struck the North Tower of the World Trade Center. At 9:03, United 175 slammed into the South Tower. At 9:37, in Washington, DC, American 77 bounced once on the west-side helipad, then careened into the Pentagon. Three strikes, three fireballs, and thousands dead, all in less time than it took to read the morning newspaper. One more aircraft was still out there: United Flight 93.

It is the nature of military operations to go wrong, in whole or in part. Osama bin Laden never attended a war college, but he knew well from his time in Afghanistan and his experiences in construction that no matter how good the plan, no matter how thorough the rehearsal, no matter how excellent the equipment, people make mistakes. The great Prussian soldier and theorist Carl von Clausewitz called this friction, like that caused by rubbing two chunks of wood together. He attributed it to the interaction of two contending bunches of humans and the influence of the surrounding environment. "Everything in war is very simple," he observed, "but the simplest thing is difficult."

Clausewitz elaborated. Friction arose from the many, many interactions of individuals and groups, including those on one's own side. Physical exhaustion and mental stresses added up. Intelligence was always poor; the enemies refused to display themselves like pawns on a chessboard. Attempts by numbers of people to coordinate their activities meant that "countless minor incidents — the kind you can never really foresee — combine to lower the general level of performance, so that one always falls far short of the intended goal." It amounts to Murphy's Law: What can go wrong will go wrong. A lot of stuff goes wrong in military operations.

So it is with sports, politics, commerce, and the construction business. What makes armed conflict even more subject to Murphy's Law, and renders even simple acts so difficult, involves the danger of sudden death or serious injury. When you bet your life and those of others, fear, bravery, and strong emotions play huge roles. Killing is easy, but dealing with the act is not. Punches get pulled, hesitations occur, and though most are aggressive in the first contact, few stand up so willingly under fire the second time, or the twenty-second, let alone the hundred and second.

Tough training, good leadership, mission rehearsals, and certain equipment (maps, compasses, radios, cell phones, and computers) help accommodate friction. Certain modern information technology proponents even claim to be able to end friction altogether through near-perfect situational awareness. No hardware or software, though, removes the tired, confused, scared human from the equation. The clash of people bent on killing each other generates friction. It arises from the very nature of warfare, as essential to it as wetness is to water. Smart commanders allow for it with backups, reserves, and alternatives. Although not a trained soldier, Osama bin Laden understood friction.

Friction came in an extra dose for the al-Qaeda team designated for United 93. Scheduled for an 8:00 a.m. departure from Newark, New Jersey, to San Francisco, the Boeing 757 did not take off until 8:42. By then, terrorists controlled or were moving to take the other three planes. American 11 was only minutes away from ramming into the North Tower of the World Trade Center. United 93's crew of seven and thirty-three of its passengers knew nothing of this. Four men aboard knew it only too well. They recognized that they were running far behind schedule.

At 9:19, after both towers had been hit in downtown New York and with American 77 turning toward its death dive into the Pentagon, United Airlines flight dispatcher Ed Ballinger sent out a bulletin via electronic chat: "Beware any cockpit intrusion. Two a/c [aircraft] hit World Trade Center." The captain of United 93, Jason Dahl, responded at 9:26: "Ed, confirm latest mssg plz — Jason." The airliner soared over eastern Ohio at thirty-five thousand feet.

Ziad Jarrah of Lebanon and Saudis Ahmed al Nami, Saeed al Ghamdi, and Ahmad al Haznawi sat in the first-class section. Jarrah, the twenty-six-year-old pilot-trained operative, had a seat in the first row, nearest to the flight deck. Nami, twenty-three, and Ghamdi, twenty-one, sat in the third row across the aisle, diagonally behind Jarrah. Haznawi, age twenty, was in the sixth row on the same side of the cabin as Jarrah. The other three provided security, or muscle, as they called it. All four had trained in Afghanistan. Jarrah had been in and out of America seven times, starting in June 2000. He held a private pilot's license earned in Venice, Florida.

Friction left the team one security man short. Sharp-eyed immigration authorities in Orlando, Florida, refused entry to its designated fifth member, Mohammed al Kahtani. The late takeoff put the terrorists behind their comrades. Although the plan called for the teams to storm the cockpit thirty minutes into each flight, on United 93, Jarrah missed his cue. It was enough time, just barely, for the flight crew to get wind of the threat.

The hijackers acted at 9:28 a.m., breaking into the cockpit. Pilots Jason Dahl and Leroy Holmes evidently resisted. On their displays, controllers at Cleveland Center watched the airliner drop seven hundred feet. Garbled transmissions included the word *mayday* (in-flight emergency) and repeated commands of "Get out of here." The security men finished off the pilots, then killed a female flight attendant. Jarrah stepped forward and grabbed the controls.

"Ladies and gentlemen, here the captain," he intoned, nervously and ungrammatically. "Please sit down," he went on. "Keep remaining sitting. We have a bomb on board. So sit." The 757 turned sharply east, toward Washington, DC. Jarrah turned off the transponder. United 93 slipped into the ether.

With the terrorists busy at the front end of the aircraft, passengers and remaining flight-crew members started making phone calls. They used their personal cell phones and the installed commercial telephones. Jarrah and his men did not stop the calls, a major error. The passengers found out what had happened at the World Trade Center. They figured out that they, too, were doomed to crash.

The passengers talked. One caller said they voted. An undetermined number of them chose to act. "Everyone's running up to first class," said one woman in her final phone message. "I've got to go."

At 9:57 a.m., a pickup team of passengers rushed to the front of the plane. The flight recorder reflected that Jarrah tried rolling from side to side and pitching the nose up and down. He appeared to be attempting to knock the passengers off their feet. The passengers persisted. One or more hijackers went down. Another held the door closed, just barely. The tumult built.

Jarrah frantically manipulated the control yoke. He asked, "Is that it? Shall we finish it off?" Amid more noises — crashing metal, glass, voices, and shouts of "Roll it!" or "Let's roll!" (from passenger Todd Beamer, perhaps, according to some who knew him) — the passengers battered the cockpit door. Another passenger yelled: "In the cockpit. If we don't, we die."

Jarrah was already there. He relaxed his hold on the controls and put the plane into its final dive. "*Allahu akbar*!" he screamed. "*Allahu akbar!*" His remaining partner chanted, "Put it in it, pull it down, pull it down, pull it down." The airliner twisted hard to the right, inverted, then smashed into an empty clearing in the woods near Shanksville, Pennsylvania, at 10:03 a.m. This concluded the al-Qaeda planes operation.

The first U.S. counterattack of the war succeeded, although at great cost. The forty American civilians who ended the hijacking of United 93 possessed no weapons but their brains, bodies, and guts. In so doing, they prevented a fourth attack on an American landmark, perhaps the White House or the Capitol building. We will never know.

The uniformed military got caught way out of position that fateful morning. On September 11, 2001, air defense of the continental United States came under the North American Aerospace Defense Command (NORAD), a U.S./Canadian organization dating back to 1958, the height of the Cold War. NORAD faced outward, scanning the skies for inbound missiles or bombers. The Soviets had gone away a decade earlier, and the much weaker air arm of the Russian Republic rarely probed NORAD's perimeter anymore. The U.S. component of NORAD was the U.S. Space Command, much more concerned with satellites and rocket science than civil defense. Some intelligence organizations had warned of terrorist use of aircraft as weapons, although exercises along those lines posited captured airliners inbound from overseas. Nobody figured on airliners seized and used as weapons inside our country. Only two alert sites, each with two Air National Guard fighters, covered the Northeast Air Defense Sector. All four fighter jets got off the ground, although too late to do anything. Two reached New York after both towers had been hit. The other pair just missed seeing American 77 crash into the Pentagon. NORAD didn't even hear about United 93 until after it went down.

The smartest U.S. decision made that morning came from the Federal Aviation Administration. Its air traffic control centers provided the first reliable reporting, which was admittedly fragmentary, especially after the terrorists shut down the transponders on all the planes but United 175. (Likely confused, the hijackers of that plane had switched the code but didn't turn off the device.) Various local radars got partial signatures as the airliners moved to their targets. With perhaps the clearest picture, and that none too clear, at 9:42 a.m., the FAA command center showed three airliners already down, United 93 hijacked, and others suspected to be in trouble. It was operations chief Ben Sliney's first day in his new position as FAA national operations manager, but it was not his first rodeo. Drawing on twenty-five years of controlling flights, he issued the bold order: land them all. More than forty-five hundred airliners were diverted safely to the nearest airports. Within a few hours, anything still flying would be either military or fair game for the interceptors.

Thanks to Sliney's directive, the air picture cleared up, although not without more scares and false alarms. NORAD struggled to police various duplicate or bogus tracks of other possibly hijacked planes. President George W. Bush left his scheduled trip to Sarasota, Florida, for Barksdale Air Force Base, Louisiana, and then Offutt Air Force Base, Nebraska, before finally heading home to Washington that evening. A threat to Air Force One proved to be nothing, but after the horrendous events of the day, no one took any chances. In Washington, with the Pentagon afire and calls coming in of an attack at the Department of State (false), Vice President Dick Cheney at the White House at one point cleared U.S. Air Force fighters to engage a supposed rogue airliner. Thankfully, it never came to that.

On the ground, death and daring shared equal billing. Firefighters, police, and emergency rescue teams saved thousands of lives. Even so, 2,973 died at the hand of the enemy. Both towers collapsed in New York, with the loss of 2,749 people, including all aboard both airliners, 343 firefighters from the Fire Department of New York, 37 Port Authority police officers, and 23 New York Police Department officers. At the Pentagon, between American 77 and those in the building, 184 died. The crash of United 93 killed 40. These numbers do not include the 19 terrorists.

It fell to President Bush to close the long, bloody day. That evening, the president spoke from the Oval Office. To the many watching on television, he looked tired but resolute. After words of condolence and reassurance, he said something that had not been heard before: "We will make no distinction between the terrorists who committed these acts and those who harbor them." He then went on to say, "We stand together to win this war on terrorism." This one would not be handled with grief counselors, lawyers, and yellow crime-scene tape. This was a war.

Bush didn't mention al-Qaeda in his speech. His CIA director, George Tenet, told the president at 3:30 that afternoon that Osama bin Laden's people had done it. Analysts identified three men on American 11 as al-Qaeda agents. Technical sources picked up a lot of congratulatory messages from recognized members of bin Laden's network. It took time to be sure, but from the first hours, the U.S. knew who did it.

Osama bin Laden kept his mouth shut, at least for the moment. His bold work spoke for itself. The Sheikh's men had taken symbols of the far enemy — the red, white, and blue planes of American and United — and used them to devastate the Twin Towers and the Pentagon, icons of U.S. might. Word spread quickly giving credit to the Sheikh and his martyrs. While Middle Eastern governments, including Iran and Libya, issued the standard condolences, thousands of their citizens gathered to celebrate the defeat of the far enemy. The Arab Street and its affiliates in other Islamic countries pulsed with excitement. If Bush was looking for "those who harbor" al-Qaeda terrorists, he faced the prospect of sorting through over a billion Muslim believers in dozens of countries, including his own.

"The Pearl Harbor of the 21st century took place today," Bush wrote in his diary late on September 11, well after his speech to the nation. At Pearl Harbor, America lost 2,403 sailors, Marines, soldiers (which included airmen at that time), and Hawaiian civilians, with the explosion of the battleship USS *Arizona* accounting for half that grim list. Yet on December 7, 1941, the enemy wore Japanese uniforms, used conventional weapons, possessed armed forces and supporting industries, and originated from a single identifiable country with a known capital. The way from Pearl Harbor to Tokyo Bay proved long and bloody, but the course was obvious from the first hour.

On September 11, 2001, the path looked anything but clear. This enemy dressed in civilian clothing and mixed easily among local populations in America and around the world. This enemy employed unusual weapons, mixing guns, knives, and explosives with commercial trucks, boats, and aircraft, individual threats hard to pick out from the general flow of life. The al-Qaeda network itself was finite in number, but it benefited from certain active supporters. These included the Taliban rulers of Afghanistan, the anti-American regimes in Iran, Iraq, and Syria, and substantial disgruntled elements in many Middle Eastern countries. It all ensured enough money and enough volunteers to keep al-Qaeda going.

Even if America went into distant Afghanistan, knocked out the Taliban, and killed a lot of al-Qaeda, including Osama bin Laden, would that do it? The existence of Egyptian Islamic Jihad, the Islamic Movement of Uzbekistan, Hezbollah, Hamas, and a hundred other similar groups reminded all that this region had a lot of hate to spare. America's "frenemies" in Saudi Arabia, Yemen, and Pakistan, among other countries, professed support for the U.S. while succoring homegrown threats, as long as those hotheads focused on the infidels. President Bush told his country they were at war against terrorists and those who harbor them. Taken to its extreme, this approach pitted the U.S. against more than a billion people. Defining the enemy defined the war. From the outset, the United States struggled with this fundamental challenge.

On September 20, with British prime minister Tony Blair in attendance, President Bush addressed a joint session of the U.S. Congress. With millions watching, Bush explained the country's strategy to his fellow citizens, to the world, and to the enemy as well. There would be turns and reverses, but this statement set the direction for what became known as the Global War on Terrorism.

Bush spoke bluntly, labeling the events of September 11 "an act of war." He named Osama bin Laden and al-Qaeda. He also identified Zawahiri's Egyptian Islamic Jihad and the Islamic Movement of Uzbekistan, the latter included because the U.S. needed staging bases in that otherwise obscure Central Asian state bordering Afghanistan. Then he turned to the Taliban, hosts of the Sheikh and his men:

> And tonight, the United States of America makes the following demands on the Taliban: Deliver to United States authorities all the leaders of Al Qaeda who hide in your land. Release all foreign nationals, including American citizens, you have unjustly imprisoned. Protect foreign journalists, diplomats and aid workers in your country. Close immediately and permanently every terrorist training camp in Afghanistan, and hand over every terrorist, and every person in their support structure, to appropriate authorities. Give the United States full access to terrorist training camps, so we can make sure they are no longer operating. These demands are not open to negotiation or discussion. The Taliban must act, and act immediately. They will hand over the terrorists, or they will share in their fate.

Neither Bush nor his lieutenants expected Mullah Omar and the Taliban to agree to this harsh list of demands. It adhered to the diplomatic niceties of offering a peaceful settlement, but no sovereign country, let alone one led by a fanatic anti-Western sect, could or would comply. Already isolated — only Saudi Arabia, the United Arab Emirates, and the midwife of the regime, Pakistan, formally recognized the Taliban's Kabul government — Mullah Omar and his disciples showed no inclination to appease the infidel Americans by turning over Osama bin Laden or his followers. Clearly, the Taliban wanted to trust God and fight it out. So be it.

Bush continued, pointing out that America's aims went well beyond bin Laden's ilk and the mullahs in Kabul. "Our war on terror begins with al-Qaeda," he said, "but it does not end there. It will not end until every terrorist group of global reach has been found, stopped and defeated." He compared the enemy's dream of an imposed, severe Wahhabi caliphate to German Nazism, Italian Fascism, and "totalitarianism," a euphemism for Russian Communism, which indicated that the U.S. wanted help from Moscow and the Chinese Communists too. Bush played it safe there.

He took a broad view, however, of the scope of the war:

> We will direct every resource at our command—every means of diplomacy, every tool of intelligence, every instrument of law enforcement, every financial influence, and every necessary weapon of war—to the disruption and to the defeat of the global terror network.
>
> This war will not be like the war against Iraq a decade ago, with a decisive liberation of territory and a swift conclusion. It will not look like the air war above Kosovo two years ago, where no ground troops were used and not a single American was lost in combat.
>
> Our response involves far more than instant retaliation and isolated strikes. Americans should not expect one battle, but a lengthy campaign, unlike any other we have ever seen. It may include dramatic strikes, visible on TV, and covert operations, secret even in success. We will starve terrorists of funding, turn them one against another, drive them from place to place, until there is no refuge or no rest. And we will pursue nations that provide aid or safe haven to terrorism.
>
> Every nation, in every region, now has a decision to make. Either you are with us, or you are with the terrorists. From this day forward, any nation that continues to harbor or support terrorism will be regarded by the United States as a hostile regime.

Most of these lines generated thunderous applause from the assembled U.S. lawmakers. The strategy sounded good. The other parts hit all the right notes too: he gave a nod to America's British ally, represented in person by Tony Blair; thanked all the other friendly states; and stated the strong desire to work with moderate Muslims and their governments. But once the clapping stopped and the televisions clicked off, the basic question remained: Who was the enemy?

The smart set already knew, of course. Elected leaders, political commentators, retired generals, former FBI agents, and intelligence men "in from the cold" agreed: it was al-Qaeda. Just go get them. That seemed simple enough. As Clausewitz warned, though, in war, even the simplest things are difficult.

Bush's inner circle included two men who had already run an American war in the Middle East: Vice President Dick Cheney and Secretary of State Colin Powell, both key architects of the 1991 Desert Storm victory.

The president also relied heavily on Secretary of Defense Donald Rumsfeld, who had held plenty of high-level posts in government, although none of them recent. These three experienced men advised Bush. In the end, though, he decided.

The first decision almost made itself. At that time, the director of the Central Intelligence Agency, George Tenet, also served as the director of central intelligence, and that was not some Monty Python wordplay. Tenet ran the CIA, America's premier clandestine organization, but he also had the duty to collate and synthesize the work of all the U.S. intelligence entities, military and civilian. When Tenet said al-Qaeda did it, defeating Osama bin Laden and company became the objective. But how do you win a war against a network?

You can track and kill the key leaders—but that's much easier said than done. A more certain route, one better suited to American military capabilities, involved knocking out the state sponsors. In September of 2001, that meant Taliban Afghanistan. Why not plop a Desert Storm on Afghanistan and be done with it? Powell, of course, thought that was enough.

Cheney and Rumsfeld, especially the latter, argued to include Iraq too. Saddam Hussein was a known enemy with recognized links to terrorist acts, if not necessarily to al-Qaeda and its latest outrage. The Iraqis had used chemical weapons in the past. Saddam openly sought nuclear and biological arms as well. Why wait to take the next shot in the face? Get ahead of it, they urged.

Narrow or broad? The strategic choice loomed. You could make a case for both, especially with the Twin Towers down and a hole smoking in the side of the Pentagon. But a truly comprehensive global effort surely promised a very long, very extensive war ranging across much of the Islamic world. In days to come, that would have to be addressed, with Saddam Hussein's Iraq as the known case in point. But in Washington in September of 2001, it wasn't time to sort that out. Not yet.

Bush did what most presidents do when given a split between trusted advisers. He compromised. They would deal with Afghanistan first, and then, if needed, Iraq. That explained why the September 20 speech was about more than al-Qaeda and Afghanistan; it opened the door for America to go after "every terrorist group of global reach." The phrase referred to Baathist Iraq, not the Irish Republican Army or the Ku Klux

Klan. Iraq went in the hold file for now, with U.S. airpower still controlling Iraqi airspace, as it had done since 1991. Going after al-Qaeda in Afghanistan looked hard enough.

Even as discussion continued on how wide an offensive to mount, Bush and his team agreed on defensive measures. The military recovered quickly from September 11. Some of the first orders for what became Operation Noble Eagle emerged from the smoke-filled Pentagon hours after the airliner struck it. Air Force, Navy, Marine, and even Canadian fighter jets flew multiple combat air patrols over U.S. cities. National Guardsmen, regular Army soldiers, and Marines posted sentries at major airports, nuclear power plants, and other designated sites. Emergency teams stood ready to react to nuclear, biological, or chemical attacks. Some of these teams moved to meet the late-September flurry of mail packets containing anthrax spores. The active, widespread presence of armed, uniformed men and women did much to restore public confidence in the wake of 9/11.

More important for the long term, Defense Secretary Rumsfeld reinforced the message that protecting the homeland came first. On September 11, while NORAD guarded U.S. airspace, the four-star commander spent most of his time on his U.S. Space Command duties. American land and sea defenses, largely theoretical, came under U.S. Joint Forces Command, the remnant of the U.S. Atlantic Command that had chased Soviet submarines during the Cold War. Rumsfeld ended this bureaucratic jumble. He told the military to create the U.S. Northern Command, which activated on October 1, 2002. The new military headquarters partnered well with the civilian sector's equally new Department of Homeland Security. It did not happen overnight, and it wasn't perfect. But it closed some of the gaps the bad guys had exploited on 9/11.

Offensive operations in Afghanistan proved much tougher to arrange. Nobody had the stomach to wait a year for action. Bush wanted something done in weeks. Powell and his diplomats worked to secure grudging help from Pakistan as well as access and overflight rights through former Soviet Central Asian republics like Uzbekistan. CIA Director Tenet had his agents in place in northern Afghanistan. Rumsfeld turned to U.S. Central Command and to General Tommy R. Franks.

While the U.S. Armed Forces plan for a very wide range of contingencies, some potential missions get a lot more attention than others. If they went back into Iraq, for example, they would benefit greatly from hard-won intelligence gathered during daily armed air missions in the north and south of that country. Doing anything in landlocked, forbidding Afghanistan hadn't generated much interest before 9/11. Shooting cruise missiles was one thing. Putting in people, other than a few CIA agents, really augured trouble unless you delivered thousands, enough to look after themselves if things went bad. The brutal Black Hawk Down battle in Mogadishu, Somalia, in 1993 reminded all what could happen when special operations elements got caught up in something bigger and worse than they'd foreseen. Afghanistan had lived in the "too hard" box until 9/11.

Pressed to do something, Franks and USCENTCOM offered three options for Afghanistan. These went before the president on September 15. The Infinite Reach cruise missile strikes on August 20, 1998, represented the type of thing that could be done almost immediately, once Navy shooter ships and submarines got into range. An expanded version added the Air Force's long-range manned bombers to weight the effort. The strongest option, built on a special operations concept tagged Infinite Resolve, added to the missile barrage and air sorties. This idea put Special Forces teams on the ground with the CIA men in order to work with the thousands of armed friendly Afghans who opposed the Taliban. Together, it was hoped, a few Americans and a lot of Afghans would use U.S. airpower to defeat the Taliban and knock out al-Qaeda. This approach could be reinforced as required with conventional Army and Marine battalions. Franks's planners called this third option Infinite Justice, keeping with the planning code names assigned to Afghan contingencies. Bush chose Infinite Justice.

The initial concept sketched out at USCENTCOM grew day by day into a real plan. The U.S. had only scant weeks to devise a campaign, coordinate with the friendly Afghans, move forces, integrate allies, carry out reconnaissance, and find the enemy. Although the effort certainly drew on the known strengths of American global operations, it was going to be very tough. Few had any illusions on that count.

The eventual plan had four phases. The third phase, Conduct Decisive Combat Operations, drove the train. From the outset, Franks and his people aimed to knock out the Taliban government and eliminate al-

Qaeda in Afghanistan. Every other aspect of the plan revolved around that objective. Franks had served in both Vietnam and Desert Storm, so he knew the difference between winning and losing. The rangy, rough-around-the-edges, plainspoken Texan focused on winning this one at maximum speed and minimal cost.

Phase one bore the title Set Conditions and Build Forces to Provide the National Command Authority Credible Military Options. In plain English, this meant getting ready to go. From the start, Franks knew he had to attack within a month or so. Even with a might of 2.1 million in uniform, counting all the National Guard and reserves, only a fraction of the U.S. Armed Forces could get to the neighborhood of Afghanistan that quickly. Even so, the lineup packed a lot of punch. The Air Force prepared its B-2A Spirit, B-1B Lancer, and B-52 Stratofortress bomber wings for launch from bases as far away as the continental United States. The U.S. Navy's USS *Enterprise* (CVN-65) carrier battle group and the USS *Carl Vinson* (CVN-70) carrier battle group moved north into the Arabian Sea. The USS *Theodore Roosevelt* (CVN-71) carrier battle group steamed from Norfolk, Virginia. From Japan, the older, conventionally powered USS *Kitty Hawk* (CV-63) was reconfigured to serve as a "lily pad" for special operations units. *Kitty Hawk* then sailed for the Arabian Sea. Other U.S. special operations forces staged into countries near Afghanistan. Army infantry battalions followed. A Navy/Marine amphibious readiness group also headed that way. Logistics, intelligence, and headquarters elements deployed, with U.S. Air Force C-17 Globemaster III airlifters in heavy use. All services prepared other units to follow.

As the U.S. Armed Forces assembled for battle, Colin Powell's State Department did some assembling too. They engineered a broad coalition, much like Powell had seen done in the 1990–91 war against Iraq. The United Nations passed Resolution 1373 supporting the American-led response. The Organization of American States invoked the 1945 Rio Treaty, pledging help from countries in the Western Hemisphere. Australia and New Zealand acted under the ANZUS Treaty, offering their assistance as well. For the first time in its history, NATO invoked article 5, which stated that an attack on one of its members was an attack on all, and solicited contributions within the alliance. Some NATO aircraft actually assisted with Noble Eagle patrols in U.S. skies. Most countries shared intelligence and granted overflight and transit authority. This one developed too fast to bring in small, unready, or symbolic contingents.

The more capable allies contributed some key elements. Great Britain's army sent Special Forces, its navy deployed an aircraft carrier and escorts, and the Royal Air Force, lacking intercontinental bombers, readied fighter aircraft once local bases became available. Australia and New Zealand provided Special Forces and warships. Canada added six ships, aircraft, and special operators. France sent a frigate and a logistics ship. Even Japan, constitutionally limited in its ability to participate, dispatched support vessels from its Maritime Self-Defense Force. All of these helped round out the force.

In arranging international assistance, the entire operation depended on the agreement of the countries fronting landlocked Afghanistan. In the usual regional style, each of those states exacted concessions from the infidels before joining the campaign. A good amount of small exercises, guest slots at military schools, and military-to-military contacts from the 1990s now paid off in a big way. To the northwest, scrupulously neutral Turkmenistan allowed only humanitarian aid to cross its territory. Uzbekistan bargained hard, granting use of its key air base at Karshi-Khanabad (K2, to Americans) but insisting on over $118 million in annual assistance funding and U.S. help against the Islamic Movement of Uzbekistan, which explained that obscure group's mention in Bush's September 20 address. Tajikistan offered three air bases. Nearby Kyrgyzstan granted the use of the large Manas air base. Each received increased U.S. aid. Except in Turkmenistan, the arrangements included military overflights and other transit approvals. America also placated the Russians, who still maintained a lot of interest in their former republics. That locked down the north. Iran to the west, of course, went its own way.

The big get, the vital assent, involved Pakistan. Battling its own viral Islamist cells, shunned by America and the West after its 1998 nuclear tests, ever jealous of its much more powerful neighbor India, Pakistan always seemed one coup away from implosion. Now, as Bush said, it was time to choose. Even among the usual Middle Eastern and South Asian double-dealers, the Pakistanis stood out for their paranoia. Friends of America (sort of) during the Soviet incursion into Afghanistan, the

Pakistanis had long nurtured and aided the very Taliban that the U.S. now wanted to destroy. But for his own reasons, Pakistan's strongman du jour, President (and General) Pervez Musharraf, elected to cooperate. Powell called him and spoke frankly: "As one general to another, we need someone on our flank fighting with us." Musharraf bought it, and the United States had a deal. But the Pakistani general had his boot squarely on the most important U.S. logistics windpipe, the one that would eventually run from the Arabian Sea up into Afghanistan. Musharraf would surely take advantage of this newfound leverage.

Phase one led to phase two: Conduct Initial Combat Operations and Continue to Set Conditions for Follow-On Combat Operations. This phase might be interpreted as "Since we can't get everything there at the start, we'll get going, feed stuff into the fight, and see how things develop." Destroying the Taliban air defense came first. Mullah Omar's half-trained air arm possessed a few elderly MiG fighters and some antiquated antiaircraft missiles and guns. There appeared to be two or three command posts tying it all together, but who could really tell? Of course, compared to the extensive array of radars, command centers, interceptor jets, and surface-to-air missiles protecting Iraq in 1991, the obsolescent bits and pieces at the Taliban's disposal didn't seem like much of a threat to U.S. air commanders. Nevertheless, they had to go.

After the U.S. blew aside the Taliban air defense, the real work started. The U.S. offensive counted on friendly Afghans, guys the CIA said would do the job with U.S. air in support. The Northern Alliance, mainly Tajik, accounted for the last shard of the old anti-Soviet guerrilla conglomerate defeated by the Taliban in 1996. Driven from the capital of Kabul, the Northern Alliance held the Panjshir Valley well northeast of the city as well as a few remote districts here and there. Its most able leader, Ahmed Shah Massoud, had died on September 9 at the hands of two Algerian al-Qaeda suicide bombers posing as television interviewers. (Their camera exploded.) Massoud's assassination had been part of Osama bin Laden's preparation in the run-up to the 9/11 planes operation and, coincidentally, a useful favor to his benefactors Mullah Omar and the Taliban. Losing Massoud greatly complicated any U.S. attempt to strike back in league with the Northern Alliance.

With Massoud dead, Northern Alliance leadership devolved on four tough, mutually suspicious commanders. Fahim Khan, seconded by shrewd Bismullah Khan Mohammedi, headed the eastern Tajiks. Ismail Khan in the west commanded a second Tajik contingent. Abdul Rashid Dostum led the Uzbeks. Karim Khalili ran the Hazara element. Together, they mustered twenty thousand fighters with AK-47 rifles, Russian-designed PKM machine guns, mortars, rocket launchers, Toyota pickups tricked out with machine guns, and a few dozen creaky tanks and armored personnel carriers. They even had some old Russian Mi-8 helicopters. The whole lash-up resembled a set for one of the Mad Max movies.

Though not lacking in numbers, the Northern Alliance labored under one big deficiency. It contained few Pashtuns. That ethnicity formed the largest group (40 percent) of the Afghan population and made up the vast bulk of the Taliban. Along with the Northern Alliance, the U.S. needed some Pashtuns. To start with, they had two. The first, an intellectual and a political organizer, was a small, neat, articulate, skinny, balding man named Hamid Karzai. The second was bluff, cruel, independent former anti-Soviet guerrilla leader Gul Agha Sherzai, a man who combined opposition to the Taliban with opium trafficking. He had a clutch of armed followers and ran a remote patch of Kandahar Province as his personal fiefdom. Compared to these two unlikely characters and their thin string of armed acolytes, the Northern Alliance looked like the Eighty-Second Airborne Division. Still, Karzai and Sherzai allowed the U.S. to check the box that stated Pashtuns were included in the anti-Taliban effort. American spokesmen dignified them as the Southern Alliance, a term unknown to any actual Afghan.

In both the north and the south, the U.S. proposed to add teams of Special Forces, each with a U.S. Air Force JTAC (joint terminal attack controller, an airman who directed air-delivered bombs to targets). To dislodge the Taliban, under this scheme of maneuver, the Afghans supplied riflemen, and the U.S. delivered accurate, devastating firepower from above. Against the seemingly ramshackle Taliban, the plan should work. But nobody knew for sure. The Soviets had bombed and bombed, killing many, but they lost in the end. Nobody could predict the Taliban's reaction. Tommy Franks had seen American airpower do the job against the Iraqis in 1991. He bet on the airmen.

Once the air got the Taliban rolling back, it was time for phase three. In this part of the campaign, USCENTCOM staff officers envisioned reinforcing the offensive in the north and the south. A few selected American conventional ground units, maybe about twelve thousand soldiers and Marines, would be inserted to finish off the Taliban, run down the last of al-Qaeda, and grab key terrain, like Mazar-e-Sharif in the Tajik north, Kabul, and Kandahar in the Pashtun south. Historically, control of these three cities and the corridor between them equaled control of Afghanistan. Everything beyond that, whether to the north and east into the Hindu Kush Mountains or to the south and west into the Registan Desert, could wait.

Phase three demanded decisive results. It all made military sense in USCENTCOM, in Washington, and to the American public. But on the ground in Afghanistan, things promised to be a lot more confusing. It circled back to the original question: Who was the enemy? Al-Qaeda executed the events of 9/11 and showed the capability to commit similar acts again; the Taliban hosted al-Qaeda but hadn't quite taken charge of its own fractious, illiterate, impoverished country. Only the former of these two very different enemies threatened the United States, even though the quasi-conventional Taliban furnished a much more appropriate target set for U.S. firepower. *Start there,* said the appointed (and self-appointed) experts, *against the guys with the tanks and howitzers, however rusty. Once we get going, we can figure out what happens next.* By painting with a pretty broad brush, Franks and USCENTCOM ensured a bold start and a much less certain finish.

What did it mean to defeat the Taliban? The military definition of *defeat* is "an enemy force has temporarily or permanently lost the physical means or the will to fight. The defeated force's commander is unwilling or unable to pursue his adopted course of action, thereby yielding to the friendly commander's will, and can no longer interfere to a significant degree with the actions of friendly forces." That suffices for doctrine. It would serve just fine should the Taliban play by the rules, surrender, go home, and stay there. But what if its members dispersed, scattered into the hinterlands, and crossed into Pakistan, their old sanctuary? What if the Taliban backed off for a while, regrouped, and returned? Then what? Nobody had thought that through.

Whatever American forces did to or with the Taliban, their actions would address only the hostiles who'd harbored the terrorists, not the terrorists who'd inflicted the 9/11 attack. What did it really mean to eliminate the al-Qaeda perpetrators? Did we have to hunt down and kill every jihadi, starting with bin Laden? Worse, al-Qaeda puttered along on a relative shoestring of money, supplies, and recruits, most of which came from various locations inside and outside of Afghanistan. Damming those trickles seemed highly unlikely. What could we do? If we could get the leaders, it might break the network and ruin the unique high-end capabilities engineered by the Contractor, bin Laden. So: nail the terrorist chain of command. Easier said than done — thus far, America's manhunting record had been kind of spotty. We weren't starting from zero, but maybe two or three on a scale of ten. Well, step one amounted to getting closer to the problem. The military's desire to go, and go now, overrode the voices raising such unpleasant matters.

Finally, what of the broader Muslim community, none too trusting of America or any Western power? Clearly, Bush, Cheney, Rumsfeld, Powell, and Franks all agreed, America must not take on the entire Islamic world. Yet could the U.S. really depend on Pakistan? What about the Central Asian republics? If they came under pressure from their coreligionists and began to choke off access, then what? Nobody wanted to bring up those difficult questions.

Oh, and what about phase four? It had a long title: Establish Capability of Coalition Partners to Prevent the Reemergence of Terrorism and Provide Support for Humanitarian Assistance Efforts. The Coalition partners mentioned weren't the French and British. The term referred to the locals, the Afghans, and maybe the people in the surrounding states too. Franks thought it might take three to five years to accomplish, as it had with Kuwait when the emirate rebuilt after the 1990–91 Iraqi invasion. He wanted to hand off as much as possible to the NATO countries, the UN, nongovernmental relief organizations, or whoever. In Franks's view, and that of many members of the U.S. military, Americans kicked down doors; let others build nations. Phase four planning didn't garner much interest.

It should have. Afghanistan wasn't Kuwait. It wasn't even Haiti, except in terms of entrenched poverty and endemic violence. Ripped apart by decades of civil war, benighted by illiteracy (85 percent of adult males could not read), devoid of modern industry, burdened by a backward view of women, Afghanistan limned the picture of a failed state. Somebody had to run the place, and when America had troops present, that somebody might well be wearing a U.S. uniform. When the Taliban toppled in accord with phase three, no current Afghan party seemed able to assume governance. Well, there was that fellow Karzai. Maybe he'd fill the bill.

One other thing popped up, itself indicative of the kind of war under way. Some in the White House began to hear complaints about the name Infinite Justice. Various experts on Islam muttered and grumbled. Only God, Allah Himself, could dispense infinite justice. The designation might neatly match the military's code-name protocol, but it grated on the ears of those in the wider Islamic world. The U.S. depended on the Muslim Afghan Northern Alliance to fight, and the Muslim Central Asian republics and Pakistan were essential to the team, so Infinite Justice would not do. Its choice demonstrated just how little the U.S. Armed Forces really knew about the region, its people, and their beliefs. More clever minds came up with Operation Enduring Freedom — much better. But starting an offensive by apologizing rarely augurs well for its outcome.

Thus, a few days after 9/11, America decided to go into Afghanistan, boots on the ground, allies in tow, frenemies dragging behind, all of it looking a lot like the front end of a much bigger deal. In all the excitement, it seemed like Desert Storm reborn. The United States would go in hard and smart and take down Afghanistan, a task that had been too tough for the British regulars during the raj and the hardened Soviets in their day. Few in U.S. uniforms — damn near none — raised a disturbing point. Isn't this just what Osama bin Laden wanted? For him, and for the U.S., the old caveat applied: be careful what you wish for.

3

THE HINDU KUSH

> I think everybody thought, "Of all the places to have to fight a war, Afghanistan would not be our choice." But we didn't choose Afghanistan; Afghanistan chose us.
>
> — CONDOLEEZZA RICE, JULY 2002

THE TALIBAN HELD the high ground. This semi-organized amalgam wore uniforms of a sort. Some had the whole getup, including snappy dark green berets. Others sported only mottled military field trousers. A few, like the great Osama bin Laden himself, favored camouflage jackets. Here and there, a bearded man in traditional woolen robes topped off his outfit with an old, rounded Soviet metal helmet. Some of their AK-47 automatic rifles and PKM machine guns were clean; others not. Just below the crest of the scrubby ridge line stood two elderly, Russian-made T-55 tanks, a couple of mortars, and one cockeyed howitzer. Disciplined soldiers they were not. But there were a lot of them.

Try as they might, the Uzbeks hadn't been able to get up there. To get into the city of Mazar-e-Sharif, a place they very much wanted to take, Abdul Rashid Dostum and his men needed to crack that position. They had tried before. On the heights, the Taliban looked down, scattering a few shots among the Uzbeks taking cover behind the rocks and dirt piles on the other valley wall. The same old game with the same old score, it appeared.

Then it changed.

A flash of white light stabbed the ragged Taliban trench line. Up went a dark brown column of dirt, rocks, bushes, and God knew what else — ammunition? Weapons? Body parts? Seconds later, delayed by a thousand-plus yards, a crashing roar sounded, a pressure wave that pushed hard against the chests of Dostum's Uzbek fighters.

It happened twice more. The Uzbeks cheered and opened fire.

Dropping his binoculars, Dostum reached for the hand mike of his field radio. The assistant tuned it to a channel used often in the country. The Uzbek leader addressed his Taliban counterpart.

"This is General Dostum speaking. I am here" — he paused for effect — "and have brought the Americans with me."

It didn't take many more big U.S. bombs to make the point. The Taliban cracked; some ran, leaving behind their damaged tanks and more than a hundred dead. An equal number raised their hands, finished. "We killed the bastards by the bushel-full today and we'll get more tomorrow," reported Captain Mark Nutsch, the American commander with the Uzbeks.

Nutsch and his men enabled Dostum to sweep all before him. Some Taliban resisted much more fiercely, but in the end, just about all quit. Within three weeks, methodically bombing from ridge to ridge, the Uzbeks joined the final push that took Mazar-e-Sharif on November 9, 2001. Other Northern Alliance bands, also with U.S. CIA and Special Forces teams, pushed on to Kunduz and Taloqan, other key cities in the area. American officers and sergeants on foot, in battered pickup trucks, and even on horseback fought side by side with the Northern Alliance Afghans. At each center of resistance, American airpower broke the enemy.

From the outset, the campaign relied on airpower. Operation Enduring Freedom began with airstrikes on October 7. That day, less than a month after 9/11, American and British jets and cruise missiles smashed up what passed for the Taliban air defenses. Every one of Mullah Omar's creaky aircraft, his handful of old surface-to-air missiles, and his few active radars were all hammered. In addition, U.S. Air Force bombers and Navy jets struck known al-Qaeda training camps. A British Royal Navy submarine fired two of the fifty Tomahawks launched that day. Within a few hours, the Afghan skies belonged to Uncle Sam.

For about two weeks, American aircraft ranged unchallenged over the country. They bombed some new sites and re-attacked previous targets. One British voice spoke sardonically about strikes to "rearrange the rubble." The ever-helpful Crown Prince Abdullah of Saudi Arabia suggested suspending bombing for the upcoming Islamic holy month of Ramadan. Taliban spokesmen began to scoff at the U.S. effort. Sure, the air show impressed. But what next? Where was the great Northern Alliance ground offensive? The guys up in the American press box started to squirm in their seats. The suits in Washington soon did likewise.

At this point, though, the only U.S. "boots on the ground" (a grossly overworked cliché of that time) belonged to a few CIA officers inserted by nondescript Agency helicopters near the end of September. Linking up with the Northern Alliance, the CIA elements brought their expertise, necessary cash in strongboxes, and an absolute guarantee of U.S. commitment. But to run U.S. airstrikes, they waited on the uniformed special operators. The first military teams entered during the third week in October.

United States special operations forces (SOF), in 2001 and now, amount to about 3 percent of the U.S. Armed Forces. Most SOF had roles in Afghanistan, but only a few kinds got much official coverage. The wide variety of names, code words, and colloquialisms associated with SOF is intentional, to some extent. These units work in the shadows, and they prefer it that way. Composed of specially selected and highly trained personnel, SOF are of two basic types: the regional forces and the others.

Regional SOF includes contributions from each service. The U.S. Army provides almost half of the total, including Army Special Forces (SF, the Green Berets of Vietnam fame), the Seventy-Fifth Ranger Regiment (superbly trained parachute-qualified light infantry), and the 160th Special Operations Aviation Regiment (aka Night Stalkers, an elite helicopter contingent). U.S. Air Force special operations squadrons fly long-range penetrator airplanes and helicopters as well as the deadly AC-130 gunships employed for close air support. Navy SEAL (sea, air, land) teams and their supporting special warfare units form the maritime component. In 2001, the Marines sent individuals only to SOF, but as the war continued, the Corps organized its own unique units tied to the service's amphibious core mission. Together, these forces organize and deploy in support of a regional commander like General Tommy Franks in October of 2001. Such regional SOF is acknowledged to exist. Given their tasks deep in hostile territory, however, the U.S. military rightly limits how much publicity they receive.

The other kind of SOF also do a lot of key missions in various regions, but they strive to avoid any publicity, and usually succeed in that effort. In the tradition of the British Special Air Service (SAS), which pioneered this line of work, these guys do the most difficult, demanding national counterterrorist tasks. They are manhunters, although they, too, had much to learn after 9/11. Nobody was supposed to talk about them, and as a rule, nobody in uniform did. Sometimes the media, fascinated with these highly capable forces and their impressive set of challenging missions, reported on their exploits. Even Hollywood got interested now and then. That mattered not at all in the SOF ranks, and so it mattered not in the wider U.S. military. You didn't speak their names. You didn't depict their emblems. You didn't acknowledge they existed. You just didn't. But you knew they'd be there. Conventional soldiers referred to them, in whole or in part, as the Task Force, and left it at that.

The men sent to help Abdul Rashid Dostum came from the regional SOF, specifically Army Special Forces. Captain Mark Nutsch commanded a twelve-man Operational Detachment Alpha (ODA), the basic SF element, with a very experienced warrant officer deputy, a senior team operations sergeant, an intelligence sergeant, and then two NCOs, each focused on the key specialties of weapons, demolitions/construction, communications, and medical aid. Each member had been through a grueling multiweek course of selection and assessment followed by overall SF and then specialist training for a year or so. Standards exceeded the demands of the most intensive college courses, with a relentless physical fitness regimen to boot. SF team medics, for example, were as skilled as some countries' licensed physicians. The men cross-trained in more than one specialty. When they joined a team, they learned appropriate regional languages, no easy task, especially along with parachute jumping, hours of skill shooting, and all the rest. Nutsch and his people, for instance, knew some Dari and Pashto. A twelve-man ODA could subdivide as required to advise a guerrilla unit of thousands.

On arrival in the mountains of northern Afghanistan, Mark Nutsch and his men met a pair of CIA officers. They were not in charge of the

SF, nor did they work for Nutsch. Rather than devise a formal command relationship, they all just shook hands and cooperated. After about a week on the ground, two highly welcome U.S. Air Force joint terminal air controllers (JTACs were the men who directed airstrikes) joined the ODA. With those useful attachments, and some borrowed horses, Nutsch and his team accompanied Dostum and the Uzbeks.

Such an ODA is indeed the basic SF element. But they do not live in a vacuum. Six ODAs form an SF company, an Operational Detachment Bravo (ODB), commanded by a major. Three ODBs make up an Operational Detachment Charlie (ODC), and three of those, plus support elements, make up a Special Forces Group. The SF mixes and matches teams to form mission-tailored organizations. To keep all the parts straight, each sub-element has its own individual number. Nutsch, for example, commanded ODA 595. Its counterparts were quite similarly configured to include the CIA ties and the Air Force JTACs.

The entire array deployed under the rubric of Joint Special Operations Task Force–North, formed by Colonel John Mulholland around the core of his Fifth Special Forces Group, augmented with U.S. Air Force, Navy, Marine, and British military elements. The headquarters were set up at Karshi-Khanabad airfield in Uzbekistan. Once established at K2, Mulholland preferred to use his command's code name: Task Force Dagger.

Large, tough, and smart, Mulholland gave his teams a mission with a lot of latitude: "Advise and assist the Northern Alliance in conducting combat operations against the Taliban and al-Qaeda; kill, capture, and destroy al-Qaeda and deny them sanctuary." He named the main enemy much more clearly than most, and it greatly helped his SF commanders stay focused. Mulholland also wanted some airfields so he could build up numbers, bring in supplies, and pave the way for conventional army battalions to follow. Mazar-e-Sharif filled the bill. Its seizure on November 9 marked the first major victory on the ground in Afghanistan.

Taking Mazar-e-Sharif started a Taliban unraveling that accelerated dramatically over the next few days. Around Afghanistan, the word went out: No men, no matter how much they prayed, could stand against the devastating U.S. bombers and the handful of sharp-eyed, sharpshooting, bearded, tough Americans that accompanied the Northern Alliance. Mulholland's Task Force Dagger had two ODAs and an ODC (battalion-level headquarters) with Dostum's aggressive Uzbeks, and eventually five teams joined the various segments that made up Fahim Khan's Tajiks. Overhead, a succession of huge Air Force bombers and the carrier-launched Navy F-14 and F/A-18 fighter-bombers continued to deliver ordnance on time, on target, and, when necessary, in tremendous volume.

The Taliban backpedaled and made for the Pashtun regions in the south and east. Up north, Taloqan, which had changed hands a few times during the ongoing 1996–2001 civil war with the Taliban, gave up on November 10, the day after Mazar-e-Sharif was taken. Mulholland's ODAs 585 and 586 were there. Victorious, energized Northern Alliance bands closed on Kunduz, the last major city holding out up north.

Urged on and accompanied by the CIA and SF teams, the anti-Taliban Afghans pressed the fight. The capital of Kabul fell to Fahim Khan's men and ODA 555 on November 14. The Taliban collapse and retreat happened so fast that the smart guys in Washington and the press color commentators began to fret that it had been entirely too precipitous. Various experts worried about the lack of Pashtuns in the streets of Kabul — except the Taliban Pashtuns running out the other end, of course. The optics were all wrong, with too many Northern Alliance Tajiks in Kabul. No, the cognoscenti intoned, this wasn't being done properly. In many cases, these American voices were those that three weeks before had worried about a developing quagmire. Some people were just never happy. It reminded one of General George S. Patton's sardonic rejoinder when his Third Army took Trier, Germany, in 1945 ahead of schedule: "Do you want me to give it back?"

Victory developed momentum even in the Pashtun belt. Afghans like winners, and it was becoming clear that Mullah Omar and his Taliban were losing. In the south, Hamid Karzai and ODA 574 catalyzed enough Pashtun militia to take the provincial seat of Tarin Kowt on November 17. No soldier, but a gifted and relentless rabble-rouser, Karzai kept the locals motivated and encouraged them to march south on Kandahar city itself, the final Taliban outpost. The much less articulate but significantly more ruthless drug smuggler and gunrunner Gul Agha Sherzai, aided by ODA 583, brought his hundreds up from Takh-te-pol, tracing the long road from the border with Pakistan to the south. Taliban Kandahar hung in the balance, its days numbered. In a few weeks, several dozen brave, tough, highly skilled Americans, a deadly cascade of aerial firepower, and thousands of invigorated Afghan partners smashed the Taliban and liberated Afghanistan. Like most successes in that sad country, it led almost immediately to more trouble.

In Afghanistan, only the dead really surrender. The living might indulge in it for a while, and many did in that momentous November of 2001. For Afghans, giving up is a time-out, not a game-over. Surrender is merely a tactic to be employed when confronted with overwhelming odds. When the opportunity arises, combat resumes. Pauses or not, scores must be settled.

In the Afghan tribal culture, especially that of the Pashtuns who filled the Taliban ranks, the death of a family member created a blood feud to be settled by lethal vengeance. Inside the community, blood money could be accepted, but not outside it, and especially not from the infidels. The Americans and the Northern Alliance had killed a lot of Taliban. Those thousands captured included many families' brothers, sons, fathers, and cousins. The Taliban remnants weren't willing to let it go.

This applied in spades to the sullen, alienated survivors of Brigade 055, sometimes called the Arab Brigade. Many of these men, volunteers raised through al-Qaeda, were graduates of bin Laden's Afghan training camps. Many came from Pakistan and were as Pashtun as any of the locals; others hailed from Chechnya, Saudi Arabia, Yemen, and Egypt. Long on indoctrination and short on useful weapons and tactical skills, the foreigners envied their comrades martyred under the rain of U.S. bombs. Now they faced shame and Allah knew what else at the hands of the Uzbeks and these infidel Americans.

Rightly focused on new business to the east, in and around Kunduz, Abdul Rashid Dostum left a mass of captives in his wake. His subordinates looked for places to put them. Any sort of enclosure would do. Among several sites they used, the Northern Alliance leaders elected to deliver hundreds of captives to the imposing brick compound of Qala-i-Jangi, about six miles west of Mazar-e-Sharif city. That place looked plenty secure.

The octagonal brick Fighters' Fortress, six hundred yards across, dated back more than a century. This massive castle had been built over the course of a dozen years, starting in 1889, at the behest of a royal lieutenant in the north, Abdul Rahman Khan. The fort's walls — ten feet wide and anchored by big corner towers fifty feet across — rose forty feet above the surrounding farm fields. Firing slits scored the upper ramparts. A smaller east-west wall, ten feet high, split the facility in half. In its center, a plain metal door limited access between the two areas. The main entrance gate stood almost in the middle on the east side, just north of the central wall. Built into the north wall, a large headquarters and a two-story barracks housed the garrison, and horse stables were provided for their mounts. The southern half was more open, with a few single-story buildings, rows of dusty stables, and dank storerooms along the central and outer walls. In the center of the wide south yard, a long rectangular structure squatted; the Uzbeks called it the Pink House, a reference to its lighter shade of mud brickwork. Just to the east, another, smaller rectangular single-story adobe building capped a warren of underground storage rooms.

Following the lead of the Taliban, which had used the fortress for years as an armory and barracks, Dostum took over the better-appointed northern half as his temporary headquarters during operations around Mazar-e-Sharif. The Taliban had left stacks of ammunition and weapons of all types in almost every room in the structure. Dostum's Uzbeks added more, so the place had a lot of value in its own right as a munitions storage site. When he left to move south, Dostum appointed militia chief Majid Rozi as the fortress commander. Rozi kept a hundred Northern Alliance people with him. They inherited all the arms. He also assumed the unpleasant job of guarding a lot of very disgruntled Taliban prisoners.

The trucks arrived within hours after Rozi took over, as dusk gathered late on November 24. Packing up to fifty prisoners each on the beds of six large, old Russian-made vehicles, the Uzbek drivers entered the main gate, turned into the southern half, and dumped their human cargo. Many of those unloaded came from Brigade 055; they were al-Qaeda types, foreigners, men with nothing to lose. The outnumbered guards barked at the new arrivals, telling them to line up and be searched. Rumors passed through the ragged crowd: Native Afghans would be freed. Foreigners would be turned over to the Americans. That did not seem likely to end well.

When one of Rozi's officers leaned in to question a man who did not look Afghan, the man pulled out a live hand grenade. The blast killed himself and the Uzbek. Another captive fighter did likewise at the other end of the mob. A second Northern Alliance leader, a Hazara, died along with the grenade wielder. The mob seethed and began to move forward.

Alarmed, several Uzbek guards fired their AK-47s into the air, long, ripping bursts. An officer stepped up and addressed the prisoners, assuring them they would be treated well. Nobody was going to be given to the Americans. In fact, he announced, the United Nations planned to take the foreign fighters. (This would have been news to that distant body.)

Accurate or not, the message spread. The detainees subsided. Some glowered at their captors. A few pulled their dirty blankets over their heads. As a concession, the sentries ended their halfhearted attempts to search the prisoners. Honor had been served. Blows had been struck, and Allah had seen fit to remove the actors and the victims. Enough was enough.

Cowed, nervous, and dealing with a surly crowd as night approached, the Uzbek sentries lowered their weapons and prodded the prisoners into the long blocks of stables and the Pink House. Some were directed into the underground passages, already hosting prisoners from earlier hauls. The outer doors clanged shut and were locked. All those piles of weapons and bullets were locked in there too.

With the south yard now cleared, the Northern Alliance men pulled back across the middle wall and slammed the big sheet-metal gate. With AKs in hand, a few Uzbeks mounted the wall to keep watch on the dark buildings. Amazingly, the night passed without incident.

In the morning, several captives asked to be released from their cells to wash up and pray. Hopeful that the previous night's episodes would not be repeated, the guards opened the doors and let the prisoners come out. Water came in large tubs, followed soon after by a rice breakfast in similar containers. In small groups, clustered here and there on the flat southern dirt yard, men washed, prayed, and ate. The Uzbek sentinels relaxed. The warm sun helped.

At 11:15 a.m., a Northern Alliance officer opened the middle metal gate. Through it walked two men dressed in a mix of Western military gear and traditional Afghan clothes. Both men were armed with AKs and pistols. The pair separated, stopping about a hundred yards from each other. The Uzbek guards were told to assemble the prisoners in the courtyard. It took several minutes to gather up the various groups of captives. The prisoners sat on the ground. Each of the two visitors stood watching, searching the faces of the grimy men below.

Johnny Micheal "Mike" Spann and David Davis (pseudonym), both CIA officers, wanted to hear what these prisoners had to say. The presence of the Brigade 055 types suggested that the people held in Qala-i-Jangi possessed important information about al-Qaeda. A prisoner or two might even be able to pinpoint Osama bin Laden, Ayman al-Zawahiri, or the other al-Qaeda senior leaders. It was worth a try.

Spann and Davis proceeded with confidence. They had been active from the outset with the SF and the Northern Alliance. Indeed, Davis was one of those who'd met Captain Mark Nutsch on the landing zone when ODA 595 came in by helicopter on October 19, just over a month ago.

The CIA pair spoke Dari and some Pashto. Each American leaned down to question a man. Spann had his bunch. Davis had another. The questions started.

"Who are you?"

No response.

"Why are you here?"

Nothing. Seated in front of Spann, a thin young man hung his head. His legs slowly uncrossed and flexed.

Meanwhile, grumbling rippled around the yard. Seated men leaned toward each other. Americans; these two were Americans. The Brigade 055 fighters knew what that meant. Several stood up even as the Uzbek sentries motioned them to sit back down. Five, ten, twenty, and then more got up and moved forward. The Uzbeks gestured uselessly. Upright men shifted toward Spann and Davis. An Uzbek raised his AK-47 auto-rifle.

Absorbed by his subject, Spann focused on the man seated at his feet. He asked again, his tone even: "Why are you here?"

"To kill you," the enraged prisoner shouted in English. He jumped up in an instant, his skinny arms reaching toward Spann's neck. The CIA officer yanked out his pistol and pulled the trigger. The attacker jerked back and went down, finished. The gunshot echoed off the mud bricks.

Now all the prisoners stood up. The frontline converged in a wild, gesticulating knot around each of the two Americans; the rest filled in, adding weight to the inchoate mass. Panicked Uzbek guards started firing. A cacophony of yelling voices filled the courtyard, interspersed with the crack of 7.62 mm AK shots. One captive wrested an AK-47 from a guard. Then another took a rifle. The Uzbeks began to fall, drilled with bullets from their own weapons.

Hands, arms, bodies, a press of hot, dirty robes closed on Spann like a woolen wave. The American fired twice more, killing an attacker with each shot. But there were way too many. Davis was separated, mobbed too. He wriggled free, beating at the swarm of hot bodies around him. The CIA officer shot two or three Taliban prisoners and ran for it, toward the middle wall.

Behind Davis, as Spann was engulfed, other Uzbek guards fell. Cheering Taliban and Brigade 055 men stripped the AKs from the dead sentries. Surviving Uzbeks ran toward the metal gate, already closing in front of them. On the walls to the north, Northern Alliance sentries opened fire. They shot at the Taliban, now dispersing, splitting into singles and pairs, seeking cover in the brick buildings. The former prisoners, now very much combatants, returned fire. More gleeful shouts erupted. Some of the Brigade 055 men discovered a stock of AKs and ammunition.

Shooting grew in volume. On the north wall, in the headquarters building, two Red Cross representatives, one of them Simon Brooks of Great Britain, looked up as Dave Davis came forcefully into the room. The CIA officer was all business. His weapons smelled of cordite. The Red Cross people before Davis provided a nice touch, albeit absurd, checking to ensure those in captivity had access to shuffleboard or whatnot. Well, Red Cross rules be damned. The Taliban weren't playing by any Western legal niceties.

"He burst in and told us to get out of there," Brooks recalled. "He was really shaken up. He said there were 20 dead Northern Alliance guys and the Taliban were taking over the fort." Davis indicated that he had to go back and get Spann. The Red Cross man and his partner didn't want any part of that. They went out on the wall and found a way down and away. A German television crew and a *Time* magazine reporter and interpreter stayed. Until then, they had figured the war had passed them by, leaving them in the boring backwash. Now they had a story, all right.

Davis called his superiors. They contacted K2 airfield, Task Force Dagger headquarters, way up north in Uzbekistan, over the mountains and under the shrouds of thick clouds. The first version of the story suggested that the prison was in enemy hands and there were two dead Americans, bad indeed. The second, more accurate but no less sobering, reflected Spann down and missing, with Davis and Majid Rozi's Northern Alliance element under siege by hundreds of rampaging, armed ex-prisoners.

For men like Colonel John Mulholland, few messages resonated like the report that one American was trapped and another American might be missing. Special operators do not leave their own, ever. A good portion of the brutal engagement in Mogadishu, Somalia, in October 1993 revolved around U.S. determination to recover every comrade, alive or not. Despite valiant efforts, the Somali foe got a few dead Americans and dragged their corpses through the dusty streets as the camera rolled. The battle spawned Mark Bowden's gripping, authoritative book and the film based on it, *Black Hawk Down,* which was just about to open in theaters back in America as Qala-i-Jangi erupted. Americans in uniform, especially in the tight-knit SOF family, knew only too well how bad this kind of fight could get.

At one point in Mogadishu, with two helicopters crashed and thousands of heavily armed hostiles swarming isolated units, the U.S. SOF commanders had only two men on hand to save a crew of downed aviators. Leaving their circling helicopter to go on the ground promised almost certain death. But it might, just might, save those trapped down below. Two senior snipers went: Master Sergeant Gary Gordon and Sergeant First Class Randy Shughart. Down in the street, they killed many Somali gunmen. But in the end, enemy numbers prevailed. Their heroism protected the wounded Chief Warrant Officer 3 Michael Durant long enough for a senior Somali to take the man prisoner — he survived and came home. Gordon and Shughart died under enemy fire. Each was recognized for his selfless valor with the Medal of Honor. Nobody up at Task Force Dagger wanted to repeat that episode.

That said, Mulholland had almost nothing to send to Qala-i-Jangi. His ODAs were spread all across northern Afghanistan, moving and fighting. He had the First Battalion, Eighty-Seventh Infantry from the Tenth Mountain Division protecting the K2 airfield. Maybe they could do something, but they'd need a lot of helicopters, of which Mulholland had only a few, to take the long flight south. They might stage to the Mazar-e-Sharif airstrip aboard Air Force C-130 transport planes, but Mulholland had none of those at the moment. Such a response took time to assemble. Task Force Dagger staff people got to work. Something could come together in a day or so. But with Spann already lost, Davis pinned down, and the Uzbeks hanging on by their fingernails, who could wait that long?

In the interim, all the big colonel could commit was a battalion headquarters: Operational Detachment Charlie 53, under Lieutenant Colonel Max Bowers. Even that little element, thirty-seven men on paper, was understrength. Bowers himself was forward with part of his team outside Kunduz, helping with the ongoing Northern Alliance encirclement of that key city, the final major Taliban position up north. The command post in Mazar-e-Sharif was overseeing all operations in support of the Northern Alliance, including looking after the city's vital airfield. Well, they were SF men, not clerks. Conditions called for immediate action. And most important, ODC 53 had Air Force joint terminal air controllers, ensuring a form of firepower not available during the Black Hawk Down battle. Mulholland gave the order.

At 1:45 p.m., the battalion XO, Major Kurt Sonntag, got the word. Sonntag had multiple situations to juggle. He turned to his counterpart and nominal subordinate, Major Mark Mitchell, the S-3 (operations officer; that is, the key planner and battle tracker in the command post). Mitchell tagged others from the small headquarters: two SF operations NCOs, the surgeon, and an interpreter. Mitchell enlisted, as the security element, the only available fighting force he owned, eight newly arrived British Special Boat Service (SBS) Royal Marine Commandos. The SBS section included two members of its U.S. counterpart formation, Navy SEALs. Most important of all, Mitchell took the two Air Force joint terminal air controllers. Staff Sergeant Michael Sciortino and his fellow JTAC had the radios and skills to ensure that the fifteen-man relief element brought a very, very heavy punch.

Within a half hour — the SF aboard two battered minivans, the SBS and SEALs in two Land Rovers — the little team drove up to the main gate at Qala-i-Jangi. Even over the engine noise, they could tell they were driving into a major firefight.

"You could hear the bullets whistling overhead," Air Force controller Mike Sciortino remembered. "Random explosions could be seen from mortar rounds landing along the northern wall that surrounded the fort." The Americans and British exited their vehicles just inside the main gate. Majid Rozi met them and, through the interpreter, explained the situation. Obviously, the Taliban held the south half of the fort and were keeping the northern half under fire.

Mitchell did not hesitate. He turned to Sciortino and told the Air Force controller: "I want SATCOM [satellite communications] and JDAMs [Joint Direct Attack Munitions]. Tell them there will be six or seven buildings in a line in the southwest half. If they can hit that, then that would kill a whole lot of these motherfuckers." Mitchell told the airman to use his long-range satellite radio and get airplanes overhead with guided bombs. A JDAM used a strap-on fin kit and a little guidance package to convert a dumb explosive device into a smart weapon that would hit within spitting distance of the assigned map coordinates. Since a thousand-pound bomb could kill out to two hundred yards and wound out to four hundred, that was more than good enough. Of course, it wasn't going to be easy steering these things into a fortress that was six hundred yards from one side to the other.

Sciortino started calling the warplanes overhead. As he did, one of the Special Forces NCOs got Dave Davis on the radio. Addressing Mitchell, the NCO summarized their conversation. "Mike is MIA. They've taken his gun and ammo," he said. "We have another guy [Davis]. He managed to kill two of them with his pistol, but he's holed up in the north side with no ammo." The sergeant then looked at Sciortino: "Tell those guys to stop scratching their balls and fly."

The SF and the Afghan commander set up on top of the northeast corner tower. It took until almost four that afternoon to get aircraft overhead. After all, the fort lay in a supposedly quiet rear area, well away from the day's major fighting to the east around Kunduz and far to the south, near Kandahar. The concept of frontlines represented a holdover from another era, the one that had favored the U.S. Qala-i-Jangi demonstrated that the enemy was learning to discard such Western conventions and go with what worked best for Afghans, which was to say, treachery and ambuscade. Fortunately for Mitchell and his embattled team, warplanes can be rerouted pretty quickly, and the experienced U.S. airmen reacted even quicker than most.

Outside the walls, Rozi evacuated his Uzbek wounded by taxicab. The Northern Alliance resupplied ammunition. Canteens and bottled water went around. The Taliban shot at them now and then, but they seemed happy to hold their part of Qala-i-Jangi. When a few Taliban tried to get outside, alert North Alliance riflemen killed them.

Now the American planes appeared overhead. The enemies were about to be reminded of how they'd ended up as captives. Sciortino talked on the radio, calmly relaying map coordinates. "Four minutes," he said. The Americans and British looked at each other.

The countdown continued. Three minutes. The Americans passed the word through the interpreter: *Take cover.*

Two minutes. Everybody got down, but some of the Uzbeks kept their heads up. They just had to see. The Americans waved them flat.

One minute.

Thirty seconds.

"Fifteen seconds," Sciortino said, calm and steady, a man doing his job.

Right on time, the guided bomb whooshed overhead, a white spear just visible as it slanted down out of the heavens and lanced into the southern wall. The thing slammed into one of the domed stables, crumpling the half sphere as if it were an eggshell. A huge V of dark brick dust shot up. Roaring air pressed the men down. Hot metal fragments spun overhead. Splinters clattered off the brick parapets.

At the south end, firing stopped. On the north wall, the Uzbeks cheered. Then a Taliban shot rang out. More followed. This was not going to be easy.

Six more times, the Air Force controllers brought down JDAMs. The British SBS commander raised Davis on the radio. After two more bombs, Davis and the German television crew planned to make a break for it. What about Spann? "From what I understand," one SF NCO said, "he was already gone before we got here." Continued bombing and darkness ensured that Davis and the television people escaped cleanly. But Major Mitchell and his people stuck with Rozi and the Uzbeks. They had to finish off this Taliban revolt. And they intended to find Mike Spann, dead or alive.

Morning brought reinforcements. Majid Rozi's people nursed an ancient T-55 Russian tank up to the battlement, its rust-streaked treads clattering slowly up the long ramp. Grinning Uzbeks methodically hand-cranked its long 100 mm cannon around to face the south. The tank began banging away.

On the nearby brick catwalk, two Northern Alliance mortar teams fiddled with their thin metal tubes, fussed with the angles, and dropped some test shots down the barrels. Each round slid down, went off with a metallic *plonk* as it struck the firing pin, and then arced out toward the south end of the fort. Once they had the range, the men took turns dispatching an occasional fat little bomblet onto the Pink House and stables. Many domed roofs showed gaps. Whole walls were gone.

Defiant Taliban gunfire continued.

About 10:00 a.m., coming all the way from K2 in Uzbekistan, four SF men and an eight-man rifle squad from the Tenth Mountain Division showed up. The Tenth Mountain riflemen represented some of the first conventional troops in country. Mulholland had no other units to send, but he delivered what he had. The SF colonel knew the airpower would keep coming.

Although hit by the tank and mortar rounds, Taliban gunmen kept returning small-arms fire. Mitchell knew it was time to bring in the bombs again. He turned to Sciortino. His partner handled the calls. These were different planes than the ones that had come the day before. These particular Navy F-14 Tomcats came off a carrier way out in the Arabian Sea.

Over the cracking and snapping of small arms, the Americans and the British near the JTACs caught snatches of the radio conversation. "Be advised," said one of the Navy guys up in the fighters. "You are danger close. You are about a hundred meters away from the target." All the strikes the day before had fallen danger close, a term defined by the military as six hundred meters from the proposed target. Hell, by that definition, any bombing inside Qala-i-Jangi equaled danger close. The Air Force controller cleared the strike.

"Making their run now," said the naval aviator far overhead. The usual countdown started. Everybody knew the drill.

Men got down, waiting. One minute.

Taliban shots kept up. Thirty seconds. Waiting.

Fifteen seconds.

A massive hot light; air sucked from lungs; brown brick dirt gushing up like gritty water; a blast wave roaring like a thousand cracks of thunder — the bomb, a two-thousand-pounder, hit almost squarely on the northeastern tower. "I went flying through the air," recalled Sciortino. "I blacked out."

Around the airman, others lay sprawled. Some rolled over. A few reared up, unsteady. Several didn't move at all. The T-55 tank turret flipped cleanly off, the hull overturned and blackened. A huge divot, hazed by drifting yellow dust, breached the thick outer brick wall. One Uzbek shouted in English, "No! No!"

An American spoke on the radio: "We have men down." Four Northern Alliance men had died. Another dozen lay wounded. So did two British and five Americans, including Mike Sciortino. It had been a mistake, no doubt. Something zigged when it should have zagged. A transposed number in the map coordinates, a bent fin on the bomb, a *1* turned *0* in the circuitry due to dust or banging around on the plane or just plain bad fortune — nobody knew for sure, except maybe Carl von Clausewitz, who'd said such foul-ups happen in every war. Sooner or later, friction collected its blood toll.

As the afternoon dragged by, shooting died down on both sides. The trapped, battered Taliban must have been hungry and tired. Americans and Uzbeks took advantage of the relative lull to move their wounded and dead. November 26 had been a long day already. It got longer.

Just before sunset, CIA officer Dave Davis returned and linked up with Mark Mitchell and the SF element. The CIA and SF leaders had a powerful reinforcement on its way in, one likely to break the Taliban's back. Davis turned to Alex Perry of *Time* magazine. "You don't want to leave here tonight," the CIA officer offered. "There's going to be quite a show."

There was. After sunset, up in thc black sky, droning engines sounded. An AC-130 aerial gunship circled overhead. It pounded the Taliban buildings with dead-on, rapid-firing 40 mm Bofors guns and a succession of 105 mm shells. The loud cracking and dull booms pulsed again and again. The aircraft had a large stock of ammunition and worked through it slowly, steadily, relentlessly. Taliban shooting died away.

When the first gunship went off station, a second AC-130 took over. The pounding went on and on for hours, all night. The succession of hits chewed up the adobe structures. Walls got knocked over. Ceilings caved in. One could imagine the effects of this on those trapped inside and helpless.

After a few hours of the relentless stream of projectiles, an ammunition stock detonated in one of the southern buildings. The explosion lit the night and initiated a fierce conflagration that burned into the dawn. The blast rattled walls and popped open doors six miles east at the ODC 53 headquarters.

On the morning of November 27, the Northern Alliance and their SF brothers decided to finish it. Taliban fire had dwindled to occasional pop shots. Subsisting on stringy horse meat from slaughtered animals, these men were low on morale, low on ammunition, and low on armed effectives. Dave Davis said to his comrades, "We're going to close in on these guys pretty hard." They did, shooting as they went.

In each of the holed, dusty storerooms, the SF men and Northern Alliance people took turns clearing. The Uzbeks proved liberal with hand grenades and bullets, dumping in a fragmentation device and following it up with an entire magazine or two (or three) of wild AK fire. Fortunately, the brick walls absorbed the 7.62 mm rounds rather well. It wasn't elegant, but it worked.

Much better trained, the Americans and British used aimed fire. Tossing frag grenades onto hardened brick floors often got you an immediate face full of hot metal. The SF and SBS men entered like lithe cats, scanned and shot, methodically cleaning each cubicle with a few precision rifle rounds. *Slow is smooth, smooth is fast,* the sergeants teach you. On it went, storeroom after storeroom.

It took time and sapped the energy of men wearing heavy body armor. But it had to be done. Smart enemy fighters raised their hands, not always an effective tactic with the enthusiastic Uzbeks. Dumb ones raised their AKs and got themselves drilled.

Resistance proved minimal. The AC-130s had reaped a grim harvest. The room-clearing added more. "On the field below lay hundreds of dead and dying," eyewitness Alex Perry of *Time* wrote. "Two embraced in death. Alliance soldiers stepped gingerly over the bodies." After crossing the killing ground, the Americans and British found Mike Spann. His body was booby-trapped with a hand grenade, which was carefully disarmed by the recovery team.

Johnny Micheal Spann became the first American killed in action in Afghanistan. Mike Sciortino received the first Purple Heart earned by an airman in the Afghan campaign. Mark Mitchell received the Distinguished Service Cross, America's second-highest award for bravery and the first one given in this war. Others also earned military recognition. It all reflected the scale of the engagement, a major one by any measure. You would have to go back to the Korean War's 1952 Geoje Island prisoner revolt to find anything similar.

Rozi's men spent the next few days cleaning out the underground rooms and tunnels. Taliban holdouts battled like cornered rats. Grenades, gunfire, and hunger did not bring them up. Burning oil didn't do it. Finally, after days of shooting and explosives, icy water washed them out. Eighty-six gaunt, filthy, beaten men stumbled up into the light. Among them, two Americans emerged: John Walker Lindh, a twenty-year-old from California, and Yaser Esam Hamdi, age twenty-one, born in Louisiana. Osama bin Laden's recruiters got around.

The Qala-i-Jangi uprising was treated as a one-off, a bloody postscript to the unorthodox, highly successful SOF/airpower/Northern Alliance blitzkrieg that blew away the Taliban. In his postwar memoirs, USCENTCOM commander General Tommy Franks didn't even mention it. His summary of the Afghan campaign in the north ends with the fall of Mazar-e-Sharif. Indeed, on November 27, the same day Mark Mitchell's men found Mike Spann, the USCENTCOM commander got a call from Secretary of Defense Rumsfeld. It wasn't about Qala-i-Jangi. "General Franks," Rumsfeld said, "the president wants us to look at options for Iraq. What is the status of your planning?" The big guys in Washington had already moved on.

What happened at Qala-i-Jangi deserved a much closer look. Inside that crumbling fortress lurked the shape of things to come. The clash at Qala-i-Jangi prefigured a few key aspects that grew to characterize the Global War on Terrorism, especially its Afghan variation.

As it so often would in the future, the trouble started with prisoners. More difficulties followed Qala-i-Jangi, especially as the Brigade 055 survivors headed to the U.S. Naval Base at Guantánamo Bay, Cuba. Clever Bush administration lawyers thought they'd evaded U.S. law by placing Camp X-Ray on a borrowed patch of territory in Cuba, an avowed U.S. enemy that nevertheless honored a long-standing prior agreement. Some of those who survived the Qala-i-Jangi battle ended up in Guantánamo. Their struggle continued inside the wire. Though the camp operated at a very high standard and admitted inspectors from international human rights organizations, the very word *Guantánamo* became shorthand in the Islamic world for "prisoner abuse."

Even the word *prisoners* was shelved. Administration lawyers in Washington didn't want to identify the men in Brigade 055 as prisoners of war. Under the Geneva Conventions, a POW denoted a person in national uniform captured under a clear chain of command via a procedure that adhered to defined regulations. Under international protocols and military law, POWs had rights, and the holding country accepted responsibilities to secure, protect, and care for enemy troops so captured.

The Taliban and al-Qaeda didn't seize any U.S. prisoners early in the war. When the Afghan foes and their Iraqi affiliates eventually did pick up a few American captives, they treated them abominably. By sawing off heads for the benefit of the Al Jazeera audience, these enemies put themselves right up there on the sadism meter, surpassing even the twisted malice of the North Koreans and the North Vietnamese.

With all of this in mind, and recognizing that many of those captured in Afghanistan were camp-trained terrorists, the suits in Washington decided that the guys in Guantánamo and other holding sites were not POWs. They were not like the bedraggled Iraqi conscripts who'd surrendered en masse back in 1991. You couldn't toss them some pork-free ration packs and send them home. No country wanted these brutal characters. So the legal people labeled the captives unlawful enemy combatants, which was certainly true, as they didn't wear uniforms or have much organized authority. They entered captivity for the duration of a war that had no defined termination point. Not entitled to POW status, they became merely detainees.

While this designation passed legal muster, it left a lingering perception up and down the chain of command that maybe the normal rules for treatment of prisoners of war didn't apply. Americans in uniform asked how to handle the captives. How were they to be housed? How about questioning? What about means of restraint? Sometimes the answers came back freighted with ambiguity; sometimes no answers came back. As the war devolved into night raids that dragged men out of bed and harsh little firefights against roadside bombers in civilian dress, U.S. soldiers developed dark thoughts. Leaving the detention situation so loosely defined would come back to bite them in ways nobody foresaw in late 2001.

Qala-i-Jangi also made it clear that people were already watching and grading the Americans. The U.S. intended to adhere to a high standard. Outsiders wanted to see how the Americans measured up. Qala-i-Jangi happened with international do-gooders, in this case the Red Cross, right in the middle of things. Those, like Simon Brooks, wanted to enforce Marquess of Queensberry rules on the side that hosted them, the Americans. The opposition got a pass. The Taliban and al-Qaeda had never welcomed international detention observers and never would, except in wholly contrived displays. It wasn't fair but it happened over and over again.

In addition to those well-meaning civilians, Alex Perry of *Time* magazine and a German television crew observed almost everything at Qala-i-Jangi. To their credit, they recorded what happened and left the editorial slants to others. The CIA and military SOF don't care for publicity, especially at the business end of their activities. Yet war generated interest, interest generated business, and the media, being run by commercial concerns, followed events that attracted customers and so sold advertising. As most of the coverage at this point was neutral or favorable, few in uniform thought much about it. Those thoughts would change as the war continued.

At Qala-i-Jangi, the spooks and the operators, the CIA and the SF, had held center stage. But John Mulholland knew that he needed reinforcements. He sent in a conventional force, admittedly just a few men from the Tenth Mountain Division. It reminded all that there were never enough SOF. Airpower had killed effectively in Qala-i-Jangi. After all the hammering, though, people had to go in on the ground with rifles, grenades, and guts. To control dirt and the societies that lived on it, you had to use live, trained, disciplined humans, and more than a few. As the war evolved, it demanded way more than handfuls of CIA officers and special operators. That, too, was a lesson from the prisoner uprising.

Those needed conventional battalions began to arrive in December 2001, even as the water froze in the abandoned dungeons at Qala-i-Jangi. In the south, near Kandahar, the First Marine Expeditionary Brigade established Forward Operating Base Rhino. Well to the north, coming in behind Mulholland's Task Force Dagger command post at K2 air base in Uzbekistan, the Tenth Mountain Division headquarters arrived. They joined the busy First Battalion, Eighty-Seventh Infantry already at K2. The Second Battalion, 187th Infantry, drawn from the 101st Airborne Division, guarded the Shahbaz Air Base in Jacobabad, Pakistan, quite a concession by Pervez Musharraf's ever-suspicious government. In Kabul itself, the headquarters of the British Third Mechanized Division set up what would become the International Security Assistance Force (ISAF), initially limited to securing the capital.

By that time, the Afghan capital wanted securing. America's chosen Pashtun, Hamid Karzai, headed something grandiosely announced as the Afghan Interim Authority. On December 5, Karzai auditioned for the job from outside besieged Kandahar via a U.S. Special Forces satellite-phone connection. After listening to him, a United Nations conference in Bonn, Germany, gave him the task.

Then, in an incident all too similar to the errant two-thousand-pound bomb at Qala-i-Jangi, a mistake in coordinates led to a strike that hit way too near Karzai, his bodyguards, and his comrades of ODA 574. Six Afghans and two Americans died. Sixty-five Afghans and the rest of the U.S. SF team sustained wounds. Karzai received a splinter cut to his face. The tough Afghans and Americans worked through this lethal friction. After their casualties had been treated and evacuated, they formed up and pressed on to take Kandahar city.

Like those who fought at Qala-i-Jangi, Hamid Karzai learned some things. The Americans owed him for stepping forward. For his part, and it frustrated him greatly, he owed his life and his new position to the Americans. And they were not always careful with their weapons . . . or their allies.

General Tommy Franks exulted. In his memoirs, he put it this way: "In Operation Enduring Freedom, Central Command and its Afghan allies had defeated the Taliban regime and destroyed Osama bin Laden's al Qaeda terrorist sanctuary in seventy-six days." In this triumphal statement, Franks echoed the conventional wisdom of that time.

True, the Taliban in late 2001 lay prostrate. In a decision reminiscent of Saddam Hussein's embrace of a conventional defense of his country in 1991, Mullah Omar elected to fight this war the Western way. "Defending the cities with front lines that can be targeted from the air will cause us terrible loss," Omar opined. But he did it anyway, and his men got plastered.

In the aftermath, local Taliban leaders went to ground. The senior ones crossed into Pakistan. There they hid, pondering the lessons of their speedy defeat.

First, they decided that they would determine the timeline for this war. Their defeat proved temporary, although it would take half a decade for them to commence a significant resurrection. Convinced they had won big, the U.S. assumed there would be no such Taliban resurgence. Naturally, few asked why the Taliban was, or if it should remain, America's enemy. The U.S. just declared victory and kept going. The Taliban, meanwhile, began to rebuild their networks in the Pashtun belt along the Afghan-Pakistan border.

Second, the Taliban saw the advantage of playing the victim. Most of the Western media outlets were skeptical, but the uneducated on the Pashtun Street and their coreligionists on the Arab Street took it all at face value. As the lesser power, not bound by inconvenient things like the truth, Mullah Omar wrapped his followers in Afghan nationalism and Islamist righteousness against the occupying infidels. Over time, inside Afghanistan and out, people forgot about the thunderous opening round and regarded instead the squalid, drawn-out morass that followed.

Finally, the Taliban went back to their preferred ways of war. Sniping, raiding, roadside bombs, trickery, and cunning, all financed by the opium trade, promised much better results for the Taliban than Western tactics had produced. Uniforms and discipline (such as it was) were abandoned. Loose tribal gangs and thuggery became the norm. Let the U.S. try to sift the fighters from the innocents in the Afghan countryside. Breaking into Afghan homes at night, rounding up both the usual and unusual suspects, dropping bombs with big blast patterns—well, the Americans couldn't help but be ham-handed. They were outsiders, infidels. Eventually, the Taliban reckoned, the locals would side with their own.

As for al-Qaeda, even at the time, many decried Osama bin Laden's escape from Afghanistan. A major operation by the Task Force in the first three weeks of December 2001 flailed at the Tora Bora cave network on the rugged border near Jalalabad. Extensive airstrikes, a handful of determined SOF, and Afghan partners gave chase through the Hindu Kush. Slipping through the snowy peaks, the remaining al-Qaeda leadership and their security details played hide-and-seek in the tunnels and bunkers of one of the Contractor's big projects from the 1979–89 Soviet war. The Task Force almost had them.

Many enemies never made it out of the air-delivered meat grinder. The thousand al-Qaeda casualties mentioned by some sources actually amounted to a conservative estimate. But the Sheikh and his immediate circle escaped. Later recriminations lambasted Franks, not to mention Secretary Rumsfeld and even President Bush himself, for not committing a Ranger battalion, the Marines, or perhaps the First Battalion, Eighty-Seventh Infantry to block the high trails into Pakistan. But in that extremely convoluted terrain, in that awful winter weather, Franks could have deployed the entire Tenth Mountain Division, and a few others to boot, without sealing off all those ratlines.

It didn't matter. The opening Afghan campaign finished bin Laden and al-Qaeda as a viable, operational entity. Never again would Osama bin Laden or Ayman al-Zawahiri select, review, and approve operations. They dared not even speak except through cutouts, couriers, and a dwindling coterie of confidants. The Afghan camps were overrun and never rebuilt. Al-Qaeda, "the Base," disintegrated. An American raiding force finally found and killed Osama bin Laden in Abbottabad, Pakistan, on May 2, 2011. But in many ways, he was already dead, and had been for years.

The smashed organization resembled a large pane of high-quality glass that had been shattered: each shard was sharp, still fine, but small and separate. Like companies in the aftermath of an antitrust corporate breakup, the lesser franchises at first very much resembled al-Qaeda. Gradually, though, they developed their own unique regional flavors: al-Qaeda 2.0. Many acts, great and small, drew inspiration from Osama bin Laden the Sheikh. But bin Laden the Contractor was no more. All that remained was an inspiration that catalyzed the Islamist fringe and haunted the West like a ghost.

Three major attacks aped the al-Qaeda approach and honored the Sheikh, who claimed them all. The Bali nightclub and consulate bombings of October 12, 2002, killed 202 people and tipped the hand of Jemaah Islamiyah (the Islamic Congregation), an Indonesian group. The Madrid, Spain, attacks of March 11, 2004, that featured the near-simultaneous bombing of four commuter trains and killed 191 civilians was traced to Moroccan Islamists. The London subway and bus attacks on July 7, 2005, that saw four detonations and fifty-two deaths was tied to a local Islamist cell in Great Britain. Other successor elements drew on al-Qaeda's name and cachet and exhibited features pioneered by bin Laden: al-Qaeda in Iraq, al-Qaeda in the Arabian Peninsula, and al-Qaeda in the Islamic Maghreb. All acknowledged bin Laden. Affiliated groups like Abu Sayyaf (Father of the Swordsmith) in the Philippine Islands, Lashkar-e-Taiba (Army of the Pure) in Pakistan, the Islamic Movement of Uzbekistan (IMU), and Al Shabaab (the Youth) in Somalia did likewise. Rhetoric aside, all ran their own shows with a nod to the ghost in Abbottabad.

It would have taken the prescience of a prophet to see all that so clearly in the heady early winter of 2001. Around President Bush, in the ranks of the senior military, the prevailing attitude amounted to relief. The fighting had gone quicker than anyone had hoped and cost far less than anyone had feared. What came next? It all went back to that old question: Who was the enemy?

Defined narrowly as al-Qaeda and its immediate Taliban patrons, that enemy was suppressed for now. In truth, classic, real al-Qaeda, Osama bin Laden's construct, lay in ruins. But in December of 2001, nobody in Washington possessed enough hard intelligence to prove that. After 9/11, going with hunches did not seem wise. Vice President Cheney and Secretary Rumsfeld advised President Bush to keep going, to finish the fight in Afghanistan (whatever that meant) and look beyond. Iraq came up, a threat still out there, a country brimming with ill will against America and presumed to have the means to act on it. Cheney argued again and again, "But if we don't get them, they'll get us." Rumsfeld agreed. That massive blackened gash in the Pentagon sure buttressed his case.

One of the only members of the highest circle who knew what battle was like warned the others. Colin Powell cautioned repeatedly against expanding and extending the war. As with Cassandra in ancient Troy, he stayed on the inside looking out, doomed to be both correct and unheeded. Years later, informed by solid hindsight, the smart set all agreed that Powell had it right from the outset, but at the time, he alone spoke these unwelcome thoughts. In late 2001, among Bush's key advisers, Powell was an outlier. To the rest, including those in uniform, pressing ahead to Iraq seemed reasonable, prudent, and probably the best way to ensure that Saddam Hussein did not strike first. Like President Harry S. Truman in the autumn of 1950 en route to Korea's Yalu River, George W. Bush decided to keep going.

To finish off al-Qaeda, America needed Afghanistan, both to mop up there and to launch strikes into neighboring Pakistan, with the grudging assent of that dubious state. Working in Afghanistan involved supporting Hamid Karzai, nursing along this ISAF business with the allies, and continuing to go after the Taliban, which threatened both. Each step made all the sense in the world. In the aggregate, however, they ensnared the U.S. in an experiment in nation-building in the forbidding Hindu Kush. Osama bin Laden might well have been out of it, but a key part of his master plan came to pass.

Was there an alternative? Bush's father had found one in 1991. Faced with Saddam Hussein on the ropes, the Kurds and Shiites in rebellion, Kuwait wrecked, and a victorious military straining at the leash, George H. W. Bush took his limited victory and pulled back. True, forces remained, patrolling the air over Iraq. But the major war ended. America could have done the same in late 2001 — declared 9/11 avenged, al-Qaeda smashed, and Afghanistan freed, and then pulled out; if the handwringers had insisted, perhaps the U.S. might have left a small residual force. One could conceive of such a thing.

Except George W. Bush's administration and the U.S. citizenry had lived through 9/11. Nobody dared chance another such horrific event. To guarantee beyond doubt the end of al-Qaeda, to knock out its affiliates, sponsors, and imitators, America had to keep up the pressure. "I will not wait on events," the younger Bush declared.

So there it stood. Al-Qaeda had been run off and disrupted, left badly disorganized, but not killed. The parts and franchises, the copycats and wannabes, took up where the Sheikh and company had left off. In December of 2001, we just didn't have the manhunters, and, more to the point, the manhunters didn't have the good scenters and able beaters to track down and tear apart a terrorist network, whether large or small, transnational or local. We could smash it — and we did — but we could not kill it, not yet.

Scattered strikes and sloppy swings left too many lingering ghosts, too much inspiration for the next wave of terrorists and their financiers and sympathizers. The work could not be left to a few hundred specialists. It took thousands, locals especially. The spooks couldn't find this will-o'-the-wisp enemy, certainly not on their own. Put another way, the early days in Afghanistan showed all too well the course of the entire war: never enough spooks but way too many ghosts.

4

ANACONDA

O gods, from the venom of the cobra, the teeth of the tiger,
and the vengeance of the Afghan — deliver us.

— TRADITIONAL HINDU PRAYER

Colonel Frank Wiercinski found the ghosts. They were right down there in the sunlight at seven thousand feet, in strength, popping up and down like rifle-range targets. The enemy shot at him and the other eight Americans in his Third Brigade, 101st Airborne Division tactical command post. That rather neat military designation might look good on a computer screen. At 6:30 a.m. on March 2, 2002, on the southwest edge of the Shah-i-Kot Valley, the term described a short, ragged line of tired, dirty men prone on the jagged rocks ducking AK bullets. Something about the radio antennas sticking up, as well as the two Black Hawk helicopters that shimmied in one by one to disgorge the soldiers a few minutes earlier, told the enemy to give this little bunch of Americans some special attention. They were getting it.

They almost got it for keeps on the landing. The lead UH-60 helicopter with Wiercinski aboard missed the initial approach to the minuscule flat spot hanging off the razorback ridge line. Threading the needle between rows of boulders, the trail bird made it in cleanly, rotors spinning a few feet from the stone outcrops. Five soldiers hopped out and squatted as the helicopter roared up and away in a hanging cloud of dust. The quintet waited for their bluff colonel and his three people.

As Wiercinski's Black Hawk looped around to try again, the radios hummed with voices. Across from the tiny brigade TAC landing zone, on the northeastern side of the valley, the Second Battalion, 187th Infantry, more than a hundred of them, got in cleanly thanks to three big twin-rotor CH-47 Chinooks. They had already met hostile contact, not serious, but enough to keep them busy. It would be almost four hours before the next three Chinooks came in.

South of 2-187 Infantry on that same eastern slope of the valley, the hundred-plus soldiers of 1-87 Infantry already had their hands full right on the landing zone. Their three Chinooks barely made it out in the face of heavy machine guns and a pair of rocket-propelled grenades (RPGs). Enemy fighters moved among the rocks above the Americans. The foe held the high ground.

On the valley floor, a column of friendly Afghan militia with U.S. Special Forces advisers was supposed to be moving to four grubby little villages to sort out bad guys from civilians. The brain trust in charge of American propaganda (or "information operations," as they called it) dubbed Pashtun commander Zia Lodin's six hundred men a key part of the Eastern Alliance — a name that reflected the wishful thinking that these Afghans might equal the very real Northern Alliance in its organizational acumen and will to fight. They showed the latter most of the time, but not today.

As Wiercinski's soldiers landed on the valley slopes, Zia's people crouched alongside a line of gaudily painted "jingle" cargo trucks and beaten-up pickups. A mistaken engagement by an AC-130 had killed one American SF officer and wounded two others and thirteen Afghans, effectively halting the advance. Now enemy riflemen and mortar crews tormented the befuddled militiamen. Casualties mounted. Pashtun leader Zia waited. He demanded to see some U.S. airstrikes, a lot of them, before he resumed his movement.

To sum up: one battalion (really a small rifle company) down and okay to the northeast; the second one (equally understrength) down but in trouble to the southeast; the Afghan ground attack stalled; and the enemy present in force, well armed and uninterested in leaving. So it stood that early morning in the Shah-i-Kot Valley as that second Black Hawk flared up to insert Frank Wiercinski and three others.

That aggressive foe sped Wiercinski's ride on its way and in so doing nearly finished them all. One RPG snaked in front of the Black Hawk's Plexiglas chin bubble and bloomed in a dirty burst, flinging fragments into the helicopter's bottom. The aircraft bucked, but the aviators knew their business and made the landing.

As they did, AK bullets stitched the tail of the UH-60. There is no good place for a helicopter to get shot, but the intricate gearbox that controlled the tail rotor is certainly one of the worst. If something broke back there, a Black Hawk without a tail rotor meant an airframe rotating down to a crash. Already wedged between those unforgiving rocks, its tanks full of volatile JP-8 fuel, the Black Hawk would have been gone in a millisecond in one searing flash. They would have been scraping Wiercinski into a sandbag if that happened. It did not.

The helicopter made it out. Wiercinski and his three partners moved to link up with the five already there. As they left the aircraft, the men heard snapping sounds, *zip-zip,* over their heads. "Guys," Wiercinski said, "I've heard this sound before. I didn't like it then and I don't like it now." There would be a lot more.

A smart aleck might have observed that with dozens of infantry battalions in the U.S. Army, it was quite a coincidence that two different units that happened to share almost the same numeric monikers were at the same place. Maybe it confused the enemy. It didn't confuse Wiercinski or his brigade headquarters people, who referred to the outfits by their radio call signs: Rakkasan Raider for 2-187, and Summit for 1-87. It demonstrated that Operation Anaconda put together a lot of pieces to chase those al-Qaeda ghosts.

The operation's name reflected its purpose. The upper leadership in Afghanistan, at USCENTCOM headquarters in Tampa, Florida, and certainly in Washington determined that Osama bin Laden had escaped Tora Bora in December 2001 because nobody had guarded the exits into Pakistan. This time, with numerous reports of al-Qaeda elements hiding in the Shah-i-Kot Valley in Paktia Province in southeastern Afghanistan, the Americans decided to encircle the area. Like the giant anaconda snake of Brazil, the operation intended to constrict its prey. Inside the ring, the SF-enabled Afghan militia would finish the opponent. Airpower served to kill off any attempted leakers.

Thus Wiercinski's soldiers endeavored to block the exits from the

Shah-i-Kot Valley. With fewer than 250 soldiers in the first wave, spread in little platoons across five miles of mountains, that seemed a tall order. But the United States intended to build those numbers as the day unfolded. Moreover, as demonstrated in the opening engagements the previous autumn, those warplanes overhead magnified the firepower available to the heliborne riflemen. Zia Lodin's militia would sweep the villages on the valley floor. The military calls this tactic hammer and anvil; Zia would hit the opposition against the solid positions of Wiercinski's air assault force. Optimistic planners named Zia's militia Task Force Hammer.

The enemy didn't read the script. According to the intelligence analysts, about 150 to 200 al-Qaeda fighters would be mixed in with ten times that number of civilians in and around the villages on the bottom of the valley. Instead, as Wiercinski noticed immediately on landing, "There were no civilians in those three towns. There were no colors, no smoke, no animals, no hanging clothes." And the hostiles gathered in three times the projected strength, most of them dug in up high among the peaks and full of fight.

Air assault units in the 101st Airborne Division, drawing on the heritage of the paratrooper and glider regiments of World War II, prefer to land right on top of the enemy. Sometimes terrain or hostile resistance doesn't allow it. Helicopters are not armored and cannot take a beating, unlike tanks. So this tactic is dangerous, high risk, but it's very high payoff when it works, as it dumps fresh, fully armed riflemen right into the foe's innards. In the Shah-i-Kot Valley, Wiercinski thought his brigade was coming in well off the target villages. In a nod to old Clausewitz and that ubiquitous friction, the enemy guessed wrong as well. They set up right where the Americans showed up.

By coming in with his men, Frank Wiercinski saw what they saw, heard what they heard, and knew what they knew. Twenty-two years in the infantry and a veteran of the Ranger jump during the Panama conflict in 1989, the colonel recognized the opportunity granted when an elusive opponent fixed himself. The goat tied itself up to its own stake. Well, this particular goat had big teeth, actively slashing hooves, and a very ornery streak. Yet here their adversaries were, with Wiercinski's people squarely atop them.

Despite the common image of a colonel back at his nice, neat command post listening to the radio, reading reports, and sticking pins in a map (checking e-mail and clicking a computer mouse these days), that kind of thing goes only so far. Some say that back there, out of the fight, people can keep their heads clear. That's true, and it's why you keep staffers there, watching and checking well out of the line of fire. Perceptions in the rear tend to run behind reality. From the rear, situations often look a lot worse than they are: contact heavier, casualties higher, enemy stronger, bad going to worse. Back a hundred and twenty miles at Bagram, the Tenth Mountain Division headquarters that morning thrummed with negative energy waves, with Zia Lodin's militia stalled and the bad guys very active right on the landing zones. Some senior staffers began to argue for pulling out the Americans.

Not Frank Wiercinski, who knew better. When things really go to hell, the commander must be right there to make the call. He also has a moral responsibility to share danger and hardship with his soldiers. Under fire, men may consent to be ordered. They will be led. That is why the motto of the U.S. Army Infantry is Follow Me, not Check Your Inbox.

Wiercinski's battlefield inbox overflowed that morning. Between wriggling up and shooting occasional hostiles, the colonel had to command his brigade. At eight thousand feet up, in bright sunlight, his perch offered quite a panorama. To the northeast, on the lower reaches of the snowcapped mountains, gray puffs of explosions and a couple of dark little attack helicopters roving above marked where 2-187 Infantry landed. Looking to the east, down below Wiercinski's TAC team, the colonel saw most of the riflemen of 1-87 Infantry strung in a loose perimeter around a horseshoe of rocks that formed the lip of a shallow bowl near the southern exit route from the valley proper. Dark-clad hostiles moved around on both the lower reaches of Wiercinski's ridge and near the 1-87 Infantry position. Others shot down from the heights to the east, the massive hulking mountain that anchored the eastern valley wall. The Afghans called it Takur Ghar, the "Lordly Mountain." The opposition knew the value of it and used that vital high ground. They were good at it too.

In the classic Western *Butch Cassidy and the Sundance Kid,* the exhausted protagonists, looking down from various vantage points, repeatedly asked about their dogged, seemingly superhuman pursuers: "Who are those guys?" The same question occurred to Frank Wiercinski and a lot of his soldiers that morning. The G-2 military intelligence types missed this one, all right.

Much later, after going through the morass of reports, overhead imagery, captured records, and transcripts of detainee interrogations, the intel people figured it all out. They were good that way. The MI folks could usually tell you the make, model, year, paint color, and license plate of the semitrailer truck that just ran over you. In this case, the bad guys were not the usual Arab Osama bin Laden inner circle. Rather, about half, the sharper marksmen, hailed from one of those al-Qaeda 2.0 regional franchises, the Islamic Movement of Uzbekistan (IMU). When President George W. Bush name-checked this otherwise obscure group in his initial public address after 9/11, it locked down the use of the K2 air base in that former Soviet Republic. But there was more to it.

The IMU was quite real and, at the time, very dangerous. The group backed Osama bin Laden to the hilt. Even when the Sheikh went over the Pakistan border, the IMU stayed. Led by Tahir Yuldeshev, Abdul Rashid Dostum's Uzbek evil opposite number, these people knew how to shoot and liked to fight. The older ones used to tangle with the Soviets. Stranded a long way from home, the IMU stood and fought on the heights above the Shah-i-Kot Valley. The local Pashtuns called them the Chechens, because many IMU men had European features, and the older ones spoke passable Russian. A few skilled snipers, a dozen capable mortar crews, and a dozen or so DShK 12.7 mm heavy-machine-gun teams — helicopter killers — came from the IMU, men who favored black clothing — a nod to the al-Qaeda banner, white Arabic script on jet black.

The other five hundred or so hostiles in the Shah-i-Kot came from the standard-issue Pashtun Taliban under Saif Rahman Mansour. Drawn from those abandoned towns on the valley floor, these men carried AK-47s and RPGs. Their logistics support came from Miranshah, Pakistan, through Jalaluddin Haqqani, a notorious fixer and guerrilla chief who'd often fought his own war in the Soviet era. Local Shah-i-Kot Taliban would fight to the last IMU Chechen. Once the IMU people went down or bugged out, the Taliban seemed likely to fade away.

Nobody was bugging out on the morning of March 2. Back in their headquarters, watching carefully as they always did, the Task Force leadership started to think that maybe, just maybe, some big game might be trapped in the Shah-i-Kot. Maybe bin Laden, Zawahiri, or Mullah Omar, one of those Big Three, was inside the thin, disjointed cordon. That could explain the ferocity of the reaction.

The Task Force used the high-technology, long-duration MQ-1 Predator drone as its eye in the sky to observe the opening moves by both sides. The Task Force's assault teams waited on the runway at an old Soviet base well north of Shah-i-Kot. Wiercinski's Third Brigade launched from the same airfield, and his higher headquarters, Tenth Mountain Division, shared the ramp side with Task Force Dagger's Special Forces and the Task Force. Colonel John Mulholland's Task Force Dagger teams and Zia Lodin's Afghan militia worked for Tenth Mountain Division, like Wiercinski's soldiers. The Task Force worked for itself, as was its wont in those early months of the war.

The Task Force had more than a Predator looking down at the Shah-i-Kot that fine day. Within their first hour on the ground, Frank Wiercinski and his team met five neighbors they did not know they had: three SEALs, a Navy explosives ordnance-disposal technician, and an Air Force controller, a reconnaissance team from the Task Force. The Navy SOF element proved their bona fides with a bright magenta VS-17 recognition cloth. The fivesome had been on that same ridge for more than two days, reporting back. Some of their information, distilled, made it through the wickets back at Bagram to Wiercinski's soldiers. Now the colonel and his people met the source.

When the trio entered the brigade TAC's little position, the lead SEAL pointed back the way he'd come. He indicated a dead body slumped on the boulders, up the ridge to the south. It was an enemy DShK heavy machine gunner. The SEALs had killed its five-man crew before sunup. The Third Brigade Command Sergeant Major Iuniasolua Savusa, forward with Wiercinski where he was needed most, said it right: "I owe my life to these guys." Had that DShK remained active, the TAC team's Black Hawks would quite likely have been blown away.

Blowing away the IMU and Taliban now became the order of the day. The U.S. riflemen did their part up and down the valley. A 120 mm mortar crew with 1-87 Infantry proved especially deadly, making all thirty-five of their high-explosive rounds count. Machine gunners reached out hundreds of yards to kill exposed enemies. Up on the brigade TAC's stone balcony, Wiercinski and Savusa joined their men and the nerveless, dead-on Navy SEALs in taking down individual Taliban fighters scrambling up the ridge side. The colonel "was in his element," remembered Air Force Captain Paul Murray. "He looked like an infantryman at some times and at others he looked like a guy in command of hundreds of men that were in a dogfight."

Murray didn't fire his rifle, the only man up there that day who did not. He did his work with the radio, bringing in airpower. He started with a danger-close mission right near the TAC site. A pilot himself, Murray talked to his brothers above in two F-16 Fighting Falcons. The men in the jets could not pick out Murray and the other Americans. Murray told the lead F-16 to drop the first bomb outside six hundred yards, beyond the danger-close ring. The fighter pilot did so.

"You see where that bomb hit? Now," Murray coached, "I want you to come up that ridge line about halfway and tell me what you see."

"I see some guys maybe running around in there," the pilot replied, "and certainly I see a little cave entrance and stuff like that."

"That's where I want the next bombs," Murray said.

"All right," the pilot came back. He paused, dead air on the radio. "How close to the friendlies?"

"Three hundred meters," said Murray. That was very close with bombs. Paul Murray then followed with "Foxtrot Whiskey," the colonel's initials. It was shorthand that meant Frank Wiercinski would take responsibility for the outcome, good, bad, or ugly. He might as well. His own hide was right there on the line.

Three bombs came down. Three flashes, dirt clouds, and thunder followed. From 1-87 Infantry below, one officer transmitted: "Thanks. You guys saved us." Murray passed it on. The F-16s moved to other targets. So did Murray in the Third Brigade TAC position, aided by the Air Force controller with the SEALs.

A tremendous amount of airpower went against the IMU and Taliban in the Shah-i-Kot on March 2. U.S. warplanes delivered 177 JDAMs and laser-guided bombs, an average of one every ten minutes or so throughout that long day. Various spotters from the Task Force, well placed on high ground like the five with Frank Wiercinski, directed some of these strikes. Air Force controllers with 2-187 Infantry and 1-87 Infantry ran their share too. So did the Task Force Dagger people with Zia Lodin's militia.

In that busy day, 162 of the engagements reflected immediate support to troops in contact. Four ongoing simultaneous air missions typified the pace of action in the valley, with multiple jets, munitions, and helicopters all churning around in the same five-mile-by-five-mile box of airspace. One Air Force B-1B bomber released 19 JDAMs on ten distinct targets over a two-hour period. Another B-1B made six attacks, placing fifteen JDAMs on six separate targets. A third bomber struck an IMU antiaircraft gun as the enemy dragged it out of a cave mouth. Two Marine Corps F/A-18 Hornets made the long run in from the carrier USS *Theodore Roosevelt* (CVN-71) out in the Arabian Sea. After expending all bombs, the Marine fighters dove down to finish the job. In three low passes, the Marine aviators strafed enemy gunners with hundreds of lethal rounds from their 20 mm Gatling guns. It was all impressive indeed, a maximum effort, and devastating to the opponent. There was more.

Well below the relentless jets, excepting the occasional strafer, the air belonged to seven AH-64 Apache attack helicopters from the Third Battalion, 101st Aviation Regiment. Trained and organized to operate alongside the 101st Airborne Division's air assault infantry, the Apache fliers prided themselves on going in hard, pressing the fight. Working in pairs, the Army aviators took to heart their orders from Frank Wiercinski: "Protect them [the infantry]. Give them everything you've got." That morning, they did.

Apaches carry three basic weapons: the onboard 30 mm cannon, 2.75-inch (70 mm) rockets, and Hellfire guided missiles. For the Shah-i-Kot mission, the AH-64s used all three, expending 540 30 mm rounds, hundreds of rockets, and a single Hellfire. The numbers do not do justice to the effort. Apaches prefer to hunt by night, unseen by their targets. On March 2, 2002, they flew in broad daylight.

In the crucial first ninety minutes after the initial landings, the Apache Aviators waded into the enemy. Swinging low, trusting in their sturdy helicopters, the aviators worked closely with the riflemen on the ground. Each infantry unit called on the Apaches. The rockets tore up IMU gun posts; the wicked cannon shredded hostiles. The IMU gunners stood up and blazed away, undaunted. They paid for it.

Air Force controller Technical Sergeant John McCabe, hunkered in the rocky perimeter with 1-87 Infantry, took a few minutes from calling in airstrikes to watch the Apaches go to work. "I do recall 2.75[-inch] rockets coming off the Killer Spades [Apache unit radio call sign]. They shot their 30 mm at them," McCabe said. The Apache fliers did not hesitate. "Again, they would fire their weapon systems, continue flying, and they received heavy ground fire from the guys," McCabe noted. Over McCabe, the attack helicopters made four separate runs. Frank Wiercinski himself later paid tribute. As usual, he put it simply and directly. Their 101st Airborne Apache Aviator teammates "clearly saved the day."

The first hour and a half of fighting tore up even the tough Apaches. Each rotor blade had at least one hole in it. Fuel tanks leaked from numerous punctures. Enemy RPGs, set for airburst, peppered the hides of every aircraft. Panes of Plexiglas shattered. Wires were cut, black boxes cracked, and key structural spars nicked. Patched up, rearmed, and refueled at a forward site, three of the Apaches went back into action that afternoon. The rest could still fly, barely, but enough to limp back to Bagram.

With the Apaches off station getting repaired, the friendly Afghan militia marking time at the north entrance to the valley, the infantry fighting up and down the east wall, and the airstrikes coming in regularly, it was time for Frank Wiercinski to decide what to do next. He had another 250 waiting near the six Chinooks at the Bagram runway. The original plan called for them to arrive in midmorning. With the enemy still shooting up the landing zones, that wouldn't work. The ordeal of the Apaches offered more than enough evidence of what could happen if a fat Chinook full of riflemen ate a few RPGs. Wiercinski told the Tenth Mountain Division headquarters to hold off for a while, to let the fight develop.

In Bagram that busy morning, division commander Major General Franklin L. "Buster" Hagenbeck put his faith in Frank Wiercinski. Hagenbeck knew the pugnacious colonel well from his own previous service as a one-star deputy commander in the 101st Airborne Division.

Wiercinski was forward, and Hagenbeck was not. Both stuck to the military adage "Trust the man in the fight." Hagenbeck did.

It accorded well with Buster Hagenbeck's unhurried approach to command. A natural coach — in fact, he had been an assistant coach for the Florida State Seminole football team in the 1970s — the general deliberated and sought consensus. He rarely jumped to decisions. Some of the more fire-breathing senior commanders would have already been in the air over the Shah-i-Kot. A few of the real hard cases would have been down on the ground trading bullets alongside Frank Wiercinski. That was not the way of Buster Hagenbeck, a command-post type who liked to stay objective, keep his head clear, and make the wise call. In a crisis, Hagenbeck's natural inclination was to wait, gather more information, poll his key subordinates, and then, if he really had to do so, pick an option. This he did as March 2 unfolded.

That said, the general recognized that the Third Brigade had definitely found the enemy and had the foe by the jugular. He and his command-post staff did what they could to get more air sorties. The Air Force and Navy pushed hard to send even more aircraft into the fray, no easy task when everything had to get into and out of the same five square miles of mountains dotted with intermixed knots of U.S. soldiers and hostile gunmen. Hagenbeck also paid close attention to messages from the Task Force's scout teams, which suggested that the enemy was getting a hell of a beating. Reports also depicted 2-187 Infantry in the northeast moving under fire, but definitely moving, toward assigned blocking positions. Above it all, Frank Wiercinski in his contested but vital ridge line skybox had a tremendous vantage to direct strikes and assess the battle. Those bits of information sounded promising.

The rest of the situation did not. In the southeast, 1-87 Infantry found itself engaged and stopped in that same compressed bowl near the landing zone, minus a few small flank elements. Worse, a succession of enemy mortar rounds wounded twenty men, including the battalion S-3, Major Jay Hall, and Command Sergeant Major Frank Grippe, both of whom declined evacuation. Of course, given the unrelenting enemy gunfire, no wounded would be pulled out by air, not for a long time. The battalion needed ammunition and a continued dose of aerial firepower. The commander, steady Lieutenant Colonel Paul LaCamera, assessed the situation and determined that his men could hold and give as well as they got. But the number of wounded, almost a fifth of the force, got the staff people back at Bagram nervous.

As for Zia Lodin's Afghan militia, they had not reached the four villages. Unable to get much air support — a huge morale issue for Afghans mesmerized by tales of the decisive strikes in initial weeks of the war — Zia's people waited for hours at the north end of the valley. Enemy fire resulted in two more dead militiamen and another fourteen wounded on top of the casualties already caused by the errant dawn AC-130 mission. Confused, irritated, and somewhat ignored in the Tenth Mountain Division's natural preoccupation with the big U.S. infantry fight to their east, Zia Lodin pulled back, immensely frustrating his able SF advisers. The Afghans were no longer part of the picture.

Finally, the damage report from the Army aviators gave pause to Hagenbeck and his staff. The six Chinooks and two Black Hawks had all taken fire on the initial lift, and the Apache gunfight underscored the hazard. Going in again in daylight risked the crucial helicopters, the only reliable, fast way in and out of the distant valley. Wiercinski already wanted to wait on the next lift. Hagenbeck went one better and decided to wait until darkness, to make it much tougher for the opposing gunners. "I didn't want a shoot-down," Hagenbeck said, "and we were pretty well stabilized by early afternoon."

Scrambling, and misreading their general a bit, a few overly zealous division staff officers issued a warning to Third Brigade to prepare for a complete pullout at dark. On the computer screens back at Bagram, too many staff people saw only gloom and doom. It was clearly the first major battle for most.

Combat veterans, and now Afghan combat veterans, Wiercinski and especially LaCamera both argued instead for using nightfall to bring in the next lift, pull out casualties, and resupply ammunition and water. The Task Force people in the Tenth Mountain Division command post agreed. Their scouts were calling in bomb after bomb. The enemy had stood, fought, and bled. Keep it up, the Task Force people argued. Reinforce success.

Hagenbeck seized on that concept, a way to satisfy both the worried staffers in the rear and the confident commanders forward. "We were being extraordinarily successful," Hagenbeck stated, "except in one place." Well, two, actually, if you counted Zia Lodin's militia as well as the 1-87 Infantry. But it reflected the Tenth Mountain Division's view. The Afghan militia was no longer factored into the equation.

The order went out, half a loaf. Come night, the helicopters would insert another rifle company to reinforce 2-187 Infantry up north. One unengaged unit, a platoon-plus from 1-87 Infantry, was to march north to link up. In the morning, the much stronger 2-187 Infantry expected to push south and clean up the enemy at that end of the valley.

The rest of the battalion, the eighty-six embattled men in the rocky bowl, were told to return to Bagram, treat their wounded, and refit. For Paul LaCamera's hard-fighting 1-87 Infantry, the order to withdraw satisfied few. Command Sergeant Major Frank Grippe, wounded but game, carried his own heavy rucksack off the landing zone that night. He spoke for many in his battalion: "Higher made the decision to evacuate us," he said. "They had these guys [the enemy] freakin' fixed."

Frank Wiercinski, Iuniasolua Savusa, Paul Murray, and the Third Brigade TAC comrades left on a Chinook about 3:30 a.m. on March 3. Calm as ever, Savusa patiently led the way to the landing zone. Savusa chose a better place than the tricky, shoot-the-gap patch used in the morning. The new area was a little flatter, although three-quarters of a mile down the black, rock-strewn slope. Behind the departing file of air assault soldiers, the SEALs moved silently back up to their outpost. "I've killed enough for today," one whispered to the command sergeant major. Well, there was always another day.

More people did get killed that day, and the next. Seven of them, unfortunately, were U.S. special operators. An attempted helicopter insertion atop the pinnacle of Takur Ghar poked a hornets' nest of heavily armed opponents. Two helicopters went down, seven Americans died, and six were wounded. Backed by more than thirty airstrikes, including a bold F-15E 20 mm strafing run, the Americans claimed more than two hundred enemy casualties. The Lordly Mountain got a new name; now it was Roberts Ridge, in honor of Petty Officer First Class Neil C. Roberts, a SEAL killed in action in the seventeen-hour firefight. The scale of that engagement caused the Task Force to agree on March 4 to place their elements under Buster Hagenbeck's tactical control for the remainder of Operation Anaconda.

Things dragged on until March 19, but the rest of the battle proved to be small change after that wild first day and the vicious SOF knife fight that night and then the long day on Roberts Ridge. Sweeps and airstrikes continued. More battalions joined the effort, including the First Battalion, 187th Infantry, the Third Battalion of Princess Patricia's Canadian Light Infantry, and the Fourth Battalion, Thirty-First Infantry. Paul LaCamera's 1-87 Infantry went back to work. Another brigade headquarters, Second Brigade of the Tenth Mountain Division, organized the follow-on effort. The Apaches of Third Battalion, 101st Aviation, reinforced by other units, including Marine aviation, flew cover. Even Zia Lodin's militia, backed by another chief's decrepit T-55 tanks, finally cleared the four empty villages. As usual, the air pounding continued. More than three hundred bombs shook the valley on March 9 and 10. When the operation ended, U.S. infantry stood on Roberts Ridge.

People who liked numbers ran them. As early as March 13, General Tommy Franks at USCENTCOM counted 517 confirmed enemy dead and about 250 probable. A week later, the Tenth Mountain Division claimed a total of 800 dead al-Qaeda and Taliban, including the latter's commander, Saif Rahman Mansour. Even the lowest estimates acknowledged at least 200 enemies killed, all on a battlefield blessedly free of civilians. The Americans also totted up eleven artillery pieces, fifteen DShK heavy machine guns, and twenty-six mortars captured or destroyed, and they'd searched and cleared sixty-two buildings, forty-one caves, and five compounds. It all smacked of similar lists from Vietnam. Clever journalists immediately made the dire comparisons.

They missed something. Attrition by itself just means dead people. Trading eight Americans (including the SF officer killed March 2) for eight hundred opponents may make some accountant happy, but as Rudyard Kipling warned his readers in the heyday of imperial Britain, there are perils to such arithmetic on the frontier. Kipling pointed out that a highly trained, expensively outfitted British soldier could be killed by a single dirt-cheap rifle bullet fired by an illiterate goat herder, and the deaths of a hundred goat herders did not make the bloody exchange acceptable in the far-off mother country. Over time, the lengthening casualty list would weigh heavily indeed on citizens in Hometown, USA. But few felt that weight in the spring of 2002 — not yet. Average Americans can tolerate a lot, provided it looks like something is getting done. Results count more than numbers.

Coming on the heels of the horse-mounted SOF and JDAM campaign of the previous autumn, Anaconda smelled like a finish fight, the last stand of al-Qaeda and the Taliban. Never again did the Afghan enemies mass, stand, and fight as they had in the Shah-i-Kot Valley. In that very real sense, the Anaconda body count mattered.

For their part, the IMU and Taliban erred grievously, and paid for it, in resorting to their Soviet war method and defending high ground. This time, the old heads like Jalaluddin Haqqani completely misread their new opponents. Holding the heights typically worked against unhappy Soviet conscripts and their glum Afghan partners. But now and then, the Afghan guerrillas ran into the really dangerous breed of Soviets, the paratroopers and Spetsnaz (the Russian SOF) skilled in close combat and excellent in using supporting fire. Those elite units routinely flensed the jihadis. A few Russian airborne companies had done just that during Operation Magistral from November 1987 through January 1988 across Paktia Province, including the Shah-i-Kot Valley. Fortunately for the Afghans, there were not many such Soviet forces. With the Americans, however, all the troops fought at the standards of Soviet paratroopers and Spetsnaz. The fearsome U.S. airpower was in a class all its own.

The Task Force's careful reconnaissance elements saw the awful results in the Shah-i-Kot and on the long, crooked mountain paths that led toward Miranshah, Pakistan. Bloody bandages, discarded field gear, and abandoned dead suggested heavy casualties and significant disorganization. When considered in light of that laundry list of heavy weapons announced by the Tenth Mountain Division, it became evident that the foe had been badly beaten.

Operation Anaconda pretty much finished the IMU, although Tahir Yuldeshev survived and kept the name alive in a few villages of northwest Pakistan. As for the Pashtun Taliban: Those in eastern Afghanistan had smirked at what had happened to their Taliban colleagues up north, around Kabul, and in Kandahar. Some thought Mullah Omar and his people hadn't fought hard enough and had folded too early. In Anaconda, those eastern Pashtuns found out what it meant to go toe to toe with the Americans. Those who escaped gained a very expensive education written in blood and fire. It fully reinforced Mullah Omar's decision to back off and rethink how to fight the war.

After the drubbing in the Shah-i-Kot, smarter Pashtun guerrilla chiefs like Jalaluddin Haqqani chose to follow the wise counsel of Khushal Khan Khattak, the great rebel leader who fought the invading Mughals in the seventeenth century:

> When you fight a smaller enemy detachment you should decisively attack with surprise. But, if the enemy receives reinforcement [or] when you encounter a stronger enemy force, avoid decisive engagement and swiftly withdraw only to hit back where the enemy is vulnerable. By this you gain sustainability and the ability to fight a long war of attrition . . . A war of attrition eventually frustrates the enemy, no matter how strong he may be.

It matched almost exactly Mao Zedong's more elegant formula:

> *Enemy advances, we retreat.*
> *Enemy halts, we harass.*
> *Enemy tires, we attack.*
> *Enemy retreats, we pursue.*

The Taliban chieftains preferred to quote their hero Khushal Khan Khattak, the scourge of the Mughals. They knew well what happened to the British and the Soviets. They knew their Mao too. All the older ones had studied in Soviet military schools, done their own reading in Western sources, and learned what happened in the Sino-Japanese War of 1937–45, the Chinese Civil War of 1927–49, the Indochina War of 1945–54, the Algerian War of 1954–62, and the Vietnam War of 1965–75. Sooner or later, a protracted war goes to the home team. No matter how many extra innings the U.S. played, in Afghanistan, the Taliban batted last.

Thus informed by the masters of the art, the Taliban chose their course. Western-style firefights were almost wholly replaced by the traditional Afghan ambuscades, snipings, and hit-and-run attacks. Every Taliban element sought to intermix with the local civilians, using them as shields to curb Yankee firepower. Future operations focused on generating propaganda leading to support—recruits, supplies, and especially intelligence tips — among Afghan villagers. If positive motivation didn't work, terror strikes and cooperation under duress would do. Each dead American counted, and each counted as more than just another U.S. rifleman removed from the battlefield. It all reflected that Kipling arithmetic, especially as time went on and such losses seemed pointless and endless back in the home country. The Taliban sought opportunities, real, half real, or entirely invented, to embarrass or vilify the Americans and their Coalition partners. Over time, the occupiers would lose heart. Their client Hamid Karzai and his clique in Kabul were damaged goods, forever stained by their reliance on the infidels. Mullah Omar and his guerrillas settled down for a long, long game. *Enemy advances, we retreat.*

The U.S. and the Coalition knew that Anaconda worked. Never shy about speaking his mind, General Tommy Franks summarized: "I thought it was a very successful operation." To Franks, this put the exclamation mark on the Afghan campaign. For those keeping score at home with their copies of the USCENTCOM plan, Anaconda ended phase three, Conduct Decisive Combat Operations. Apparently, the remaining enemy had pretty much left the battlefield.

Now came phase four, the one that hadn't been all that well thought out: Establish Capability of Coalition Partners to Prevent the Reemergence of Terrorism and Provide Support for Humanitarian Assistance Efforts. It meant messing around with the embassy and Karzai, taking care of ISAF/NATO stuff, digging wells, and teaching Afghan policemen how to shoot pistols. Franks had bigger things in mind. He intended to keep the U.S. troop number low, its footprint light, and to outsource the mop-up to the allies. Let America do what it did best: high-end, joint service offensives to knock out hostile states. Tommy Franks was already looking ahead to Iraq, the next big thing. Afghanistan faded in the rearview mirror.

The military term for Franks's plan is *economy of force*. It has nothing specifically to do with managing money, although it typically results in a lot less being spent than in the main show. In an economy-of-force effort, you do what you must (and no more) while you put your major forces against the bigger task. In war colleges, they teach you that it's a big mistake to shovel extra resources into a secondary undertaking. The

U.S. Central Command had an entire unstable region to superintend. With Afghanistan quiescent after Anaconda, Franks and his staff treated it as an economy-of-force theater.

That's not to say they ignored it. Operation Anaconda made it clear that a two-star headquarters lacked the experience and prestige to handle a theater in the absence of daily attention from U.S. Central Command. Divisions fight battles. They don't supervise theater campaigns, at least not well. Asking Tenth Mountain Division to coordinate SOF, airpower, Afghan militia, allies, the U.S. embassy, the CIA, the humanitarian relief, the United Nations, the news media, and all the other day-to-day military activities in-country pretty much tapped out Buster Hagenbeck and his people. Well, the guys in Tampa couldn't keep watching things perk along in the Hindu Kush. The U.S. Central Command had other challenges, starting with Saddam Hussein's Iraq. They needed somebody to look after Afghanistan, keep it under control, and do it without additional resources.

Hagenbeck's actions post-Anaconda made USCENTCOM's decision easier. He had done well enough running the Shah-i-Kot lash-up. Afterward, he got into a messy contretemps with the Air Force over the planning and execution of the bombing. He groused to his own Army service staff over the small size of his headquarters and lack of cannon artillery batteries. He even created a minor flap when he mentioned the possibility of hot pursuit into Pakistan to finish off the withdrawing Taliban. "I was out of my mind at the time," said Hagenbeck later. In Tampa and Washington, eyebrows raised.

The Tenth Mountain Division had been there only a few months. But no division had run a fight like Anaconda since Vietnam. This sure wasn't Desert Storm. In this one, General Tommy Franks didn't have more troops to spare. Under similar circumstances in World War II, General George C. Marshall had dealt with the China-Burma-India (CBI) theater — a secondary effort supporting a secondary effort, the military equivalent of the dark side of the moon — by sending in his sharpest, most aggressive infantry general, Joe Stilwell. Marshall couldn't spare any soldiers, planes, or tanks. But he could send the right guy to keep the CBI going and off the main radar screen. Like Marshall, Franks needed someone to hold the bag, someone who wouldn't spill it, bust it, or complain.

Lieutenant General Dan K. McNeill and the Eighteenth Airborne Corps headquarters were told to take over in June of 2002. Major General John R. Vines and the Eighty-Second Airborne Division took the tactical fight that same month. They led the Army's premier expeditionary fighting formations. Both commanders had real combat experience: McNeill in Vietnam, Panama, and Desert Storm; Vines in the latter two campaigns. Both had jumped under fire in Panama. They were problem solvers, carnivores, out and about with their paratroopers. They were the kind of senior officers who could keep a lid on Afghanistan.

The idea of a higher commander to look up and out and a tactical commander to run the fight set the standard for the rest of the war. It freed up USCENTCOM's four-star commander to look after the entire regional fight against the Islamists, especially with Iraq kicking up. And things happened in other countries beside Afghanistan and Iraq. The two-tier structure worked, especially as pioneered by two soldiers who knew each other as well as Dan K. McNeill and J. R. Vines. Along with the four-year lull beaten into the Taliban, this eminently practical command arrangement became one of the most important outcomes of Operation Anaconda.

One other community had some soul-searching to do after Anaconda. The entire offensive employed a wide range of special operations units, including the SF advisory teams and elements of the Task Force, all right in the main fight. Another element coordinated an outer ring of U.S., Australian, Canadian, Danish, French, German, New Zealand, and Norwegian special units. They killed many foemen, and killed them well. But despite all of the tasks done and teams involved, the SOF-o-rama broke no terrorist networks, caught no big fish, and, in the end, could not tell where the hostile leaders went. Man-killers they were. Manhunters they were not.

So far, the Afghan campaign had featured local militias and some U.S. Army rifle companies stomping around here and there to kill the Taliban and any al-Qaeda they could find in the bargain. The aerial attacks slaughtered a lot of bad guys. But it all amounted to a better-executed version of the old Soviet tactics: round them up, kill a lot, and let God sort them out. To continue like this was — well, you could pick your cliché; they all got used. It was like Whac-A-Mole, or chasing smoke, or picking pepper out of fly excrement, or hunting for a needle in a haystack—it all amounted to just operating and hoping something turned up.

Going all the way back to the Vietnam War, the last major U.S. attempt to deal with this problem, the doctrine writers at the U.S. Army Infantry School at Fort Benning, Georgia, came up with their own little formula, sort of like the Mao Zedong strategy jingle: Find, fix, fight, finish, they said. The SOF people followed that mantra as they built up capabilities throughout the 1980s and 1990s. The special operators deleted *fight* from the sequence. If you used SOF for direct action, it was a targeted raid, not some drawn-out tank battle. The Task Force, in particular, focused on *finish.* Indeed, the Task Force were the best finishers in uniform. Put them next to terrorists, and that opponent went away. The problem involved finding the enemy.

The Task Force tried. In numerous previous operations in the 1980s and 1990s, some well known and others never discussed, the Task Force fumbled and stumbled, trying to find an enemy hiding in plain sight in a hostile populace but very determined not to be found. In the Shah-i-Kot, while scouts did yeoman work finding the enemy, the bulk of the Task Force sat on the runway, waiting for the call to swoop in and snatch the top targets. Of course, that summons never came.

It was as if a superb baseball hitter—Ted Williams in his prime—with an eagle eye, matchless reflexes, and a perfectly turned bat, were compelled to play every game in stygian darkness, flailing at pitches whizzing in from all points of the compass. Hell, you couldn't even see the pitcher, and it seemed like the other team was using more than one, some of them chucking chin music from the dugout and even the box seats. In this blacked-out stadium, the Task Force needed to turn on the lights, find the pitcher—or pitchers—see the ball, and hit back.

Notably self-critical, the SOF people evaluated their progress to date. Few were happy. Men like John Mulholland and others asked the hard question: How do we find the enemy? The answer came back, as it had for years: with actionable intelligence. And where did that arise? Now the intel guys looked into their ties and mumbled, *Well, it depends . . .*

The U.S. collected information superbly and everywhere, from space to dirt. They tracked all kinds of events and things and people. For long-lead-time matters, like the order of battle for the Chinese fleet, that sufficed. For short-fuse needs, it got much, much more excruciating. Of the mass of data gathered, only a small percentage (50 percent? 10 percent? 5 percent?) ever got analyzed. Only a tiny fraction of that produced the specificity to allow action. The whole machine squatted back in the Washington area. Meanwhile, out on a bare ridge line in the Shah-i-Kot, tough men burrowed into cold-rock hide sites saw the real enemies but didn't know which, if any, of eight distinct clutches of bearded males wandering in and out of sight included the key terror-cell leader. So they were back to the usual plan: Try to kill them all. Hope for a break.

The field manuals told you how to do it. It seemed slow and painful: Take captives and interrogate them, not the way it had been done in the hastily arranged and ultimately fatal terrorist town hall at Qala-i-Jangi, but real questioning, in depth and consistent. Pick up documents and electronic media and go through them methodically, detecting subtleties. Figure out which Muhammed knows which Ali, who pays the bills, and who carries the messages. Draw up the links, like urban police chasing gang lords. The rudiments were known. It usually took a lot of time.

In their haste to act, to do what they did best, to *finish*, the Task Force operators counted on others to do the finding. They trusted far too much in the experts back in Washington. As one senior SOF officer put it, you had a huge funnel in DC pouring intelligence down to the battlefield, and a massive vacuum out in theater policing up prisoners, documents, and local tips. And they both came together at a narrow little soda straw, a few hot tips dripping through from back there, but most of them ignored, and a couple of useful nuggets percolating up from the frontlines, handy for press conferences at the Pentagon, if little else. The giant factory in Washington didn't provide much from day to day. The part in the field hardly ever went through all the stuff picked up. At both ends, detainees went unquestioned, documents and equipment waited unchecked, and relationships got missed. Dots didn't get connected.

To address this, to turn on the stadium lights, the Task Force needed to push those smart Washington-area analysts right up against the theater. The hard work of sorting through all the taken materials, talking to all the detainees, and rescrubbing both as often as necessary had to be done. It had to go quickly; hours, not days, and minutes if possible. To see the patterns, perceive opportunities, and act decisively took extraordinarily skilled and dedicated people. The brilliant work of the recon teams in Operation Anaconda demonstrated that the Task Force had just these kinds of people. They excelled at seeing openings and taking them. Rather than hang around back at base waiting to get directed toward the bright spot on the horizon, the Task Force would turn on its own lights.

Crusty old Secretary of Defense Don Rumsfeld agreed. In a directive on July 1, 2002, the secretary asked: "How do we organize the Department of Defense for manhunts? We are obviously not well organized at the present time." This matter was going to get resolved.

While the Taliban went to ground, Tommy Franks sorted out the command structure, and the SOF figured out how to bust ghosts, the Bush administration again faced an opportunity. The style of the British on the North-West Frontier Province circa 1897, with young Winston Churchill of the Fourth Hussars and the Malakand Field Force out swatting Pashtuns, gained the nickname "butcher and bolt." The British went in, killed a lot of local troublemakers, and left. Each incursion kept things quiet for years.

After six months in country, the U.S. had butchered. After Anaconda, to some it seemed judicious to bolt, to declare victory and leave. The military definitely thought so. Lieutenant General Dan McNeill believed the bolt was imminent. Before he headed over to Afghanistan, the vice chief of staff of the Army, hard-bitten General Jack Keane, told the Eighteenth Airborne Corps commander, "Don't you do anything that looks like permanence. We are in and out of there in a hurry."

That wouldn't please the allies. Those worthies were already invested; they had a few thousand ISAF troops hanging around Kabul teaching some preliminary courses to a nascent Afghan military. As usual, too many Coalition countries stood ready to fight to the last American soldier and dollar. The British, who should have known better after their three difficult wars in Afghanistan, argued vigorously for a continued strong U.S. commitment. Many other voices in many other capitals concurred.

So did Hamid Karzai and his coterie, the few educated people in the country now flying in loose formation trying to make a go of running the government. They were just getting started in Kabul. The U.S. Department of State, notably the embassy, saw this attempt at governance as eminently fragile. If not supported, the country faced sinking into the vale of sloth, violence, and illiteracy that had characterized its long, troubled history. Through heroic efforts by many dedicated civilians and more than a few in uniform, the Afghan people made it through the winter without mass starvation. If the Kabul experiment went under, the next cold season promised no such happy outcome.

The U.S. military might have been ready to wrap it up. That decision, though, belonged to the president. Those in uniform did not feel too strongly one way or the other, as long as America capped its troop strength at around ten thousand, maybe two brigades' worth—that wasn't much, was it?—and left the nation-building to the U.S. embassy, the allies, the United Nations, nongovernmental organizations, and Karzai's Afghans. Few noted that the continued presence of thousands of infidels, fewer than the Communist Russians, but a lot nonetheless, offered the most likely catalyst to a Taliban resurgence. In other words, over time the outsiders' staying might actually engender more instability and end up overturning the whole shebang. In the happy interlude after Anaconda, hardly anyone saw that coming.

President George W. Bush explained the policy in an April speech at the Virginia Military Institute, George C. Marshall's alma mater:

> We will stay until the mission is done. We know that true peace will only be achieved when we give the Afghan people the means to achieve their own aspirations. Peace will be achieved by helping Afghanistan develop its own stable government. Peace will be achieved by helping Afghanistan train and develop its own national army, and peace will be achieved through an education system for boys and girls which works. We're working hard in Afghanistan. We're clearing mine fields. We're rebuilding roads. We're improving medical care. And we will work to help Afghanistan to develop an economy that can feed its people without feeding the world's demand for drugs.

"Stay until the mission is done." (And what mission might that be?) It all seemed rather ambitious and open-ended. The high-flown words that followed, a nation-building litany that Bush and his cabinet had previously disdained, didn't sound very limited either. The president wrote out a very large promissory note for Dan McNeill, J. R. Vines, and a few thousand paratroopers. In London, Paris, and especially Kabul, ministers listened and nodded. So, too, did the Taliban elders in the dusty villages of northwest Pakistan. The Yankees would not go home. For disenfranchised Pashtuns, that way led to a Taliban resurrection.

The same old question floated out there: Who was the enemy? Al-Qaeda members seemed, and were, pretty damn thin on the ground. Would America continue to take on the Taliban to prop up Hamid Karzai? It was easy to agree to that after Anaconda, with the Taliban dissipated. Over time, it might get very hard and demand a lot more than ten thousand soldiers. There was a reason the old British regiments butchered and bolted.

Not so America, Bush argued. The U.S. was a different kind of power; it was not seeking empire, but doing good, liberating, helping. And killing terrorists too, he added. Britain and Russia had their histories in country, not good ones either. Bush made reference to the usual course of events for outlanders in the Hindu Kush. "It's been one of initial success followed by long years of floundering and ultimate failure. We're not going to repeat that mistake," Bush said. We did not. Instead, we tried something new. Iraq beckoned.

Like a gambler on a lucky streak at roulette leaving his marker on the red seven for a few more spins, the Americans stayed in the first game even as they strolled down the street to the next casino. The Afghan debt came due years later. But it came.

5

A WEAPON OF MASS DESTRUCTION

Fliers with premonitions are not healthy people.

— TOM WOLFE

SHOOTING DOWN A jet fighter flying straight and level on a clear afternoon takes a lot of concentration. The damn things go fast. A surface-to-air missile (SAM) can do the job, zipping right up there, tagged to a solid radar spot. If the aircraft slides down to avoid the SAM, then a good collection of antiaircraft guns can nail him. It's a nice high-low squeeze play. Running that scheme in the dark gets really hard, more so when turning on the radar dish may earn you a homing projectile right in your face. So you glance at a quick on-off flick of the radar, then shut down. You throw stuff up there, missiles and exploding shells and hardened slugs by the gross. Volume counts.

Just after midnight, in the first few minutes of December 18, 1998, Iraqi missile men and gunners around Baghdad ran the high-low drill on an F-16C Fighting Falcon piloted by Captain Kevin "Grace" Kelly. (The wise guys in his squadron named him after the movie star.) That night over Baghdad, he needed every kind of grace, divine and otherwise, to get through.

His 522nd Fighter Squadron commander, Lieutenant Colonel Tom Jones, was up there too, piloting another F-16 on the same mission. As he put it later, "It got pretty scary because they were shooting pretty close to us." In World War II, the pilots called the ground fire *flak*, a contraction of the German *fliegerabwehrkanonen* (air-defense cannon). Even today, antiaircraft cannon offer serious threats. Around Baghdad that night, Russian-made 23 mm, 57 mm, 85 mm, and 100 mm batteries stood in for the vaunted German 88 mm emplacements that had surrounded Berlin in 1944. But they worked just the same. Jones continued: "If you watched the news from Baghdad and saw those green rounds of anti-aircraft artillery being fired, that was when they were shooting at us."

To the pilots, the green tracers seemed to float up like hot baseballs, each one you saw showing the way for the four you didn't. Tracer or not, if a round or some hot chunks off the airbursts holed your fast-moving plane at the wrong angle, the laws of physics became unforgiving. The fliers called it the "golden BB." So hit, the F-16 would, as the engineers put it, "depart controlled flight." Then the best outcome you could hope for was a jarring ejection, a short parachute ride followed by hitting the desert floor like a sack of potatoes, and a stint of harsh captivity at the hands of Saddam Hussein's henchmen. Other scenarios did not merit speculation.

In Iraq in 1998, the guns ruled up to about ten thousand feet. The SAMs owned the space above, at their best over twenty-five thousand feet. SAMs flying toward you looked like awkward telephone poles with fire coming out of the bottom. The ones you didn't see were the ones that got you.

For Tom Jones and Kevin Kelly and the other hundred or so Americans (and a few from Britain's Royal Air Force) flying over Iraq that night, the shroud of darkness protected them. Air defenders usually countered lack of visibility with the electronic pulses of radar. By this time, the Iraqis, fearful of the deadly U.S. homing missiles, rarely dared to turn on their radar sets. Lacking tracking data, Iraqi gunners could barely make out the darting shapes far above them. "There was a very low moon," Kelly said. "It was pitch-black coming into the target area."

The single-engine F-16 Falcons were officially lightweight fighters, at least relative to the larger, more sophisticated twin-engine U.S. Air Force F-15 Eagles. *Lightweight* was definitely one way to describe the plane its pilots called the Viper, the Electric Jet, and, at times, the Lawn Dart. Although elegant and cleanly designed with its needle nose, gaping under-chin engine scoop, and big bubble canopy, the F-16 was not small.

Kelly, Jones, and company each flew an aircraft that, when loaded for a mission, weighed more than twenty tons, about half of that fuel and ordnance. By comparison, a World War II B-17G Flying Fortress weighed about thirty-two tons and lugged four tons of bombs. F-16s were built as nimble air-to-air fighters but did a lot of ground attack. The pilots were good at it.

That night, Kevin Kelly and his wingman made the four-hundred-mile push up from Ahmed al Jaber Air Base in Kuwait. His F-16 was carrying external fuel tanks, some self-defense Sidewinder missiles, and a full load of six five-hundred-pound Mark 82 general-purpose bombs. The Mark 82s, three under each wing, were destined for a Special Republican Guard target cluster in greater Baghdad. A plane ahead of Kelly dropped the first load. "After the first bomb detonation," Kelly noted, "the skies really lit up with triple-A [antiaircraft artillery] fire." Kelly pressed on.

The green dots of fire rose up. Kelly's F-16 flew above them, exploiting that useful envelope between ten thousand and twenty-five thousand feet. That was the idea: stay well up from the gunfire and take chances with the less common and far more radar-dependent air-defense missiles, especially at the lower end of the SAM engagement range. In a true integrated air-defense system, a network of radars linked to a few key regional ground command posts sort out the data. Those regional centers then feed the national command post to convey a complete picture of the developing air raid. Defending commanders can then choose to engage attackers with guns, SAMs, and even interceptor jets. Adding defending fighters to the mix made it very hard for attackers to fly serenely above the cannon barrage or simply jam the SAMs, as the home team could be up there ready to bounce. Once in a chase with hostile fighter planes, if the dogfight didn't do the job, intruders might be forced down into the flak or pushed up into the ideal SAM range. At a minimum, the attackers would have to dump external fuel tanks and bombs to get out of the fur ball.

Most countries used this same approach. For example, in America, the rudiments of the old NORAD Cold War array still remained on 9/11, and it allowed for some tracking of the hijacked airliners and the belated scrambling of fighters. Saddam Hussein's military had had a much better, well-armed system in place until January 17, 1991. It went away overnight once the U.S. opened Desert Storm with a massive initial series of airstrikes that tore apart the Iraqi integrated network. A lot of uncoordinated pieces survived the 1991 U.S. air onslaught. Though the Iraqis spent the 1990s painstakingly reassembling a huge number of working guns, radars, SAMs, and fighter planes, the entire system did not tie together anymore.

Iraqi jet interceptors, none too active in 1991, proved no-shows in 1998. Saddam Hussein judged his air force ineffective, hobbled by inferior aircraft and inadequate training. He was right in this assessment. Saddam's solution was not to retrain or reequip his air arm but to clear the skies and give his guns and missiles free rein. This he did. Iraqi MiG and Sukhoi fighters stayed out of action, skulking around the edges at best.

Though he kept looking, Kevin Kelly saw no enemy fighters. What about the SAMs? The F-16 pilot craned his neck in the clear bubble canopy, looking down into the black ether for the hot tail flame and rising shadow of the long, slim, wide-finned V-75 type of SAM known to NATO as the SA-2 Guideline, the great plane cracker of Vietnam. He watched for the shorter, fatter, faster, smaller-finned S-125 Pechora (named for a Russian river), called the SA-3 Goa by NATO and known for its prowess in downing Israeli jets in the 1973 war and Iranian intruders in Iraq's 1980–88 conflict. That night, Kelly and his teammates saw none of these high-altitude threats. The lethal suppression and electronic jamming must be working, Kelly thought.

The exploding antiaircraft shells, however, kept tracking his aircraft. They burst by the hundreds below his racing plane. "It's no kidding for real," Kelly said. "It looks exactly like what you're seeing on CNN, except I'm looking down on it." Kelly punched off his bombs.

Each Mark 82 unlatched and began the long fall, about three miles. Kelly's bombs were "dumb"; that is, not IT-enabled, not the precision JDAMs later used in Afghanistan. That said, Kelly had plenty of experience dropping ordnance, having carried out similar attack profiles on many training runs. In addition to having a good pilot, the F-16 itself was very smart indeed. Its onboard computer calculated well. Against the row of buildings below, six dumb five-hundred-pounders worked just fine. The Iraqis admitted that thirty-eight had been killed and a hundred wounded, but who knew the truth? Some Iraqi spokesmen alleged thousands of civilian casualties, of course. To listen to the government ministers in Baghdad, every U.S. bomb showed an unerring nose for a hospital, orphanage, elementary school, or mosque.

As Kevin Kelly, Tom Jones, and their 522nd Fighter Squadron mates turned away for the long run back to Kuwait, more U.S. warplanes continued the attack. About a hundred miles southeast of Baghdad, the Fifth Baghdad Mechanized Division of the Republican Guard got a lot of attention. Two B-1B Lancer bombers bore in at 550 miles an hour, twenty-six thousand feet up. They had come all the way from Thumrait, Oman, headed for the Republican Guard compound at Al Kut, near a bend in the Tigris River.

The thousand-mile run to the northwest represented the first live combat mission for the graceful four-engine bomber known to its four-man crews as the Bone (from "B-one"). The B-1 fleet sat out the 1990–91 Gulf War. At that time, the Bones focused on the nuclear mission and endured a series of engine and avionics problems. All of that had been fixed by 1998. With swing wings, supersonic top speed, and the ability to execute aerobatics akin to an F-16 fighter, the Bone also featured a degree of stealth thanks to its shape and special skin coating. The bomber had no guns, but it packed a jammer suite second to none. It could thoroughly confuse enemy radars and missiles. All of this allowed each B-1 to deliver a huge amount of ordnance. In those dark minutes around one in the morning on December 18, 1998, the bombers Slam 11 and Slam 12 each carried sixty-four Mark 82 five-hundred-pound munitions in their massive bomb bays. In terms of punch, Slam 11 equaled almost eleven F-16s. In other words, a single Bone could drop as many bombs on one pass as the entire 522nd Fighter Squadron.

As they crossed the Persian Gulf, Slam 11 and Slam 12 formed up with a dozen U.S. Navy F/A-18 Hornets off the aircraft carrier USS *Enterprise* (CVN-65). An EA-6B Prowler also linked up. The Prowler had the ability to blank out hostile radars with jamming and kill them by launching HARMs (high-speed anti-radiation missiles). It reflected what Lieutenant Colonel Jeffrey Taliaferro, who flew on the B-1 mission, called "a very healthy pairing." The night's mission demonstrated the close joint service cooperation developed during many training evolutions.

The American aircraft pivoted to avoid one active SA-2 SAM site. Aboard Slam 12, Lieutenant Colonel Gordon Greaney watched his display. He saw a second Iraqi SAM position, an SA-3, briefly activate its radar. The Iraqis shut down as fast as they lit up. Had the enemy persisted, the EA-6B would have gone to work. Either way, the U.S. attack would go on.

The SAMs stayed on the ground. The antiaircraft artillery opened up and began banging away. But the barrage flowered well below the big U.S. aircraft. The two crews noticed but stayed focused on their instruments, determined to make the trip count.

About seven miles out from the target, both B-1s opened their double bomb bays. In five seconds, all sixty-four five-hundred-pounders separated from each Bone. After the mission, Lieutenant Colonel John Martin noted that the fliers "put steel on target like we had trained to do so many times while at Ellsworth [Air Force Base, South Dakota]." The pair of B-1s slid into a steep turn and headed back toward Oman.

Below and behind the bombers, the Al Kut complex and the flat desert around it erupted with 128 explosions. Although some of Slam 12's bombs came off the B-1 a fraction of a second late, photo interpreters assessed a good pattern for the mission. The two crews got credit for twelve direct hits. The strike photo was released to the news media as an example of an effective attack.

The Al Kut mission received a little extra interest because it featured the combat debut of the Navy's first group of women trained to fly fighters. One of them, Lieutenant Kendra Williams of Strike Fighter Squadron 105, flew her F/A-18 off the USS *Enterprise.* Various press people wanted to make a big deal about it. Williams answered like the naval aviator she was: "I was just doing my job." As were they all.

So were the Iraqis. In the typical languid local fashion, *inshallah* and all that, they arrived a bit late to the fracas. Almost three weeks after Kelly, Jones, Taliaferro, Greaney, Martin, and Williams completed their night missions, the Iraqi air force crept back out into the sunlight. Southwest of Baghdad at about 10:15 a.m. on January 5, 1999, a pair of twin-engine MiG 25s crossed the thirty-third parallel of latitude into the U.S.-patrolled zone in southern Iraq. About eighty miles to the east, another pair of Iraqi jets, swing-wing MiG-23s, loitered near the fringe of the American area.

Vectored by an overwatching U.S. radar plane, two Air Force F-15C Eagles turned toward the MiG-25s. The Iraqis used their onboard radar sets and illuminated the Americans. That constituted hostile intent. The aggressive Americans immediately fired a spread of three Sparrows and one AMRAAM (advanced medium-range air-to-air missile), all radar-guided weapons. The Iraqis turned tail and lit their afterburners, fleeing north. The speedy MiG-25s outran the missiles.

In the east, about fifteen minutes later, the two MiG-23s tried the same thing. Two U.S. Navy F-14D Tomcats of Fighter Squadron 213 from the USS *Carl Vinson* (CVN-70) fired back. The Iraqis spun around and screamed north, ahead of the trailing Phoenix missiles. Neither appeared to hit, but one MiG-23 went down. It may have exhausted its fuel. Still, a kill is a kill.

The military called the four-day December air offensive Operation Desert Fox. Marine general Anthony Zinni, the commander of U.S. Central Command at the time, suggested the name himself when he proposed to "outfox" Saddam Hussein. It followed the usual USCENTCOM naming convention: Desert Shield (the defense of Saudi Arabia, August 1990 through January 1991), Desert Storm (the liberation of Kuwait, from January to March 1991), and Desert Strike (the response to an Iraqi incursion into Kurdistan in September of 1996). The planners even had two other names on the shelf: Desert Badger featured measures in response to a downed U.S. aircraft; Desert Thunder addressed the defense of Kuwait if the Iraqis tried to invade again.

Shakespeare once asked what's in a name. Evidently, the Bard did not anticipate the dynamics of spin and nuance always at work inside the Beltway. In December of 1998, more than a few of the government's publicity people in Washington didn't care for the name. The journalistic smart set connected the title to Field Marshal Erwin Rommel, commander of the German Afrika Korps, known since the 1941–43 campaign in North Africa by the nickname Desert Fox. How could America call an operation after a Nazi general? they asked. The fact that Rommel was no Nazi, and indeed paid with his life for his opposition to Adolf Hitler, didn't matter. Instead, at their headquarters in Tampa, Florida, Zinni and his people had to interrupt their duties to offer repeated assurances that they did not mean *that* Desert Fox.

This Desert Fox came about because Saddam Hussein had celebrated Halloween on October 31, 1998, in his typically perverse way. The Baghdad regime evicted United Nations inspection teams trying to check on suspected Iraqi nuclear-, biological-, and chemical-warfare programs. President Bill Clinton directed preparations for air operations to degrade Iraq's suspected chemical-weapons stockpiles. At the same time, the Americans attempted to line up diplomatic support.

That proved rather problematic. The longtime no-fly zones above the thirty-sixth parallel (Operation Northern Watch over Kurdistan) and below the thirty-third parallel (Operation Southern Watch guarding the Shiite Arabs) had frayed a bit since the heady days after Desert Storm. Various regional states permitted overflight and basing rights. Others limited their support. Turkey grudgingly let U.S. and British warplanes patrol the Kurdish region but declined to host any of the Desert Fox strike aircraft. In the south, France went its own way, as always, and backed out of Southern Watch on the eve of Desert Fox. China and Russia chafed. It all frustrated the Clinton team.

There was more from the usual direction. In standard ambivalent mode, Saudi Arabia exceeded the French example, allowing continued Southern Watch defensive flights but refusing to permit the use of its bases for any Desert Fox offensive sorties. The Saudis then helpfully reminded the Americans and their British partners that the Muslim holy month of Ramadan began on December 20, so wrapping things up by then would be much appreciated. The situation had changed a lot since the anxious days of August 1990 when Saddam Hussein's armored brigades squatted on the kingdom's northern frontier. This time around, the Saudi princes shrugged. Those Westerners had to know their place.

For the Americans and the British, and their reluctant allies dragging behind, it all amounted to a continuing war against Saddam Hussein's Baathist Iraq. Ever since General Norman Schwarzkopf walked out of the truce tent at Safwan in March of 1991, the U.S. and its friends had lived with the consequences of leaving the Iraqi dictator in power. Within a few months, all U.S. ground forces left Iraq.

A U.S. Army caretaker contingent stayed in Kuwait. During periodic Intrinsic Action exercises, often timed to coincide with perceived Iraqi threats, a U.S. Army brigade practiced deploying, drawing pre-positioned equipment, and then carrying out live-fire training in Kuwait. Marine and Navy amphibious ships practiced landing operations. Kuwaiti units participated in all of these war games to the extent that their limited capacity allowed. It all contributed to deterrence, but neither Kuwait nor America nor the neighboring Saudis wanted a lot of boots on the ground.

After 1991 in the Persian Gulf region, America trusted the really heavy lifting to airpower. Britain had showed the way. After the Great War ended in 1918, the British army reduced its massive wartime order of battle to its regular regiments. That shrunken ground force had to guard numerous colonial holdings. For those who appreciate the irony of later developments, two particularly tough ones were the frontier between British India and restive Afghanistan and the newly acquired League of Nations Mandate of Iraq. In both places, the British found more fighting than they had ground units. The new Royal Air Force provided the answer.

The British called it air policing. The head of the Colonial Office at the time, Winston S. Churchill, announced at a regional conference in Cairo on March 12, 1921, that the Royal Air Force would defend the new country of Iraq. The energetic colonial minister was just the man to direct such a bold move. When the Ottoman Empire collapsed in the wake of the Great War, Churchill, advised by the Middle Eastern expert Gertrude Bell and the ever-restless Colonel T. E. Lawrence, approved the conjunction of three former Ottoman Empire vilayets (administrative divisions): Mosul in the north, Baghdad in the center, and Basrah in the south, on the coast of the Persian Gulf. Having played midwife to the new state, Winston Churchill named its new protector: the Royal Air Force. He judged that by not sending more ground troops, he saved his country twenty-five million pounds annually, the equivalent of $1.3 billion in today's U.S. currency.

The RAF assumed the duty with alacrity. When villagers shot at British government facilities, ambushed officials, or waylaid Western travelers, the mandate power had to do something. Army columns took weeks to assemble, provision, and dispatch. The RAF could act swiftly, within hours, one of the huge advantages of airpower. This matched very nicely with the requirement. As one 1924 RAF order noted: "Hesitation or delay in dealing with uncivilized enemies are invariably interpreted as signs of weakness." Air policing ensured a rapid response.

The RAF followed a protocol that would be familiar to us today. First, the British officers made demands, such as the surrender of certain rebel leaders. The Iraqis usually complied, especially when they had already experienced a prior dose of RAF persuasion. But sometimes, the locals held out. In most of these cases, a leaflet drop on the offending village warned the Iraqis of the upcoming air attack. Of course, given that most of the Iraqi people did not read Arabic, let alone English, the bilingual pamphlets often had no effect. Those Iraqis who knew what was coming got out of town. Most stayed, defiant, trusting in Allah.

The RAF might follow up with a demonstration air raid in a nearby open desert to overawe the Iraqis. That often did the job, especially in the early days. As the years went by, though, and the Iraqis learned by experience, such displays served the RAF less and less well. Confronted with noncompliant Iraqis, the RAF attacked. The British timed their raids carefully "to ensure the greatest concentration of men and animals around the village." RAF doctrine was clear on that point.

When the appointed day came, a dozen or so obsolescent biplanes swept down on the designated Iraqi hamlet. Machine guns sufficed to kill men and livestock. RAF pilots observed that twenty-pound bombs "will only make a small hole in the huts of tribesmen, 112-lb. will be expected to blow off the roof, and 230-lb. and larger bombs should destroy large houses." If that wasn't enough, the RAF senior commanders advised more: "Incendiary bombs should be used for good effect against villages and crops." Loose talk even mentioned using chemical weapons, but that didn't happen. Bullets and bombs proved enough. A few hours after each attack, another set of leaflets were dropped, summarizing the lesson. It all amounted to the old British butcher-and-bolt tactic, done from the air. It worked.

For the U.S., it worked too. In its day, the Roman Empire dominated due to the land power of its marching legions. Winston Churchill's British Empire relied on the Royal Navy and sea power. For America, airpower characterizes its strength. Since 1945, punctuated by the atomic bombing of Hiroshima and Nagasaki, U.S. airpower has undergirded the country's global influence. It shapes how we keep the peace and fight our wars.

Since the end of World War II, the strongest air arm in the world has consistently been the U.S. Air Force, followed closely by the second-strongest, the fleet aviation of the U.S. Navy, supplemented by the U.S.

Marines. The U.S. Army's helicopters give it unprecedented mobility and firepower on land. Wry commentators often complain about the fact that America sustains four air forces (five with the Coast Guard), failing to grasp that all modern U.S. combat capability relies on aircraft. They are as integral to American fighting methods as cannon were to nineteenth-century armies and navies. The U.S. brings it all together better than anyone else. The Russians and Chinese sometimes claim more warplanes and occasionally field impressive technologies. But those countries do not integrate their air wings very well, train them to high standards, or deploy them around the world to exert influence. Coupled with equally dominant U.S. sea power, a vibrant aerospace industry, and now the burgeoning information technology sector, airpower enables America to range the world.

That tremendous capacity can destroy any fixed target on earth. And that's a crucial weakness. Killing and burning eliminate rivals. Taken to the extreme, airpower can finish off an entire nation. But unless you're willing to go all the way, and the enemy fully knows and accepts that, airpower cannot control ground. It cannot make a determined people submit to U.S. control. To do that, you have to enter their streets and stand there, rifle in hand. Like the imperial British with their ships of the line, America prefers to rely on its masterful technology, to stand far off and pound. We don't like messy, bloody, dirty ground wars.

Sometimes we can avoid them. In certain conflicts, destruction is enough. Firebombing and atomic-bombing imperial Japan in 1945 finally broke that proud state without ground invasion. In a lesser, though still impressive, case, the American-led NATO air campaign of 1999 caused Serbia to back out of Kosovo, ending a decade of Balkan civil war. Punitive raids in Libya in 1986 curtailed that terrorist regime. If you narrowly define your goals — the measured advice of Colin Powell comes to mind — air raids suffice.

Yet too often, destruction is not sufficient. Conventional bombing badly weakened but did not crack Hitler's Nazi Germany from 1944 to 1945. With little to lose, North Korea rode it out from 1950 to 1953. It did not cow the implacable North Vietnamese in that long war. It hollowed out and fatally weakened Saddam Hussein's divisions in and around Kuwait in 1991, but the Iraqis held their ground even as they writhed under the Desert Storm air bombardment. Foes can choose to rely on international sympathies for the weak, to fan Americans' moral qualms, to play for time, and so ride out air attacks.

Airpower zealots, like Marxist ideologues or religious fanatics, explain away shortcomings by pointing to improper or incomplete application. Certainly, despite incredible valor by thousands of airmen, the instruments weren't yet up to the task over Germany in World War II. Inordinate strictures hobbled the Vietnam air campaign. As for Desert Storm, a few more weeks of bombing might well have seen the Iraqis fold their tents and decamp. That one was hard to call.

Yet the arcane micromanagement of the air campaign over North Vietnam reflected America's innate uneasiness with death from above. It ill accords with the values of a democratic republic. In a world war, with almost no possibility of hostile retaliation, Americans agreed to impose fire raids on places like Hamburg, Dresden, and Tokyo. We didn't like it, but we did it. We'd never try it against the Russians or Chinese, who can hit back with equivalent ferocity. And against a lesser foe, such extreme measures make most Americans very queasy. For smaller threats, we'd rather use smaller tools. Killing for the sake of killing is not our way. This, too, is the heritage of Hiroshima.

We strive for precision in our use of airpower. Even in World War II, the U.S. Army Air Forces in Europe flew by day and tried to use the sophisticated Norden bombsight to hit military and arms industry targets. American losses mounted in the face of German fighters and flak, but U.S. airmen refused to shift to outright city destruction. Having endured the German Luftwaffe's blitz of Britain's cities, RAF bombers showed no such qualms. They flew by night and relied on firebombing in what they euphemistically called "de-housing" attacks. Eventually, faced with the highly dispersed Japanese war economy, the Americans resorted to the same grim approach in 1945, which culminated in two atomic bombs. From then on, though, chastened America put strict limits on its fearsome airpower. In the Korean War, U.S. bombers never attacked China, the main enemy. Over North Vietnam, the Americans plinked and chipped around the edges for so long that the Communists were well prepared by the time a few serious strikes happened. Enemy propaganda and antiwar-movement allegations aside, the U.S. scrupulously — perhaps too much so — spared most of the North Vietnamese populace.

The great Linebacker II B-52 offensive in December of 1972 killed 1,318 civilians in Hanoi, hardly numbers that reflected the supposed carpet-bombing so often claimed. It does demonstrate, however, that explosive aerial bombs and missiles, no matter how carefully aimed, are very blunt instruments. Using airpower means killing people. Killing fewer may make Americans feel better. But they are still just as dead. And more are left to seethe and fight back.

In Desert Storm and other post-Vietnam air attacks, including Desert Fox, America tried hard to hit the hostile leadership and its armed components. Civilian casualties were minimized, often at great risk to the attacking U.S. fliers. In public statements in the 1990s, President Bush and then Clinton made it clear that the U.S. quarrel was not with the people of Iraq but with their ruler. Unlike the Germans or Japanese in World War II, by the final decade of the twentieth century, the Iraqis and others were considered victims of their dictators, awaiting succor and, maybe, liberation from America. Instead of using its full might, hopeful to curry some favor even in the countries attacked, the United States deterred itself.

Our enemies see this as a weakness. The dictators who have had such power used it. Hitler, Stalin, and Mao killed millions under their sway, employing what they had at hand. If they had possessed America's degree of atomic airpower and were undeterred by us, they would assuredly have run up that hideous score at the United States' expense. Saddam Hussein never shied away from using jets and helicopters against his own and, when he could, on neighbors like Kuwait, Saudi Arabia, Iran, and Israel. He did not understand why the Americans flew over his country daily but did not destroy him.

The U.S. saw it as containment. The military talked of keeping Saddam Hussein in the box, much as the RAF had suppressed his Iraqi ancestors in the 1920s. Operations Northern Watch and Southern Watch together constituted a huge air campaign. American, British, and, for a while, French and Turkish airmen carried out nearly ten times the number of sorties flown in the much more compressed, violent Desert Storm offensive. U.S. Navy ships fired hundreds of Tomahawk cruise missiles as well. Although the December 1998 Desert Fox episode was the largest series of raids after Desert Storm, there were other major strikes in 1993 and 1996. The Iraqis postured and growled. Over time, they found the flights and strikes less and less impressive. But the battered Iraqi armed forces remained fixed in place.

Constant daily skirmishes occurred. From 1998 to 2001, the Iraqis switched on their tracking radars, sent up jet fighters, fired air-defense guns, and launched SAMs some sixteen hundred times, well over one incident a day. The U.S. and British shot back, too, 286 times in that three-year stretch. Operation Desert Fox came on top of all that.

The ongoing U.S. air campaign kept thousands of Air Force crews and hundreds of aircraft, as well as an average of two U.S. Navy carrier battle groups with thousands of sailors and more than 150 planes, engaged. One estimate pegged the cost to America as $12 billion. Driven by the continuing Iraq mission, the U.S. Air Force undertook a significant reorganization into ten expeditionary air wings. One wing would form, train, head overseas as a cohesive team, fulfill its assigned duties in theater, then return, and the cycle would restart as the next wing took over. The Air Force modeled it on the time-tested Navy/Marine deployment model. America's airmen carried out many other tasks in the 1990s, deterring ever-combative North Korea, running the 1995 Balkan air attacks, and winning the large-scale 1999 air campaign against Serbia, not to mention supporting U.S. interventions in Somalia and Haiti. Always, however, the airmen focused on their key mission: keeping Saddam Hussein's Iraq in check.

The hapless Iraqis never knocked down a Coalition plane, though not for want of trying. But the daily grind of getting shot at took its toll on the Coalition members. The French and Turks dropped out. The Saudis grew ever more testy. The British scaled back. For Saddam Hussein, every day in power amounted to another victory. To an extent, the U.S. had been fixed in place too. America joined Saddam in the box.

By Operation Desert Fox in 1998, Americans wanted out. Channeling popular opinion, both political parties looked for a way to eliminate the loathsome Iraqi strongman. Seven years after it had occurred, Desert Storm looked a lot less like a victory. The daily U.S. military reports of dodging SAMs and dropping bombs coincided with constant bellicose Iraqi rhetoric. George H. W. Bush's great success became Bill Clinton's headache.

Operation Desert Fox responded directly to one well-accepted problem. After the 1991 war, protected by American airpower, United Nations inspectors entered Iraq. They found extensive evidence of nuclear, biological, and chemical research. They located dozens of long-range missiles and stockpiles of nerve gas. Most alarming, a good number of these sites had not even been targeted. Analyst Anthony H. Cordesman summarized what went wrong in Desert Storm: "We did not identify over 80% of the actual Iraqi facilities, our most successful strike against nuclear facilities were an accidental hit by a diversionary strike against an uncategorized target, we hit no major missile or biological warfare site, and over 95% of Saddam's biological and chemical weapons and missile forces survived until the end of the war." Clearly, the Iraqis wanted these deadly arms, the infamous weapons of mass destruction (WMDs).

That term, beloved of Soviet propagandists during the Cold War, lumped together three very different kinds of devices. Nuclear bombs featured massive blasts, blazing heat, and lethal radiation. Even the small ones could wreck an entire city and kill thousands in an instant. Biological weapons, natural and manmade, were intended to spread diseases. By their nature, once delivered, germs take their own course, subject to winds and weather. As for chemicals, used extensively in World War I and now and then in other conflicts, they too degraded if the weather wasn't right, and most militaries had the proper protective gear. Only the nukes really merited the WMD hype. Bioweapons might live up to it someday, but they had too many weird vagaries. Chemicals would not kill masses unless they were aimed at panicky, defenseless civilians. Not surprisingly, Saddam's shabby science and industry focused on chemicals, tried-and-true in the 1980–88 war with Iran and used in several reprisals in Kurdistan. The Iraqis wanted the nuclear and biological toys, but they just couldn't create them. Of course, in 1998, we didn't know that. Things got a lot more transparent once we poked around in all of Saddam's hidey-holes in 2003. Until then, it suited the dictator's purpose to bluff all, inside Iraq and out.

When Saddam tossed out Richard Butler and the United Nations weapon inspection teams at the end of October 1998, their work was incomplete. Butler estimated the Iraqis had not shown him about 32,000 individual chemical munitions, 550 mustard-gas-filled bombs, and 4,000 tons of industrial chemicals used to make more. He also thought the Iraqis had up to a dozen concealed Scud rockets, possibly outfitted to carry chemical warheads. If you wanted to deliver nerve agent on a lot of unready civilians in Israel or Saudi Arabia, that would do it. As shown in 1991, Iraq knew the proper trajectories only too well.

In addition to the WMD issue, the U.S. had to consider Iraq's persistent threat to Kuwait. While no match for the U.S. military, the Iraqis could still overrun Kuwait. Even after the beating Iraq's military had taken in Desert Storm, American intelligence credited Saddam with seventeen army divisions and six Republican Guard divisions, about 350,000 troops armed with 2,200 tanks and 4,000 artillery pieces. The emir of Kuwait's 23,000 soldiers wouldn't last long without help from the United States. When the Iraqis massed two Republican Guard divisions just north of Kuwait in October 1994, America quickly reinforced the emirate. Within a week, two Army mechanized infantry brigades arrived in Kuwait; multiple Air Force fighter squadrons deployed to bases in the region. Offshore, a brigade-strength Marine expeditionary unit waited aboard amphibious assault ships. The USS *George Washington* (CVN-73) carrier battle group took up station, joined by British and French warships. The Iraqis backed off. The entire episode, though, showed how quickly bad things could happen, even with Southern Watch aircraft overhead.

Inside Iraq, Saddam Hussein and his Baathist clique ruled with a heavy hand. Over the years, the dictator's secret police, internal security forces, Republican Guard, and regular military murdered 250,000 Iraqis, up to 180,000 of them Kurds. In an especially cynical effort, Saddam's forces intervened in an inter-Kurd clash in August of 1996 and took the opportunity to inflict numerous casualties on his hated rivals. In the southeast, Iraqi engineers drained the marshes, driving the local Shia Arabs out of their homes and shooting many in the forced relocations. Whatever he did to surrounding countries, Saddam never let up on his own unfortunate countrymen. Iraqis did not have to search for a weapon of mass destruction. He ran the show from Baghdad.

Finally, Saddam Hussein wholeheartedly backed terrorists of all stripes. In a direct action, his intelligence operatives tried to stage an assassination of former president George H. W. Bush during a Kuwait visit in April of 1993. Saddam trained, financed, and offered moral support to various entities. His people ran camps for and cooperated closely with several terrorist groups, with a particular interest in attacks on Egypt, Iran, Israel, Kuwait, Saudi Arabia, and the United States. Saddam hosted his own Fedayeen Saddam, Abu Nidal's Fatah (Conquest) Revolutionary Command, the Palestinian Liberation Front of Abu Abbas (perpetrators of the 1985 seizure of the Italian cruise ship *Achille Lauro* and the murder of an American passenger), the notorious Palestinian group Hamas (Zeal), Egyptian Islamic Jihad (Ayman al-Zawahiri's network), and Ansar al-Islam (Partisans of Islam), an al-Qaeda offshoot. That last one, hidden in the mountain fastness of Kurdistan, aided Saddam's cause by attacking various Kurdish targets. A key leader, Jordanian Abu Musab al-Zarqawi, fought in Afghanistan against the Soviets in 1989 and against the Americans in 2001. Zarqawi didn't have an official al-Qaeda identification card, but he sure thought of himself as part of the gang. He had more to contribute after the Americans invaded in 2003. Zarqawi was one among many in Saddam's broad swath of terror agents.

Confronted with all of this, a bipartisan consensus emerged in the otherwise fractious U.S. political landscape. In 1998, the Democratic president and the Republicans in Congress agreed on little else — indeed, the Republicans moved to impeach Bill Clinton even as Desert Fox occurred. Despite the poisonous partisan atmosphere, they joined together on the need to rid Iraq, and America, of Saddam Hussein.

On October 31, 1998, the same day Saddam Hussein threw out the UN inspectors, Clinton signed the Iraq Liberation Act. The bill passed the House by a vote of 368–30 and received unanimous consent in the Senate. It listed ten charges against Saddam: making war (including chemical strikes) on Iran in 1980; forcibly relocating and slaughtering thousands of Kurds in 1988; using chemicals against the Kurds in that same year; invading and despoiling Kuwait in 1990–91; failing to dismantle WMDs; attempting to orchestrate a terrorist attack on former president Bush in April 1993; pushing 80,000 men to the Kuwaiti border in September 1994; attacking Kurdistan in August 1996; interfering with UN weapons inspectors; and, finally, evicting the inspectors altogether. The list focused squarely on the four big problems: WMDs, external threats, internal repression, and support for terrorism.

The law's text included these words: "It should be the policy of the United States to support efforts to remove the regime headed by Saddam Hussein from power in Iraq and to promote the emergence of a democratic government to replace that regime." If airstrikes like those in Desert Fox could do the deed, good enough. But the law clearly sought a homegrown Iraqi resistance to back with American training, money, and air support. It sure beat using U.S. soldiers on the ground.

The Kurds up north were more than ready. They had their own factions, of course, but the CIA kept good connections with them all. The Kurds already had U.S. air cover, thanks to Northern Watch, and they would fight. Although the Afghan campaign still lay in the future, long before 9/11, the Kurds looked like good candidates to play the part of the Northern Alliance. They also shared the Northern Alliance's liability. The Kurds, like Afghanistan's Tajiks and Uzbeks, formed a minority in Iraq. The country's Arab majority was not likely to follow a Kurdish lead. Just as the Americans needed Pashtuns for the 2001 Afghan campaign, so the U.S. sought some Iraqi Arab equivalents of Hamid Karzai, Gul Agha Sherzai, and their erstwhile Southern Alliance.

Arab opponents of Saddam were numerous enough inside and outside Iraq. Chasing and executing them kept Saddam's people busy. But the most capable potential rebels, those of the Shia Arabs in the south, already had a patron in Iran. Remembering only too well America's inaction during the failed 1991 southern uprising, the Shia Arabs proved a very hard get. One Shia Arab expatriate, however, made himself all too available.

Ahmed Abdel Hadi Chalabi offered to organize the Shia Arab resistance. He had a good pedigree; too good, in retrospect. His wealthy father had rebuilt the elaborate Kadhimiya shrine in Baghdad, which marked the last resting place of two of the venerated Shia imams. With his family, young Chalabi emigrated to the West at age twelve, just ahead of the 1958 Baathist takeover in Iraq. He earned an undergraduate degree in mathematics at the Massachusetts Institute of Technology and got his doctorate at the University of Chicago. After Chalabi graduated, his banking business in Jordan collapsed under allegations of fraud. He went on to organize the Iraqi National Congress in 1992, a self-designated umbrella group for opponents of Saddam Hussein. Although his family had money, Chalabi always seemed to have more than his share. The rumor mill placed him on the payrolls of various intelligence services, and some supposed he served the will of others. The shady

Chalabi, though, worked only for himself. He lobbied long and hard for the 1998 Iraq Liberation Act, portraying himself as the potential George Washington of a free Iraq. Well, maybe, but first he had to deliver something besides self-aggrandizing schmoozing. Though many American politicians feted him, the CIA and Department of State had no faith in Chalabi. In Iraq, there would be no Southern Alliance.

Thus the 1990s ended. Bill Clinton and most Americans wanted Saddam gone. Most Iraqis did too. The no-fly zones kept him down. Taking him out, though, meant using U.S. men and women on the ground. The Iraqi resistance remained a long shot. Beset by many, many other foreign and domestic problems, not the least a rancorous impeachment attempt in the U.S. Congress, Clinton left Saddam to the ministrations of his airmen and the hopes of the CIA. You could think of worse outcomes.

Worse outcomes seemed all too possible, even likely, in the anxious days after 9/11. As Saddam played footsie with various UN inspection delegations, American officials worried. "Keeping Saddam in a box looked less and less feasible to me," said President George W. Bush. "He had used weapons of mass destruction in the past. He has created incredible instability in the neighborhood." And there was a personal angle. "He tried to kill my dad," Bush said. After 9/11, Bush had no intention of waiting to act.

Vice President Cheney and Secretary Rumsfeld agreed with him completely. Even as the Afghan campaign achieved some early successes, the defense chief put General Tommy Franks to work in late November of 2001. The extant USCENTCOM plan envisioned a repeat of Desert Storm with months of buildup and 380,000 troops. Well aware of what was transpiring in Afghanistan, Rumsfeld and Franks chose to look at faster options that relied more on airpower and SOF. It would take a while to nail down international and domestic support. Several small units packed up and moved into the theater. As 2001 became 2002, Central Command planned, revised, and reconsidered. Meanwhile, Northern Watch and Southern Watch kept picking apart the Iraqi air defenses.

The domestic consensus to depose Saddam Hussein grew stronger after 9/11 and in the face of the Iraqi maximum leader's dogged intransigence regarding UN weapons inspections. Bush moved slowly, as many in his own party and in the Democrat ranks insisted on UN and allied support. Amassing such support took time and wasn't wholly successful. The British lined up, of course, at substantial political cost at home for Prime Minister Tony Blair. The Saudis agreed to it, but only if America promised to pull out most of its forces after the operation. Other countries concurred, more or less. The Turks couldn't, or wouldn't, commit to letting U.S. units transit on the ground to enter northern Iraq. They left it hanging. Some allies, notably the French and Germans, showed little enthusiasm for going after Saddam. Strong domestic political objections, not to mention ongoing business dealings with Iraq, kept them out of any potential coalition. China and Russia, of course, demurred. The UN gave a mixed blessing to the effort, not a ringing endorsement but not a disapproval either. Unable to gain agreement for "all necessary measures" (military means), America sponsored a rather vague resolution promising "serious consequences" for Saddam's Iraq. That loose verbiage garnered a 15–0 vote in the U.N. Security Council, a tally that included the votes of China, Russia, and Syria. So there it stood. If the U.S. campaign worked, the usual suspects would pile on the bandwagon. It took most of 2002 to line up enough foreign cooperation to mount an offensive.

In America, Bush asked for a congressional resolution authorizing the use of military force. The one from September 2001 didn't mention Iraq, and the clearly expressed and recorded support of the citizenry through their elected representatives was essential for a new campaign. The Democrats barely controlled the Senate, and the Republicans ran the House. As the voting began on October 10, 2002, the entire House and a third of the Senate faced an election the following month. The American people, to believe the pollsters, wanted to finish off Saddam Hussein. Thus the vote ran true to form. It passed 296–133 in the House and 77–23 in the Senate. The margins far exceeded those for the 1991 war with Iraq.

Given the strident partisan political recriminations that followed in the second half of 2003 and afterward, it's noteworthy that the text of the 2002 authorization consciously emulated the 1998 act, referencing it extensively. Twenty-two findings walked down the familiar charges, including WMD threats, conventional threats against neighbors, internal repression, and support for terrorists. Many Democrats who voted for the bill later claimed they had been misled, perhaps intentionally, by skewed intelligence estimates, particularly those regarding potential Iraqi nuclear, biological, and chemical weaponry.

Party politics in America has never been for the faint-hearted. As the war soured in late 2003, the debate grew quite heated. The former consensus vaporized. A new one arose: Saddam had no WMD. He was in that oft-cited box and was no danger to surrounding countries. He was rough, in fact brutal, to his population, but that described many cruel dictators in other countries — Syria, Iran, and North Korea, not to mention China. And Saddam had no ties to al-Qaeda, none at all. Facts to the contrary were ignored. The opposition boiled it down to a pithy, angry slogan: Bush lied, thousands died.

All of that unhappy political strife lay well in the future as Tommy Franks planned the end of Saddam's regime. Drawing on the Afghan playbook, USCENTCOM devised an intriguing mash-up of the old Desert Storm and the new air/SOF idea pioneered in Enduring Freedom. Steadily pressed by the robust U.S. no-fly efforts, the Iraqis were weaker and hence more vulnerable, provided the United States moved quickly. A long buildup would not serve anyone well. This time, the air and ground offensives would go off together. One good hard blow, carried through to Baghdad, should topple Saddam.

As with the Afghan campaign, Franks and his planners proposed four phases: staging; seizing the initiative with air and SOF; decisive operations; and then transition. Franks focused most strongly on phase three, decisive combat operations. Employing less than half the ground strength of Desert Storm, the new scheme emphasized speed and firepower, much of that from the air. The U.S. would use a land component of 140,000, plus 30,000 allies (mostly British), with six divisions (four Army, one Marine, one British) and some separate brigades, supported and working in concert with a massive air offensive. Franks directed air and SOF contingents to handle the sparsely populated open deserts in the west, where Saddam's Scud rockets had operated in 1991. The United States wanted to bring one Army division through Turkey into the friendly Kurdish north, but at the last minute the Muslim Turks, despite being longtime NATO members, just couldn't agree to it. The Turks did permit some air transit and ground movements. Denied a Turkish avenue into northern Iraq, the Americans elected to rely on the sturdy Kurds, their CIA and SF advisers, and airpower, reinforced with a parachute drop by the 173rd Airborne Brigade. It all presented Saddam Hussein with multiple, simultaneous attacks.

As the Iraqis dealt with challenges on several fronts, the Army would make the main effort, charging up the Euphrates River. To the east, the Marines and British intended to go up the Tigris River, with the British holding back to focus on the coast, the major southern oil fields, and the key port of Basrah, Iraq's window to the Persian Gulf. The soldiers and Marines, paced by airpower, were going all the way to Baghdad. The planners figured on months of fighting. Tommy Franks strongly believed it would be over much faster.

As before, Franks did not spend much time on that phase-four stuff. His staff dutifully cranked out plans and charts aplenty, but they received little attention. As recounted by Major General William G. "Fuzzy" Webster, a key deputy in the campaign, "There was seriously not anything but a skeleton of Phase IV until very late." Maybe, as in Afghanistan, the follow-up could be done on the cheap. The key thing was to blow Saddam off the map. The rest might well take care of itself.

For his part, Saddam Hussein assumed another iteration of Desert Fox, a menu of cruise-missile strikes and air raids, at worst accompanied by a limited U.S. ground push into southern Iraq as in 1991. Well, they were all Shiites down there anyway. Saddam explained to his generals, "The Americans will use their airstrike methods, which they prefer and used recently, rather than sending troops, based on their horrific experience in Somalia." Rarely has a strategic leader so misread his opponent.

The United States, too, exuded certainty of victory, albeit on a much sounder basis than anyone in Baghdad. Taking down Saddam Hussein looked pretty likely. Then what? General Eric K. Shinseki offered a prescient warning that was largely ignored. The U.S. Army chief of staff was a Vietnam veteran badly wounded in that conflict. He had also served in the NATO peacekeeping operation in the Balkans from 1997 to 1998. During testimony to the Senate on February 25, 2003, Shinseki suggested that "something on the order of several hundred thousand soldiers" were needed to control Iraq, a country the size of California with a population of twenty-seven million. In public, an annoyed Donald Rumsfeld replied: "The idea that it would take several hundred thousand U.S. forces I think is far off the mark." He expected the Americans to be hailed as liberators. With Rumsfeld and a great many others banking on that happy thought, Operation Iraqi Freedom began.

6

APOCALYPSE THEN REDUX

We're going to grab him by the nose and kick him in the ass. We're going to kick the hell out of him all the time, and we're going to go through him like crap through a goose.

— GENERAL GEORGE S. PATTON JR.

AMERICAN SOLDIERS BEGAN their attack on Baghdad by digging all night. Dozens of sweating men with D-handled shovels pushed into the loose sand, turned it over, and cursed in the darkness. They weren't digging a tunnel. Having rolled through two linear packed-dirt border berms and then crossed two antitank ditches filled by the engineers, the men of the Second Battalion, Seventh Infantry found the going slow on the enemy side of the Kuwait border. Their thirty-three-ton M-2A2 Bradley infantry fighting vehicles and seventy-ton M-1A2 Abrams tanks kept bogging down in the viscous Iraqi sand. Each brief sprint across a hard patch led to a grinding halt in softer dust, some stretches of it gooey from recent rain. At each stop, the tank crewmen dismounted, the rifle squads piled out of the back end of the Bradleys, and the shoveling commenced. At one mile per hour, it was going to be a long, painful three-hundred-mile push to Baghdad.

The 2-7 Infantry and the rest of the Third Infantry Division had chosen this tough trudge through the desert west of the Euphrates River. Hardtop Highway 8 ran right along the river, with Highway 28 to the west. That was the straight shot, the obvious route. Looking at the known Iraqi unit locations, guessing about the unknown ones, the intelligence people predicted Iraqi defenses strung like four pearls on a necklace, anchored on the key river cities: Nasiriyah, Samawah, Najaf, and Karbala. Seventy-five miles north of Karbala stood Baghdad.

The Third Infantry Division expected to avoid most of those defenders by pushing well to the west, through the open desert. The opening round came off nicely, with Kuwaiti and U.S. engineers opening the clear lanes and the lead units heading north on schedule. To speed them on their way, at 9:00 p.m. on March 19, 2003, five battalions' worth of preparatory artillery fire pummeled the Iraq border guard towers. The 458 155 mm artillery shells and numerous 227 mm rockets upended eleven towers, killing many hostiles and scattering the rest. As the initial artillery barrage ended, scouts forward of 2-7 Infantry crept ahead to check things out.

Their thermal sights showed the glowing images of tanks. Iraqis? Must be, since there were no Americans this far north. The scouts maneuvered carefully, expecting contact. Two American Bradleys shot at four enemy vehicles. The Iraqis did not return fire. Daylight explained why. The targets turned out to be burned, rusted T-55 hulks, left over from the 1991 campaign. A good portion of the Desert Storm tank fighting happened just forward of the advancing Americans. H. R. McMaster's old 73 Easting position was just to the northwest.

Laboring across the desert, the Americans passed occasional dead enemies, crumpled bodies left behind when the surviving border guards fled. One particularly unlucky Iraqi died when hit directly by a box of surrender leaflets dropped from an Air Force C-130. The U.S. soldiers hoped the rest of the surrender papers were used as intended. Intelligence anticipated mass Iraqi capitulations. But first, the battalion had to get to hard, dry, flat ground. That took all night, and a lot of shoveling. Meanwhile, airplanes and helicopters passed above them, heading north, and the artillery kept banging away.

By morning, the battalion had cleared the morass. The next three days went by in a blur as the battalion's sixty armored vehicles (fourteen tanks, thirty Bradleys, and a collection of older M-113 series armored personnel carriers and M-9 armored bulldozers) clanked north. The battalion's tracked fighting vehicles shielded over thirty big ten-ton wheeled fuel, ammunition, and repair trucks, thirty more older five-ton types, and another forty or so flat, squat Humvees, the modern equivalent of the jeeps of World War II. Later in the war, all of those wheeled vehicles would feature armored cabs. But in 2003, the wheeled fleet had only thin metal and fiberglass between them and the enemy. Most of the Humvees and a number of the big trucks had ring-mounted machine guns. In the prevailing Army doctrine of the time, wheeled support trucks stayed well behind the frontlines. The idea was that they might race up under cover of darkness to refuel, rearm, and repair, then scuttle back to the safety of the rear echelon. In a three-hundred-mile attack, there was no rear area. So the trucks moved with the tanks and Bradleys.

On this mission, the Second Battalion, Seventh Infantry constituted what the Army calls a mech-heavy task force. The Army refers to this as task organization, building a unit from familiar, somewhat interchangeable subunits to meet a particular battlefield need. Each part of the Army has strengths and weaknesses. Tanks can move, shoot, and survive heavy hostile fire but cannot hold ground. Infantry takes and retains ground and clears buildings, trenches, and bunkers, but the riflemen will take a beating from enemy fire, especially artillery. Engineers breach obstacles (like the border berm in Kuwait), dig field fortifications, enable river crossings, and improve roads. Artillery can kill people and damage things, but it isn't all that accurate; its shells spread a lot of fragments but don't always destroy hardened targets like buildings, bunkers, and tanks. Helicopters can deliver troops or devastating gunfire and rocket volleys, but they fly only a few hours at a time and can't take much punishment. Air support is immensely lethal but must be aimed carefully; it's like hunting mice with a pistol. If your aim is off, with big bombs, your side can suffer as much as the foe's. It all amounts to a giant military version of rock-paper-scissors. In combat, you want some of each.

Task organization is a longtime practice in armored formations, hearkening back to the Second World War German panzer *kampfgruppen* (armored battle groups) and U.S. combat commands. It seems like a lot of switching around, but the units train together regularly. Battalions swap companies. Companies trade platoons.

For Iraqi Freedom, 2-7 Infantry gave one rifle company to another battalion in First Brigade. In return, 2-7 Infantry received a tank company (Team Knight) and kept two of its own rifle companies (Team Rage and Team Bushmaster). After the usual back-and-forth, the tank company got two infantry platoons. The rifle companies each picked up a tank platoon. The battalion retained its headquarters company, with the commander and staff, the medical platoon, the service elements, scouts, and mortars. Company B of the Eleventh Engineers was attached to breach obstacles, build barriers, blow things up, and do some light road and fortification work. All of the units knew one another well from previous training at Fort Stewart, Georgia, and the preparatory weeks in Kuwait. They'd run numerous live-fire exercises together. In short, they were a cohesive team.

The battalion's high state of training showed the high standards of its commander, Lieutenant Colonel Scott E. Rutter. A twenty-year infantry officer, Rutter had already fought the Iraqis. In 1991, he commanded a Bradley rifle company with the First Infantry Division. He earned the Bronze Star with V device for bravery under fire. Despite his prematurely white hair, Rutter had a great deal of stamina and fitness. He was smart, energetic, and extremely perceptive, with a real sense for getting ahead of trouble. He exuded an unearthly steadiness under fire, the type of guy who grows calmer under pressure and so keeps those around him focused on the task. There was another thing about the 2-7 Infantry commander. "He had an approved retirement when we got a deployment order," remembered one of his sergeants, "but postponed it rather than leave his troops as they were getting ready for a fight. That says something about a man."

Rutter expected contact at the initial border breach. For once, his prediction didn't pan out. After the passage, other units pulled in front of his. That had been the plan, to use Rutter's battalion at the border, then pass others through and save 2-7 Infantry for other firefights to come. Soldiers from the Third Brigade were already tangling with the Iraqi Eleventh Infantry Division at the first Euphrates River city, Nasiriyah. If you tuned to the right networks, you could hear snatches of the battle on the radio, and soldiers did so. Only one in fifty of them had been in a firefight before. Over the radio, combat sounded like training. Except when they called for artillery or brought in a medical evacuation helicopter; those were real shells and real wounded. The action wasn't that far away. But it was not 2-7's fight.

An army advancing across the desert in an armored column may sound romantic and exciting. It conjures images of George Patton up in a tank turret, goggles down, mouth in a grimace, holding on to the grips of a .50-caliber machine gun while a hundred tanks spread out behind him and dust rises in great yellow plumes. That scene certainly played out. Soldiers took it in, realizing they were doing something big, something powerful. It got the adrenaline going, all right.

Yet those exhilarating runs never lasted long. The column slowed or stopped. A vehicle began to falter. Or the radio crackled with an order to change speed or direction or to announce the next halt for fuel. An armored column on the move defies the second law of thermodynamics as long as it can. Sooner or later, though, entropy takes over. Complex modern military vehicles, especially tanks and Bradleys, have needs. Over time, those needs become issues. If those issues aren't addressed, the machines grind to a stop. Carl von Clausewitz saw friction in an era of horses and muskets. Modern armored forces intensified that friction dramatically.

Consider a cross-country ride in a Bradley fighting vehicle. To an unschooled eye, it looks like a little tank topped off by a small turret with a thin cannon protruding. It is not a tank. Rather, a Bradley is designed to keep riflemen under armor, moving up alongside tanks. It's that rock-paper-scissors thing. When the tanks need infantry, the Bradley ensures they're right there. The ride is rarely pleasant, but it's safer and faster than walking in the open.

The torsion-bar suspension allows a Bradley to cross uneven ground, but those inside feel every bump and jump. To the front, the big six-hundred-horsepower diesel engine drives the treads; on a hard-surface road, the thing can reach forty miles per hour. The driver is crammed into a narrow left front slot, shoehorned between the 25 mm cannon turret to his right rear and the engine to his right front. A driver can open his hatch and suck dust from the vehicles to his front or keep it closed and try to peer through thick armored glass blocks, each about the size of a brick. To the driver's right, the turret has the gunner and Bradley commander, both standing in a space equivalent to the area inside an old folding-door phone booth. They share that spot with a thermal sight, a 7.62 mm coaxial machine gun, and the back end of the deadly 25 mm chain gun that provides the Bradley's primary firepower. They too can keep the hatch open or drop down and use the superb thermal sight. The jouncing ride bangs them back and forth, hour after hour. The body armor helps some.

In the back, crammed knee to knee, sit six riflemen. They can't see much through the little vision blocks. They wear helmets and torso body armor, and they carry their rifles and machine guns, plus ammunition, water, first-aid kits, and small radios. Every rifleman uses eye protection, as in a chemistry lab, and gloves. At the very back, a big ramp forms the rear of the Bradley. When the ramp drops, ready or not, the riflemen unpack themselves and get out. Thanks to the flexibility of youth, they can do it pretty fast.

The first night at the berm, each Bradley was fueled and flush with ammunition; every tool, spare battery, smoke grenade, ammunition can, and extra water container was neatly stowed. Lined up and tied down along the outside on both flanks, on each vehicle, the soldiers' rucksacks formed an outer layer of padding. Any inbound round would hit the rucks first — there go the clean socks and razor blades. The neatly executed load plans matched the fondest dreams of exacting platoon sergeants. It was all dress-right-dress, neat and orderly, a place for everything and everything in its place. Then the Bradleys started moving, and entropy ensued.

At the first soft spot, the ramp went down. The riflemen tumbled out, and then in the dark they yanked the shovels out of their tightly strapped hold-downs. Some were put back, but most got tossed into the troop compartment. The riflemen knew they'd be needed again. Over time, people got thirsty. Canteens and water bottles came out. Some ate parts of the bagged rations, the MREs (meals ready to eat), dozens of packs and subpacks all neatly wrapped in plastic-covered foil. The tracked vehicles needed various oils. Cans were opened; empties were tossed overboard or went underfoot. Trash bags filled up. Some overflowed and were ground into the dirty metal floorboards. The ramp went down again. The riflemen got out. More shoveling, and some careful scans in the dark, hunting the absent enemy. So it went for hours, start and stop, as night became day.

Soaking up the brilliant Iraqi sun, the Bradley warmed up. Bradleys had fine heaters, but no air conditioning. The men in back, sporting Kevlar-and-ceramic helmets and vests, learned how a chunk of roast beef felt in a Crock-Pot. Various parts began to act up. A road wheel started to shed its rubber rim. A shock absorber seized up. A thermal sight started to fuzz out. Then came a halt. Ramp down. Soldiers tromping around clearing a culvert. Ramp up. More dirt got on things. The battery box spilled. The radio needed a different channel set. That disturbed the repair tools. A socket wrench escaped onto the floor.

Then came a fuel stop. Bradleys needed fuel daily. Tanks were even worse, sucking fifty-six gallons an hour. The ramp went down, and the riflemen got out to provide a security ring. The driver and gunner made some adjustments to a faltering track shoe. After a half hour, all mounted up. On it went, day and night, for three days. Some Bradleys broke down and had to be towed. None were left behind, as even the break-downs might become a source of spare parts. The accompanying five-ton trucks carried a lot of spares, but often they lacked the one thing you needed most. Stop and start. Entropy ate its way through the column like slow rust.

The heat, the roar, the dust, the grime, the rollicking garbage, and the stink of overloaded bodies pressed too close together — it wore on you. You didn't really sleep. You dozed, pressed to your mates, drifting in and out. The trash bounced around at your feet. Armored warfare in a Bradley resembled living in a dumpster. Except that when that ramp went down, you fought.

•

The ramps went down at dusk on March 22 as 2-7 Infantry halted along Highway 8, twelve miles short of Samawah, a city of 150,000. An RPG whizzed past a scout Humvee. The scouts reported a vehicle fleeing into the little town of Al Kindr, just ahead.

Scott Rutter checked his computer display. Remnants of the enemy's Eleventh Infantry Division were sixty miles to the southeast. The Republican Guard's Second Medina Armored Division was thought to be well north, up between Karbala and Baghdad. Looking at Samawah on the bright screen, Rutter could see the current markers for the Third Brigade battalions up ahead, along with the division's Third Squadron, Seventh Cavalry. They were fighting somebody; now it looked like 2-7 Infantry was too.

Rutter sent Team Rage into Al Kindr. The army calls this "developing the situation." It amounts to hunting with very big guns. The rifle company commander led with his infantry, riflemen spread out on foot, covered by Bradleys, backed by a platoon of four tanks. It took a while. The light faded away. Whether by chance or by choice, the power in Al Kindr cut out, dousing the town's few lights.

In the rear of the stopped column, Team Knight's First Sergeant Wilson Rodriguez saw an overturned SUV on the side of the road, its wipers scraping the windshield. That looked odd. Rodriguez directed a rifle squad to dismount. Together, the men checked an Iraq gas station. In the gloom, they saw the glint of expended AK cartridges scattered on the cement near the disused pumps. In the office, a teapot was still warm. As the Americans moved around the building, rifles up, two Iraqis jumped up, tossing aside a corrugated metal roofing sheet, and raised their hands. They then pointed to the garage. Rodriguez and the riflemen moved slowly. As they neared the second building, six Iraqis walked out, hands up. All eight prisoners wore uniforms and gave up their AK-47s.

At the front of the column, the scouts and Team Rage found no such cooperation. Out in the waving swamp grass between the town's low buildings, two armed Iraqis crawled. They figured the Americans couldn't see them in the long grass. Thanks to their vehicular thermal sights and individual light-intensifying goggles, the U.S. soldiers saw both Iraqis clearly. The scouts hesitated, waiting for instructions. On the radio, Rutter calmly ordered: "Engage, over."

The scouts and Team Rage's riflemen opened up. A few shots knocked down both hostiles. "Stop shooting," directed First Lieutenant Stephen Gleason. "They're dead."

Silence followed. In the inky darkness, ten more Iraqis, AKs in hand, rose up like wraiths from the tall rushes. Evidently, they didn't know that the Americans could see them plainly. They began to step forward through the gently rustling stalks. The enemy moved slowly, gingerly, as if tiptoeing. Gleason's scouts knew what to do. Team Rage joined in. Another series of single shots, and then a Bradley's machine gun ripped off a long burst. Six Iraqis fell over as if struck with baseball bats. The other four stumbled, waving their arms. They had had enough. The scouts secured the four prisoners, all wounded to some degree. Not one wore a uniform.

That got some attention at the 2-7 aid station when the Iraqis arrived under guard for treatment. The battalion's intel people, the S-2 section, sent a man over to take a look. So the Iraqi soldiers didn't fight, but the Iraqi civilians did. Well, that made sense. Saddam Hussein had decided after Desert Storm that in the next round, his people would fight their way, tribal-style, in the hit-and-run Bedouin tradition. Selected outfits, like the Eleventh Infantry Division, stood — and failed — Western-style. But these wounded guys, eyes burning with hate, weren't soldiers. They were irregulars, Fedayeen Saddam, Quds (liberators of Jerusalem) Force, and Baathist Party militia. On paper, the three groups should have fielded hundreds of thousands; instead, they turned out hundreds. Unlike Saddam's regulars, though, these people very much wanted to fight.

Saddam Hussein thought they all served him. In the dozen years after the 1990–91 war, his subordinates had passed weapons around liberally, especially in 2002, when American action became imminent. Baathist Iraq never lacked for arms and ammunition. The local Shiites in and around Samawah didn't take the AKs and RPGs in order to protect Saddam, whom they detested. A number of Shia Arabs had another motivation, much closer to home, thanks to their coreligionists, friends, and active backers in Iran. The ayatollahs' agents advised Iraqi Shiites to gather all the implements and use them to battle the infidels, the double-dealing Americans who had encouraged the uprising in 1991 and then stood back and watched Saddam crush it. If the Shiite militia stopped the U.S. invaders, fine. Saddam would owe them, and they might get a deal. Even if Saddam somehow endured, Shia Arab irregulars would possess the necessities to repeat the 1991 uprising and win this time. Enough took arms to make the reception hot for the invading Americans. It came as a nasty surprise to Rutter and his soldiers, who had been told over and over to expect a much kinder reception.

Following the two engagements, the battalion gathered up fifty detainees, mostly militiamen. As each hour passed, though, the ever-present need for fuel grew. Rutter's soldiers marked time, popping at occasional miscreants, real or imagined, out in the high grass. Finally, at 11:00 p.m., the column cranked up again and rumbled toward Samawah. When he spotted a few armed enemy soldiers kneeling on an overpass, Team Knight's forward observer called in 155 mm artillery rounds. The big shells hit dead-on, bowling over the Iraqis.

The artillery impacts did the job and distracted the next unit in the order of movement. Perhaps fixated on the dead opponents, maybe looking down inside the tank or just plain exhausted and confused, the tank platoon soldiers from Team Knight turned right when they should have gone left. All the global positioning systems in the world can't reverse four seventy-ton M-1 Abrams tanks gone astray. The lieutenant knew it almost immediately, the key word being *almost.* It was too late.

Within minutes, the first tank wedged itself into an alley, its big hull stuck into adobe walls on both sides. Despite the late hour, Iraqis poured into the streets. Some wore uniforms, most had AKs, and a few had PKM machine guns. From a rooftop, a man started zipping out RPGs, one after another. The fat explosive projectiles smacked into the American side armor, too close for the fuses to arm. Enemy bullets rang off the massive tanks. To the rear, Bradleys and dismounted riflemen inched up, trying to provide cover. The tank company commander thought he counted hundreds of Iraqis in the parallel dirt streets, darting in and out. A bunch gathered right in front of the stalled vehicles. Muzzle flashes sparked all around. The guy on the roof kept shooting RPGs. Team Knight couldn't move forward and couldn't easily back up. And the soldiers dared not shoot the 120 mm main guns, because on the tracking displays, it looked like friendlies ahead in the town center.

Still out on Highway 8, Scott Rutter judged carefully, estimating based on the map and experience. He cleared his men to fire. Use the machine guns, he told them. Those bullets shouldn't reach the nearby U.S. units. With that, the tanks began to fire back, lengthy, insistent bursts, three-quarter-inch-long 7.62 mm bullets and inch-and-a-half .50-caliber slugs. The guns chewed strips out of the adobe and divided the mob like Moses parting the Red Sea. Engines roared. One by one, now unmolested, the tanks backed out, treads clanking. Part of a wall collapsed. The streets were full of rolling dust, with bodies sprawled here and there in the dirt. Refueling went quickly at the next stop. Nobody said much.

Rutter's battalion ended up outside Najaf for almost a week. It wasn't a pleasant stay. During the battalion's first full day patrolling west of the city, an unseen Iraqi marksman killed Specialist Gregory Sanders. The soldier was on top of his M-1 tank, trying to tie down loose equipment. The local farmers seemed exceptionally well armed. Scott Rutter ordered his riflemen to search nearby farmhouses. Many floors featured discarded Iraqi army uniforms. The patrols saw very few adult men, just an elder hobbling here or there.

On March 25, a line of violent thunderstorms blew in from the west, bringing lightning, rain, and an intense sandstorm. Skies turned dull red and visibility dropped to a hundred feet, sometimes less. Incessant rains carried the blowing dust down to the ground in fat, muddy drops. The foul deluge coated the already grimy tracked vehicles and big trucks with a layer of slimy sand. Soldiers slipped and slid as they tried to refuel tanks, change blown tires, and clean off gritty ammunition linked belts for the machine guns. Even the drinking water tasted like dirt.

When the storm blew out, the militia shooters returned. The battalion's patrols cleared various compounds. Struggling to keep the tracked vehicles running, Rutter's soldiers jerked and started over pocked roads and open, bumpy desert. Refuel stops dragged out as the mechanics and drivers banged and twisted to remove stuck parts and jury-rig fixes. The units depleted the battalion's stock of common spare parts: fuel filters, batteries, track shoes, road wheels, shock absorbers, and the like. Absent a major resupply — not scheduled, with the bulk of the repair items still way, way down the route near Nasiriyah — more than one 2-7 Infantry tank or Bradley would be limping into Baghdad.

When night fell, movements became treacherous. Yet the battalion had to keep pressing north to make up for time lost in the sandstorm. Individual companies hopped their platoons one by one through a patchwork of farms, berms, and drainage canals; several vehicles lurched into wide irrigation ditches. One of the tanks in Team Rage nosed into a wide depression. A trailing Bradley's crew wasn't so lucky. Trying to skirt some abandoned mud-brick buildings, the thirty-three-ton tracked hull tipped into a wide, dry well sixty feet deep. The Bradley crumpled some, bow buried in the hard dirt, the smashed turret nearly vertical against the far wall. Heroic efforts saved the men who'd survived the jarring impact. They could not reach Sergeant Roderick Solomon, who died in the crash. It took the engineers and mechanics, working various winches on two enormous fifty-six-ton M-88 tank retrievers, well into the next morning to recover Solomon and pull up the wreck. The smashed vehicle was stripped for parts.

Two days later, soldiers at a roadblock north of Najaf got a bitter taste of the future. When four Americans approached a man sitting on the side of the road, he explained by gestures that he could not walk. He pointed to his ankle, just below the hem of his dishdasha (an ankle-length one-piece robe), the biblical-style ensemble that the U.S. troops called a man-dress. Almost on cue, an orange-trimmed white Iraqi taxicab drove up. The cabby spoke English. That wasn't too odd. Many Iraqis spoke some English. Schools taught the language, and some Iraqis picked it up, more or less, by watching the flood of pirated Hollywood movies smuggled into the country. The talkative cabdriver indicated he had come to pick up the lame civilian.

Rifles leveled, the Americans told the driver to get out and open all the doors for a search. Typically, Iraqis given such instructions drag their feet and grumble. This guy sprang right out, Mr. Helpful. He swung open the two doors on the driver's side. Then he went to the trunk.

When he pulled on it, the car exploded in a hot, loud fireball.

Both Iraqis died instantly. So did all four Americans: Sergeant Eugene Williams, Specialist Michael Curtin, Private First Class Diego Rincon, and Private Michael Creighton. From that point on, Rutter's men stopped searching civilian cars and trucks. If they approached, 2-7 Infantry security teams fired a warning burst. Then they shot to kill.

Over the next few days, the battalion column pushed north, mile after mile. Other Third Infantry Division battalions, the Air Force, and the Apache helicopters smashed up the Republican Guard Medina Division, leaving nothing but scraps for Rutter's people. The long string of tanks, Bradleys, M-113s, and trucks passed Karbala and finally crossed the Euphrates River early on April 3. Ahead lay Baghdad.

The Third Infantry Division turned to the First Brigade. None of the division's three maneuver brigades was fresh — far from it. Second Brigade, Third Brigade, and the division cavalry squadron did most of the hard fighting around Karbala. When First Brigade grabbed the Euphrates crossing southwest of Yusufiyah, it put a lot of American combat power thirty-five miles south of Baghdad. Up higher in the chain than the division, nobody really wanted to get into a knife fight with these swarms of fedayeen types in the Baghdad neighborhoods. But the international airport was out on the western edge of the city surrounded by a lattice of hamlets and open farmland. With the Medina Division gone, the airport looked like low-hanging fruit ripe for the plucking. The map showed it as Saddam International Airport; the dictator didn't miss many chances to exalt himself. An American brigade on those runways, within easy artillery range of downtown Baghdad, might just trigger the airport namesake's final collapse.

Tired they might have been, but the First Brigade soldiers saw the finish line. The Third Battalion, Sixty-Ninth Armor would make the main thrust into the facility. Rutter's 2-7 Infantry, in relatively good shape, got the task to support the other battalion by securing a key four-way intersection just east of the airport. Both battalions moved out as the sun set on April 3.

The people in 3-69 Armor found a pretty good path, at least part of the way, and made it to the airport in a few hours. Not so Scott Rutter's battalion, which found itself in another convoluted rat's nest of canals and walled farms and roads just wide enough for a cart. It was just like the area where that Bradley had taken a fatal plunge a few days earlier. With a slim crescent moon, there wasn't much ambient light. The night-vision goggles blanked out. Thermals worked, but the rural landscape didn't offer much temperature contrast. It all looked like a hazy curtain. So the men moved forward the hard way. Riflemen and engineers on foot were often out front, leading the column slowly north. Scott Rutter was right there with them, in his Bradley. Soon enough, he was out walking along the narrow cart track too.

The 3-69 Armor waited south of the fifteen-foot airfield perimeter wall, ready to breach and enter when 2-7 arrived. They waited. And they waited. The 2-7 Infantry move slowed and stopped. Tanks creaked across elevated farm trails, hanging over on both sides. The churning treads barely gripped the crumbling, sandy soil. One after another, slow-moving Bradleys scraped house walls and knocked off overhanging roofs. Some M-113s got stuck. Ammunition trailers jackknifed. Shovels went to work, men sweating as they turned stubborn dirt. The column lurched up a little. Then a canal bridge collapsed. Entropy, friction, exhaustion, you name it — it enveloped the battalion like a stifling miasma. The entire effort seized up. Rutter and Team Bushmaster were cut off from the rest of 2-7 Infantry. Up ahead, 3-69 Armor pressed to get going on its own. Just before eleven that night, the brigade commander pulled the trigger.

The armor battalion blew open the wall and rumbled onto the airfield proper. The Iraqis melted away. Well to the south, hearing the commotion, knowing nobody held the vital blocking position, Scott Rutter decided to go with what he had. Team Bushmaster kept picking its way along. Behind it, the battalion S-3 (operations officer and third in command), Major Rod Coffey, and the rest of 2-7 Infantry slowly, painfully, backtracked and began looking around for another route. Laconic radio messages suggested minimal progress. Hour after hour, the drill continued. Drop the ramp. Get out. Find a way. Get in. Raise the ramp. Sporadic gunfire from 3-69 Armor echoed in the distance.

At last, at 3:30 a.m. on April 4, Rutter's lead element reach the key intersection, a huge highway interchange. The battalion called it Four Corners. That's how it appeared on a flat map display, but in the small hours before dawn, the men saw that it clearly had a significant vertical aspect. The cement-and-steel conglomeration loomed up out of the blacked-out city like a slice of the Los Angeles freeway plunked into western Baghdad. Even in the darkness, Scott Rutter could see that this interchange had already been ill-used: guardrails were peeled back, shallow craters dimpled the road, and pieces of twisted metal debris lay here and there. One four-lane divided street ran north to south, parallel to the two main airport runways. The intersecting four-lane highway ran east, right into the heart of Baghdad. Across that east-west avenue, about five hundred yards east of the airport boundary, a stout four-lane overpass carried another north-south highway. Access ramps trailed off that overpass into the east-west roadway. To the north and south lay darkened walled compounds. *Whatever the hell they might hold,* thought Rutter.

The battalion's job was to block Iraqi counterattacks onto the airfield proper. To the east, 3-69 Armor now had a hot firefight going. Evidently, the Iraqis had figured out that the U.S. had gotten on the runways. Somebody on the enemy side decided to do something about it. With only one company, in the dark, surrounded by God knew what, Scott Rutter kept Team Bushmaster together. He could always move the unit. Some engines or generators or something were running off to the north. Rutter put Bushmaster there, athwart those roads. A very long hour passed. Red U.S. tracers and green Iraqi ones laced the sky to the west out on the airfield.

About 4:30 a.m., Major Rod Coffey dragged in the rest of the column.

A few vehicles pulled others on tow bars. Some listed to one side. All wore streaks of dried mud. Attached equipment dragged on the ground as the vehicles rolled by. But the soldiers were up and alert despite the long, frustrating night movement. They were in Baghdad now. Bushmaster stayed in the north. Team Knight went to the east. Team Rage extended the ring to the south. The 120 mm mortar platoon and scouts covered the west. Engineers with small M-9 armored bulldozers and trusty D-handled shovels worked in each sector, knocking down walls to clear fields of fire. In the middle sat the command post, the aid station, and the service vehicles. The books say infantry exists to hold key ground. There was none more key than this.

Dawn brought trouble. As the advantage conferred by American night-vision goggles and thermal sights went away, 2-7 Infantry found out what was in all those walled courtyards to the north and south. The occupants announced their displeasure with a ragged series of mortar rounds. The enemy shells hit near the battalion command post. The explosions got everyone's attention.

Well, maybe not everyone's—on the access ramp to the overpass, a flat-topped, lightly armored six-wheel M-93 Fox chemical reconnaissance vehicle slowly rolled upward. It was unclear why these soldiers thought this seemed liked a good time to look for WMDs. A loud crack sounded, and something whipped by the Fox. The vehicle immediately lurched into reverse. There was an Iraqi tank on the opposite access ramp.

Team Bushmaster First Lieutenant Paul Mysliwiec immediately pushed his Bradley up the ramp to do that now-familiar deadly dance known as developing the situation. Along with its 25 mm cannon, which might or might not damage a tank, depending on where the rounds hit, the Bradley turret also had a pair of TOW (tube-launched, optically tracked, wire-guided) antitank missiles. One of those would destroy a tank. The TOW needed about sixty-five yards to arm, and it trailed thin little wires, like metal fishing lines, back to the launcher. Electric impulses along the wires served to direct its fins as the missile sped to its target. It could go more than two miles. The tank that took a whack at the Fox was a lot closer than that. Mysliwiec's Bradley climbed deliberately up the ramp.

The enemy T-72 tank sat on the opposite access ramp, just over the crest of the overpass road. Using the lay of the land to hide all but the tank's cannon and turret was known as *defilade.* This Iraqi tanker knew his craft. The hostile tank fired again, a sharp metallic crash.

The Bradley bucked as the enemy round glanced off its side. The unkempt, sand-covered rucksacks peeled open, dumping boots, dirty shirts, socks, and water bottles onto the pavement. The impact pitched Mysliwiec out of the turret, popping his thick communication cord like a shoelace as he tumbled off the reeling Bradley. Private First Class Wendell Gee threw the tracked vehicle into reverse and backed down the incline. The lieutenant followed on foot, shaken but still in the fight.

How do you get at an angry enemy tank in defilade? Any U.S. tank that went up the western ramps might well take a 125 mm T-72 round right in the flat belly, a vulnerable place, and so not good. But the U.S. infantry had a new weapon, used many times in training but just now getting its shakedown in combat. Called the Javelin, the fat cylindrical launcher looked like a golf bag with a box—the sight—stuck to it. It weighed about the same as a full load of golf clubs, and had the same awkward heft. The missile inside, however, had a useful feature. If you clicked the right switch on that sophisticated sight, the Javelin would zip toward the target and then, obeying its little onboard computer, pop straight up and come down like a spear, right through the thin armor on top of the enemy tank turret. Much newer than the bigger TOW, the Javelin had no trailing wires. You aimed, fired, and then grabbed the next missile in its throwaway launcher and stuck it on that clever sight box. The manufacturer, Raytheon, promised one shot, one kill. That morning in Baghdad, 2-7 Infantry tested the advertising.

Private First Class John Davis and Private First Class Jefferson Jiminez each grabbed a Javelin and angled up the access ramp. As they got near the top, they could see over to the diagonal approach road—no T-72. It had moved somewhere. The riflemen scanned slowly, using the bridge's height to peer into the walled compounds. There! Just over a tall wall to the south, they saw three T-72s in a tight bunch. Davis reared up and fired.

Scott Rutter and the soldiers below watched the missile race across the roadway, soar up, then drop behind the wall. A blast roared, a huge fireball erupted, and a tank turret flipped into the air. Secondary explosions went off, tank ammunition blowing up, tossing up more dark T-72 chunks. Davis fit another Javelin onto his sight and shot again. Straight out, then up and over, then down, killing a second T-72. Jiminez also fired, nailing the third tank.

To the west, the mortar platoon dealt with another pair of probing T-72s. Captain Matthew Paul got separated from his platoon and took cover in a palm grove within forty yards of the two Iraq tanks, both firing machine guns to kill the American officer. The Iraqi bullets went high. Even as the 120 mm mortar gunners ducked incoming small-arms fire and pumped out rounds to silence Iraqi mortars, Sergeant First Class Robert Broadwater ran over to an M-1 tank towing an inert, disabled mate. He pointed to the two T-72s, just barely visible sitting in the palm-fronted road. The Iraqis were consumed with trying to find and kill Captain Paul.

Dragging its useless partner, the U.S. M-1 tank demonstrated surprising quickness, shifting forward ten yards or so and then swinging its big 120 mm cannon right onto the first Iraqi T-72. Hugging the dirt under the Iraqi machine-gun bursts, Captain Paul heard the Iraqis yelling in Arabic. Two breath-stealing cracks, and two more frying-pan turrets spun up into the air. The hulks burned. His tormentors eviscerated, Paul stood up and sprinted back to his busy platoon. He got there just as another serious threat appeared.

To the north, through a hole punched earlier by the U.S. engineers, two hundred or so uniformed Iraqis began forming to push through. Major Rod Coffey pulled together the command post and service troops. At the medical aid station, corpsmen looked up from working on the wounded to fire back at approaching Iraqis. Mortar men alternated dropping 120 mm projectiles down their hot tubes and shooting their M-4 carbines. Iraqis stumbled and fell, but not enough of them. They kept coming.

The main enemy thrust aimed straight through the broken wall. The Iraqis hadn't performed to Fort Benning Infantry School standards, but the combination of mortars, tanks, and now Iraqi foot soldiers had put together just about enough rock-paper-scissors to get to the center of the 2-7 Infantry ring. One good push might do it. And these Iraqis were pushing hard.

With tanks and other dismounted Iraqis keeping the line companies busy all around the perimeter, Scott Rutter turned to the engineers. In the U.S. Army, engineers breach and build but always retain a secondary task to fight as infantry. That time had arrived for Company B, Eleventh Engineers. Led by Sergeant First Class Paul R. Smith, a platoon rolled up. To cork the yawning opening, a single M-113 headed directly into the hail of enemy bullets. The eleven-ton M-113 was old, a Vietnam-era design with thin aluminum armor and an unprotected .50-caliber machine gun on its top. But that was all they had, that and some twenty dismounted engineers who had dropped their shovels and leveled their rifles. For 2-7 Infantry, the entire three-hundred-mile push to Baghdad came down to this: an understrength platoon of tired, determined men squaring off against hundreds of equally determined Iraqis.

Paul Smith coolly directed traffic as if he were on a rifle range. He placed two other M-113s to cover the lone gap-closer vehicle. He motioned a headquarters Bradley into a good firing spot. He showed individual engineers where to fire into the open wall. Then he set the example — do as I do, in Army parlance. An engineer recalled: "He calmly stood in the open picking off attacking Iraqis one by one." An Iraqi RPG struck and ignited the focal M-113, the one holding the middle. Smith went in three times to pull out wounded soldiers. Iraqis clambered atop the walls to shoot down at the Americans. "It was non-stop shooting," remembered Private Michael Seaman. "The Iraqis just kept spraying fire over the walls, hopping up to fire RPGs and dropping mortar rounds on us." Smith glanced over his shoulder. The entire perimeter of Four Corners was engaged; he heard staccato small-arms fire, mortar bursts, and, intermittently, the flat boom of tank main guns, U.S. and Iraqi. The engineers had to solve this one themselves.

Smith dropped his left arm and pointed backward, signaling his men back to the next low wall. He himself moved to the smoldering M-113, squeezed into the driver's seat, and drove it right through the center of the hole. The boxy thing braked hard and rocked back, its blunt bow protruding onto the Iraqi side. There Smith could see and engage the Iraqis sheltering behind the stretches of wall on either side of the gap. The engineer NCO got out of the driver's seat and climbed up on the .50-caliber machine gun. He waved for help. Private Seaman ran forward and got into the driver's seat. The other engineers, now with some of Rod Coffey's headquarters people integrated, continued to blaze away at any Iraqis in the opening or atop the flanking walls.

Smith ripped off long, steady bursts with the .50-caliber. Pauses were short. The big machine gun hammered away. It dated back to 1933, but its eminently reliable, simple design had ensured results in earlier wars. It did the job for Paul Smith that morning. Five times, Seaman passed up cans of linked rounds, the gleaming belts of golden cartridges and wicked, copper-toned inch-and-a-half slugs. If one of those brutal items hit you anywhere, even clipped your fingertip, it would bowl you head over heels. If it punched through you, you were done. The Iraqis lacked body armor. After five cans of .50-caliber, the attack lacked Iraqis.

Inspired by Paul Smith's heroism, the engineers and the 2-7 headquarters soldiers advanced and finished the foe. There were few prisoners. When they counted the bodies later that day, they found about a hundred dead around the damaged M-113. It was hard to be sure of the exact number, as it depended on how you tallied up the gory partial remains. Paul Smith died that day, and he would receive the Medal of Honor posthumously. That critical morning in Baghdad, at the crucial juncture, he and his resolute men stopped the enemy.

Scott Rutter and his people knew the worm had turned. You could feel it. After Smith's stand, 2-7 Infantry had the momentum. The men spent the rest of April 4 cleaning out each of the other neighboring compounds in turn. As he had done all morning, Rutter stayed out on the ground with his riflemen. He joined Team Rage in clearing one barracks complex, the usual room-to-room dirty work. Along with rifles, machine guns, and grenades, the battalion used its tanks, Bradleys, mortars, supporting artillery, and close air support. The Americans knocked out two more tanks plus eight other armored vehicles, and they took about forty prisoners, a fraction of the hundreds of enemy they had engaged that day. The number of Iraqi dead, in addition to those spread around Paul Smith's M-113, could only be estimated. There were a lot.

The intelligence folks confirmed a few days later that 2-7 Infantry had taken on and destroyed one of four brigades of the elite Special Republican Guard. Rutter's battalion continued to fight over the next few days. Building on the foothold at the city airport, the Second Brigade, Third Infantry Division mounted two thunder runs right into downtown Baghdad, sending a battalion on April 5 and the entire brigade on April 7; 2-7 Infantry got into a very sharp clash backing up that second one. Two days later, cheering Iraqi citizens and helpful Marine tankers took down Saddam's statue in Firdos Square in east Baghdad.

To the many Desert Storm veterans, like Tommy Franks and Scott Rutter, the first three weeks of Iraqi Freedom served to restore the deleted scenes from the 1991 production. In this director's cut, the Americans pressed all the way to the Iraqi capital and deposed Saddam Hussein. By any standard, it was an impressive performance, a three-week blitzkrieg that tore through an opposing army of some 350,000. Famed British military historian John Keegan summed it up: "The Americans came, saw, conquered."

In the thrill of victory, Americans, especially soldiers, made a little more of the campaign than it was. *On Point,* a pretty good official U.S. Army history of the war through May 1, 2003, appeared within a year. Drawing on a very thorough survey of documents and lots of interviews, the authors told as complete a story as they could. They did not shy away from recording ugly, bloody mistakes, such as the wrong turn through Nasiriyah that resulted in the flaying of the 507th Maintenance Company and a botched night helicopter attack on the Medina Division. But such incidents were exceptions to the narrative — and, of course, to the highly successful campaign.

The curious part of the book involved the maps. In this, the official Army reporting mirrored the initial press coverage, which portrayed Iraqi Freedom as being similar to George Patton's sweep through France in 1944: battalions outflanking hostile positions, penetrations of fixed defenses, and tank-on-tank smashups. Looking at the standard military symbols on the maps, a reader would get the impression that the Iraqi military fought as formed units and that the various militias followed a neat organizational wiring diagram and executed traditional ground maneuvers. Sometimes that happened, as at Four Corners on April 4. But usually, the only forces maneuvering were the Americans and British. The Iraqi cycle of order-counterorder-disorder made many enemy movements and tactics as inept and incoherent as lice hopping on a hot griddle. Keegan pointed out that the Iraqi army "faded away" and that "their resistance had simply been without discernible effect."

Yet despite clear and rapid success, the advance proved anything but easy. For the Third Infantry Division, there had been spots, stretches, and successive incidents of vicious resistance from the Kuwaiti border all the way to Baghdad. The Marines and British faced it too in the east. Speaking to reporters in the wake of the sandstorm of March 25–27, Lieutenant General William Scott Wallace, who commanded the Army forces moving up the Euphrates, observed, "The enemy we're fighting is different from the one we war-gamed against." Wallace meant the irregulars.

As the Fifth Corps commander, Scott Wallace started the campaign with three major outfits: the Third Infantry Division, the 101st Airborne Division, and a brigade-plus of the Eighty-Second Airborne Division. He ended up committing both airborne contingents to secure the routes. On the Tigris avenue of approach, the Marines left the British to clean up the area around Basrah, Iraq's second city. As the Marines went north, they too dropped off security forces. Saddam's regulars fought harder than in 1991. There were few major surrenders. That said, the half-trained, ill-led, Saddam-directed Iraqi military, lacking any air support, really had no prayer against trained, disciplined American and British soldiers working in concert with overwhelming airpower.

The more dangerous Iraqi threat turned out to be the militias. Saddam guessed right after Desert Storm. Iraqis made excellent ambushers, raiders, and deceivers, melting in and out of the civilian populace. The conventionally educated Iraqi commanders did not know how to harness these tactics and the consequent enthusiasm, and thus most Iraqi battalions and brigades tried to fight in the Western style. As one captured Iraqi general put it, "Most commanders understood the nature and theory of modern warfare but in Iraq it was in conflict with the tribal nature." Focused on what they considered legitimate foes like the Eleventh Infantry Division and the Medina Division, the Americans shot up, ran over, and slaughtered the militia. That worked to get to Baghdad and remove Saddam's central government.

It had been a terrific campaign, one for the military textbooks. Yet as the dust started to settle in Baghdad, thoughts circled back to the war's basic question: Who was the enemy? The Iraqi army was gone. The Republican Guard, too, had disintegrated. Yet there hadn't been many cheers for the liberators. Worse, that long, awful series of knife fights with citizen irregulars did not bode well. Somebody out there was still shooting at Americans. And most of them had nothing to do with al-Qaeda.

American citizens rightly measured the outcome of Operation Iraqi Freedom against the announced purpose of the campaign. Saddam Hussein remained at large, but he ran nothing. He joined Osama bin Laden in oblivion, and Saddam enjoyed little, if any, affection from the region's Islamists. True, the enemy of one's enemy can be one's friend. But in the Middle East, there seemed to be only different shades of enemies. Americans recognized that two of the war's goals had been achieved. Saddam would never again menace nearby countries or his fellow Iraqis.

Nor would Saddam Hussein himself bankroll, train, or direct any more terrorists. The Americans and British found plenty of documentary evidence of Saddam's terrorist efforts. Up north, the SOF and Kurds attacked and destroyed an element of Ansar al-Islam, a known al-Qaeda affiliate. Nobody found an underground lair with Osama bin Laden in it, but for those hunting terrorists, Iraq offered fertile pickings. Saddam's ties to al-Qaeda were not dissimilar to the tenuous links between Nazi Germany and Imperial Japan in World War II. They were not close, but there were some connections, and they were clearly on the same side.

Of course, Saddam's terrorist inclinations would have been a lot more threatening had those operatives been given weapons of mass destruction. After April 2003, Saddam's WMD plans were clearly at an end. But as days became weeks, and weeks became months, and multiple search teams scoured Iraq, it became evident that Saddam's nuclear, biological, and chemical programs were decrepit and damn near illusory. In a country that had buried its air force fighter jets, docked its few pathetic navy patrol boats, and maintained factory after factory that produced nothing, a mostly nonexistent WMD stockpile accorded well with prevailing practice. It wasn't accurate to say there were no WMDs. There were some chemical munitions scattered here and there, leftovers from years before. Certain laboratories were screwing around with nuclear notions, biological ideas, and chemical possibilities, but nothing substantial had appeared. Diligent U.S. survey elements even turned up some yellowcake uranium at the old reactor site at Tuwaitha, southeast of Baghdad. And certainly, the Iraqis had Al Samoud and Ababil-100 missiles that exceeded UN guidelines in range; the Iraqis fired seventeen of them during the three-week campaign. One Ababil-100 almost wiped out the Second Brigade, Third Infantry Division command post during the thunder run of April 7, 2003. There wasn't exactly nothing to the Iraqi WMD lineup. But those expecting to find long rows of gleaming, combat-ready munitions were sorely disappointed.

Iraq's weak ties to al-Qaeda and almost nonexistent WMDs both became very divisive domestic political issues in Britain and the United States. Supporters of the invasion pumped air on the embers. Detractors refused to see anything there. It all made minimal difference in the end. The greatest Iraqi terrorist and weapon of mass destruction departed Baghdad on April 9, 2003.

The American victory celebration took physical form on May 1. Like an updated version of an ancient Roman triumphal procession, President George W. Bush donned a flight suit and climbed aboard a U.S. Navy S-3B jet. On a brilliant sunny morning, off the coast of California, he flew out to the aircraft carrier USS *Abraham Lincoln* (CVN-72). He became the first serving president to make an arrested fixed-wing landing on a carrier. A one-time Air National Guard jet interceptor pilot, Bush clearly relished the moment. He stripped off the flight gear, shrugged into his suit jacket, and strode to a podium on the gray flight deck. Above and behind him, on the carrier's island superstructure, a long banner read MISSION ACCOMPLISHED. It was there because the carrier was finishing a ten-month deployment, four months longer than the normal stint. The sign wasn't for Bush, but it fit the mood.

Bush launched into a short, bold speech full of superlatives. He praised the U.S. military and its allies. Near the outset, he stated, "Major combat operations in Iraq have ended." He reminded all of the earlier victory in Afghanistan and the continuing American mission there "to help the Afghan people lay roads, restore hospitals, and educate all of their children." He promised the same in Iraq. "Our coalition will stay until our work is done," he vowed. "Then we will leave, and we will leave behind a free Iraq." The president offered cautions about more hard work ahead and no certain finish date. Those caveats got lost in the applause and the bright sun. In the overall war on terror, with the removal of Saddam Hussein, Bush said, "We have seen the turning of the tide."

He concluded with a tribute to those killed in action and then thanked the sailors for their service.

White House officials Andrew Card, Michael Gerson, and Karl Rove clearly had orchestrated a dramatic scene. But in more than one way, the real producers were Aeschylus, Euripides, and Sophocles. You just sensed where this thing was headed. The imperious nineteenth-century German statesman Otto von Bismarck said it well: "You know where a war begins but you never know where it ends." Bismarck understood. Rhetoric, signage, and fond hopes aside, the Iraq campaign did not end on a carrier deck.

PART II

HUBRIS

THE IRAQ CAMPAIGN

APRIL 2003 TO DECEMBER 2011

Mene, mene, tekel, upharsin.

— DANIEL 5:25

7

"MISSION ACCOMPLISHED"

But I suppose you're not too keen on what we've now got to do, Glatigny. Afraid of getting your hands dirty, perhaps?

— JEAN LARTÉGUY, *THE CENTURIONS*

EVERYBODY NEAR THE Balad Canning Factory knew Fouzi Younis, but nobody knew where to find him. At least, nobody was willing to talk about it. Fouzi Younis was the sort of fellow who motivated his neighbors to remain silent. Indeed, he gruesomely silenced a few just to encourage the others to stay quiet. When he told Iraqis to shut up, they believed he meant it.

They did not believe the Americans meant anything they said. Oh, yes, the infidel riflemen possessed the mighty thirty-three-ton Bradley fighting vehicles with their all-seeing thermal sights, stout armor plate, churning treads, and deadly 25 mm cannon. Moreover, the men themselves were superbly fit, consistently alert, well trained in battle drills, and excellent marksmen. As one former Iraqi general put it: "The American soldiers are very disciplined. They fight like robots and engage and kill everything on the battlefield." Well, not quite everything — with a very few very sick exceptions, the U.S. troops confined their work to armed opponents. There was no shortage of those around Balad, Iraq, in January of 2004. As for the tough Americans, they were spread pretty thin, and it didn't seem like they'd stay long, maybe a few months or a few years. Fouzi Younis wasn't going anywhere.

As a Sunni Arab, Younis benefited from plenty of local sympathy. The families around the canning plant were uniformly Sunni, and they looked after their own. Balad city proper was mostly Shia. The two Muslim sects distrusted each other, to say the least, and Saddam Hussein fanned those flames. He'd always favored his own bunch, the Sunni, who made up about 18 percent of the Iraqi populace. Under the Baathists, the majority Shia (60 percent) had always taken the pipe in the head. Now, with Saddam gone (captured on December 13, 2003), the Shia flexed their muscles, although not too much in Balad, a Shiite island of one hundred thousand surrounded by a million Sunni in the Salah ad Din Province. It was Saddam's home turf. Indeed, he had been taken not far from Tikrit, sixty miles north. Shia Arabs trod lightly outside Balad. As well they should. The Sunni Arabs cared little for the fallen dictator. But they hated the Americans who had displaced the Sunni from power, shut down their military and civilian sinecures, and installed the hated *sheruggis,* the easterners, a reference to the strong Iranian influence on Shia Arabs. Fouzi Younis preferred killing Americans, but slaughtering Shia would do.

The Americans in Balad definitely wanted to nab Fouzi Younis but didn't know how or where to find him. Although well trained and fully armed, the soldiers of the Second Battalion, Third Infantry were new to this area. They were on loan from a different U.S. Army brigade, destined to head way up north to Mosul, near the Kurdish territory. By taking over Balad for a few weeks, 2-3 Infantry freed up the experienced First Battalion, Eighth Infantry to go twenty-five miles north to clean up the restive city of Samarra. The 2-3 people were renters, not owners, and they had more pressing business in Mosul. So when an IED (improvised explosive device, a roadside bomb) blew up on December 18, 2003, and killed seven Iraqi police (IP), the temporary tenants needed some help to sort it out.

That help arrived on the afternoon of January 2, 2004. Company A, 1-8 Infantry rolled in aboard their dirty, banged-up Bradleys. Now they returned to their former compound in eastern Balad city. The small fort, surrounded by sandbags and dirt-filled wire-and-canvas Hesco containers, still hosted their affiliated engineers, Company B, Fourth Engineer Battalion. The relationship between the two companies was strong, built on eight-plus months of daily patrols and raids. The engineers had been assisting 2-3 Infantry. They were glad to see the return of their familiar teammates.

With Company A came two more scarred but ready Bradleys carrying the battalion tactical command element and the commander, Lieutenant Colonel Nathan Sassaman. The charismatic lieutenant colonel had a reputation, and he lived up to it. Eighteen years earlier, tall, broad-shouldered Nate Sassaman had quarterbacked the West Point football team to wins over Air Force and Navy and then on to its first-ever postseason victory against Michigan State in the 1985 Cherry Bowl in Pontiac, Michigan. It's part of U.S. Military Academy lore that during World War II, Army chief of staff General George C. Marshall (himself a graduate of the Virginia Military Institute) said: "I have a secret and dangerous mission. Send me a West Point football player." Nate Sassaman was just the kind of guy Marshall wanted.

Sassaman was very close to his soldiers. He was out with them constantly. He rode to the sound of the guns and led the way under fire. When Staff Sergeant Dale Panchot was killed by a direct hit from a rocket-propelled grenade that shattered his chest, Sassaman was right there. He was often the man dragging his wounded to safety and killing marauding insurgents. On all-night patrols, Sassaman walked with his squads. He got his hands dirty.

The battalion commander emphasized action to preempt hostile moves, and swift, immediate reactions to enemy attacks. Sassaman's riflemen rounded up suspects by the dozens. When Sunni sheikhs threatened the battalion, Sassaman detained twenty of the elders and held them for weeks. The 1-8 Infantry ringed one recalcitrant Sunni village with barbed wire and systematically went through every hovel, removing weapons, arms, and tens of sullen Arab militiamen. Enemy mortar shots prompted return 120 mm mortar fires within a minute, followed by a thorough search of firing areas and all nearby structures. An IED explosion resulted in an immediate roundup of all adult male Iraqis in the area. A gunshot would be returned by accurate, aimed fire, plenty of it. In the words of the battalion S-3, Major Darron Wright, "In our eyes, nothing was too extreme. We were at war." The battalion's soldiers had an attitude, aggressive and uncompromising. They reflected the commander. Not every man loved the demanding lieutenant colonel. But they all recognized that in a fight, he'd be there.

Getting Fouzi Younis might well lead to a fight, a big one, so Sassaman arrived with his Company A. Captain Matt Cunningham, also a former West Point football player, although much younger than Sassaman, would lead his riflemen on the raid. Engineer Captain Eric Paliwoda, a hulking six-foot-six former West Point discus thrower, would command his men on their usual supporting tasks: blowing open locked doors and dealing with any IEDs. And fighting as infantry, of course—the engineers always stood ready for that eventuality.

For Sassaman, Eric Paliwoda was more than the engineer commander. All the firefights, the daily grind of 120-degree heat, the grime, the monotonous food, the week after week of eighteen-hour days, and the costs imposed by his own driving personality took a toll on Nate Sassaman. Killing people is corrosive, as is seeing your own killed. It eats at you. Most battalion commanders confide and let loose with their majors or their command sergeant majors. In 1-8 Infantry, that happened, but as time went on, Sassaman got more tired, and more frustrated. He pulled back into himself. The senior guys couldn't reach him anymore. Sometimes, a battalion commander can relate to the colonel above him, the one who commands the brigade. That didn't work for Sassaman. He had little in common with his brigade commander, a much more deliberate officer. Despite a fourteen-year age difference, the lieutenant colonel and the engineer captain hit it off. Paliwoda became Sassaman's sounding board. The battalion commander summed it up: "Eric ended up being my closest friend in Iraq." Indeed, big Eric Paliwoda was well known and well loved across 1-8 Infantry.

Now Paliwoda, Cunningham, and Sassaman each huddled with his men to plot a portion of the night's raid. It had taken two weeks to finger Fouzi Younis. The IPs dead in the IED explosion had included the able deputy chief Lieutenant Colonel Nasser Hamid Ahmed. Popular and well-respected by the police and the Americans, Nasser ran his own string of local snitches. He had survived by having eyes and ears in and out of town. With Nasser dead, the other IPs fumbled some trying to figure out who did it. It had taken a couple of shakedowns and some squeezing and a few things the Americans probably did not want to know about to pin the tail on this donkey, Fouzi Younis and his ten followers. By January 2, the Iraqi police were pretty sure Younis and his men lived in the nondescript company apartments at the Balad Canning Factory south of the city. It was Sunni territory, badlands all the way.

Founded in 1974, the Balad cannery was a "state-owned enterprise," which was a euphemism for a fake factory. For years, it had made tomato paste. But in Saddam's latter days, it just fell into disuse. Workers gathered for Baathist-sponsored meetings or to get their minimal salaries and food rations. Baathist Iraq had three green stars on its flag, for unity, freedom, and socialism. Saddam's boot guaranteed the first; the second was honored in the breach; but the third existed all over. *We pretend to work, and they pretend to pay us,* as the joke went in the old Soviet Union. So, too, in Iraq. Plentiful oil money ensured a food basket and a fuel ration for every family, both suitably greased by bribes and side deals. As a rule, the Sunni got more than the Shiites. Baathists got more than non-Baathists. Saddam's insiders got the most. By 2004, no tomato paste came out of Balad. But the thousand or so Sunni workers passed the days in their shabby government-built apartments, getting by on odd jobs and handouts from the Americans and their Iraqi collaborators. Evidently, spurred by Fouzi Younis, some had found other things to do.

The Iraqi police identified the apartments of Younis and his ten followers. At 11:00 p.m., the 1-8 Infantry planned to lock a cordon around the housing area. Then riflemen, engineers, and IPs would kick in the doors and drag out the targets. An Iraqi informant, his face hidden by a ski mask, would identify the men taken. It was an old drill to 1-8 Infantry. In their manhunts, the special operators in the Task Force increasingly used technical means, drones and electronic sniffers. But at the infantry battalion level in 2004, manhunting was done the old-fashioned way, methods familiar to the Roman legionaries in Jerusalem in A.D. 33. You got a source, rounded up the suspects, and pawed through them until you found the one you wanted.

These heavy-handed measures worried some in the U.S. chain of command. Blowing through houses made America no friends, and even in Sunni parts of Iraq, not everybody was an enemy (yet). Sassaman's brigade commander quizzed a company commander before a December operation in Samarra: "How many are you planning on knocking on the door," asked the colonel, "and how many are you planning on kicking in the door?" The captain replied, "Uh, we have been kicking in the doors . . . or following five pounds of C-4 [plastic explosives]." Caught in their nightshirts, surrounded by armed, armored Americans, Iraqis rarely fought back. The women wailed and the children cried. A lot of blankets got torn, and rows of crockery got broken. It wasn't pretty, but it worked.

So the riflemen cleaned their weapons, the engineers assembled their breaching charges, and gatherings of officers and sergeants and their IP counterparts examined computer-printed overhead photographs and marked the key buildings. It should go just fine. And the source was certain he had Younis located. The enemy counted on 2-3 Infantry being confused, and it had been two weeks since Younis disappeared. It was unlikely Younis and his group realized that Sassaman's people had come back for the raid.

It was about four thirty in the afternoon, six and a half hours to go, when — *wham!* The command post shook; dust sifted down. *Wham! Wham!*

"Incoming! incoming!" The Americans and the Iraqi police dropped to the floor. Those outside hit the dirt. Fourteen mortar rounds hit, one after another. If 1-8 Infantry had been running the area, the 120 mm mortars would already be firing back. Under 2-3 Infantry, nothing happened. Nate Sassaman immediately ordered his Bradley readied to go out and look for the enemy shooters. A platoon of Matt Cunningham's Company A with some attached engineers were already out the gate.

As the shelling ceased and Sassaman ran toward his tracked vehicle, he saw men down in the open vehicle parking area. Four soldiers sat up, dazed, bloodied, cut up, but okay. But two didn't move. His helmet and body armor rattling, Sassaman raced over to check on the casualties. One of the prostrate men was Eric Paliwoda.

The large man was wearing his flak jacket, but a hot splinter had gotten through. Blood seeped from under the Kevlar vest. Paliwoda wasn't conscious. Sassaman tried to find and stop the bleeding until two medics shouldered him aside. Rank be damned, expertise counted more, and the lieutenant colonel deferred to them. He leaned close to Paliwoda: "Hang in there, man." The medevac helicopter was already overhead. Sassaman mounted his Bradley and moved out, following Company A.

Three hours later, in the dark, the riflemen returned. They had found a dug-in mortar base plate in an open dusty field, but no firing tube, no ammunition, and no insurgents. As the column motored back to the small base, the radio call came. "Eagle Six [Sassaman], this is Eagle X-Ray [the battalion command post]. We have some bad news, over." Paliwoda was dead.

Sassaman did not take it well. When he got back to the compound, he exploded. He stripped off his Kevlar vest, sweaty and streaked with blood — Paliwoda's blood — and kicked it. He cursed the 2-3 Infantry commander, who stood there mute and took it. What else could he do? Sassaman raged and roared, a strong, emotional man racked with grief. As Darron Wright put it, "He reached his breaking point."

It happens thus in every war. Thousands of years ago, the rage of Achilles led to the climax of Homer's *Iliad*. When the Trojan Hector killed Achilles's friend Patroclus, the Greek warrior sought out Hector, slew him, and tied the Trojan's body to his chariot. Achilles then dragged Hector's corpse through the dust around the walls of Troy. Even among the cruel Greeks in a much more primitive age, this was over the top. Yet rage can do that to a soldier. And it did it to Nate Sassaman. Achilles sought his own personal revenge. Sassaman knew better, and he backed off. But certain of his loyal 1-8 Infantry myrmidons got the word, and they did not.

The raid went off late, at two in the morning. Fouzi Younis was nowhere to be found. His family allowed that they had heard of him but of course stated that they had no idea of his location. Eight men were detained. Two fought back — odd, that — and were killed by U.S. gunfire. Some of the myrmidons might have evened up the score already.

By midday, Nate Sassaman, Darron Wright, Matt Cunningham, and all of the tactical command post and Company A moved out for Samarra. Operations there had not let up during the raid interlude. People knew about the mortar attack, Paliwoda's death, and the severe wounding of Staff Sergeant Todd Moyer. Captain Todd Brown's Company B patrolled aboard their Bradleys, moving slowly along Power Line Road, on the southern boundary of Samarra. Rifle squads walked forward to each intersection, then brought up the vehicles. *You cover, I move;* a common tactic.

Just after noon, the enemy chose to spice things up. A blue car halted about two hundred yards away from the company, right near the long line of one-story storefronts that flanked Power Line Road. A man shouldering an RPG launcher jumped out. How the guerrillas loved that thing! It featured in many of their recruiting photos and videos, maybe because it seemed to offer some hope against the American armored fleet. Each RPG made two sounds in rapid succession: the sharp, flat pop of its rocket motor, and the duller, louder blast of its warhead striking home. Todd Brown and his men heard two RPGs go off. The rounds zipped by — you can see them — and kept going. Back a block in the column, mounted in his tracked vehicle, Brown watched the grenades pass by. He couldn't see the shooter.

Staff Sergeant Ashton Legendre could see just fine and immediately returned fire. The 25 mm chugged its rhythmic bursts, the high-explosive incendiary rounds tearing into the car. The blue sedan lit up, dirty black smoke rising. Two other nearby cars, uninvolved but too close for Legendre to avoid, were also pummeled. The RPG shooter was cut to pieces, as were his four partners. Some of the nasty little exploding shells whizzed into the backs of small shops, tearing strips out of the concrete and scattering the wares. The incendiaries started some small fires in the buildings. Iraqi civilians ran away or ducked behind cover. They had learned all too well what to do.

Todd Brown had heard 104 RPGs before these two. He knew what to do also. Another of 1-8 Infantry's tough young West Pointers (and former first captain, the top-ranked military cadet in his class), Brown ordered up another rifle platoon to reinforce the Americans returning fire at the intersection and to block enemy escape. After giving that direction, he took off his combat-vehicle-crewman helmet with the built-in headphones, grabbed his Kevlar dome, a backpack radio, and his M-4 carbine, and loped out the back ramp of his Bradley. He had to get to the fight, to join his riflemen and Bradleys near the RPG engagement.

As usual, Nate Sassaman was on his way too. He and Darron Wright and their security team got there after the shooting stopped. Legendre's well-placed return fire had put paid to the enemy. When Sassaman arrived, Brown's riflemen had secured the intersection and were busy questioning Iraqi bystanders. Iraqi police showed up to assist. One of the insurgents was still alive, although just barely. Sassaman's medic tried to keep him that way, but the man died. The locals helped to douse the fires. Many Iraqis stared at the three shredded automobiles and the sprawled dead. As things settled down, a few Iraqi elders came forward. They alleged that two uninvolved civilians had died in the firefight.

Maybe they had. The enemy wore civilian clothes. Most of the Sunni locals sympathized fully with the bad guys, so who was truly clean in Samarra? Brown summarized: "We did hit two innocent bystanders but smoked the RPG team—that is how it goes in the city fight. What are you going to do, not shoot back?" The myrmidons shot back.

In Matt Cunningham's Company A, one platoon did more than shoot back. The night after the Balad raid, in the wake of Paliwoda's death, well aware of the ongoing fighting in town, First Lieutenant Jack Saville, Staff Sergeant Tracy Perkins, and their platoon were patrolling northern Samarra. The curfew descended at 11:00. All Iraqis had to be inside. Anyone outside could be detained or, if he or she acted at all off, shot dead.

Around midnight, the Americans stopped a beat-up white truck. It carried two male cousins, Marwan and Zaydoon Fadil, both in their twenties. The truck had no weapons, just some plumbing fixtures in the cargo bed. What were these two guys doing out after curfew?

The U.S. riflemen at first waved at the Iraqis to go on home. Then the platoon sergeant motioned them to hold up. He told his men to get the Iraqis out. That done, he had the soldiers put plastic flex cuffs on each of them. Perkins turned to the lieutenant. "Somebody is going to get wet tonight," he said.

Orders stated that those detained after curfew were to go to the U.S. base. But Perkins, Saville, and the platoon chose not to follow that direction. Soldiers pushed the cuffed prisoners into the back of a Bradley. The ramp closed with a clank.

When it opened again, the Fadil cousins were prodded out into the chilly January night. The Bradley was stopped right near the bank of the Tigris River. It was about forty-five degrees or so, pretty cold for Iraq. Armed Americans escorted the pair to the base of a big dam that also served as a bridge. The water was a few feet deep and looked calm. A soldier cut off the hard plastic cuffs. The Americans pointed to the water and leveled their rifles. The message didn't require translation: *Get in.*

The Iraqis struggled, to no avail. They tried begging. No luck there, they saw. The black rifle muzzles pointed right at them. They both waded into the cold water. The Americans drove away.

In the ensuing rounds of investigations, hearings, testimony, and courts-martial, the truth, or a generally accepted version, came out. Marwan survived. Zaydoon drowned. Perkins and Saville had resorted to this method before to teach a lesson, they said. As Specialist Ralph Logan put it later, "It was like giving someone a swirly." He meant that it resembled the high-school bully's tactic of sticking someone's head in a toilet and flushing just to let him know who was boss. It was a way to demonstrate the pecking order, to show the Fadil cousins who ran Samarra.

The rumor mill cranked up immediately. The Military Police Criminal Investigation Division (CID), the Army's detectives, got involved as early as January 4. Perkins and Saville at first clammed up, claiming nothing had happened. When Matt Cunningham told Sassaman about it on January 7, the battalion commander told him and Saville, "Don't say anything about the water." Exhausted, heartsick, and frustrated, Sassaman spoke hastily. Paliwoda's dried blood was still on his flak vest. The battalion commander had shot and killed enemies and sent his own soldiers to their deaths. To make a big deal about this thing, maybe just a couple of Iraqis lying to get a compensation payment, well, he just could not do it. Sassaman backed his men. Much later, his division commander, Major General Raymond T. Odierno, told him, "You just wanted to be one of the boys." Not wanted to be — he was. In the band of brothers that was 1-8 Infantry, Sassaman and his men, his faithful myrmidons, were one and the same.

Within days of the incident, under CID questioning, Ralph Logan and some others talked. They admitted that they had put the two in the water but said both got out fine. When the Fadil family produced a body two weeks later, the American soldiers became adamant. Nobody drowned. Nobody died. It was just a nonlethal punishment. The Iraqis never permitted an autopsy, so no definite proof existed that Zaydoon perished. Iraqis had been known to fake deaths to get U.S. compensation money or even to conceal the survival of a terrorist-cell leader. Still, it became quite clear that something bad had happened that cold night at the Tigris River. Although it was not yet public knowledge, the U.S.

senior generals already knew of a very ugly October 2003 incident at Abu Ghraib regarding detainees. The 1-8 Infantry case got attention all the way up to the U.S. headquarters in Baghdad. Soon enough, the American press got wind of it.

Perkins and Saville went to courts-martial, were convicted, and served short prison sentences. Five others, including Sassaman and Cunningham, received nonjudicial punishment under article 15, mostly for concealing evidence or obstructing the investigation. For the lieutenant colonel and the captain, the proceedings finished them off in a professional sense. Both left the Army.

It is easy to sit in a climate-controlled room and wag your finger at Sassaman, Cunningham, Saville, and Perkins. One of the perils of fighting alongside your soldiers, of leading from the front, is that you climb aboard a potentially deadly emotional roller coaster. In the era of Achilles, that was well understood and accepted. Yet even that Greek hero, blinded by his volcanic sorrow, went too far. In 2004, few got it, even those in uniform. Not enough senior people were out on the ground, seeing the elephant. The war in Iraq happened at a thousand tiny crossroads, in a thousand flyblown villages, frequently at night, in searing heat or drenching cold rain, and was often over in a few random minutes. As for any civilians trying to empathize, well, forget it. Of the few citizens who sought to understand, too many saw Sassaman and his soldiers as ill-used victims, pathetic dupes of some neoconservative conspiracy hatched in the wood-paneled conference rooms of the White House, Halliburton, and Enron. The problem wasn't just that 1-8 Infantry crossed a dark, twisted line, as they surely did. Confronted with the same messy, dangerous situations in other villages, in other battalions, a great many American soldiers found themselves balancing precariously along that same grim line.

That line looked pretty faint and hazy in Samarra and Balad, but it wasn't all that clear back in Baghdad, let alone Washington. Nate Sassaman hadn't just wandered into trouble. His commanders put him there, and wanted him to stay. The great liberation had gone awry within months. It all came down to the usual question: Who was the enemy? Around bad areas, as 1-8 Infantry discovered, the answer was only too simple: everybody.

It may well have seemed that way. There were gradations, of course. The Sunni Arabs in the center and west, 18 percent of the population, clearly backed what they liked to call the honorable resistance. The Kurdish north, about 20 percent, went with the Americans strongly, as did the trace elements, such as the fire-worshipping Zoroastrian Yezidi, the Turkmen, and the Assyrian Christians. The Shia majority of 60 percent, concentrated in the south, tolerated the Americans but chafed as the occupation persisted over the years; Iranian-backed anti-American militias reared up now and then, with bloody results. As a whole, neither Sunni nor Shia Arabs wanted large numbers of infidel outsiders in their towns. Both cut deals of convenience, the Shia more often than the Sunni. But regardless of sect, Iraqi Arabs wanted to run their own country their way, even if that didn't exactly correspond to the fondest hopes of the United States. Winning over a majority of twenty-seven million Iraqi people to U.S. goals was no easy proposition.

Along with the sullen, unhappy general public, none too thrilled by thousands of foreign troops in their midst, there were active enemies. Like Iraqi Arabs as a whole, the bad guys came in two basic flavors, Sunni and Shia. Estimates of hostile strength ranged up to sixteen thousand. They were a quarrelsome, shadowy bunch, burrowed deep into the civilian populace.

Month after month, year after year, most of the opposition stemmed from Sunni Arabs, those who rejected the American occupation and the consequent Shia-dominated government. Across Iraq's eighteen provinces, the consistent hostile action germinated in the Sunni strongholds of Anbar, At Tamimi (the Kirkuk area), Baghdad, Diyala, Ninewah, and Salah ad Din, and it sometimes spilled over into Babil south of Baghdad city. Sunni rejectionists made up most of the attackers going after Americans like the 1-8 Infantry. A man like Fouzi Younis was typical of this element.

In the aftermath of Saddam's initial departure from Baghdad in April 2003, hopeful U.S. intelligence officers opined that "former regime elements" made up the bulk of the Sunni resistance. Units received multiple sets of playing cards embossed with Baathist biggies, the enemy brain trust conveniently depicted on a deck of fifty-two cards, to help soldiers recognize the men if they saw them. Saddam, of course, was the ace of spades. Though the SOF diligently played 52-pickup throughout 2003, that didn't curtail attacks against U.S. forces and their allies. True, most of the Sunni opposition had lived well under Saddam. Many had served in the old Iraqi army, the Republican Guard, or the Baathist ranks. But the deposed dictator, bouncing from one hiding place to another, did not direct or sustain them. When his sons Uday and Qusay died under attack in Mosul in July of 2003 and then Saddam himself got fished up out of his spiderhole in December of that year, the various Sunni insurgent networks slowed down not a whit. They didn't work for Saddam. They barely worked with each other.

Sunni Arabs did have one ally, although not by choice. Cooperating with the Sunni and sheltering under their wings was a more virulent threat, the rapidly coalescing cells of al-Qaeda in Iraq (AQI), a 2.0 model franchise headed by the implacable Jordanian firebrand Abu Musab al-Zarqawi. The average Sunni cells laid IEDs, fired RPGs and rifles, and shot mortars and rockets. Zarqawi's AQI followed the Osama bin Laden template and developed periodic spectacular attacks. They brought in foreign fighters, not many, but enough to stage mass-casualty suicide car bombings, an AQI specialty. Zarqawi was happy to slay Americans and other Coalition members but was even more interested in slaying Shiites, whom he considered apostates and thus not true followers of Islam. Zarqawi longed to ignite a major Sunni-Shia conflict, bigger and bloodier than the one already percolating. Iraq's very name meant "the shore, the edge, the split point" — geographic (desert to mountain, marked by the two great rivers), ethnic (Arab to Persian), and religious (Sunni to Shia). Zarqawi intended to crack the country wide open.

The Sunni guerrillas and the AQI were ever present. The other kind of enemy, the Shia militants, came and went. They responded to the wildly anti-American cleric Moqtada al-Sadr, a sometime tool and full-time recipient of Iranian largess. When he got excited, the militias did too. Shiite areas usually tolerated the Americans, not out of love, but because the U.S. backed the Shia-dominated government that had emerged in Baghdad by mid-2004. Whenever Sadr called, though, the streets filled up in Shiite areas of Baghdad and the southern cities of Karbala, Al Kut, and Najaf. The British in Shiite Basrah tangled regularly with Sadr's Jaysh al-Mahdi (JAM, the Mahdi army, named for the great Shia hope, an avenging messiah who would supposedly arise to end time and save the devout). Iranian Pasdaran (Revolutionary Guard) and intelligence-service agents stirred the pot steadily and kept up a flow of weapons, money, and expertise. Over time, with Iranian help, JAM and its offshoots proved quite difficult opponents.

Had the Americans simply pulled out in the summer of 2003, Iraq would likely have collapsed into an even more brutal civil war than it did. Backed by Iran, with the Baathist superstructure destroyed by the American-led invasion, the Shia would probably have prevailed, and the Sunni would have knuckled under, fled, or been exterminated. The already autonomous Kurds looked to go their own way, tied loosely, if at all, to the authorities in Baghdad. Even with the Americans staying in Iraq for eight and a half years, this is essentially how it played out. Maybe that unsatisfying yet realistic result was adequate, an outcome acceptable as good enough. In the summer of 2003, though, it wasn't good enough. The U.S. wanted more, a democratic republic in Mesopotamia that backed America, particularly when it came to hunting down terrorists. President Bush had said as much on the deck of the USS *Abraham Lincoln* on May 1. That high-flown objective put Nate Sassaman and the 1-8 Infantry in a very bad place by January of 2004. They weren't alone.

Despite well-articulated public aspirations to establish a Jeffersonian government on the Tigris, the United States took a while to reorganize its war effort. The ill-planned phase four arrived right on schedule. In the absence of rehearsed, agreed-upon arrangements in country, improvisations arose. The first almost scuttled the entire enterprise. It had to do with the most basic military consideration: Who was in charge?

Almost from the start of planning for Operation Iraqi Freedom, General Tommy Franks and his USCENTCOM staff assumed the follow-on headquarters structure to have two levels above the various combat divisions. On top would be a civilian strategic headquarters, the Office of Reconstruction and Humanitarian Assistance (ORHA), under retired Lieutenant General Jay Garner, a Vietnam veteran who also oversaw the Provide Comfort relief effort in Kurdistan in 1991. ORHA prepared to go in, get the Iraqi army and police back to work, and put the Iraqis in charge of their country. "Coordinating" — a vague idea, not proper command authority — with Garner's ORHA was Lieutenant General Scott Wallace's Fifth Corps headquarters. Under Fifth Corps, U.S. troops would act as a shield while the Iraqis got back on their feet. Franks guessed the entire thing would take a few months, six at the most. It would be like the aftermath of Desert Storm, some cleanup and a U.S. withdrawal, perhaps with a follow-on advisory effort. The old Northern Watch/Southern Watch air policing was over, of course. What better leader to run all that than Jay Garner, who had done it well in the Kurdish north in 1991?

Iraqis made themselves available to take over. Characters like the oily Ahmed Chalabi, his cousin and fellow Western expatriate Ayad Allawi, former Iranian "guest" Ibrahim al-Jafaari, and another prior resident of Iran, Nouri al-Maliki, all hovered on the periphery, only too ready. Each of these men and his supporters were viewed by Sunni Arab Iraqis as pure *sheruggis,* Persian pawns, and, worse, U.S. collaborators. It would have been news to Iran and the United States that they were backing the same team. Well, Bush said you were either with us or against us. Little did he realize that in Iraq, a good number of Shiite leaders checked both boxes.

That key point became subsumed in a greater transition, one that surprised most of the senior uniformed Americans. Ten days after Bush's speech aboard the USS *Abraham Lincoln,* Ambassador L. Paul "Jerry" Bremer III took charge of a new strategic headquarters, the Coalition Provisional Authority (CPA). After less than three weeks in country, Garner's ORHA was no more. Bremer was in charge. In the ambassador's words, "I had the requisite skills and experience for that position." He did not speak Arabic, although he had served in Kabul, Afghanistan, from 1966 to 1968, which was something. His most notable assignment had been as ambassador to the Netherlands from 1983 to 1986. Bremer enjoyed close connections to the Bush White House. Now he was the president's man in Baghdad.

He acted decisively. Within days, Bremer issued CPA General Order Numbers 1 and 2, dissolving the Baath Party and all its social groups, the entire Iraqi army, air force, and navy, the assorted intelligence services, the Saddam-era militias, and the Republican Guard. None of this was coordinated with the U.S. military; indeed, Franks's plan had always counted on the Iraqi military to keep order. De-Baathification was announced as a fait accompli. It guaranteed Sunni outrage.

Bremer defended the measure as essential to convince Iraqis, especially the Kurds and Shiites, that times had changed for good. Of course, while many Baathists and military officers were Sunni Arabs, Shiites and Kurds also held such party membership. So did a wide variety of minor officials in education, commerce, and local government. Bremer argued that it was similar to removing Nazis in postwar Germany. That was true, except for three problems: bad as they were, Baathists weren't Nazis; Iraqis definitely were not Germans; and the war in Iraq wasn't over.

With Bremer's arrival, Fifth Corps also changed command. Newly promoted Lieutenant General Ricardo S. Sanchez stepped up from command of the First Armored Division to take over Fifth Corps in its expanded role as Combined Joint Task Force 7, a new military operational headquarters directed to coordinate — that slippery word again — with Bremer's CPA. Bremer wasn't officially in charge, and Sanchez didn't officially report to him. Given Bremer's assertive personality, that would not be a pleasant setup.

There had been another option, the one that most expected. Franks could have chosen to leave the Combined Forces Land Component Command (CFLCC) headquarters under Lieutenant General David McKiernan. An experienced three-star general, McKiernan had just run the Army/Marine offensive that took Baghdad. His headquarters featured numerous handpicked senior officers with a lot of combat time, a "dream team," in the words of General Jack Keane, the Army vice chief of staff. But Tommy Franks did not think the Iraq deployment would last long. McKiernan's CFLCC might well have business in another area of the region, such as Iran or Syria. Afghanistan continued to bubble. The dream team moved on, preparing for a game that never came.

Keane spoke bluntly at the time. "Let me get this right," he said. "We are going to take the last arriving division commander, who just got here a couple weeks ago, and put him in charge of the war in Iraq. That is what we are going to do?" It was.

The forty-nine-year-old Sanchez might have been the youngest lieutenant general in the Army, but he was not unqualified. A tank battalion commander in Desert Storm, a key joint staff officer in U.S. Southern Command, and a general officer commanding in the Balkans, Sanchez had just deployed his First Armored Division into Iraq. Dealing with Jerry Bremer looked to be difficult, but Sanchez had seen worse. He took over for what many, including General Tommy Franks, told him would be a speedy drawdown and departure.

Franks would not be around to see how his prediction played out. Wrung out after the Afghan and Iraq campaigns, he retired. General John Abizaid took over U.S. Central Command. A Lebanese American West Pointer and infantry officer, Abizaid spoke Arabic. He led his Ranger company on the combat jump into Grenada in 1983, served as a United Nations observer in Lebanon in the middle of the 1980s, and then commanded the U.S. paratrooper battalion that served in Kurdistan during Operation Provide Comfort in 1991. When Abizaid commanded the First Infantry Division in Germany from 1999 to 2001, one of his brigadier general deputies was Ric Sanchez. Sanchez referred to Abizaid as "my friend and superior officer." The men were close, and together they would see this thing through in Iraq.

In later years, many heaped scorn on Sanchez and Bremer. The self-assured diplomat and the prickly three-star got along poorly at best, although both men tried hard to work together. Bremer overpromised and underdelivered. Many complained that he just could not figure out how to get the Iraqis to run their own country. For his part, Sanchez found himself acting from day to day, trying to set the right standards and get out with his troops. He got shot at more than most at his level, and that mattered to the people under him. Sanchez was both the operational Fifth Corps commander and the strategic CJTF-7 commander, each of which was a full-time job. Neither was done very well. Critics referred to a lost year.

Maybe they were right. But it is difficult to see how different personalities might have changed that year much. Replace Bremer with Henry Kissinger and Sanchez with Dwight Eisenhower, cancel the de-Baathification orders, and the stark facts on the ground still sat there, oozing pus and bile. With Saddam gone, any voting would install a Shiite majority. The Sunni wouldn't run Iraq again. That, at the bottom, caused the insurgency. Absent the genocide of Sunni Arabs, it would keep it going.

John Abizaid saw this hard truth early on. It explained why he was not in a hurry to dump Ric Sanchez. The USCENTCOM commander recognized the intractable nature of the problem. He was among the first to refer to the situation in Iraq as "a classical guerrilla-type campaign," and he chose the words carefully. With that in mind, Abizaid went to the next logical step. The Americans could not defeat the Iraq insurgency. They could at best assist the Iraqis. "I believe in my heart of hearts," Abizaid said, "that the Iraqis must win this battle with our help."

Defining the size, scope, and nature of that help took about a year. Far from being "lost," for the Americans in Iraq, the twelve months after Saddam's fall became a steady struggle to define the problem and put solutions in place. It all happened in the face of growing enemy resistance and consequent Clausewitzian friction. By the end of that stretch, things were set for a long-term campaign.

Strategic decisions came first. With the *Lincoln* speech and Bremer's appointment, the U.S. committed to a lengthy effort in Iraq. That required a decision on how long individuals should remain in country. The Army went with a year. The Marines chose seven months at battalion level and twelve months for higher headquarters. The Air Force, Navy, and SOF used four to six months. But the question arose: Should they go one by one or by unit rotations?

In World War II, the rule had been to send units, then backfill casualties with single replacements. Once in theater, you stayed until the war ended. Korea began with units on an open-ended timeline, then shifted completely to individual fillers on one-year tours. Vietnam followed suit. The fixed individual tour, one man or woman at a time, is the basic U.S. military assignment method, aligning neatly with personal enlistment contracts and service obligations. About a twelfth of a unit departs every month, with some upticks or drops over the year. While the method is beloved of military personnel officers, as moving people one at a time offers maximum flexibility to meet immediate service needs, schooling standards, and promotion gates, the use of individual replacements makes a mockery of team training, unit cohesion, and close ties between leaders and led. Individual assignments treat people like interchangeable parts, widgets or screws to be shuffled from drawer to drawer. In combat under this system, new guys showed up one at a time, and often got shot all too quickly. That was bad enough. For units in contact with the enemy, it was worse. Paper strengths remained high, yet beneath that serene number was a constant churn, what Lieutenant General Lesley J. McNair called "the invisible horde of people going here and there but never seemingly arriving." It led to the constant need for retraining; discipline challenges among officers, sergeants, and junior enlisted people who didn't know each other well; and more casualties. Individual replacement was supremely efficient — every man in the right place, every place with the right man — and crudely ineffective.

The Vietnam veterans running the U.S. military in 2003 wanted no part of it. In Desert Storm, Somalia, Haiti, the Balkans, and Afghanistan, the military started the campaigns by turning off the personnel system (an indictment there, to be sure). The official terms were *stop-loss,* which suspended discharges, and *stop-move,* which locked people into their units. Battalions trained, deployed, and returned together. The Navy and Marines had used this method for years for fleet rotations. Pushed by the constant requirements of Northern Watch and Southern Watch, the Air Force followed suit with their air expeditionary forces. The British, who had relied on battalion swap-outs for decades in their imperial and post-imperial days, swore by it. In the summer of 2003, facing a lengthy Iraq counterinsurgency, the U.S. Army bit the bullet. One-year unit rotations became the rule.

A few hiccups marred the transition. The pushy personnel types insisted on changing out a good number of Army battalion commanders, as well as some one-star generals, in May and June. That explained why Lieutenant Colonel Nate Sassaman took over 1-8 Infantry in theater on June 17, 2003. At Army headquarters, uncompromising General Jack Keane put a stop to that. Unit rotations would be the standard. Stop-loss and stop-move became the norm. Once alerted to deploy, units would stay together until they came home. It permitted the Army in particular to sustain a long war.

That long war demanded more troops in order to swap out every year. In the summer of 2003, the U.S. Army had four divisions in Iraq: the 101st Airborne Division up north, the Fourth Infantry Division (Sassaman's outfit) in the center, the Eighty-Second Airborne Division out west, and the First Armored Division in Baghdad. The Tenth Mountain Division was in Afghanistan, and the Second Infantry Division in Korea. That used six of the ten regular Army divisions, an unsustainable number. Come the spring of 2004, the Army planned to swap out three divisions in Iraq. The Marines would handle the fourth, heading into Anbar Province to supplant the Eighty-Second Airborne. Another

Army division would go into Afghanistan, and the Second Infantry Division had to stay in Korea to deter the threats from the Communist north. So four Army divisions a year had to rotate, maybe for a decade. This arithmetic argued strongly for bringing in the U.S. reserves and drumming up some more allies. Both efforts bore fruit.

Former Air National Guardsman George W. Bush emulated his Army National Guardsman predecessor Harry S. Truman in the Korean War. Both presidents wisely called up the Guard and Reserve. Although their wars proved immensely unpopular, by bringing in men and women from every county in America, the presidents guaranteed those in uniform support month after month. American citizens may turn on a faulty policy, and most eventually do. But they won't turn on their neighbors. In contrast, President Lyndon B. Johnson did not call up the reserves during the Vietnam War, and the resulting defeat blew back on the veterans of that war and all of America. In the Vietnam War, activating the reserves probably wouldn't have brought victory, but not doing so helped ensure loss of public support and final failure.

In addition, unlike Lyndon Johnson, who tried halfheartedly to bring a few "other flags" into Vietnam, but very much in the spirit of Truman in 1950 and his own father in 1990–91, Bush sought and employed many foreign contingents. This demonstrated hard work by Secretary of State Colin Powell and the diplomats, nudged as necessary by Bush's personal appeals. Eventually, thirty-seven countries sent troops. The British, Koreans, and Poles headed multinational divisions and contributed multiple brigades. Other brigades came from Georgia, Italy, the Netherlands, Spain, and Ukraine. Battalions were sent from Albania, Australia, Bulgaria, the Czech Republic, Denmark, the Dominican Republic, El Salvador, Honduras, Hungary, Japan, Lithuania, Romania, and Thailand. Companies joined from Armenia, Azerbaijan, Bosnia-Herzegovina, Latvia, Macedonia, Mongolia, New Zealand, Nicaragua, Norway, the Philippines, Portugal, Slovakia, and Tonga. Platoons arrived from Estonia, Kazakhstan, and Moldova. NATO eventually contributed a small military training mission in Baghdad. The United States supplied a lot of equipment and support for the smaller participants, but those countries came, and most stayed for a good stretch. About 24,000 Coalition troops served in addition to an average of 130,000 U.S. personnel in 2004. In later days, pundits carped that the Bush administration didn't create a real coalition in Iraq. Canada, France, and Germany loudly stayed out. Those who came, though, held areas in the south and north and so freed up the Americans to tackle the embattled Sunni heartland.

The winter of 2003 was a tough one for Nate Sassaman's battalion and the Iraqis around them. Across the country, however, the clammy, rainy months appeared to indicate that maybe, just maybe, things were settling down. Saddam Hussein sat in confinement. The number of attacks dropped in the Sunni regions. The Shia militias were behaving. In his memoirs, Lieutenant General Ric Sanchez refers to this period as "a window of opportunity lost." He must have seen something, but few others noticed. After a few more Iraqi winters, it would become obvious that the enemy just fought less in the cold months.

The opposition remained maddeningly opaque. Intelligence people scratched their heads. A January 25, 2004, count of detainees totaled 9,754. Casualty estimates ran to at least that number. So, essentially, the U.S. and its partners had captured and killed the entire insurgency. Yet the bad guys were still out there.

The Task Force stayed at it. The SOF continued to refine their manhunting but still spent a lot of time waiting for actionable intelligence. They nailed Saddam and continued to tick through the deck of prominent Baathist figures. But it was slow and uncertain work. They just could not find the hostile leadership. Few of those in American custody were big fish. Zarqawi and his crew moved freely through the Sunni villages.

The American Special Forces and the British, among others, suggested a solution that had always worked in counterinsurgencies. Nate Sassaman relied on it at the Balad Canning Factory: Get Iraqi partners with you out on patrol. Build a new Iraqi army. Revitalize the Iraqi police. That effort began top down from Bremer's CPA in Baghdad, and bottom up in the various Coalition battalions. The CPA approach promised success in ten years; it was like hand-tooling a hundred Ferraris when the field wanted a hundred thousand Yugos. But the grassroots version out in the units started to get some response. This was what Abizaid had meant by the Iraqis winning it for themselves with U.S. help.

Iraqis on the ground could help sift the bad guys from the good. Massive roundups just didn't work, except to inspire more young Sunni Arab men to join the insurgency. The detention camps already contained enough pipe swingers, bomb planters, and pop shooters. The SOF teams and the conventional battalions all understood that they had to find the key hostiles, the spiders at the center of the network webs. Those spiders hid well.

Dealing with daily IEDs and a steady string of killed and wounded, small units grew impatient. Any attack began to generate prisoners, guys swept up in the reaction. If the enemy engaged the Americans, Iraqi males were taken. Some were questioned and released. Many were not. The population at the Baghdad Central Confinement Facility at Abu Ghraib rose steadily. In January 2004, during a relatively quiet week, an average of eighty detainees a day arrived. It overwhelmed the MP guards and swamped the MI interrogators.

The captive Iraqis proved anything but docile. They fashioned shanks and shivs from scrap metal, strong-armed cooperative prisoners, defiled their spaces with feces, spat at guards, flung human waste at the MPs, and twisted, bit, and bucked during cell transfers. Escapes were attempted. The leaders among the prisoners, often portraying themselves as low-level nobodies, insisted on continued resistance behind bars. A good number of detainees supported such actions. The military police had to endure all of this and limit their responses to the denial of privileges and solitary confinement.

At Abu Ghraib, several prisoners mixed it up with guards on October 18, 2003, led by a detainee with a smuggled pistol. A few of the MPs chose their own countermeasure, not unlike the 1-8 Infantry soldiers at the Tigris River. That night, five enlisted MPs pulled twelve Iraqi prisoners from their cells. They stripped the captives naked and then piled them in sexually humiliating positions. A week or so later, the same guards put a hooded man on a box with fake electrodes clipped on his fingers; the prisoner was told the wires were real, and if he stepped off the box, he'd be electrocuted. Three days later, the same MPs again stripped prisoners and put them in sexually embarrassing poses. This incident also involved K-9 police dogs. A trio of military intelligence soldiers participated. These abuses were not linked to any interrogation. The soldiers later explained that they were teaching the Iraqis a lesson, the same reason offered by the soldiers in 1-8 Infantry. The MPs, however, took a lot of pictures.

By January 14, 2004, Sanchez and Abizaid knew about it. They informed Secretary Rumsfeld and immediately initiated a series of investigations. The casual cruelty and gross poor judgment demonstrated by those involved in the incident particularly frustrated Sanchez, who had spent a lot of his personal time at Abu Ghraib and on detainee matters in general. Much later, when the story and the graphic photos broke in the press on April 28, 2004, confused critics — some not so confused but going down the trail for their own reasons — conflated the misconduct at the prison with the Bush administration's description of prisoners as "unlawful combatants" and rumors of terrorist captives spirited away to be tortured in unaccountable "black sites." It became accepted among many that the U.S. routinely tortured detainees, rivaling the severe methods of French paratroopers in the 1956–57 Battle of Algiers. The Arab Street ate it up, pointing to the photos as proof beyond doubt of infidel perfidy. It definitely didn't help the Americans' cause that Abu Ghraib was a notorious Saddam-era prison, which allowed opponents to tar U.S. soldiers with that brush too. A few dumb young MPs inflicted a lot of damage to the U.S., an egregious "own goal," as soccer enthusiasts might say.

To counter all the hyperventilation about torture, some perspective is in order. The U.S. military did not torture anyone. None of the subjects of the Abu Ghraib photos were killed or injured, although they were certainly denied basic human dignity. Interrogating a detainee requires some manipulation of his environment, and that can be done without denying a captive food, rest, or hygiene. To a civilian, being awakened at odd hours, having a leashed dog bark at you, being bombarded with loud noise, or enduring harsh white lighting might seem very threatening, as indeed these approaches are meant to feel to a detainee. But they are not torture. Much standard military training includes these same sensations. The CIA apparently has some different rules, but even they very rarely apply any enhanced interrogation techniques, none that are life-threatening. There have been crimes dealing with detainees, as at Abu Ghraib. We must be careful in ascribing criminality to the mere acts of detention and interrogation. Sadly enough, it all traced back to too much loose talk about unlawful combatants. At the outset, Vietnam veteran Colin Powell had warned the senior Bush people to be careful about all matters dealing with prisoners, but as in so many other strategic aspects, he was ignored. After Abu Ghraib, Bush and Rumsfeld belatedly adopted Powell's view. Orders went out: treat detainees as standard prisoners of war, no more and no less.

Abu Ghraib festered beneath the radar as the squads of investigators came and went during the chill Iraqi winter. The Americans, the British, and the other Coalition troops, now side by side with some new Iraqi partners, manned checkpoints and patrolled. At night, they raided. In later years, the American intelligence community in Iraq would look back and choose January through March 2004 as the benchmark months, "as good as it gets," in assessing hostile activity. In Vietnam, the metric of choice was the infamous body count, the number of dead Communists. In Iraq, while bodies were counted, the favored statistic became enemy attacks. Those averaged about twenty a day in the first three months of 2004. T. S. Eliot wrote that "April is the cruellest month." In Iraq, his words rang true.

8

WHAT HAPPENED IN FALLUJAH

> We are not only fighting hostile armies, but a hostile people, and must make old and young, rich and poor, feel the hard hand of war, as well as their organized armies. I know that this recent movement through Georgia has had a wonderful effect in this respect. Thousands who had been deceived by their lying newspapers to believe that we were being whipped all the time now realize the truth, and have no appetite for a repetition of the same experience.
>
> — MAJOR GENERAL WILLIAM TECUMSEH SHERMAN

"MEN, WE'RE ORDERED to withdraw from the city."

The company commander said it quietly. Still, he said it. They all knew what those words meant.

Corporal Ethan Place looked up from the scope atop his M-40A3 sniper rifle. He lay with his belly flat on the uneven cement floor, his tan desert camouflage gone gray with dust and stiff with sweat as well as some of his own vomit and some blood, thankfully not his. For six days, the corporal, his spotter, and a platoon of Marines had held this key vantage point, a three-story abandoned house overlooking the cemetery in the Jolan District of Fallujah. During his time there, Place thought he'd killed thirty-two insurgents. He'd hit more but couldn't be sure what became of all of them. And then there had been the five-hundred-pound bombs, the Cobra attack helicopters hosing the other buildings down the street with 20 mm cannon shells, all the other Marines banging away, then the two tanks that came up, and even some expert SOF snipers — killing machines, those guys — all in all, a lot of firepower. Yet even in that maelstrom, the aimed single shots told. At one point, Place and his spotter counted twenty-two enemy bodies lying stiff and broken in the dirt street. At odd hours, gaunt dogs slid from the shadows and crept up to gnaw on the remains. The lurking Iraqi foe quit coming out. They'd learned not to try to recover their dead. Place and the other snipers educated the enemy, one 7.62 mm bullet at a time.

Now here was Captain Doug Zembiec, the skipper of Company E, Second Battalion, First Marines, an officer fighting alongside his men from the start, announcing that after all of this, after seventeen Marines wounded of thirty-nine in the shooting house, after all of the dead suffered by the U.S. and inflicted on the enemy over the last month, it was over. Some big guys up the chain evidently had had enough. Marines must follow orders, but they don't have to like them. They hated these.

They called the April 2004 operation to clean out Fallujah Vigilant Resolve. To the tune of forty dead as they crunched and smashed to within a few days of securing Fallujah, Ethan Place and his brother Marines had proven plenty vigilant. But in Washington and Baghdad, others ran out of resolve.

The Marines pulled back. As is their tradition, they carried out every one of their dead and wounded. Behind them, Fallujah still belonged to the insurgents.

Several rungs up the chain of command, Major General Jim Mattis, the First Marine Division commander, gave voice to the frustration of men like Ethan Place and Doug Zembiec. "First we're ordered in, and now we're ordered out," Mattis growled to his chief of staff. Then he quoted Napoleon: "If you're going to take Vienna, then by God, sir, take it." Something had clearly gotten very fouled up in Iraq that dark April of 2004. The problem was bigger than Fallujah.

For the U.S. war effort, the spring of 2004 was the eve of key strategic transitions. Ambassador Jerry Bremer's Coalition Provisional Authority had worn out its welcome — if indeed it had ever been welcome — among the senior Iraqi civilians, already jostling for power as soon as the CPA went away. After much back-and-forth, which included strong

Iraqi demands for immediate, sweeping nationwide elections, Bremer and his CPA team crafted a reasonably rational hand-over sequence. The American-selected interim Iraqi government prepared to take over on July 1. An election in early 2005 would create a transitional National Assembly to write a constitution for approval by plebiscite later that year. Finally, by the end of 2005, an Iraqi government would be elected under the new constitution. It all depended on enough security to permit the rule of law rather than the rule of the AK-47.

The second major transition enabled security. After months of the military burdening Lieutenant General Ric Sanchez and his overworked, undermanned CJTF-7 headquarters with both strategic and operational responsibilities, mid-May would see the long-awaited creation of a four-star theater command to handle matters of strategy and policy and a three-star corps headquarters to run day-to-day operations. This would match the long-established practice in Afghanistan. Sanchez would initially run the four-star headquarters, Multi-National Force–Iraq (MNF-I), although he expected to hand over command to a successor four-star officer by summer. At the three-star level, Lieutenant General Thomas F. Metz and his Third Corps at first filled in behind the departing Fifth Corps in CJTF-7. Metz and his Phantom Corps from Fort Hood, Texas, looked forward to their operational role as Multi-National Corps–Iraq (MNC-I) come May 15, 2004. These top-tier reorganizations came on top of the overall force rotations across the country, as the initial invasion divisions, brigades, regiments, and battalions handed off to their successors. It promised to be a very busy spring.

Their enemies, of course, had their own timeline. During the cool, wet Iraqi winter, they'd planted fewer IEDs and dropped fewer mortar rounds, but the warming trend of spring saw more hostile actions. Perhaps misreading the winter lull as more of an opportunity than it was, Sanchez directed a series of local offensives under the rubric Valiant Saber. The idea was to keep up the pressure.

One particular thorn in the American side remained Moqtada al-Sadr, the Shiite cleric son of a beloved grand ayatollah martyred by Saddam Hussein. The sprawling rectangular Thawra low-income housing district in northeast Baghdad, home to two million Shiites (more than a quarter of the city's population), proudly called itself Sadr City in tribute to Moqtada's father, Muhammed. Young Moqtada did not have the scholarly devotion to inch his way up the ladder of the Shiite ayatollah hierarchy. But what he lacked in academic skills, he more than made up for with charisma. Already blessed with a famous name that gained him immediate credibility among Shia Iraqis, Sadr found his issue: the infidel occupation. "I'm just striving to apply the Sharia law," Sadr said earnestly. "Death to America!" he added. Sadr attributed Saddam's departure to Allah's will, which had at long last brought the Shia Arabs to power in the Land of the Two Rivers. Now, apparently, the great deed having been done, Divine Authority wanted the occupiers gone, and their collaborators too. Sadr's Jaysh al-Mahdi (JAM) kept up their own pressure, using press releases and targeted violence. Sadr himself was indicted by an Iraqi court in August of 2003 for murdering a rival Shiite cleric. When Jerry Bremer asked Sanchez to grab Sadr that month, the general chose not to do so. At the time, Sanchez dismissed the proposal as a potential "strategic blunder for us." The Coalition units in the south, like the Spanish in Najaf, the Ukrainians in Al Kut, and the British in Basrah, were surrounded by large, suspicious Shiite communities. The allies much preferred to negotiate with local JAM leaders. So Sadr remained at large. Whatever else it did, the conscious choice to avoid confronting Sadr in 2003 allowed Sanchez and his people to focus on the growing Sunni insurgency. The Shiite flank remained cranky but contained.

With the reduced Sunni troubles in winter, it looked like the early 2004 Valiant Saber initiatives could include not only operations in Mosul and Samarra but also a long-overdue reckoning with Sadr. Bremer and the intel guys thought maybe some squeezing might work. Sanchez agreed. "We know who his lieutenants are," Sanchez noted, "and whenever we have the opportunity, our forces will be more than glad to launch these kinds of operations." The general expected a reaction, a JAM upsurge in every Sadr stronghold, but he figured his forces were ready. "We can handle that," he concluded. Orders went out.

On March 28, the Sadrist newspaper *Al-Hawza* (meaning "the Islamic schoolhouse") was shut down by U.S. soldiers of the 759th Military Police Battalion. On April 3, Sadr's associate Mustafa al-Yaqoubi was detained. As expected, JAM trouble began to spin up in Sadr City and Shulla in Baghdad, as well as in Karbala, Basrah, Al-Kut, Najaf, and

Nasiriyah. The Shiites were stirring big-time, but with the Sunni still quiet, the new challenge could be met.

But then came the bolt from the blue, the type of incident that Carl von Clausewitz rightly called "the kind you can never really foresee," friction on a platter. On March 31, in Sunni Fallujah, four Blackwater security contractors, all U.S. military veterans, were ambushed, dismembered, and burned. A howling, grinning, celebratory mob strung up two of the charred bodies on the green-painted iron bridge over the Euphrates. Media outlets worldwide carried the shocking pictures. In the Fallujah neighborhoods, a strong U.S. response was considered unlikely; the enemy was *shwaretek:* "gutless."

So now CJTF-7 faced twin challenges: the expected one with Sadr's JAM, and the one that wasn't supposed to happen in Fallujah. Indeed, a few weeks before, the outgoing American commander, Major General Chuck Swannack of the Eighty-Second Airborne Division, had told reporters, "I'm discounting a very serious insurgency ongoing here right now." The Marines, newly arrived in Anbar Province, hadn't been so sanguine. Now came the desecration of the Blackwater foursome.

The graphic images seemed to demand a strong reaction. In Baghdad and Washington, senior leaders pounded the table. In country, the Marines, led by Jim Mattis and his three-star higher commander, Lieutenant General James T. Conway, argued for doing this right, not just dropping bombs and killing people. Anbar was the most Sunni-populated (97 percent) province in Iraq, the largest of the eighteen provinces, with most of its million-plus people living in the thirteen major cities clustered along the Euphrates River. Fallujah, the most eastern city, held two hundred and eighty thousand people. Sometimes called the City of a Hundred Mosques (forty-seven inside the city limits, another fifty or so in surrounding villages), Fallujah had long been a hotbed of antigovernment sentiment. The 1920 revolt against the British Mandate originated in Fallujah. Many imams in those hundred mosques preached the Wahhabi line in a style that might have brought comfort to Osama bin Laden himself. For many reasons, including lack of manpower, the Eighty-Second Airborne had manned no permanent posts inside the city. Now the Marines owned the problem.

The U.S. Marine Corps has a long heritage of fighting counterinsurgency campaigns. Between the world wars, Marines led U.S. efforts in Haiti and Nicaragua. Like most successful campaigners, the Marines learned to develop and rely on local forces and native intelligence sources to gain enough popular support, or perhaps acquiescence, to find the bad guys. The Marines collected their successful methods in the *Small Wars Manual*, published in 1940 and used ever since. In Vietnam, the Marines applied these kinds of techniques in their effective Combined Action Platoon program. The Communists greatly feared these local forces and had worked diligently, and with increasing effect over the years, to divert the Marines from the villages. Now the Marines brought this approach to the landlocked desert and river towns of Anbar Province. They intended to train and accompany Iraqi soldiers and police on patrol. They would help with civic action, building schools, wells, and clinics. And when enemies resisted, the Marines would fight, but always arm in arm with local Iraqi units. As Mattis put it, "No better friend, no worse enemy."

The slow, smart Marine methods might have worked if not for that gruesome tableau at the Fallujah bridge. In Washington, Secretary of Defense Donald Rumsfeld declared, "We've got to pound these guys." President Bush agreed. In Baghdad, Ambassador Jerry Bremer said to Sanchez, "I encourage a vigorous attack." Sanchez concurred. His operations deputy, Brigadier General Mark T. Kimmitt, told the press on April 1 that "U.S. troops will go in. It's going to be deliberate, it will be precise, and it will be overwhelming."

Overwhelming it was, but it turned out to be the Coalition command that felt the pressure. The JAM uprising arrived right on schedule but exceeded the intelligence predictions. Sadr's people outdid themselves. In Baghdad, a horrific street fight erupted in congested Sadr City. Newly arrived troopers of the First Cavalry Division went toe to toe with crowds of fanatic JAM militiamen. The Americans killed hundreds at the cost of seven of their own soldiers killed and sixty wounded. In Al Kut, the Ukrainians were swamped by Sadrists who grabbed the radio station and threatened the CPA compound. In Najaf, the Spanish backed away as Sadr's gunmen took over the key religious shrines. At the encircled Najaf CPA center, Ric Sanchez himself got involved in a rooftop firefight. Fighting in Karbala, Nasiriyah, and Basrah flared as JAM shooters roamed the streets. Except for the British, who dealt with matters in Basrah, the embattled allied contingents in the south pleaded for U.S. help.

With the Marine offensive under way in Fallujah, Ric Sanchez needed additional forces to clean up the Shiite south. He was compelled to turn to a former command of his, Major General Marty Dempsey's First Armored Division. A third of Dempsey's soldiers were already en route home to their bases in Germany. Sanchez had to hold them, turn them around, and use them to clean up the JAM uprisings, starting with Najaf and Al Kut. He also brought down reinforcements from Mosul to help in Baghdad.

For Marty Dempsey, it was a hard decision. An officer who had been in plenty of fights himself during a tough year in Baghdad, Dempsey gave his soldiers the news straight: "I know you are eager to get home. I am too. But not if it means allowing one thug to replace another. We've worked too hard here to watch that happen." In one case, soldiers who had just gotten off the airplane in Germany reboarded and went right back. The American campaign teetered. Most military operations reach such a point, when only the will of the commander and the valor of the troops can win through. With Sunni surging in Fallujah and Shiites rampaging at multiple sites, that juncture had arrived.

The enemy knew it and jumped on it. Sunni insurgents in Fallujah played on this crisis atmosphere. The Marine attack began on April 5 with four battalions and immediately made solid progress. In a very shrewd move, far more important than acquiring another crate of RPGs or better car bombs, the insurgents invited reporter Ahmed Mansour of Al Jazeera into the city. He and his camera team camped out at a crowded city hospital just west of the bridge where the Blackwater men had been strung up. Ahmed Mansour's cameramen sent out a steady stream of images of dead babies, mangled elders, bloodied children, wailing mothers, and thin, pathetic, still corpses in civil dress. Given that every hostile in Iraq wore civilian clothes, that only made sense. But it did not play well in Peoria, let alone in Baghdad, Riyadh, Amman, or Washington. Joined with the television footage of fighting in the Shiite areas, it got immediate attention from residents of the ever-fevered Arab Street and the wider, troubled Muslim Boulevard. Those worthies expressed loud and extensive sympathy for the "honorable resistance" in Fallujah, referring to it as another intifada (uprising), akin to the Palestinians battling the Israelis. Always mindful of their fiery followers, Arab potentates also objected, many in strong public statements. Even the British complained that the Americans were too heavy-handed. Grumbling escalated: too much firepower, too many civilian casualties, too many bombs, too much violence, too much, too much, too much. It looked a lot worse than it was, but sober military judgment cannot easily trump a parade of bold images. The Americans persisted. Fast-moving First Armored Division units cleared Al Kut and trapped Moqtada al-Sadr himself in Najaf. Some of the senior U.S. commanders, though, started looking over their shoulders. As well they should. In this prizefight, too many in the audience had brought their own white towels.

The tragic Al Jazeera footage resonated in Iraq. Not much brought Iraqi Sunni and Shiites together, but the April combat did the trick. Sadr's people immediately dispatched a symbolic aid convoy to the beleaguered Sunni; the Sunni insurgents likewise sent some trucks with arms and a few fighters into Shia enclaves in Baghdad. A vociferous Shia mob in Baghdad convinced a new battalion of Iraqi soldiers not to proceed to Fallujah. The unit disintegrated; the men climbed off stopped trucks, shucking their uniforms as stunned U.S. advisers watched, powerless to stop it. Other enemy cells and networks began to blow holes in bridges on the major highways leading down to Kuwait, snarling Coalition troop and supply movements. Finally, the senior Iraqis in the governing council went wobbly. Several resigned in protest over Fallujah, Najaf, or both. One Sunni elder, Adnan Pachachi, said, "We consider the actions carried out by U.S. forces as illegal and totally unacceptable. It is a form of mass punishment." Even Shiite members like Ayad Allawi and Ahmed Chalabi expressed reservations. If the governing council fractured, there would be no Iraqis to take over from the CPA.

Ignoring their many critics, dismissing complaints fueled by insurgent propaganda, American and Coalition commanders kept up the pressure. The military situation improved. By mid-April, the Sunni and Shia uprisings had been contained. The Marines were within days of finishing off the enemy in Fallujah. The First Armored Division flying columns restored order in Al Kut and cornered Sadr in Najaf, awaiting only a final okay to nab him. Baghdad had been quieted, to some degree, by the First Cavalry Division. Coalition elements restored security on the highways. Yet in April of 2004, the propaganda front proved to be the only one that counted. As Marine commander Jim Conway summarized, "Al Jazeera kicked our butts." With Iraqi senior civilian officials objecting, the British chafing, the publicity all wrong, and the U.S. 2004 presidential campaigns well under way, Bremer and Sanchez got the word: Stand down. Placing a cherry on the crap sundae, the Abu Ghraib detainee-abuse story broke worldwide on April 28.

Fallujah lapsed into a negotiated settlement that left insurgents running the city under the fig leaf of a Fallujah Brigade, members of which were incapable of cooperating with one another, let alone with the Marine cordon fronting the city. Moqtada al-Sadr squatted in Najaf, still a free man, declaring victory. His JAM continued to stage events, but the heavy casualties suffered among the faithful took their toll. When Sadr ran the same drill again in August, he faced a different lineup. By that time, the Iraqis were (sort of) in charge.

That day had come. On June 28, 2004, with little fanfare, Jerry Bremer completed the CPA mission. He handed authority over to Ayad Allawi and the Iraqi interim government. It was a small event held in the Iraqi prime minister's office. A photographer took a few posed pictures. The big ceremony had been scotched for fear of an enemy attack. That told its own tale. Bremer left Iraq after a thankless year.

His military counterpart followed him three days later. On July 1, Lieutenant General Ricardo S. Sanchez turned over command of Multi-National Force–Iraq to General George W. Casey Jr. Sanchez departed a bitter man. He was rightly proud of his Coalition and U.S. subordinates, lauding their bravery and sacrifice. But he knew things had not gone well. "The last week of March and the first two weeks of April 2004," he wrote later, "were a strategic disaster for America's mission in Iraq." The American military view is clear: the commander is responsible for everything his unit does or fails to do.

Few American commanders have inherited a worse situation than George Casey. The Sunni enemies were active in all of their home provinces, and Sadr's Shiite JAM seemed determined to pop up when and where least wanted. Casey's MNF-I was engaged in a counterinsurgency, yoked to an unelected Iraqi government that featured too many former expatriate slicksters like Ahmed Chalabi. The Iraqi army and police existed here and there. But by and large, those facing the insurgents with firearms were foreigners and so, by definition, irritants to the native folk. The Iraqi population was at best tolerant of the occupiers, and that was only in the Kurdish and certain Shia regions.

A student of history, Casey knew that successful counterinsurgencies prosecuted by an external power demanded a long-term commitment, up to a decade of actual fighting and several more decades of follow-up and follow-through. Like smoldering forest fires, even after suppression, insurgencies tended to flare up now and again. They were rarely really and truly over. Think about America's long road in the Philippines (since 1898) or South Korea (since 1945), both of which endured hot wars and cold, consistently backed by lengthy U.S. troop deployments, formal mutual-defense treaties, and firm political and economic support. For counterinsurgency to work, the enemy had to be convinced that he couldn't outwait you. Regardless of how the upcoming 2004 U.S. elections came out, Casey hoped he could count on that long-term commitment.

Like his friend and superior John Abizaid at USCENTCOM, Casey understood that a long-term U.S. effort needed to focus on helping the Iraqis. Hours after taking the flag from Sanchez, the new general questioned his MNF-I staff. "Okay, who's my counterinsurgency expert?" A major general dutifully answered, but to be fair, the question was rhetorical. If the U.S. and the Coalition intended to win this thing, they were all going to have to become counterinsurgency experts — starting now.

For observers expecting a driving, imposing, forceful general, George Casey didn't look like much. Short, stocky, with a shock of graying hair and reading glasses he pushed up onto his head when thinking, Casey wasn't a West Pointer. He'd gone to Georgetown and got his commission through ROTC. His West Point graduate father, George Senior, a major general commanding the famous First Cavalry Division (Airmobile), had been killed in Vietnam months after young George earned his infantry commission. Army insiders had pegged George Senior to go all the way. Now his son had done it, four stars, and he'd stepped into the toughest wartime command since Vietnam.

George Casey didn't look like the general you'd get from Central Casting, the one with a colorful personality who slapped backs and barked memorable one-liners. He didn't say a lot to the press, nor did he grandstand for the troops. To be honest, rather than a dashing commander, he seemed more like an insurance salesman or a small-town storekeeper. Clever types had once said much the same thing about a rumpled, nondescript former shop clerk from Illinois named Ulysses Simpson Grant. They were wrong then too.

Like Grant, there was a lot more to George Casey than met the eye. As a Georgetown student, he had served as an intern for Coach Vince Lombardi during the football legend's tenure with the Washington Redskins. Like Army chief of staff General Pete Schoomaker, Casey was one of fewer than twenty out of an initial hundred handpicked men to gut his way through one of the brutal initial selection courses for the Task Force. Schoomaker stayed; Casey chose to return to his conventional infantry unit. Yet the achievement showed the measure of the man. As had John Abizaid, Casey served in the Middle East as a United Nations truce observer in the Sinai. He commanded a mechanized battalion in the Fourth Infantry Division, a mechanized brigade in the First Cavalry Division — his father's legacy loomed large — and, finally, the First Armored Division in Kosovo in 2000. It was unusual for an infantryman to command an armored formation. But George Casey was an unusual infantryman.

Along with the customary succession of line-officer duties, Casey had some other key assignments that provided crucial preparation for Iraq. After commanding the battalion, he served as a special assistant for the Army chief of staff, as close as Casey ever came to joining the AAA club; not the West Point Army Athletic Association, but the other one, the careerist self-promotion society that hung out near the military throne rooms: Aides, Adjutants, and Assholes. Casey eschewed that stuff. He held a key junior general post in the Joint Staff at the Pentagon, pulled together the homeland defense plans at U.S. Joint Forces Command after 9/11, stepped up to be the Joint Staff J-5 overseeing strategic plans and policy, and then took over as director of the Joint Staff, the chairman's chief of staff overseeing worldwide military operations. In those latter two jobs, Casey followed his old teammate John Abizaid. Just before coming to Iraq, Casey received his fourth star and served as General Pete Schoomaker's vice chief of staff for the Army. Casey understood the formulation of strategy, including the conundrum that had become

Iraq, as well as any American general. He probably grasped it better than almost all of those who thought they did.

There was another thing. Although he hadn't been in combat before he arrived in Baghdad, George Casey shared another characteristic with Ulysses S. Grant. Watching the stolid Grant ignore the strike of nearby bullets and even a shell burst during the vicious, hard-fought 1864 Battle of the Wilderness, a soldier of the Fifth Wisconsin Regiment remarked: "Ulysses don't scare worth a damn." So it was with George Casey. He evinced an extraordinary calm even under dire circumstances, constantly watching for opportunities and options. He looked always up and toward the horizon, searching for what might be done next rather than glancing back and down into the void.

That summer of 2004 appeared bleak indeed both inside and outside Iraq, a path strewn with grim milestones: Fallujah, Najaf, Abu Ghraib, rising casualties, and strong partisan dissent in the United States. The great, liberating march upcountry of 2003 had degenerated into the heat, squalor, and blood of Iraqi resentment, prisoner abuse, and baby-killing in 2004. Many Americans wanted out, pure and simple. Ever the strategist, even at this nadir, Casey saw an opportunity.

He started with the mission. Recognizing the erosion of relationships between the CPA and CJTF-7, Casey reached out early to John Negroponte, the first U.S. ambassador to post-Saddam Iraq. Negroponte was willing and interested in close teamwork. Together, the two crafted a joint mission statement: "To help the Iraqi people build a new Iraq, at peace with its neighbors, with a constitutional, representative government that respects human rights and possesses security forces sufficient to maintain domestic order, and deny Iraq as a safe haven for terrorists." The words would change, but in essence, this remained the U.S. mission right to the end. It wasn't overly ambitious. As Casey liked to say, "The more we do for the Iraqis, the more we do for the Iraqis." It all rode on them, not us.

That said, Casey and his team worked with the ambassador and his people to come up with an operational construct, a campaign plan. They issued it within weeks of Casey's arrival. It, like the words in the mission statement, changed over the years. But the basics stayed true until 2011.

With chai cups in hand, General Colin Powell (chairman, Joint Chiefs of Staff) and General H. Norman Schwarzkopf (commander, U.S. Central Command) meet in Saudi Arabia before the beginning of Operation Desert Storm in 1991. These two Vietnam veterans crafted an overwhelming offensive that liberated Kuwait. As secretary of state, Powell was also involved in the Afghanistan and Iraq campaigns that followed the 9/11 al-Qaeda attacks.

The guided missile destroyer USS *Cole* anchors in Aden harbor following the October 11, 2000, terrorist attack that blew a hole in the port side and killed seventeen U.S. sailors and wounded thirty-nine others. Valiant damage-control efforts saved the ship, which continues to serve today.

A blackened hole marks the Pentagon three days after the hijacked American Airlines Flight 77 struck the western side at 9:37 a.m. on September 11, 2001. Six flight crew, 53 passengers, and 125 in the Pentagon were killed by the attack. Less than a month later, U.S. forces directed from this building began operations in Afghanistan.

Unless otherwise noted, photographs are reproduced courtesy of the United States Department of Defense.

NATO advisors gather for a short briefing at Qala-i-Jangi fortress on September 10, 2012. The officer at the far right is Air Force Captain Michael A. Sciortino, who served as an enlisted terminal air controller during the Qala-i-Jangi fight in November 2001. By 2012 he was back in country as an advisor to the Afghan national army.

The Pink House was the center of the prisoner uprising at Qala-i-Jangi fortress in November 2001. The dome in the background marks a memorial to CIA officer Mike Spann, killed in action during the fighting.

A CH-47 Chinook pulls away after dropping troops in Afghanistan in 2009. U.S. Army Chinooks provided reliable lift year-round in the Hindu Kush Mountains.

Colonel Frank Wiercinski and his tactical command post team under fire during Operation Anaconda on March 2, 2002. The colonel is the soldier to the lower right of the X-shaped antenna.

The M-2 Bradley Fighting Vehicle raises dust as it churns through Iraq in 2004. These armored infantry carriers played a key role in the initial 2003 invasion and follow-on counterinsurgency operations.

Lieutenant Colonel Scott Rutter, commander of the Second Battalion, Seventh Infantry, poses in his M-2 before the March 2003 invasion of Iraq. Rutter and his soldiers advanced all the way to Baghdad in a lightning campaign.

In July 2003, General Richard B. Myers, chairman of the Joint Chiefs of Staff, visits the First Battalion, Eighth Infantry, in Balad, Iraq. At the far left stands Lieutenant Colonel Nathan Sassaman, the battalion commander, who proved to be a highly aggressive combat leader. Behind Myers is Major General Raymond Odierno, then the commander of the Fourth Infantry Division, later both the corps and force commander in Iraq. At right is Colonel Fred Rudesheim, Sassaman's brigade commander, later a major general and also a veteran of a subsequent Iraq deployment.

General John Abizaid (commander, U.S. Central Command) and Secretary of Defense Donald Rumsfeld listen to General George Casey (commander, Multi-National Force–Iraq) at Baghdad headquarters in 2004.

A U.S. Marine M-1 Abrams tank fires its 120 mm main gun to clear a hostile position in Fallujah in November 2004. The enemy could not match this kind of firepower, but eventually found ways to attack and destroy even the heavily armored M-1.

On the highway between Ramadi and Fallujah, Iraq, an IED detonates against a U.S. truck patrol on March 28, 2006. The moment was captured by an onboard video recorder. In this explosion and the following firefight, one American soldier was killed and three were wounded; at least three insurgents were killed. For many U.S. troops, this is how they met their enemy.

Lieutenant Colonel Muhammed Salman Abbas, Iraqi army, and Major Kevin Hendricks, U.S. Army, talk prior to a May 2005 operation in north Babil Province, Iraq. At the time, Abbas commanded the Second Battalion, Fourth Brigade, Sixth Iraq Army Division. Hendricks was the S-3 (operations officer) for the Second Squadron, Eleventh Cavalry.

Captain Alan Simpson, U.S. Army

Staff Sergeant Danny Laakmann stands watch in Ramadi in the hot summer of 2005. The enemy has left a message on the wall: "Slow Death."

Captain Alan Simmons, U.S. Army

Colonel Michael Steele, commander of the Third Brigade Combat Team, 101st Airborne Division, accompanied his soldiers on operations in mid-2006 in Salah ah Din Province, Iraq. Steele led from the front on mission after mission; he'd been wounded leading his Rangers on the so-called Black Hawk Down battle in Mogadishu, Somalia, in October 1993.

Captain Travis Patriquin sent this photo to the U.S. Army *Military Review* to accompany an article he submitted in 2006. In Ramadi, Iraq, he grew a mustache, spoke Arabic, and was a catalyst in the Anbar Awakening that brought Sunni Arab militamen into the fight against al-Qaeda. Patriquin was killed in action in Ramadi on December 6, 2006. His article was published posthumously.

In Ramadi, the governor of Anbar Province, Mamoun Sami Rasheed (left), and Sheikh Sattar Abu Risha look skyward. The sheikh led the Anbar Awakening, organizing tribal militiamen in 2006 to battle al-Qaeda in western Iraq. He was killed by an insurgent bomb in 2007.

Captain Frank Rodriguez, commander of Troop A, Fifth Squadron, Seventy-Third Cavalry.

Paratroopers of the Fifth Squadron, Seventy-Third Cavalry, led by Captain Rodriguez, advance along a narrow street in the congested Fadhil neighborhood of Baghdad on February 28, 2009. Down the street, an Iraqi woman walks with her children. Fadhil was very difficult to patrol and secure.

Casey summed it up: al-Qaeda out, Sunni in, Iraqis increasingly in the lead.

That little formula reflected Casey's shrewd assessment of the opposition. To help the Iraqis win, the war effort must focus on the Sunni insurgency. It meant an unpalatable but necessary accommodation with Moqtada al-Sadr. That agitator antagonized the Americans — and killed too many — but the reality of 2004 Iraq was that the Shiites held sway, and they preferred to deal with Sadr's JAM in the political arena, not to fight them. Sadr's evident Iranian connection made this very hard. After all, Iran was a charter member of the Axis of Evil named by President Bush in January of 2002, and Iranian agents and their terrorist auxiliaries had spilled a lot of American blood over the years. Yet Casey recognized that for Ayad Allawi and his Shiite-dominated government, and any likely successors, Shiite Iran was a neighbor. The Americans came from an ocean away. The Americans might or might not stay. Iran was forever.

The one enemy that could be destroyed was obvious: al-Qaeda in Iraq, Zarqawi's network. The Baghdad Shia elite always showed energy for going after "foreigners," especially these militant suicide bombers who so hated Shiites. The Sunni worked with AQI, allies of convenience wearing the same sectarian colors. But you could see an Iraq where the Sunni had local power in places like Anbar and Salah ad Din Provinces. AQI wanted no deals, no quarter, simply Americans gone and Shia brought to heel, all of it run according to stringently force-fed shari'a law. Casey and his team saw a potential split here. If they could drive in a wedge, get AQI out and Sunni included, it might well defuse the entire insurgency. George Casey understood this in 2004. For seven years, the Americans pursued it, with three successive Iraqi governments more or less in agreement.

Getting the Iraqis into the lead, especially in security, was the means to push and keep AQI out and bring the Sunni inside the fold. Here the somewhat disjointed activities of 2003 left a lot of room for improvement. The CPA-led effort was fine as far as it went, just too small scale and too slow to field new Iraqi forces. Most U.S. units, like Nate Sassaman's battalion in Balad and Samarra, did good work raising local Iraqi elements variously labeled as Iraqi Civil Defense Corps and the Iraqi National Guard. As the Marines learned in Fallujah in April, though, these new Iraqi battalions proved fragile as well as few. This effort needed a jump-start, a top-to-bottom shakeup, billions of dollars, and quality advisers. Casey got all of that.

The jump-start came in the form of newly promoted Lieutenant General David H. Petraeus, the most ambitious, energetic, and articulate officer in the U.S. Army. Petraeus had led the 101st Airborne Division during the 2003 invasion and did a creditable job in and around Mosul, including some excellent recruitment and fielding of new Iraqi police and army units. Now he came to activate and command the Multi-National Security Transition Command–Iraq (MNSTC-I, pronounced "minsticky"), with the mission to build numerous and competent Iraqi forces as quickly as possible. A charter member of the more dubious version of AAA, Petraeus was the polar opposite of Casey in terms of personality and press profile. He was exactly the sort of guy you'd think would be a general, and a top one, a brilliant one, a Douglas MacArthur in every way. Indeed, as Petraeus took over, *Newsweek* ran a story with the new three-star "warrior-scholar" (the editors' term) on the cover behind huge, urgent letters: "Can This Man Save Iraq?" That kind of thing grated on Casey, but he was willing to give "This Man" his shot. And Petraeus was the guy to take it. Every future operation after midsummer of 2004 would feature more and more Iraqi forces.

Casey's other two key subordinates enjoyed no such adulatory press, nor did they seek it. Tom Metz of Third Corps would take the Iraqi army and police generated by MNSTC-I and employ them in battle. Another long-term Casey comrade, Metz was all about work. He had no shortage of that. The other key MNF-I element, the Task Force, sought no limelight whatsoever. Yet its members were refining effective manhunting techniques they would soon share with their conventional brothers and sisters. Casey wanted al-Qaeda out, and the Task Force would shortly enjoy the capacity to exterminate AQI's leadership cadre.

Along with getting key U.S. generals in place, one other organizational step was necessary. George Casey led a multinational command, with twenty-five thousand troops from thirty-two countries in addition to the United States. The turbulent Iraqi spring of 2004 shook many of the Coalition countries. Some, like the Netherlands, announced their departure in early 2005. Others left sooner, spurred by domestic events.

The 3/11 terrorist train bombings in Madrid resulted in a government change and the subsequent withdrawal of the Spanish brigade. The Dominican Republic, Honduras, and Nicaragua followed suit. Grateful for U.S. help in its own 1980s counterinsurgency, doughty El Salvador stayed and even increased its forces. Wary Iraqi insurgents in Babil and Najaf Provinces greatly feared "the little people," the tough Salvadorans led by highly capable, veteran American-trained officers. But the Salvadorans were the exception. Many participating countries expressed reservations at how the American-led campaign had gone. They did not think their voices were being heard in the headquarters.

As the largest contributor after the Americans, the British took the lead in arguing for an expanded Coalition role in the high command. The British made a strong case. Three of Casey's seven divisions — the British in the southeast, the Poles in the central south, and the Koreans in the northeast — were Coalition. These contingents allowed the U.S. to focus its troops on the Sunni regions. The Coalition commitment had value and needed to be nailed down. Multi-National Force–Iraq could ill afford any more defections or discontent.

As he shaped his team, Casey consciously and carefully integrated senior Coalition officers into MNF-I and its subordinate organizations. The British, ever ready to play the sage Greeks to the bumptious American Romans, immediately supplied deputy commanders at force, corps, and MNSTC-I, as well as other well-placed subordinates on each staff. The British generals and colonels were not shy about offering opinions and advice. Other countries also sent key leaders. Casey's operations chief came from Australia. Italy assigned a two-star deputy to the corps headquarters; Poland also provided a senior officer to Third Corps. NATO sent a training mission to serve under Petraeus in MNSTC-I. It became common to see people in non-American uniforms in the major headquarters. Other voices had their say and were often heeded. Casey knew it all strengthened MNF-I for the long haul.

His mission set, his plan developed, his team formed, Casey turned to his able diplomatic partner John Negroponte to get the Iraqis involved. Prime Minister Ayad Allawi proved quite amenable. He was the typical refined, well-educated, English-speaking expatriate Shia leader, a close colleague of the ubiquitous Ahmed Chalabi. Allawi, however, had survived a personal brush with death that kept him wonderfully grounded as he wrestled with the situation in 2004 Iraq. In 1978, an ax-wielding Baathist assassin broke into Allawi's home in Surrey, England. Allawi was left for dead, and it took him a year to recover. When Negroponte and Casey came to him, they found an Iraqi grateful for the gesture of respect and willing to cooperate in strengthening the country. Allawi agreed wholly with Casey: al-Qaeda out, Sunni in, Iraqis increasingly in the lead.

Moqtada al-Sadr gave them the first opportunity to try the new approach. In August in Najaf, outraged by continuing U.S. patrols, JAM elements took over police stations and Iraqi government buildings. American Marines and soldiers responded rapidly, accompanied by three of the brand-new Iraqi army battalions. Allawi publicly directed and authorized the action, a bold step for a Shia Arab leader in Iraq. He also drastically limited Al Jazeera's access, which cut off a key propaganda avenue for the insurgents in Najaf. This response worked. His JAM ranks shredded once again, trapped in Najaf, Sadr pleaded for negotiations. Of course, his request was granted, and he slipped out of the noose, as usual. Yet the message was clear. The new Iraqi government would fight insurgents, even Shiite militiamen.

Having secured the Shia side by the now-familiar combination of killing JAM and shaking hands with Sadr, Allawi moved on to a tougher problem in Sunni Arab Samarra. Things there had not improved since Sassaman's time. The U.S. Army's First Infantry Division, joined by Iraqi army forces, mounted a long, methodical, and ultimately successful campaign to reclaim that city. Again, Allawi spoke up strongly, authorizing the operation. The Iraqi battalions fought credibly. Petraeus and MNSTC-I delivered.

Fallujah remained. There, with the Marines effectively besieging the city since April, the Sunni rejectionists created an insurgent sanctuary. Marine and SOF strikes and raids kept the enemy guessing, nervous, and off balance, but inside the ring, what was theirs remained theirs. Zarqawi moved key segments of AQI into this "liberated" zone. Enemy strength swelled to forty-five hundred and included hundreds of foreign fighters sponsored by al-Qaeda in Iraq. The local citizenry grew less and less supportive of AQI as Zarqawi and his confederates imposed strict shari'a law. Western music compact discs, magazines, and DVDs were eradicated. Barbers were run out of town. Male beards became mandatory. Females learned to stay scarce and covered. Any use of alcohol—a guilty pleasure for many Fallujahans, as it was for many citizens in Iraq—became a cause for beatings. Commercial and private truck drivers stupid enough to pass through or near town "contributed" to the defenders. Civilians left Fallujah by the thousands. Maybe thirty thousand, less than a seventh of the April head count, remained by November of 2004, most of them anxious for relief.

It was coming. George Casey and Tom Metz looked carefully at what had been done in Najaf and Samarra. Together, with Allawi and Negroponte orchestrating the political aspects, the Coalition military leadership developed an integrated plan that set conditions for a successful offensive. The Third Corps called it Phantom Fury, intentionally using the corps nickname and so indicating that the operation was bigger than the Marines. Iraqis referred to it as al Fajr (New Dawn), and Casey saw the benefit of that Arabic title: it emphasized Iraqis increasingly getting involved in suppressing the insurgency. Getting the word out—telling the Iraqi citizens first and often just what kind of hideous enemy was running Fallujah—formed an important aspect of the MNF-I and Third Corps planning.

The Marines would again lead. Last time, there had been four Marine battalions, a regiment-plus. This was a full-scale divisional attack. Two Marine regiments, each reinforced with a U.S. Army mechanized/tank battalion task force, intended to advance abreast from north to south. Another U.S. Army brigade served to isolate Fallujah, preventing enemy reinforcement or escape. At Casey's request, the British sent up the Black Watch (the Royal Highland Regiment) from Basrah to augment the outer security ring. Six Iraqi army battalions prepared to join the assault, following right behind the Americans. Ayad Allawi declared a national emergency in all Iraqi provinces except Kurdistan and agreed to curfews, press limits, and full cooperation in clearing the insurgents out of Fallujah. Emboldened by Najaf and Samarra, Allawi echoed Casey and Negroponte: "Start together, stay together, finish together."

The Jolan District in northwestern Fallujah had not gotten any better since April. The Third Battalion, Fifth Marines drew the unglamorous duty of clearing Jolan, building by building. Infantrymen hate this kind of task. It is exhausting, dangerous, and grinding. Lieutenant Colonel

Patrick J. Malay and his Marines leaned into the job, well aware that it had to be done. The Army tankers, riflemen, and other Marine battalions had pushed right on south to Highway 10, shooting as they went. Now, just as their forefathers had once pried Japanese holdouts from bunkers on Peleliu and Okinawa, young Marines in Jolan slowly cleared out the enemy, room after room, floor after floor. The sequence rarely wavered: toss in a grenade, let it blast, go in shooting, check the bodies, remove any documents or computer media, leave the dead for the Iraqis.

Sometimes the enemy made it sporting. AK fire or an RPG might spurt from an upper floor as Marines approached. At that, some Marines flattened against the walls and others slid behind junked trucks, clearing the street. The AK bullets pinged into the hard dirt. Then the Marines waved up a big, flat-topped M-1 tank. The tank would trundle down the street. A few AK shots or another RPG might annoy the behemoth, scarring its metal hide. Still, on it came, slowly elevating its 120 mm cannon, centering on the offending window. Then the massive metallic shots rang out, a slow-cadence drumroll, ten or more, tearing open the top floor and its feisty occupants. What was left wasn't too feisty.

In the Jolan, each Marine carried seventy or more pounds of weapons, body armor, ammunition, water, and medical items. The Corps strove to make each piece weigh as little as possible, but seventy pounds of lightweight gear still weighs seventy pounds, which amounts to fighting with a medium-size baboon clinging to your back. Water and ammunition just didn't compress well. Short on sleep, slurping up individual food packets stripped from random MREs at dusty halts, sweating constantly, the Marines pressed on, house after house.

As hunters compare their various quarries, thoughtful Marines started to categorize their foes. The hit-and-run types they called guerrillas; these were the ones the G-2 intel people identified as Sunni rejectionists, combatants who followed the old Maoist dictates of *Enemy advances, we retreat; enemy halts, we harass,* and so on. The stay-and-hold characters the Marines called martyrs, and they were often non-Iraqis, Zarqawi's AQI recruits from other countries. The guerrillas sometimes slipped away. The martyrs died in place.

The Marines didn't see any civilians. Well, yes, the individuals shooting at them all wore civilian attire, but they weren't innocent townsfolk. Company K ran into one bunch who hollered down in English from a third-story window: "Mister, mister, help us! Family! Family!" Sheets of hostile RPK machine-gun and AK fire belied the pleas for help. The tanks clanked up and ripped open the apartment's façade. A Marine rifle squad found only adult male bodies.

Sometimes it got much tougher. A Marine colonel, himself wounded in an earlier Fallujah encounter, explained what happened in one such situation: "There's four guys going into a courtyard. The first guy that goes in, he's killed. The second guy is wounded. The third guy, he's wounded. The fourth guy is [Private First Class Christopher] Adlesperger, Adlesperger takes charge. By the time he's done, he's killed eleven insurgents. He's safe. He's saved the lives of the two wounded . . . by dragging them upstairs onto a roof and defending it." It was that kind of fighting.

In close-quarters infantry combat, Marine leaders are expected to be well forward. That acknowledged, there are never enough officers or staff NCOs, good as they are. In the Marine Corps, twenty-year-old corporals give the life-and-death orders. That happened in Fallujah. On the 229th birthday of the Corps, November 10, 2004, Corporal Michael Hibbert was running his part of the show as he led his tired, dirty Marines toward a battered, metal-sided warehouse near the river, on the far west edge of the Jolan. Nobody had shot at the Americans for a while. Hibbert's point man found three 122 mm artillery shells in a drainage ditch alongside the building. Marine engineers clipped the wires and then slapped C-4 demolitions on the nearby wall. A ragged hole blew open. The dust blossomed and then subsided.

Hibbert and his men slipped quickly into the gloomy hall and found stacks of RPGs and artillery rounds. Evidently, this was a big cache for the enemy. The Marines began to search the side rooms. In one crawlspace, Hibbert uncovered a bony, chained man, slowly starving to death. The Iraqi had been held since August. In a larger room, Hibbert's Marines found a makeshift movie studio. The green-and-black Islamic flag on the wall and dried blood on the floor matched the video made when Zarqawi had filmed the awful beheading of captured American contractor Nick Berg back in May. There was even a nice script in English and helpful instructions on how to get snuff tapes to Al Jazeera's Baghdad office.

There was worse half a mile away. Near an abandoned children's playground, complete with a metal merry-go-round, a small Ferris wheel, and a red, white, and blue pressed-metal moon lander marked *USA*, the Marines smelled . . . well, they smelled something they wished they did not recognize. But they did.

After entering a two-story house right near the Ferris wheel, battalion commander Malay saw a dead man with his legs chopped off, twisted in rigor mortis, blood all over. Another room had a similar corpse, also mutilated. The odor in the abattoir was overpowering.

Thanks to very smart preparatory work by Casey, Metz, the Marines, and notably Allawi's Iraqis, these horror shows were widely publicized. In America, most of it got lost in the backwash of another very close, highly acrimonious domestic election. But for the first time on the Arab Street, especially on the Iraqi Street, the locals got a long, hard look at what an al-Qaeda caliphate really portended.

Four days later, Lieutenant Colonel Pat Malay and his command security team walked through the ravaged Jolan souk. The market was empty, trashed: metal sliding doors punched aside, tables upended, power lines dangling. Stall after stall featured various stacks of ammunition and weapons, organized not quite to Marine Corps standards, but close enough. All of it had been abandoned, the would-be users dead or gone. A single thin dog ducked into an alley.

The international news media, led by Al Jazeera, decried the April attack for its supposed excess, which included 150 airstrikes that put holes in or collapsed a hundred buildings. The November assault saw 540 airstrikes, more than 14,000 artillery and mortar shells, 2,500 tank main-gun rounds, and almost half of Fallujah's 39,000 structures destroyed or damaged. It wasn't exactly Stalingrad, but swaths of Beirut in the mid-1980s came to mind. Yet this time, shaped by thoughtful military advance work, the press narrative was much more balanced. It included lurid tales of what Zarqawi and his cronies had wrought, as well as their die-hard, bloody defense. The operation cost the lives of seventy Americans and wounded 609 others; in addition, six Iraqi soldiers were killed and fifty-five wounded. Outside the city, the British Black Watch lost four men, and eight others were wounded. More than twelve hundred insurgents surrendered. Some two thousand died. Fallujah was back in the fold.

Now, in the late-afternoon sunshine on November 14, Pat Malay moved slowly up the incline toward the green-painted bridge where the entire Fallujah thing began. In white flowing Arabic script, dimmed since it had first been daubed months before, the bridge proclaimed for all to see: *Fallujah — Graveyard of the Americans.* Beneath it, Malay's Marines used black paint to write their message: *This is for the Americans of Blackwater murdered here in 2004. Semper Fidelis 3/5 Dark Horse 9/11.*

9

THE COLOR PURPLE

> The ground was always in play, always being swept. Under the ground was his, above it was ours. We had the air, we could get up in it but not disappear in *to* it, we could run but we couldn't hide, and he could do each so well that sometimes it looked like he was doing them both at once, while our finder just went limp.
>
> — MICHAEL HERR

THEY SHOULD HAVE known better. But human nature is ever thus. The slugging match in Fallujah worked. The heroism and determination of the Marines and soldiers — among them Iraqis — seemed definitive. With the smoke clearing in the smashed-up city, Marine lieutenant general John F. Sattler offered his opinion: "I think that in the west, we have broken the back of the insurgency." The Marines rightly added the battle to their annals, alongside the Battles of Belleau Wood, Iwo Jima, the Chosin Reservoir, and Hue City. They even hosted tours of the quiet, repopulating city, taking gawking congressmen, wide-eyed reporters, and well-scrubbed staff types from Baghdad to see the sights: the initial breaching lanes in the north, the Jolan merry-go-round death house, Zarqawi's studio, Highway 10, and, of course, what was now known as Blackwater Bridge. In Anbar Province, the U.S. victory lap went on and on.

For their part, the various Sunni insurgent groups thought differently. True, they'd failed spectacularly in Fallujah. In retrospect, the opposition's months-long Islamist enclave represented its greatest long-term success of the entire American counterinsurgency campaign. Everything afterward would be small, local, and incremental. The fractured nature of the enemy, hampered directly by MNF-I's comprehensive countrywide pressure and the rising tide of Iraqi security forces, prevented the type of serendipitous concerted action that had paid off for them so well in April. A sympathetic Sunni rising in Mosul ran off some Iraqi police, but other Iraqi forces, well backed by Americans, stood and fought. Mosul held. Targeted Sunni attacks in Baghdad and Samarra floundered. Sadr's Shia JAM sat on the sidelines, more interested in the January 2005 voting than taking another bloody whipping. Sunni Arab guerrilla leaders pulled back too. *Enemy advances, we retreat . . .*

George Casey and MNF-I had accomplished something, clearing the decks for the January 30, 2005, vote to elect delegates to start writing a new Iraqi constitution. The normal enemy pullback in the colder months coincided nicely with balloting, although there were definitely major attacks leading up to the nationwide vote. Undaunted by the car bombings and sniping, eight million Iraqis walked proudly to more than fifty-two hundred polling stations. Photos of ecstatic men and women waving purple index fingers — the indicator of completed balloting — told their own story. The big guns in Baghdad, America, the Coalition, and Iraq calculated — well, more like hoped, or prayed — that, chastened by Fallujah, Sunni Arabs would vote too.

They did not. "This did not bode well for our efforts to defeat the insurgency," wrote Casey. There would be no quick end to this war. Zarqawi's AQI wasn't backing down, and the Sunni Arab rejectionists weren't either. Contra Casey, al-Qaeda was still in and the Sunni were still out. Worse, the January 30 election elevated the colorless, diffident Shiite expatriate Ibrahim al-Jafaari to prime minister. Openly beholden to Moqtada al-Sadr in particular and the Shia Arabs in general, he would be much tougher to influence than the energetic, cosmopolitan Allawi. The October plebiscite and the December parliamentary election loomed. And more than 160,000 MNF-I troops sat in Iraq, tied to the entire undertaking of voting, building Iraqi forces, and trying to improve public services in a thousand suspicious Sunni villages. Continuous operations in Baghdad, Fallujah, Al Kut, Mosul, Najaf, and Samarra, and many other hot, decrepit, contested warrens, wore on the Coalition too. Another major troop rotation began, bringing in fresh faces and, by now, more than a few veterans of 2003 back for another round. Spring would come soon enough. That always got the hostiles going. *Enemy halts, we harass . . .*

"The top two, we can get," said Lieutenant Colonel Muhammed Salman Abbas. "My scouts confirm they are either at the Diyara house, near the school, or at the farmhouse out at Highway 1."

Lieutenant Caesar nodded. In front of the two uniformed Iraqi army officers stood three thin young men with scraggly beards, mops of black hair, and dark eyes blazing with interest. The three wore off-white man-dresses that hadn't seen the washbasin lately. Outside the one-story, cement-block orderly room, the men had parked their vehicles: a battered white Toyota pickup truck and the oldest Opel taxi in Iraq, ancient white door panels with dull orange fender highlights. The three were Muhammed's scouts. They looked like farm hands, dull ones at that. They were, in fact, the three sharpest sergeants in the Second Battalion, Fourth Brigade, Sixth Iraq Army Division.

Caesar's real name was never said. It was like his beat-up plastic right hand — a gift from a mortar attack, or an IED, or God knew what; you just didn't talk about it. Caesar was a forty-year-old lieutenant, Muhammed's man from the old army, and that was that. They called him Caesar because he'd learned English from watching the movie *Gladiator* over and over. The name proved apt, as the locals learned to render unto him the things that were his. And he had many needs. Caesar served as Muhammed's battalion S-2, his intelligence officer. But both men followed Soviet Russian practice. In the U.S. Army, the S-2 is an MI staffer, an adviser. The old Soviet system, the one taught to the old Iraqi army, made him much more: the chief of reconnaissance, commanding the scouts, running the informants, interrogating prisoners, the entire deal. That was Caesar.

Now the scouts brought the word. By midnight, Thamer Bedrani and Majeed Mohammed Ghuraim al Janabi would be at the house of the latter's sons in the little farm hamlet of Diyara. Alternately, they might be at Thamer's farmstead just west of Highway 1. Lieutenant Colonel Muhammed and the Americans of the Second Squadron, Eleventh Cavalry had been after these two for months. One detainee just yesterday had fingered Thamer's farm, and that break led the scouts to Majeed and his sons. The faux taxi driver had dropped off a fare next door just before returning for this conference.

Caesar smiled, rubbing his plastic hand with his real one. "We'll get both tonight," he said. "You help."

Major Kevin Hendricks, S-3 of 2-11 Cavalry, agreed. Together, the American and Iraqi soldiers set the time of the raid for 12:10 a.m. on Thursday, May 12, 2005. The men spent the rest of the day and early evening lining up all of the moving parts.

Babil Province was a backwater in George Casey's MNF-I campaign. An irregular, sandy quadrilateral sixty miles south of Baghdad, Babil enjoyed great fame in antiquity as the site of fabled Babylon. That was then. In 2005 Iraq, Babil was notable only because through the sparse clusters of flat, irrigated farmlands and expanses of open desert ran the main U.S. logistics artery to Kuwait: Highway 1. Keeping Highway 1 open and free of hostile IEDs was important. Doing it without a lot of forces was preferred.

The threat to Highway 1 originated in the northern third of Babil Province. The Shia Arabs predominated elsewhere in the province, and unless Sadr was on the warpath, they remained quiet. In the north, though, Sunni Arab tribes followed the rejectionist line. The Janabi and Obeidi clans fought each other with machine guns and RPGs when they weren't assailing American truck convoys, shooting at the new Iraqi army battalions like the one led by Muhammed and Caesar, and taking cracks at Shia Arabs who'd wandered onto the wrong turf. The Sunni and Shia faced off along a rough east-west frontier. The little Sunni town of Diyara lay just north of that split.

Centered right where the Sunni-Shia demarcation crossed Highway 1, at ground zero, stood Forward Operating Base Kalsu, the home of the Second Squadron, Eleventh Cavalry Regiment and their Iraqi partners.

The targets for May 12 were all Janabis or Janabi running buddies, Sunni gangsters who had long profited from the smuggling ratlines in north Babil. In Diyara, the Janabi network operated a basic protection racket. Local shops and farmers paid dues to Majeed Mohammed Ghuraim al Janabi. His two sons collected. When a shopkeeper got behind on payments or a farmer seemed too interested in an American-sponsored well project, Thamer Bedrani and his henchmen would pay an evening call, crowbars in hand, pistols in their belts. People paid up and shut up. And no wells were dug.

During 2004, the Americans killed a few of Majeed Janabi's pipe swingers and foot soldiers but never got close to him. That is, until Muhammed and his Iraqi soldiers arrived. They were almost all Shiites from Nasiriyah, the wrong religious sect for Sunni north Babil, but a hell of an upgrade compared to English-speaking Americans. The lieutenant colonel himself came from near Basrah, and he'd fought the British and Marines in March of 2003 before switching sides. Muhammed learned English working as an interpreter and security man for the United States. When the new Iraqi army sought officers in 2004, Muhammed reported to the Marines at Kalsu. Within weeks, he and the Marines formed and trained a few hundred Iraqis, pumped up with weapons, trucks, radios, and ammunition supplied by David Petraeus and MNSTC-I. When 2-11 Cavalry arrived to backfill the Marines, Muhammed found very willing partners. For an energetic commander like Muhammed and a fixer like Caesar, American combat power and technology gave them the edge they needed to establish a foothold in north Babil.

For the Americans, Lieutenant Colonel Bill Simril and his S-3 Kevin Hendricks knew exactly what they wanted to accomplish. It's unlikely that either of them, or any of their cavalry troopers, had ever read a counterinsurgency field manual or heard of George Casey's "al-Qaeda out, Sunni in, Iraqis increasingly in the lead" formulation. The 2-11 Cavalry, though, planned and executed right out of the books they'd never studied. Simril knew it all rode on Muhammed's Iraqis, not the Americans. Once they worked together to clear away the Janabis, they could place locally recruited Iraqi police to hold places like Diyara and then let the Iraqis build something like a village economy and civil society. Later, various counterinsurgency theorists claimed they had derived this clear-hold-build practice and acted like they'd invented it. Common sense and mission focus all over Iraq brought it boiling up from the bottom throughout 2005.

The key impediment in north Babil was not the Sunni people themselves but the thin thread of opposing leadership, the brains, the motivators. That was a finite enemy resource and could be not only limited, but eliminated. Bill Simril's Americans came in country seeing that their contribution to the fight would be manhunting. That job was not just for the Task Force anymore.

Here, the craft had certainly advanced dramatically since Nate Sassaman's night raids in Balad eighteen months earlier, which had been dependent on the good offices of a masked snitch. The Task Force find-fix-finish tactic expanded to F³EA: find, fix, finish, exploit, analyze. For the SOF, the goal was to latch onto an enemy network and never let go, never blink, never stop watching and pulling it apart, piece by piece. Each raid led to the next, drawn by the rapid on-site evaluation of detainees, documents, and electronic media. It created a diagram of the foe's network, sketched on the move, under fire, dot by dot. When dots connected, more actions followed. Actions got bad guys scurrying, talking, and making mistakes. And when they did, the SOF were right there to capitalize. Many nights in 2005, the Task Force struck more targets in one night than it had over an entire month in 2003 or early 2004. The Task Force tracked Abu Musab al-Zarqawi after Fallujah. He and his AQI were the prey, and the Task Force concentrated all of its intelligence resources right there. Systematically, nightly, by day when necessary, the best finishers in the world punched into building after building. Each detainee weakened the AQI network. Each document or computer disk set up the next set of raids. Momentum mattered. The key was to feed the fusion of intelligence, to build the AQI picture. As this happened, the entire enemy array, both AQI and the Sunni rejectionist cells, slowly came into view.

Fusion of intelligence was the key to finding the enemy in Iraq, argued Lieutenant General John R. Vines, commander of Eighteenth Airborne Corps and successor to Tom Metz as Casey's key operational subordinate. Vines had been in as many fights as any American airborne commander—the Eighty-Second Airborne jump in Panama, Desert Storm, and Afghanistan—and his early years in the Seventy-Fifth Ranger Regiment ensured strong ties to the SOF community. The corps commander well understood that, as they said on *The X-Files,* the truth was out there—if the various communities compared notes. The intel guys tended not to do that. What was theirs was theirs. But the operators, SOF and conventional, needed results. And J. R. Vines was first and foremost an operator. He asked all parties to dismantle the walls that kept information from flowing between them. Hungry for the mass of intelligence out there among the landholding corps units, the Task Force agreed.

Encouraged by J. R. Vines, the Task Force recognized that U.S., Coalition, and Iraqi conventional forces on the ground had a role to play, like safari beaters batting the tall grass to get the lions moving. In the past, the SOF guys had arrived out of nowhere, did their mysterious business, and then departed like smoke, as if they'd never been there. As 2004 became 2005, the Task Force emphasis on exploiting and analyzing raid results showed that sometimes the beaters became the hunters, and vice versa. Those in the Task Force ceased to be monks and became missionaries, not simply perfecting their techniques but sharing them. Line units like 2-11 Cavalry learned how to work closely with SOF on operations. Both sides shared vital information, fusion at the working level just as Vines intended.

It led to a practical division of labor. The SOF sought the big game. Conventional MNF-I units and their Iraqi army and police affiliates went after the Sunni rejectionists. Both teams, however, focused on the hostile leaders. Low-level militiamen were killed and captured, of course, but now the opposing leadership felt the heat too. On the night of May 11–12 in Diyara, those leaders would be the Janabis.

The Janabis wouldn't be easy marks. They were vicious bastards, fond of taking a power drill to a kneecap or cutting off a hand. They were not given to Abu Ghraib underpants-on-the-head silliness but real, hard-core Baathist torture. Janabi gunmen hit back hard at both 2-11 Cavalry and Muhammed's Second Battalion. Come the warmer days of spring, right on schedule, Forward Operating Base Kalsu ate a clutch of mortar shells and rockets every few days. Deep-buried IEDs flipped over and charred the new American up-armored Humvees. The sturdy, slab-sided trucks offered a huge step up from the thin fiberglass versions of the Scott Rutter/Nate Sassaman era, but even an M-1 tank couldn't eat eight 122 mm artillery rounds wrapped in red detonation cord and blown sky-high by a hidden watcher. The U.S. armored Humvees had no chance. In the previous two weeks, the squadron had suffered three dead, all from IEDs. The makeshift bombs left four wounded. Incoming mortar rounds hurt another pair. When you went against the Janabis, you had to be ready for the worst.

Well aware of their foe, the Americans and Iraqis put together a mission scheme with a lot of options. Iraqi Company C planned for the three Majeed family houses in Diyara. Iraqi Company D was going after the Thamer farm near Highway 1. An American platoon would take four more Diyara houses belonging to Majeed cousins who'd planted IEDs and carried AKs for the cause. Each of the three units prepared to take any of the three objectives, or all of them. As contact looked likely, the assault forces included two advanced trauma life-support teams, the squadron's surgeon and his party going to Diyara and an attached Navy doctor and his team following the Iraqi Company D to the Highway 1 house complex. A military police platoon at FOB Kalsu would be mounted up in its armored trucks, engines running, as a reaction force. The targets were about five miles north of the base.

At the two Diyara blocks and the farm compound, the Americans and Iraqis would do their best to set a cordon. But at night, with the various canals and underbrush and goat trails, you could never block every possible escape route. Accordingly, Kevin Hendricks coordinated for Apache attack helicopters to watch for and engage squirters, the ones trying to get away. With its superb thermal sights, an Apache could find anyone. As for someone outrunning the helicopter's 30 mm gun — well, that couldn't happen.

Hendricks added another layer of air cover: two F-16C fighter jets. Each of the Air Force planes packed a pod called a LANTIRN (low-altitude navigation and targeting infrared for night), which allowed the pilot to see a lot on the ground. The LANTIRN could target the F-16's smart weapons, designate impact points for the Apaches, and even illuminate a location for the ground soldiers. For riflemen wearing night-vision goggles, a spot would suddenly light up with twinkling green brilliance. The Viper pilots called it "sparkling" an objective.

Then the cavalry S-3 added yet another layer. He asked for an EA-6B Prowler, a Navy radar-jamming jet usually used to break through enemy air defenses. If the gizmos onboard the jet were retuned, the electronics could suppress the little walkie-talkie radios and cell phones popular with the foe. You had to be careful with a Prowler. It would blank out the friendly Iraqi radios too. But if you prepared for that and put the right kind of U.S. radio with each Iraqi unit, you could work under the electronic interference. The enemy could not. It would cut off the hostiles' early-warning net.

Hendricks wanted two more capabilities he couldn't get. He longed for a Predator drone overhead, to broadcast real-time video of the target houses. If you had a Predator, you could watch your prey hang out, and if they moved, you could follow them or launch early and nab them. (Or kill them, if that was required.) There were about 220 available hours of Predator orbits in Iraq in May of 2005, and the Task Force used 215 every day and night. The other five hours weren't going to a mission in north Babil Province. Caesar's scrounging scouts would do surveillance the old-school way.

The second thing Hendricks wanted was an MI technical team that could sniff bad-guy electronic signatures. A year or two later, every squadron and battalion had one. But not in 2005. Again, Iraqi soldiers and Caesar's instincts would fill the gap.

The Americans and Iraqis spent the afternoon practicing room clearing, getting on and off the trucks, and shooting at their twenty-five-yard range. The Iraqis liked banging away with the AK-47s. "In the old army," said Muhammed, "we only got to shoot a few bullets a year for training. You have taught us how to aim . . . and hit."

About eleven that night, the Iraqis clambered onto their new Nissan bongos, the tan flatbed trucks Babil farmers used to move their crops. Muhammed's men and the U.S. squadron mechanics had welded big steel plates on the sides, a form of armor that warded off bullets and might shrug away a glancing IED explosion or RPG hit. A few of the Iraqi infantrymen were artistically inclined. When they'd gotten the trucks months back, they'd decorated the plates with images of the Iraqi flag: red, white, and black, with the three green Baathist stars and *Allahu Akbar* on the white strip. About twenty armed Iraqis, a platoon or so, fit comfortably in the truck bed. Three cargo trucks made a company, with an unarmored Toyota for the Iraqi company commander and first sergeant.

Some of the U.S. advisers climbed into their own up-armored Humvees, which would be interspersed with the assembling Iraqi column. A few Americans got in the back with the Iraqis. It would be helpful at the objective houses. The U.S. night vision and radios were pretty important in keeping things moving, especially if—when—the Janabis pulled something ugly out of their hat. The ten advisers were officers and sergeants pulled "out of hide," drawn from the ranks of Bill Simril's 2-11 Cavalry. Even at its height, and they were not there at all in May of 2005, MNSTC-I never fielded more than about 40 percent of the unit advisory teams. Units always created the rest themselves.

The standard Dave Petraeus model required three parts. First, MNSTC-I and the Iraqi Ministries of Defense (army) and Interior (police) recruited, trained, and fielded a new battalion. About half were converted from existing units, like Muhammed's bunch. The rest were new builds. As the Iraqi unit grew, the second key part linked up. A ten-man advisory team joined the Iraqis for leader and collective training, then accompanied the rookie battalion into its assigned region. When the Iraqi unit deployed to fight, it passed from Petraeus and his institutional guys to Vines and the corps operators. That brought in the third component, the partner unit, the side-by-side teammates. With Muhammed's Second Battalion, both the advisers and the combat partners came from Bill Simril's 2-11 Cavalry.

Muhammed's hidden scouts reported in to Caesar. The targets were definitely onsite in both Diyara and the farmhouse. It wasn't clear exactly who was where. Hendricks confirmed the Apaches, the F-16s, and the Prowler overhead. At 11:45 p.m., the column rolled off FOB Kalsu.

Iraq Company C led: a Toyota pickup, a U.S. adviser Humvee, and three Nissan bongo trucks. They turned onto Highway 1 and eased into a long, dark gap between the seemingly endless sequence of MNF-I supply trucks rolling north. Next came the U.S. element—eight up-armored Humvees—and then Iraq Company D and its adviser Humvee. They moved along quickly. Iraqi drivers have only two speeds: stopped and floored.

Company C turned off to the east, down the long thin country road to Diyara. The lights were on in town, a good sign. As the Americans slid into the turnoff, the black road behind them lit up. You couldn't hear anything over the engines roaring, but something happened back there. Up in the Humvee turret of the eighth U.S. truck, the rear-facing gunner reported: "India Echo Delta [IED] on the trail element, over."

The white flash spanked and rolled the second of Company D's three bongo trucks. The blast tore off the back axle and scattered men and equipment onto the blackened, cratered highway. It had been a big one.

Simril and Muhammed, together in the American lieutenant colonel's Humvee, did not hesitate. "Eaglehorse X-Ray, send Watchdog to link up with Tiger Delta at Checkpoint 18 Alpha for casualty evacuation. All other Eaglehorse and Tiger elements continue to first set of targets. Tiger Charlie will take the western objective as a follow-on. Acknowledge, over." Muhammed issued orders simultaneously in Arabic over his radio set. In sum, the two commanders called an audible. They would have the MPs clean up the IED strike, finish business in Diyara, then let Muhammed's Company C raid the farmhouse.

Kevin Hendricks came up on the radio next. "The Air Force reports two men onsite at the farm, over." Out in the dark sky, the twin Apaches backed away, overwatching from the horizon. The aviators did not want to spook the twosome.

It took about twenty minutes in Diyara. Muhammed's soldiers dismounted quickly. Files of Iraqi riflemen moved like cats along the house fronts, slipping over courtyard walls. Glass smashed. Doors opened. In the dark buildings, shadows and light alternated as the Iraqis moved through with flashlights. All three of their houses were empty.

The U.S. soldiers had better luck. Three of their four houses yielded detainees, both Majeed sons and a cousin, plus some weapons and documents. But Majeed and Thamer were not in town tonight.

Outside, notebook in hand, Caesar quizzed the sons. They offered nothing. The cousin dummied up, head down. Stymied, the Iraqi lieutenant turned to another group standing to the side—the wives and children. Without a female interrogator, Caesar knew better than to address the women. That just wasn't done in a traditional Muslim setting. Rather, Caesar counted them and looked them over closely. Even in the dark, with the women covered to their eye slits in wide black chadors, the lieutenant recognized two. "She is Thamer's wife. This other one is one of Majeed's wives," Caesar said. The women didn't talk, and the Iraqi officer did not ask them to. In the gloom of Diyara's shabby, cramped main street, a few feet beyond the still-lit houses, you could almost see the gears clicking in Caesar's head. Thamer and Majeed were still around; they just weren't here.

"The farmhouse," Caesar said. "We must go now." Simril and Muhammed mounted their Humvee, leaving the other Americans and one Iraqi platoon to clean up Diyara. Company C's commander and his two other platoons crawled into their bongo trucks, fired up the engines, and followed. Over the radio, Kevin Hendricks confirmed what Caesar hoped. "The Air Force still has two adults at the farm, over."

Out west by themselves, Thamer and Majeed figured the IED had done the job. Because of the Prowler's jamming, nobody from Diyara could contact the men at the farmstead. They did not know that the Americans had other tricks up their sleeves, far overhead, out of sight but by no means out of range. Muhammed's Iraqis raced west as quickly as they had bounced east, juddering over the old, crumbling asphalt. The trucks didn't come quietly, though. As the Iraqi vehicles approached, Hendricks relayed: "Air Force reports both men running to an outer building. Will sparkle, over."

At this point, Bill Simril could have unleashed the Apaches. He could surely have done it with a clean conscience after the IED strike on Highway 1. But killing Thamer and Majeed didn't move the ball. They were Janabi honchos, the ones who showed the trail to the rest. If possible, they had to be taken alive.

As the bongo trucks and Humvees swerved into the open area in front of a low adobe stable, everybody tensed for the burst of machine-gun fire that seemed likely. Nothing happened, just brakes squeaking and hard metal sides banging. Iraqis soldiers scrambled over the steel plates even before the dust settled. As they hopped out of their Humvees, American soldiers looked up through their night-vision goggles to see a vision, a glittering cone hanging over a garden house, the air sparkling like bright green snowflakes wafting down under a streetlight.

Within minutes, Iraqi riflemen had both men cuffed and stuffed into the trucks. Neither had been armed. Caesar confirmed their identities as Thamer and Majeed.

The Diyara element was already back at FOB Kalsu. Twenty-five minutes later, Muhammed and Simril rolled in with their prizes. The MP quick-reaction force was back too. The Navy doctor and his people had done superbly out on Highway 1. Six seriously wounded men, five Iraqis and one American, all made it out alive and conscious by medevac helicopter, bound for the Eighty-Sixth Combat Support Hospital in Baghdad. The U.S. soldier, Sergeant John M. "Pudge" Smith, had been conscious and talking. That was Pudge, all right. He liked to talk.

At 2:20 a.m., as Caesar chatted up Thamer and Majeed, and the wrecker dragged in the carcass of the smashed Iraqi bongo truck, the rest of the night's bill came due. Simril got a phone call from Baghdad. Smith hadn't made it. He'd bled out.

Muhammed was beside Simril. "I have lost a brother," the Iraqi said quietly.

Muhammed Salman Abbas lost nineteen more American brothers that year, plus some ninety wounded; one-third the casualties of a Fallujah suffered by a single cavalry squadron a long way from Fort Irwin, California. For those doing the arithmetic on the frontier, what Abraham Lincoln once called the awful arithmetic, what did America get for the loss of twenty-three good men like John Smith? For a year, 2-11 Cavalry and Muhammed's Iraqi soldiers patrolled by day, raided by night, cleared the roads of IEDs, and resupplied themselves, all at cost. They left behind a different region, not yet better, but just maybe getting there.

When 2-11 Cavalry turned over north Babil to the next unit, the Janabis were far less active, and their Obeidi rivals were also beaten up. With many of their most aggressive leading members dead or captured, both tribes reverted to easier prey—each other. There were U.S. platoons salted across the landscape, each living with Iraqi soldiers and, more and more, locally raised Iraqi police. They held outposts all along that contested Sunni/Shia dividing line, in Iskandariyah, Eskan Village, the Mussayib Power Plant (Lieutenant Colonel Muhammed's official headquarters), Jurf-al-Sakhar bridge, and even lowly Diyara. Local Sunni farmers, hesitant at the start but more willing every week, took a few U.S.-funded bridge-repair and well-drilling contracts. When the constitution was ratified in October and the first Iraqi government elected in December, north Babil Sunni Arabs voted. They didn't get the results they wanted—few Sunni did—but they showed up and earned their purple fingers.

It was an especially impressive performance in a secondary effort, what the Army calls an economy-of-force area by an economy-of-force organization. The Second Squadron, Eleventh Cavalry was not supposed to be a deployable unit. It was, in fact, a superbly conditioned and skilled training aid, a full-time enemy for stateside exercises. In the Fort Irwin desert, Bill Simril's squadron formed part of a unique opposing force that matched up against units coming to train at the Army's National Training Center. The 2-11 Cavalry troopers used their own versions of Soviet T-72 tanks and BMP infantry fighting vehicles. In their time, they had portrayed the Soviet Russians, the Iraqi Republican Guard, and even Afghan and Iraqi guerrillas, all to hone the tactical expertise of units across the Army. Nobody considered the Second Squadron or its parent Eleventh Cavalry Regiment fighting units. They were school troops. But in 2005, that changed. The Fort Irwin Opposing Force went to war.

Bill Simril's troopers deployed as an expedient, to gain time for the rest of the U.S. Army to reorganize, re-equip, and retrain to fight the long war shaping up in Iraq and Afghanistan. When he took over as chief of staff in August of 2003, General Peter J. Schoomaker brought a decidedly fresh perspective to the position, starting with a fierce urgency to make overdue changes. A charter member of the SOF elite, Schoomaker was pulled from retirement to take over the Army. Secretary of Defense Donald Rumsfeld endeavored to shake up the service most heavily involved in both Iraq and Afghanistan. As well as the U.S. Army did in the opening campaigns, it was evident that for senior soldiers, the idea of rotating units in and out of two lengthy, bubbling counterinsurgencies just didn't match the preferred methodology. After Vietnam, the Army configured itself for short, intense blitzkriegs, such as Desert Storm or the initial weeks of Iraqi Freedom. It would take a massive overhaul, top to bottom, to rig the Army to carry on a long, grinding counterinsurgency, a lengthy war of attrition. The change had to happen while the Army kept fighting. There would be no time-out. In choosing Schoomaker, Rumsfeld got just the man to pull it off.

Commissioned through ROTC at the University of Wyoming, Schoomaker alternated his early SOF assignments with conventional armor/cavalry duties. Thoughtful and perceptive as well as tough and forceful, Schoomaker got an early education in combat during special operations missions into Iran, Grenada, Panama, Haiti, and Iraq. Building American SOF had been long and painful, but the Task Force chewing its way through Zarqawi's AQI in 2005 directly reflected the vision and leadership of Pete Schoomaker. Now this most unconventional soldier ran the entire Army. And he had ideas.

Schoomaker started his tenure by dying. The unwieldy Army finance system made no provision for reactivating a four-star general. The solution was to have Schoomaker officially pass away and then restart his accounts. The new chief of staff drew a lesson from that. In essence, his assignment, and indeed the entire war, was being treated by the institutional Army as a one-off, a workaround, a temporary unpleasantness soon to pass. General Schoomaker knew better. With his spur, soon enough, the Army did too.

Reorganization came first. The Army's divisions (twenty thousand soldiers) were too big and too few for repeated unit rotations. Why not organize by brigade combat teams (BCTs), about four thousand strong? By Schoomaker's figuring, without adding to its overall end strength, the Army could raise seventy-eight BCTs (forty-five regular Army, twenty-eight National Guard), and they could be plugged in as required rather than tied down to one of the divisions. From now on, divisions would supervise two to six brigades drawn from across the Army. The cohesive team would form and deploy at the brigade level. The new chief of staff called it a modular approach. Various entrenched bureaucrats quailed. As usual, all the brainiacs demanded more analysis, and *anal* was the key syllable. That wasn't Pete Schoomaker's way. "The field commanders want five-dollar and one-dollar bills," he said. "We keep handing them twenties."

In this, Schoomaker drew on his own SOF background as well as his cavalry experience. He also read and circulated a persuasive book that strongly advocated a brigade-based army. A lot of what Schoomaker wanted to try was discussed in *Breaking the Phalanx* by acerbic retired Colonel Douglas Macgregor, a veteran of the 1991 clash at 73 Easting. Macgregor himself rubbed many generals the wrong way, but given the demands of a long war, his ideas made a lot of sense. The fact that Schoomaker endorsed Macgregor's book to his generals sent a message all its own: *Get over personalities and egos. Let's fix this thing.* Like Schoomaker, Macgregor came from the cavalry perspective. Cavalry units routinely mix tanks, scouts, mortars, and engineers at very low levels, within hundred-man elements. Form the team, train the team, and deploy the team. Push the key tools to the lowest level. Don't hoard all the goodies up at division while you wait to put together some grandiose crossing of the Rhine River.

Realizing he wouldn't get more soldiers, Schoomaker told his subordinates to squeeze more out of what they had. Each of ten regular Army divisions raised a fourth maneuver brigade, adding ten more deployable BCTs to the pool. Divisions shut down long-established but now extraneous headquarters: the division engineer brigade, the division artillery, the division support command, the MI battalion, and the signal battalion. All of their subordinate battalions and companies got divvied up and assigned to the new BCTs. Short-range air-defense battalions converted to cavalry squadrons — every BCT got one, yet another reflection of the critical importance of finding the enemy in this war. Along with the new cavalry squadrons, brigades cut to two infantry or armor battalions, giving up their old third-maneuver battalions to help create the new BCTs. Inside the heavy battalions, the ones with tanks and Bradleys, the model became two tank and two Bradley companies, plus an armored engineer company, a formidable array. The light battalions (airborne, air assault, and light infantry) also kept four companies: three rifle units and a weapons company. Cold War air defense, heavy artillery, chemical defense, and headquarters went away, cashed in to create the new BCTs.

This reorganization spread root and branch, Army-wide, but it took time. With General George Casey's agreement in Iraq, as well as acceptance in the much smaller Afghan command, Schoomaker bought the necessary months in 2004–05 by sending all the Army could send. School troops like the Eleventh Cavalry Regiment prepared and deployed. Their counterparts from Fort Polk's Joint Readiness Training Center did too. A large number of National Guard units, some thirty thousand, organized on the old setup but already mobilized, also trained up and deployed. In May 2005 in Iraq, for example, the Army National Guard provided the Forty-Second Infantry Division (New York), the Second Brigade, Twenty-Eighth Infantry Division (Pennsylvania), the 256th Infantry Brigade (Louisiana), and the 278th Armored Cavalry Regiment (Tennessee). The higher headquarters and partner units for 2-11 Cavalry came from the 155th Armored Brigade (Mississippi). Trained extensively and employed in tough areas, the Guardsmen performed well, and brought something extra. As citizen-soldiers, Guardsmen came from every profession and trade. A Guard rifle company included in its ranks businessmen, farmers, teachers, mechanics, well drillers, law enforcement officers, and medical specialists. It made every Guard contingent a de facto civic action team as well as a troop of fighters. That went a long way in 2005 Iraq.

Schoomaker's reorganization went beyond the line-and-block charts. He demanded, and got, authority to re-equip the force wholesale. In SOF, the leadership has long enjoyed independent procurement authority not tied to the slow, cumbersome acquisition laws that often make it hard to get a new tank or even a new pistol in less than a decade. Schoomaker asked for similar authority for the conventional forces. He got most of what he asked for. Every Army truck got up-armored, with thousands of vehicles rebuilt or produced new on an industrial scale. To defeat IEDs, the Army organized what became the Joint IED Defeat Organization (JIEDDO), a skunk works of technical and tactical countermeasures to reduce the impact of the enemy's favorite weapon. Although 2-11 Cavalry could not borrow a Predator surveillance drone for the May 11–12 mission, its successors would bring their own, little ones, called Shadow at the BCT, and the hand-launched Raven in each company, battery, and troop. Even the Army's desert-camouflage uniform underwent a major makeover; now it was a practical gray-green, pixelated-pattern army combat uniform with Velcro easy-on/easy-off patches, reinforced elbows and knees, and comfortable, unshined (and unshinable) desert-tan boots. The SOF brothers recognized the family resemblance.

Reorganization and re-equipping solved the physical needs of an army in a long war. But Schoomaker knew well that it took more. The old trade-union mentality — if you need fighting, call an infantryman, but don't expect clerks or mechanics to shoot machine guns — might have made some sense in some far-flung, massive, linear conventional battlefield. It made zero sense in Iraq. The fuel convoys on Highway 1 were knee-deep in IEDs and RPGs every night. Every soldier had to be ready to fight, navigate, patch the wounded, deal with prisoners, and keep going. The Marines had long embraced this ideal, every Marine a rifleman. Under Pete Schoomaker, the Army trained and indoctrinated in its soldiers the same kind of belief.

Schoomaker referred to it as the "Warrior Ethos." He himself wrote the four key lines, right out of the Ranger/SOF worldview:

> *I will always place the mission first.*
> *I will never quit.*
> *I will never accept defeat.*
> *I will never leave a fallen comrade.*

More than a few Army insiders snickered. *We're not all Rangers. You can't ask female finance clerks or tubby supply sergeants to live up to that.* In every war, there were carnivores and herbivores, hunters and hunted. You can't teach the cows to fight. So said the crusty old NCOs.

Schoomaker thought different. All his time in SOF had taught him that most people can rise to challenges. Set high expectations, and most would strive to meet them. Starting in basic training, Schoomaker forced use of weapons, not just rifles, but machine guns, by all soldiers. He re-instituted hand-to-hand combatives. He emphasized tactical thinking and endurance. From the highest Army level, Schoomaker dictated basic warrior tasks — marksmanship, first aid, land navigation — that every soldier, from private to general, had to master before deployment. Shown the high bar, young American men and women did just as the Marines had done for years. They stepped up and delivered. As one sergeant major put it, referring to the elite black headgear long tied to the Rangers, "General Shinseki gave us all the beret in 2000. But Schoomaker made us earn it."

Schoomaker's Warrior Ethos directly affected the nature of combat in both Afghanistan and Iraq. In any casualty situation, American units fought tooth and nail to recover their dead and wounded. Americans left no one behind. Accordingly, and despite hundreds of thousands of squad-level dismounted operations performed day and night by every kind of unit, including those composed of supply clerks and wrench turners, the U.S. Army accounted for every man and woman. It could not bring all of them home alive. But it brought all of them home. Thousands would not go missing in this war.

Sunni rejectionists saw the American solicitude for casualties as a weakness. They often massed fire on evacuation efforts, trying to kill and wound more Americans and derail ongoing operations. But given that finding the enemy was the usual U.S. problem, a bunch of hostiles jumping on a medevac amounted to a convenient target-rich environment. The bad guys never figured out that backing off might have saved them. In this sense, they should have hewed more closely to Chairman Mao's dictum. *Enemy advances, we retreat . . .*

Others watched and learned too. The Iraqi soldiers and police loved the fierce American commitment to recover their dead and wounded, including the Iraqi partners, of course. Saddam's old army had never cared about privates, but Iraq's new army certainly did. It all greatly aided recruitment. Knowing you would come home, dead or alive, meant something in a small village like Diyara. The more Iraqis in uniform, the more they did for themselves. In 2005, even in the Sunni districts, that was starting to resonate.

The hard work of 2-11 Cavalry resembled many similar efforts across Iraq, thanks to the various National Guard formations, the new Schoomaker-approved modular Third Infantry Division in Baghdad, the next set of Marines in Anbar Province, the British in Basrah, the Koreans up north with the Kurds, and the other Coalition partners doing their work in the southern provinces. It all added up as 2005 went on. Behind this security shield, David Petraeus handed over MNSTC-I to Lieutenant General Marty Dempsey, late of First Armored Division, and the factory kept right on churning. By late 2005, Iraq fielded ten Army divisions, two special police divisions, and almost a hundred thousand local police. Not all the Iraqi units were completely manned, not all were fully equipped, and there were major holes in terms of combat support (they were missing engineers, signal troops, intelligence detachments, and logistics). But every fight now included Iraqis.

The largely unsung achievements of 2-11 Cavalry and the rest did much to clear many Sunni areas. One high-profile effort up north garnered a lot more interest. Colonel H. R. McMaster, arguably the most well-known alumnus of the 1991 tank battle at 73 Easting, orchestrated the clearance of Tal Afar, west of Mosul. The city of 150,000, including eighty thousand Turkmen, also held squabbling minorities of Sunni and Shia Arabs. The Sunni Arab rejectionists in the city ramped up intimidation against both Shia Arabs and Turkmen. From May to July of 2005,

10 percent of all attacks in Iraq happened in and around Tal Afar. McMaster decided to do something about it.

McMaster's cavalry troopers applied all the experience others had gained in Najaf, Samarra, and Fallujah. They went slowly in June, July, and August, gradually clearing the surrounding villages in partnership with the Third Iraq Army Division. Together, the American and Iraqi soldiers then installed police stations and ensured they stayed manned. American and Iraqi forces encircled the city and encouraged the citizens to depart. They drummed hard on certain themes, especially the cruelty of the Sunni Islamists trying to run neighborhoods under their version of bloody shari'a strictures. Having lived under this gory regime, Tal Afar's citizens didn't need much encouragement. The trickle of intelligence tips from inside the city became a torrent. In August, Iraq's Prime Minister al-Jafaari appealed to the people of Tal Afar to support their imminent liberation. An expert at using propaganda, McMaster brought all of these messages together. His squadrons and the Iraqi battalions then walked the walk and made it so.

The steady summer pressure led to a fairly fast final sweep in early September. In the words of one U.S. adviser NCO, in his Iraqi battalion's area, there was "not a hostile shot," and "only one blackened corpse, left rotting for days." Just 137 enemy bodies were found, less than a tenth of those expected. Most of the enemy was gone — not dead, just dissipated, broken, out of the fight; they'd quit, given up. In the end, that served just as well. The Third Iraq Army Division and its American partners immediately established outposts throughout the city. Police followed. Clear-hold-build was in effect.

Some skeptical senior officers in Baghdad discounted the Third Armored Cavalry Regiment's Tal Afar operation as nothing more than what other units were doing, just better packaged and oversold. Maybe so. Few U.S. Army officers matched McMaster in engaging and winning over journalists, and the press flocked to Tal Afar to soak it up. His McMastery of the American press rivaled that of his mentor Dave Petraeus. But you couldn't argue with success. Following the clearance, Tal Afar remained firmly under Iraqi government control. Even President Bush cited Tal Afar as an example of progress in Iraq.

Above it all, carrying out what he and General John Abizaid at USCENTCOM remained convinced was a sound approach, George Casey assessed the campaign. Whether publicly acclaimed like Tal Afar or known only to those in country like north Babil, the counterinsurgency moved ahead. Al-Qaeda out, Sunni in, Iraqis increasingly in the lead.

With regard to al-Qaeda in Iraq, the SOF proved relentless indeed. They shredded Zarqawi's network, killing and capturing each newly elevated subordinate in turn. Had he been a U.S. Army division commander, Abu Musab al-Zarqawi would be operating with the equivalent of captains as his brigade commanders and lieutenants as his division staff, if that. Hounded, hunted, one step ahead of the black helicopters, Zarqawi kept going. His suicide bombers struck less frequently, but when they did, they slaughtered Iraqi civilians by the dozens. On November 9, Zarqawi's teams hit three hotels in Amman, Jordan, killing 60 civilians and wounding 115. It was an intentionally bold stroke, serving notice that the AQI chief was still in the game. Zarqawi wasn't prospering, but he and way too many of his pieces were still on the board. Squirming under the pressure, Zarqawi wasn't making much headway on pitting the Sunni against the Shiites. Indeed, the elections and the American-led counterinsurgency were pushing the Iraqis together. Sunni were starting to come in, not many, but some. It wasn't a good sign for AQI. Frustrated, hard-pressed, Zarqawi sought a game changer, but Amman was not it.

As for "Sunni in," the U.S. embassy did fine work behind the scenes. Ambassador Zalmay Khalilzad, a personable Afghan American who'd previously served in Kabul, took over from John Negroponte. The new ambassador continued the embassy's tight coordination with MNF-I. Khalilzad and his foreign service officers nudged Ibrahim al-Jafaari and his Shia colleagues into accommodating the Sunni in the draft constitution. This did not come easily. Credible Sunni were thin on the ground in Baghdad governing circles, but to defuse the Sunni insurgency, a political path had to exist. It became accepted that Sunni would hold a certain number of key posts, as would the Kurds. The Shiite leaders appointed some Sunni to key offices. For example, under al-Jafaari's transitional administration, Sunni Arab Saydoun al-Dulaimi of Ramadi in Anbar Province served as the minister of defense, an absolutely critical post.

Another initiative also glimmered. Encouraged by Minister of Defense Dulaimi, approved by George Casey, the Abu Mahal tribe out on the Syrian border asked permission to be armed and trained as a militia to fight al-Qaeda. There were not enough Marines or Iraqi soldiers way out there, and Zarqawi's AQI depredations had slain men, women, and children in Abu Mahal villages. At Casey's direction, Lieutenant General Marty Dempsey and MNSTC-I trained this new security force. The Ministry of Defense supported it fully with instructors and weapons. Dulaimi even gave them their name on September 8, 2005: the Desert Protectors. They represented the first time Sunni Arab tribal sheikhs had asked for government help to defend themselves. Here, al-Qaeda out, Sunni in, and Iraqis increasingly in the lead really seemed possible. If it went anywhere, that is. And if it lasted.

The fall elections went off on schedule. Sunni Arabs joined the rest of Iraq in voting for the constitution in October and then electing the first real government in December. In the largest of the year's three turnouts, on December 15, 2005, some 11.8 million Iraqis raised their purple fingers. If the Iraqis quickly formed an inclusive government, this thing just might work. The normal winter lull helped. After all the death and blood, you could almost see it, the light on the far horizon. In Baghdad and Washington, a few even began making noises about U.S. troop withdrawals. It could happen.

Only one number sat wrong. At year's end, the MNF-I organization running the detention camps, Task Force 134, reported holding almost fifteen thousand captives. That amounted to the entire insurgency as estimated by military intelligence, and it didn't include the thousands known killed. Purple ink went only so far. Somebody was still out there shooting back.

10

IMPLOSION

Have you ever been experienced? I have.

— JIMI HENDRIX

IT'S NOT OFTEN that you can pinpoint exactly when a massive undertaking fails. For the U.S. in Iraq, though, the moment was obvious. At 6:44 a.m. on Wednesday, February 22, 2006, two bombs detonated inside the al-Askari mosque in the city of Samarra in the heart of Nate Sassaman's old area of operations. The Shiite shrine had stood since 944, its shining golden dome a bright spot in the tawdry old Sunni city. The mosque housed the venerated remains of the tenth and eleventh imams, revered by Shia believers inside and outside Iraq. The al-Askari mosque had hosted millions of pilgrims over the centuries. It had stood through wars, earthquakes, and storms. But modern demolitions proved its undoing. The well-placed charges cracked the north wall. The golden dome sagged, shrugged, then collapsed inward in seconds, casting up a vast cloud of dark dust, a lingering shroud over the devastation. Witnesses later alleged that five to seven men in Iraqi military uniform had been seen in and around the building. There were no casualties onsite that morning. Despite frantic immediate local searches and various follow-up investigations by Iraqi security forces, no viable suspects were ever apprehended.

It was Abu Musab al-Zarqawi's greatest attack. For years, his suiciders had struck Shia markets and neighborhoods, running up the casualty toll. Zarqawi longed to pit Sunni against Shia in a death battle, one that would see the Allah-favored, educated Wahhabis triumph once and for all over the American-backed *sheruggi* bumpkins. By blowing the roof off the Samarra mosque, Zarqawi finally found the right catalyst to ignite the Iraqi sectarian inferno. Barely a step ahead of the Task Force, its ranks savaged, its leadership cut to ribbons, Zarqawi's AQI still pulled it off, better late than never. American pursuers caught up with Zarqawi scant months later, on June 7, 2006, and vectored in a pair of Air Force F-16s to give the AQI chief two five-hundred-pound bombs up the gut. It was a clean kill, but too late. The demons were loose.

Within hours of the news of the February 22 mosque explosions, members of Sadr's JAM were in the streets. They couldn't find al-Qaeda agents, but they could locate Sunni just fine. Sectarian killings began. The violence escalated, with shops grenaded, mosques shot up, and houses burned. Sunni Arabs — already in rebellion — immediately initiated reprisals. Hundreds of Iraqi civilians died. Caught flat-footed, American troops stood back as the two sides went at it. The U.S. troops had their own problems, with both factions taking time out from slaughtering each other to hit the occupiers too. The winter troop rotation was ongoing, coming off the three successful Iraqi elections of 2005. The plan seemed on track: al-Qaeda out, Sunni in, Iraqis increasingly in the lead. Now there was this. One mosque bombing — not like there had not been others or would not be more — and the place went crazy.

Sadr's people stoked the insanity with rumors. The Americans did it. The Israeli Mossad did it. They both did it. In an opportunistic move, AQI propagandists eagerly parroted the Shiite line. Zarqawi didn't claim the attack. He didn't need to. The societal damage exceeded his fondest hopes.

The Samarra mosque implosion found the Iraqi government in transition. Shia, Sunni, and Kurdish parties met and argued, endeavoring to split up the key ministerial posts after the December 2005 election. As this horse-trading persisted, Prime Minister Ibrahim al-Jafaari remained in nominal charge, never too sure of himself even on good days. With armed sects shooting it out in cities and towns, al-Jafaari appealed for calm. "There are no Sunnis against Shiites or Shiites against Sunnis," he intoned. As usual, this ineffectual man was ignored.

Here was the real Iraq, the age-old Sunni-Shia divide scored in blood. The mosque bombing was just the latest pretext. It was always something, and it always ended the same way: beheadings, shootings, stabbings, bombings, death retail and wholesale, violence as language. Death came pretty easily in Iraq. The propensity for murder and mayhem in Islamic Iraq made the gunfights of the American Old West or the Mafia gang clashes of New York City look like playground pranks. Purple-inked fingers; forming an inclusive government; al-Qaeda out, Sunni in, Iraqis increasingly in the lead — all gone, vaporized, swirling around in the crater that used to be the al-Askari golden dome. Clear-hold-build was reset to zero. Well, the Iraqis were definitely in the lead, Shia and Sunni. They would start by clearing each other.

Most Americans in uniform once more asked that perennially urgent question: Who was the enemy? In February of 2006 in Iraq, the answer came back real ugly: Everyone. And they all had guns.

Colonel Michael D. Steele had long ago told his troops what to do. Before they even left Fort Campbell, Kentucky, he'd made it clear to the soldiers of the Third Brigade, 101st Airborne Division: "The guy that is going to win on the far end is the one who gets violent the fastest." He continued, "I'm talking about the moment of truth," he said, "when you're about to kill the other son of a bitch." The soldiers' training was sound. They knew the deal. "Men, it is time to go hunting. You're the hunter. You're the predator. You're looking for the prey."

Now they were out among the prey. Steele was with them on the helicopter assault that morning, knee to knee in the dusty cabin with the tired young riflemen of Company C, Third Battalion, 187th Infantry. They were going to a very bad place, Zarqawi country, the abandoned al Muthanna chemical weapons complex thirty miles south of Samarra. Steele's intel guys predicted a hell of a shootout when the seven Black Hawk helicopters landed. A few even guessed that Zarqawi himself, the evil genius, bête noire of the Task Force, might be onsite. Steele didn't think so. But he was pretty sure there would be plenty of Sunni insurgents, and a lot of resistance.

That was the colonel's way, to set the example, to go where he expected trouble. Mike Steele had been in action before. He'd commanded the Ranger company in the center of the October 1993 Mogadishu battle and earned the Bronze Star with Valor device for his leadership under fire in that gruesome street fight. Tall and imposing, a former University of Georgia football player on the 1980 national championship team, Steele looked every bit the part of the infantry colonel at war. Because he spoke with a southern accent and used folksy terms, people who didn't know him mistook his directness for ignorance. In fact, Mike Steele was among the most astute, educated men wearing an Army uniform in 2006. He had studied this war and this enemy with rare passion, even for a busy brigade commander, and prepared his air assault soldiers accordingly.

Echoing the approach of General George S. Patton Jr., Steele put it plainly to his soldiers:

> I have been in more third world countries than anybody in this room, and I tell you most of them do not speak English. They all speak food chain. And from the time you set foot in their country, they're checking you out, from top to bottom. They've figured out where you are on the food chain. Because if you look like prey, what happens? You get eaten. If you stand there and look people dead in the eye, you have your weapon at ready, and you don't flinch. You look like you're not scared. Even if you are scared, you look like you're not scared. You send the message that *I am the dominant predator on this street, and if you mess with me I will eat you.*

Soldiers understand that kind of blunt talk. It scares the military lawyers and gives palpitations to the public-affairs types, but Steele cared little for either. His brigade combat team, the same organization that Frank Wiercinski had led into the fire in Operation Anaconda four years before, was designed and fielded to find and kill the enemy. At 5:00 a.m. on May 9, 2006, under cover of the last hour of darkness, that was what Steele's men were about to do.

Steele's air assault force was headed for a low, marshy patch west of the derelict al Muthanna compound. Informants, technical means, and even the Task Force analysts all agreed that the few mud huts there hosted hostiles. One source described training with shoulder-fired antiaircraft missiles. That could get interesting for Black Hawk helicopters. Another report indicated that AQI might be playing with residual chemical munitions hidden in the debris of an al Muthanna storage bunker. The fear of hidden Iraqi WMDs may have long evaporated among readers of the *New York Times,* but those in country were not so trusting.

Steele requested an airstrike to prepare the landing zone. There wasn't much point to eating a shoulder-fired rocket on the final approach. Steele had lived that nightmare in Mogadishu in 1993. So the colonel asked for a good plastering of the two flat adobe huts near where the helicopters would set down. The 101st Airborne Division commander, Major General Thomas R. Turner II, turned it down. Deliberate and cautious by nature, Turner did not want to risk spreading potential chemical contamination if the rumors proved true. To make up for the missing aerial firepower, Turner ensured that Steele had a pair of AH-64D Apache attack helicopters accompanying the initial assault.

Aboard one Black Hawk, First Sergeant Eric Geressy looked down as the aircraft swung toward the landing zone. He saw tracer bullets coming up. "This is for real," he recalled thinking. "The intel isn't bullshit." Then the chopper bounced down, twisting a bit on its big rubber side tires. The riflemen tumbled out, weapons level. And the aircraft was off, dust settling behind it in the darkness.

Night-vision goggles down, searching for enemy movement, Geressy's soldiers moved toward the two mud huts. Some fired while others slowly edged forward. There was no return fire. The buildings sat there, windows blank, empty. "Like something maybe Jesus lived in," Geressy said. The objective was abandoned — no enemy, no resistance, no civilians. But who was shooting those tracers?

Still focused on finding out who'd shot at them before the helicopter landing, Geressy's commander, Captain Daniel Hart, directed a machine-gun team to the roof of one of the two abandoned adobe structures. As soon as the pair got up there, they reported a building a few hundred yards distant that had a light in the window. The M-240 machine gunners opened fire, cranking through a dozen 7.62 mm rounds. Geressy immediately ordered a cease-fire, and Hart was right with him on that.

"We don't know what the fuck is in those buildings," said the first sergeant. "We didn't get fire from those buildings, and the enemy on the ground is not the enemy we were briefed on. You gotta be able to switch gears right away. Worst thing I want is we're shooting up a fucking building with women and kids." When the men reached the isolated hut, they found nothing but an oil lamp.

But the Apaches circling overhead found something. They saw a skiff with a motor puttering down the canal. The aviators guessed that the three men on the craft had exited the house with the lamp in it. Maybe they were the tracer shooters. But nobody was sure. Dan Hart, another former West Point first captain and already all too experienced under fire, told the Apache crews to pull the trigger. They killed all three as the Iraqis tried to exit their little boat on the far bank. Hart's soldiers searched later but didn't find the bodies.

So the great air assault turned up nothing: no insurgents, no weapons, no documents, no chemicals, certainly no surface-to-air missiles, and no Zarqawi. It was a dry hole, a bust. Well, at least the Apaches had gotten three. Steele stayed with his men for a few hours. Then, with the hot sun well up and the riflemen beginning to spread out to search other areas, the colonel called in a helicopter and moved on to check out other units working nearby.

One of Captain Hart's other platoons hadn't joined the air assault. The men came by road on armored Humvees to search the villages south of the al Muthanna complex along Highway 23, which ran southwest to Fallujah. Their target was Abu Abdullah, a known Sunni rejectionist cell leader. The brigade S-2 analysts tied Abu Abdullah (whose name meant "father of Abdullah," a common sort of Arabic nickname) to the deaths of at least ten U.S. soldiers. He had links to other Sunni trainers and weapons suppliers in Fallujah. Highway 23 was his place of business.

Lieutenant Michael Horne's platoon drew the Abu Abdullah mission. For this task, Horne's platoon added three snipers and a squad of Iraqi soldiers. Horne's men arrived at their first objective about dawn. The village was abandoned, although a few buildings had red handprints painted on them. Sometimes, that marked an insurgent area. Except for two Arabs digging in a field, the Americans found nothing.

From here, accounts differ regarding what happened next. Horne said later that they talked to the two Iraqis and moved on. According to several NCOs, however, Horne directed the snipers to shoot the two males in the field. The snipers, led by Sergeant Geoffrey Kugler, demurred. With the help of the Iraqi troops, the Americans instead approached and questioned the two civilians, who knew nothing. The younger one appeared to be mentally disabled.

Meanwhile, another of Horne's squads had better luck. The men unearthed a sheikh, his wife, and three children. The sheikh told the Americans that he had been dragooned into housing and feeding insurgents and al-Qaeda fighters as they passed up and down Highway 23. The sheikh hadn't seen any enemies for days. But he'd heard there was some kind of major meeting a few miles to the south that morning, at a gas station on Highway 23. Horne and his sergeants saw their opportunity. The sheikh agreed to come with them.

Horne called Hart on the radio. The Company C commander okayed the impromptu raid and shifted two Apaches to cover the operation. Horne left part of the platoon to continue the search at the first site, and he took four Humvees with a squad, the snipers, and a few Iraqi soldiers to check out the gas station. The Apaches would suffice for backup if it turned out to be a mess.

As the attack helicopters circled on the horizon, awaiting the call, Horne's trucks rolled toward the gas station. They got there around 11:00 a.m. Although the locals appeared unhappy and suspicious, nobody opened fire. The Americans parked, got out, and surrounded the gas station. There were plenty of people there, mostly men, a lot of stink-eye but no AKs in sight.

Then the sheikh motioned to a man sitting on a curb and told Horne the person was an insurgent. That was enough for the lieutenant, who was pretty keyed up. "Shoot that man," he ordered. Taken aback, the nearby riflemen did not. One NCO, Sergeant Nathan Beal, flatly said, "We're not going to kill a guy just for sitting there." They detained the man instead. He squirmed some, but gave up.

The scuffle with the detainee got the crowd agitated. People began moving away from the gas station. Some Iraqis got in their cars. Engines started. That demanded attention, as even without bombs aboard, vehicles can run down a soldier on foot. And they certainly allow a speedy getaway for those intending to leave.

Seeing clearly that he was about to lose a bag of potentially useful prisoners, Horne, already emotional, ordered his men to fire on the fleeing Iraqis. The American riflemen immediately opened up, as did the Iraqi soldiers and two of the U.S. snipers. Horne then got on the radio and attempted to bring down the Apaches, but the pilots couldn't sort out the scrum on the ground and declined to engage. A Humvee gunner began shooting 40 mm grenades onto the roof of a building across Highway 23. It went on for about a minute, a mad minute, some seven hundred rounds flying out, ninety degrees of mayhem tearing up the road and its fronting buildings. The Iraqi people in the area crouched against walls, slithered down low in their car seats, or froze prostrate on the dusty ground.

Steady and clearheaded, Sergeant Beal gained control of the shooting. He and the other NCOs shut it down. "What the hell were you shooting at?" he asked Horne. The lieutenant, clearly confused, didn't say much. Told that some civilians might be down, the officer used the term *collateral damage.* One truck driver was dead. A woman and two other men were wounded.

In the search and cleanup, the Americans took sixty-four detainees. Several hailed from Fallujah. Three were part of al-Qaeda in Iraq. One was Abu Abdullah, who had been walking to a car carrying a baby—getting away cleanly—when the shooting started. No Iraqis carried weapons; there had been no hostile fire. For the price of one dead Iraqi civilian and three wounded, Horne had gotten his man. But the Americans knew well they had done it all wrong.

Lieutenants make mistakes in war. With enemies intermixed with civilians, all dressed alike, it's amazing they don't make more. The Army knows that, and it is part of the duties of sergeants like Geoffrey Kugler and Nathan Beal to speak up and intervene when young officers give dumb orders. The "death blossom" at the gas station was bad news. Thanks to the NCOs, it was fixed quickly, although too late for four Iraqis.

Things did not work out even that well on the west side of the canal. There, about the time it all went haywire at the gas station, Captain Hart's other platoon moved to a secondary site aboard two Black Hawks. The helicopters deposited two squads near a single hut with a roof of thatched reeds.

Staff Sergeant Raymond Girouard led his squad right at the brown mud building. When the NCO saw a male in an off-white man-dress appear at a window, he fired. His men followed suit. In infantry small-unit tactics, soldiers are taught to do as their NCOs do. If the sergeant takes a knee, you take a knee. If he moves right, you move right. If he shoots, you shoot. They shot.

When they entered, Girouard found a splayed, skinny, white-haired seventy-year-old man, Jasim Hassan Komar-Abdullah, with a hole blown in his chest, blood everywhere. His final breaths rattled out. Two teenage males and two women huddled in the corner, the women being held in front of the men. Another middle-aged man stood there. The Americans saw no weapons. With one combat deployment behind him already, a graduate of the Army's tough Ranger course, Girouard had taught himself Arabic to be more effective this time in Iraq. As his men tried to help Jasim Hassan, the NCO questioned the Iraqis. An Iraqi soldier joined in to help.

They figured out that the teenagers were grandsons and one woman was a daughter. The other man was in the wrong house at the wrong time, as was the other female. While this was being established, Jasim Hassan died.

One of the Iraqi soldiers objected to how the old man had died. He spoke English. "You guys know the situation," Sergeant Mohammed said. "This incident makes the people, the citizens, hate us." The Americans made no reply. Instead, one photographed the three other men. A few U.S. soldiers prepared the Iraqi threesome to be flown out for questioning. The Americans bound the detainees' wrists with zip ties, heavy-duty versions of the plastic strips used to close trash bags. All three Iraqis were blindfolded.

Meanwhile, Girouard and his squad moved to the next house. They thought they heard somebody inside. Unsure, the Americans popped a few warning shots over the roof. After the shots, in loud Arabic, Girouard ordered whoever was inside to come out. Seconds later, an adult male came through the low door opening. He wore a white head wrap and a white man-dress and carried a baby girl at arm's length, held out like a peace offering. Or a shield. Or a bomb. Girouard directed the man to put down the girl. The man continued walking slowly forward. The Americans converged on him. Soldiers removed the baby. The idea that the Iraqi was using a baby in some way enraged the soldiers. Girouard's men got very rough with the Iraqi, wrestling him to the ground and dragging him back inside. They punched him several times before Girouard ordered them to stop. A search of the hut revealed nothing of interest. They didn't even detain the Iraqi involved.

What happened next became the subject of numerous investigations and courts-martial proceedings. Girouard initially reported that the three detainees had tried to escape, that one had grabbed a knife, cut the zip ties, and tried to poke his guards. Then the Iraqi trio took off. Two American riflemen, Specialist William Hunsaker and Private First Class Corey Claggett, opened fire and killed the fleeing Iraqis. In the official verbiage of the first of many investigations that followed, a major stated: "I find that the Soldiers clearly acted in self-defense fearing for their safety as they were physically assaulted by the detainees." Hunsaker had knife cuts on his hands and face; Claggett had been belted in the head.

The truth, established by the military courts, proved much more sordid: Girouard had directed his soldiers to kill the detainees. They did so, helped reluctantly by another soldier, Specialist Juston Graber. Then the NCO cut Hunsaker's face and arms with a Gerber knife and punched Claggett in the head, producing injuries appropriate to their agreed-upon cover story. But the men erred in one big way. They let the three bodies be flown out—the intel guys wanted to see who they were, to check them against the list of known insurgents. All three Iraqis still had bound hands and blindfolds partially in place. As Major Stephen Treanor, the brigade S-2, asked: "Why would we shoot somebody, then put a blindfold on them?" This stunk. Steele immediately ordered an investigation.

We will never know what dark thoughts or feelings motivated Ray Girouard, William Hunsaker, Corey Claggett, and Juston Graber to kill three Iraqi prisoners on May 9, 2006. It was the entire Clausewitz friction list—bone-deep exhaustion, Murphy's Law gone wild, an enemy hidden in the fog of an opaque population, and fear. Especially that last one. Every time those three Americans went out, they rolled the dice. In the heat and dust west of Highway 23 on May 9, 2006, all three decided to change the odds. But in the Army, as in Las Vegas, the house always wins in the long run.

Steele's suspicions took a while to confirm. The officer heading the first investigation, finished within Steele's Third Brigade by May 21, bought Girouard's cover story. Statements were taken and proper prose prepared. Digging in this graveyard, though, brought up very grim rumors. As early as May 22, Private Jonathan Porter made a statement to the Criminal Investigation Division that placed Colonel Steele himself at the center of the killings. Porter, facing discharge for cocaine use and other infractions, wasn't a very credible source. But his jungle telegraph had picked up some vibrations, and CID was not slow to follow up.

Word quickly got to the Multi-National Corps level. It reached a very anxious ear in Lieutenant General Peter W. Chiarelli. A smart, charismatic armor officer, Chiarelli had gained his commission through ROTC at Seattle University. His brains and skill got him a position teaching in the prestigious social science department at West Point, and Chiarelli benefited immensely from that experience and the insights he gained. In an army where the blitz-job crewcut or even the shaved head was de rigueur, Chiarelli cultivated a shock of dark hair and the free thoughts that came with it. He'd picked up some gray streaks over the years, but his mind remained young, questioning. Not old enough for Vietnam, left out of Desert Storm, Chiarelli got all the combat he wanted commanding the First Cavalry Division in Baghdad from 2004 to 2005, tangling with Sadrists and Sunni rejectionists. But ever the guy who looked past the obvious, Chiarelli became fascinated by the possibilities inherent in civic action. He believed that if average Iraqis received essential services, the insurgency would be deflated.

A senior combat commander interested in governance was rare indeed. Just as most NFL coaches specialize in either offense or defense, so most Army infantry, armor, and artillery commanders in Iraq focused on targeting or, maybe, developing partner forces to do likewise. Chiarelli was like an NFL coach who fixated on the punting game. General Tommy Franks's lack of interest in phase-four reconstruction was typical. Coaches had to think about punting, and commanders had to consider governance, but it was a side matter, special, tertiary, something for reservists, contractors, and earnest young State Department types. The real war was about finding and killing the enemy. That's what soldiers did. Colonel Mike Steele might be an especially bold example, but he was a lot more emblematic of U.S. Army leadership than Pete Chiarelli.

Yet now Chiarelli commanded the Multi-National Corps headquarters in Iraq. His predecessor, Lieutenant General John R. Vines, a manhunter to his core, would have dealt with the Highway 23 incident as a crime and left it at that, an aberrant act by undisciplined soldiers. In the wake of the February 2006 Samarra mosque bombing and the subsequent sectarian warfare, just keeping after the enemy and staying alive in Salah ad Din Province took everything Steele's soldiers had. But Pete Chiarelli could not leave it at that. He knew Steele, and he did not think much of him. In one spring encounter with Steele at the colonel's command post outside enemy-infested Samarra, Chiarelli called the brigade's plans "unacceptable." "You are going to go around conducting operation after operation," said Chiarelli, "but you don't give these people some reason to hope their life is going to get better." Steele didn't get the value of reconstruction. Whether anyone in Samarra, one of the most consistently hostile cities in Iraq, wanted anything reconstructed by American occupiers went unasked.

Even in the face of mounting Sunni/Shia violence, and long before the May 9 killings, Chiarelli had gotten worried about how often soldiers were shooting at the locals. When the general's staff pointed out that American soldiers were killing an Iraqi a day at checkpoints or near convoys, Chiarelli ordered that every shot fired in an "escalation of force" was to be investigated by a field-grade officer under the provisions of Army regulation 15-6. The corps staff reported a few weeks later that such events had dropped by 85 percent. Chiarelli considered it progress, but he shouldn't have been so optimistic. Out on the roads, embattled Americans had kept right on shooting. To avoid the endless burden of written investigations, units reported only the encounters that resulted in confirmed Iraqi casualties. American soldiers didn't inquire too closely about those either.

It all circled back to the same damnable question: Who was the enemy? In the Kurdish north or Shia south, or even in Shiite neighborhoods in Baghdad—provided Sadr wasn't acting up this week—you could find friendly Iraqis. In shattered Samarra and the wasteland around it, and in all the Sunni Arab regions, friendly Iraqis were rare indeed. Polls conducted in Sunni areas showed 85 percent support for the "honorable resistance," especially for attacks on Americans. How did you categorize an Arab man in such an area? Enemy? Sympathizer? Potential sympathizer? Every man shot by U.S. soldiers wore civilian clothes. If he had an AK-47, was he getting ready to shoot you or merely defending his family? If he was talking on a cell phone, was he tipping off the insurgency or setting off an IED, or was he phoning his wife? Multiple times a day, in hundred-degree heat and gritty dust, young Americans bet their lives on answering these questions correctly. Chiarelli had agonized over the fact that his people were killing one Iraqi a day in roadside encounters. Steele's men could well have replied that seventy-nine Americans had died in May 2006 alone, nearly three a day. The troops did their own math.

Under the pressure of questioning, Steele's accused soldiers tripped over their lies. Cornered, implicated, they began to claim that their orders on May 9 were to "kill all military-age males." Of course, no such orders had been issued, in writing or verbally. Smart defense attorneys, however, saw an opening to mitigate their clients' guilt. The atmosphere in Company C got a hard look. But that was an appetizer. The defense counsels really keyed on Mike Steele's fighting words. The colonel supposedly set the tone of "kill them all." It all echoed a similar incident in Sicily in 1943 when soldiers in the Forty-Fifth Infantry Division had killed seventy-three Italian prisoners and then justified their acts by reference to the pre-battle "blood and guts" exhortations of their Seventh Army commander, Lieutenant General George S. Patton Jr. Patton's superior, General Dwight D. Eisenhower, dismissed the entire episode as a desperate court-martial ploy, a distraction from winning a world war. But this wasn't World War II.

It was tragic to watch the colonel and general heading toward their certain collision. They were two strong-willed, passionate commanders. Mike Steele didn't want to kill every Iraqi. He was just struggling to find some who didn't want to slaughter him and his soldiers. For his part, Pete Chiarelli had been under fire more than enough, and his First Cavalry Division had slain thousands of insurgents who'd needed to be eliminated. But he knew enough to realize that manhunting wasn't solving this thing. Maybe civic action would work. It might help, for sure. In some ways, Pete Chiarelli would have been the perfect corps commander for Iraq in 2016 or, better, 2026, many years after the blood-soaked dust settled. But in 2006, with Sunni and Shiites at war with each other and with the infidels too — well, it was like trying to fix a slashed throat with a Band-Aid.

In late May, Chiarelli got word of what had happened on Highway 23.

He was already absorbing a lengthy, dire account about a Marine element that had killed twenty-four Iraqis at Hadithah in November 2005, another confused, awful event with junior people pushed too far. In addition, word had percolated up of a murder-rape in Yusufiyah, south of Baghdad, perpetrated by another brigade of the 101st Airborne Division in March. Now came word of the prisoner deaths in Steele's brigade. If the Marines and the 101st Airborne — some of the military's most select, well-led units — were doing such things, Chiarelli had to wonder what it indicated about the overall morale and discipline of the corps. The general knew how the U.S. military had disintegrated as Vietnam wrapped up. This stuff had to stop immediately. Chiarelli was inclined to act.

The early line coming from Samarra fed Chiarelli's bias. The "kill all military-age males" phrase alarmed the corps commander. True or not, it was an extant perception, reported by multiple witnesses. As CID investigators probed further, they heard things that raised other serious questions for Chiarelli: Dan Hart's Company C maintained a scoreboard listing the number of enemy deaths credited to each platoon. Mike Steele gave out kudos and commander's coins for good kills. Accounts of Lieutenant Horne's emotional kill orders surfaced, as did the solid work done by NCOs in keeping things under control. Then the videos of Mike Steele's pre-battle speeches turned up. It wasn't hard to connect the dots: an overly aggressive chain of command, a lot of emphasis on killing, and an enemy who wore civilian clothes. It seemed like a rerun of the worst episodes of the Vietnam War.

Chiarelli left the criminal matters to the CID investigators, who proceeded to deal with them apace. But he wasn't content to leave suspect commanders in place in Samarra. He could have gone after the company or the battalion leadership, but Chiarelli knew where he wanted to look. The MNC-I commander asked a general officer to investigate Steele. Brigadier General Tom Maffey, an experienced infantry officer who knew Steele well from the Rangers, was detached from the Fourth Infantry Division fighting in Baghdad. Maffey was told to examine Steele's command climate. Chiarelli wanted to fire the brigade commander.

Maffey did a thorough job. His report clarified a lot. He stated that Steele "did not condone or attempt to cover up detainee deaths," and in fact, immediately after the event, he'd commenced an investigation. Despite the rumors, Steele had nothing to do with the prisoner killings beyond determining they looked fishy, a positive reflection on the colonel's desire to sort out the incident. Although aggressive, Steele had consistently emphasized discipline and lawful treatment of captives. Even his pre-combat speeches, the ones alternately enthralling and horrifying civilian audiences on YouTube, included clear instructions to adhere to discipline. Steele had done nothing wrong. This was about a few soldiers, not the brigade as a whole.

Chiarelli didn't see it like Maffey. He issued a memorandum of reprimand on July 11, 2006, long before the CID wrapped up its work. In the view of the MNC-I commander, Colonel Mike Steele had played fast and loose with the rules of engagement, resulting in the deaths of "five unarmed people." In a favorite phrase, one he used often in daily discussion, Chiarelli noted that Steele had not considered the "second- and third-order effects" of his words, actions, and decisions. Chiarelli stated: "Your acts, omissions, and personal example have created a command climate where irresponsible behavior appears to have been allowed to go unchecked." When Steele reported to Baghdad to receive the dressing-down, some of Chiarelli's personal staff worried the big colonel might react violently. To the end, they misjudged Mike Steele. He took it like an officer and a gentleman.

Afterward, bowing to the characteristically judicious advice of Major General Tom Turner at the 101st Airborne, Chiarelli left Steele in command. The reprimand ensured that the colonel would never make general. But one of Turner's deputies, Brigadier General Mike Oates, pegged the real issue. Chiarelli's intervention arose from "a fundamental difference of opinion about how to prosecute the war in Iraq." In the U.S. Army, three stars always trumps an eagle. Uninterested in such intramural infidel scrapes, the Iraqi enemy fought on.

One level up from Chiarelli, General George Casey at MNF-I recognized that the situation in Iraq was sliding into a bad state. The U.S. four-star didn't waver. He hewed to the strategy long ago agreed on with his USCENTCOM superior, John Abizaid. Help the Iraqis take over; al-Qaeda out, Sunni in, Iraqis increasingly in the lead. But after Samarra, that mantra rang very hollow. The Iraqis were not ready to take over, and conditions were deteriorating. The end, or even a reduction in U.S. effort, was nowhere in sight. Three years into the war, Casey and those back in Washington stared at what a counterinsurgency really meant: an unending war among a hostile people going on as far as you could see.

Casey had been in place two years, Abizaid three. They were hanging in there, and they recommended that America do the same. In their view, if the U.S. remained committed, a good result might emerge. But if the U.S. bailed out, failure was certain. Thanks to the hard work done by Pete Schoomaker's U.S. Army, the Marine Corps, and the other services, the rotational forces would keep coming, and they would be ready. This wasn't the crumbling U.S. force of 1971. The volunteer military was in it for the long term, deployment after deployment.

Still, the prospects for success were poor. Casey's intelligence people saw that "sectarian violence had become self-sustaining." Casey identified "an almost predictable cycle of al Qaeda suicide attacks, followed in a few days by Shia death squad attacks against Sunni areas in Baghdad." Civilians were now the primary victims of both Sunni and Shia insurgents. According to MNF-I estimates, there had been more than a thousand civilians killed in sectarian slaughter every month since the Samarra mosque bombing in February. American casualties remained high, averaging more than seventy dead a month. Enemy attacks ran more than a hundred a day, a third higher than in 2005. It was getting worse.

American soldiers saw the awful evidence daily in the Iraqi capital. Each morning revealed bodies floating in the Rustamiyah water-treatment plant a half mile from the Iraqi military academy. Random corpses decorated markets and gutters, usually on the borders between Sunni and Shia blocks. "Every patrol, you're finding dead people," remarked Staff Sergeant Ian Newland of the Fourth Infantry Division. "It's like one to twelve a patrol," he continued. "Their eyes are gouged out. Their arms are broken. We saw a kid who had been shot ten to fifteen times." It was brutal, with little relief in sight.

The Iraqi government wasn't improving either. The promise of the 2005 election cycle soured into the reality of a very weak Iraqi partner. The new Iraqi government in Baghdad, finally seated by June, reflected sectarian dealmaking, not meritocracy. Kurd Jalal Talabani served as president, nominally ceremonial, but in fact, a player, perhaps the most able and certainly the most pro-American voice in Baghdad. The Sunni got one of the two vice presidencies, the Speaker of the parliament, and the vital defense ministry. The Shia, however, held most of the power, notably the interior ministry, the key social welfare ministries, and, of course, the prime minister. Sadrists held several key posts, including the leadership of three ministries, as integral parts of the ruling coalition. Nouri al-Maliki succeeded Ibrahim al-Jafaari as prime minister.

Maliki's last name meant "king" in Arabic, but his résumé screamed "pawn." This slight, quiet man was an unlikely pick, clearly a compromise choice plucked from obscurity over several stronger candidates. He had spent twenty-three years outside Iraq, much of that time in Syria and Iran, hardly positive recommendations. Maliki appeared even more hesitant and inept than Jafaari had been. Tellingly, he refused to shake hands with U.S. Secretary of State Condoleezza Rice. Maliki didn't deal with women.

He also didn't want to deal with the Shiite insurgency. When General George Casey and Ambassador Zalmay Khalilzad appealed to Maliki to cooperate in cleaning out the Sadrist/Iranian rat's nest in Sadr City, Maliki flatly refused. He preferred to chase "Baathists," the various Sunni rejectionists, and the al-Qaeda network. Exerting his authority, the Iraqi prime minister often refused to commit his forces against the Shia side.

Unshackled, Sadr's JAM and its Iranian facilitators went hard against the Americans in Baghdad and the British in Basrah. The Iranians had been experimenting with a new kind of IED, a neatly machined concave copper warhead that, when fired, formed a molten metal jet capable of spearing through an M-1 Abrams tank like a hot nail through butter. The Americans called it an EFP, for explosively formed penetrator (or explosively formed projectile). In the summer of 2006, EFPs turned up over and over again in and near Shia districts. American soldiers died. Beholden to his Sadrist supporters and, perhaps, his Iranian backers, Maliki prevented the Americans and their Iraqi partners from going in to address the threat.

There was another problem. In the early days of MNSTC-I, Dave Petraeus and his team gave in to Ministry of Interior requests to raise two divisions of special police, complete with armored vehicles. Such separate armies are common in the Middle East as regime-protection troops. The Saudi National Guard and Saddam's far more sinister Republican Guard come to mind. In 2004, hungry to aid any Iraqi who wanted to get into the fight, Petraeus had supported the formation of the two divisions. Since that time, both had become Shia-dominated. They often had their own agendas, thoroughly anti-Baathist, in Maliki terms. As early as November 2005, Brigadier General Karl Horst of the Third Infantry Division found fifty-six prisoners, all but three Sunni, being horrendously maltreated in Site 4, a Ministry of Interior facility near central Baghdad. Briefed by Horst, Casey and MNSTC-I cut off some support. For their part, the Iraqis took no action beyond releasing those particular captives and reassigning a few officers. The Site 4 follow-up got lost in the undertow after the February 2006 Samarra mosque bombing. The Americans needed Iraqis to fight. And the Iraqis would do it their way, more and more as the U.S. turned areas over to them.

The Iraqi way in mid-2006 proved utterly inadequate, especially in Baghdad. Two major American-Iraqi operations, Together Forward I and II, both fizzled. Americans and Iraqis cleared Baghdad districts and set up checkpoints. But the Iraqis couldn't hold the areas, and the Americans just didn't have enough troops to spread out into hundreds of outposts. Unless dragged along by Americans, Iraqi soldiers did not patrol on foot to secure neighborhoods or raid to root out militias. Iraqi soldiers and police stood impotent on their little street-corner checkpoints while each night, gunmen waylaid sectional opponents. Al-Qaeda suicide bombers hit Shiite markets and housing blocks. Sadrists retaliated by shooting Sunni Arab men and evicting the surviving families from their homes. Much of east Baghdad became wholly Shiite in a sectarian cleansing that often resembled the Balkan excesses of the 1990s. Turning over an area to Shiite-dominated Iraqi units meant running off the Sunni . . . or worse.

As Maliki's Shia-led Iraqi forces took over, they also began limiting U.S. actions in "their" Baghdad districts. In an especially egregious incident, on October 23, Shiite militiamen snatched an Arab American soldier, Ahmed Kousay Al-Taie, outside a U.S. compound in central Baghdad. Unable to find the missing man, the Fourth Infantry Division established an extensive grid of checkpoints, focused on known Sadrist hangouts. Maliki objected to placing all of these new checkpoints in Shiite districts. After extensive face-to-face discussion, Casey got the prime minister to agree to allow the checkpoints to be manned at night. It was half a loaf, and not enough. Al-Taie wasn't found until 2012, after the U.S. withdrawal. By then, of course, he was long dead, executed by his captors.

It all led to questions about the basic Casey plan of turning the war over to the Iraqis. With the three 2005 elections over, MNF-I wanted to begin passing cities, districts, and entire provinces to the Iraqis and commence the long-awaited U.S. drawdown. But Maliki was too weak to take the baton. Iranian influence was growing. The Iraqi forces, while numerous, varied widely in reliability. Clearly, the Shiite-dominated Ministry of Interior special police battalions could not be trusted. Confronted with all of this, Casey and Abizaid canceled all talk of American troop withdrawals. They had to reclear too many areas, starting in Baghdad. In addition, the U.S. and British, without a lot of Iraqi military help, were tangling with the Shia militias and their deadly EFPs. To meet the growing crisis, Casey requested a four-month extension of the 172nd Infantry Brigade from Alaska, one of the Army's medium-weight BCTs equipped with the wheeled Stryker series armored vehicles and a lot of quality infantry, particularly suited for service in a city like Baghdad. As Casey ruefully wrote years later: "In retrospect, I waited too long to make the decision to cancel the drawdown and to extend the Stryker brigade." With Abizaid's full support, Casey persisted in working to turn over matters to the Iraqis. It would go more slowly, but go it must. To the MNF-I commander, it was the only viable strategy.

The numbers actually supported George Casey. Al-Qaeda was reeling, still striking back, but badly beaten up by the Task Force. The Sunni rejectionists continued to suffer at the hands of Casey's forces and the Iraqi military and police. American casualties ran slightly below the 2004–2005 level, and the Iraqis had more than 250,000 men alongside on operations. Iraqi soldiers and police were in the thick of the fight, enduring 2,545 killed in 2005 and 2,091 throughout 2006. Provinces, cities, and districts shifted to Iraqi control — some were shaky indeed, but many passed smoothly in the south. In the war of attrition, MNF-I was more than holding its own. The unknown involved the Sunni-Shia bloodletting and the impressions it created within and beyond Iraq. Few wanted to risk two years, five years, ten years, or even more of this. Whatever resulted would never look like a victory.

In a way, Casey and his generals in Iraq resembled the British commanders on the western front in the Great War. They saw the problem, realized their methods were not enough, yet recommended the same old solution. From 1914 through 1918, confronted with consistent failure to breach the sodden German trench lines, the British generals plodded on, adding shells and manpower to each big push, all with little result beyond more casualties. It took tanks, airpower, major tactical changes, and some key new leaders, not to mention the shock of a near victory by the more flexible Germans, to change the calculus of the trenches. Casualties in Iraq never mounted as high as those in the vast slaughterhouse of World War I, but they were bad enough. As 2006 went on, Americans in Iraq and at home began to lose patience and stomach for the fight.

Those at home had been grumbling, louder and louder, since the first Iraqi summer in 2003. The U.S. domestic consensus held through the 2004 presidential election, but President Bush remained maddeningly inarticulate and stubborn in dealing with the Iraq campaign. Partisan griping grew strident. Certain figures within Bush's own party also began complaining. Bush said and did nothing differently. Abizaid and Casey stayed in command. Their strategy continued. The enemy persisted. In November 2006, the American electorate changed the equation.

The repudiation of the Bush administration at the polls wasn't unusual in American history. Second-term presidencies usually lose steam in the midterm voting. But this "thumping," as Bush called it, placed the Democrats in power in both houses of Congress. Iraq wasn't the major issue, but it was on the table, another indication that it was time for a change. With the Democrats now holding the purse strings, business as usual in Iraq wasn't going to work. Change came swiftly.

Secretary of Defense Donald Rumsfeld went out on November 8, 2006, the day after the election. His replacement, Robert M. Gates, was a Vietnam-era Air Force veteran with a doctorate in Russian history, a former CIA officer, and a deputy national security adviser and CIA director under the president's father. Along with his duties as president of Texas A&M University, in the autumn of 2006, Gates was a key member of the congressionally mandated bipartisan Iraq Study Group, one of several high-level committees reassessing the war effort. The panel had not yet put together its report, but clearly, it wouldn't endorse more of the same.

The Iraq Study Group also included former secretary of state James Baker, former representative Lee Hamilton, former New York City mayor Rudy Giuliani, former Clinton administration adviser and civil rights attorney Vernon Jordan, former attorney general Edwin Meese, former Supreme Court justice Sandra Day O'Connor, former representative Leon Panetta, former secretary of defense William Perry, former senator Chuck Robb, and former senator Alan Simpson. This panel of worthies talked to leaders in Washington and went to Baghdad and interviewed many key figures there. Over the course of eight months, they wrote a fine, erudite document with seventy-nine recommendations and a lot of footnotes. It could be summarized in two words: *Get out*.

Simultaneously, though with considerably less star power, the National Security Council, the Joint Staff, the U.S. Central Command, Multi-National Force–Iraq, and Multi-National Corps–Iraq all carried out their own assessments. They looked at the Baker-Hamilton option (get out) but also identified two others, with variations. The U.S. could stand pat for a while, two years or so, then draw down to a residual troop commitment; this was the Casey-Abizaid hand-over scheme. Or America could reinforce, actually add troop strength, blow back the Sunni and Shia insurgents, and give the Iraqis one more chance to unscrew themselves in Baghdad. This idea became known as the surge.

Baker and Hamilton and their bunch, nudged by Vietnam veteran Chuck Robb and Cold War Army veteran Ed Meese, also offered a surge as a bridging effort to set up a clean U.S. withdrawal. Hadn't President Richard Nixon concluded the U.S. effort in Vietnam with massive B-52 strikes on Hanoi and Haiphong in December of 1972? It didn't win the war, but it did force Hanoi's hand and got its leaders to release U.S. POWs and sign an agreement. It had also provided a "decent interval" for the Saigon government to try to become functional. Maybe a surge in Iraq could do as much, or more. In the Bush White House, national security adviser Stephen J. Hadley began to think that way too.

A loud discordant note came out of Baghdad. Casey didn't want the surge; neither did Chiarelli, nor Abizaid up at U.S. Central Command. To a man, the senior field commanders saw a surge as going the wrong way, re-Americanizing a war the Iraqis had to win for themselves. All were willing to hold the line for two years or so. But they wanted no more U.S. forces. This campaign had to be passed to the Iraqis. Led by their chairman, General Peter Pace, the Joint Chiefs of Staff squarely and unanimously supported the generals in theater.

President George W. Bush, a man who'd once labeled himself "the decider," now faced a critical choice on Iraq, the biggest since the initial invasion. He could continue to defer to the strategic acumen of George Casey, as he had done since July of 2004, and stay. He could go with his new secretary of defense, Bob Gates, and withdraw. Or he could try the minority position held by some of the Baker-Hamilton team and his own national security adviser, Steve Hadley, and surge. To send more troops would go against the stark election results, the polling data, and the conventional wisdom. Surging did not promise a win — nothing did at this point — but leaving or standing pat looked to Bush like sure paths to immediate defeat.

One more voice weighed in at this point, a soldier no longer in uniform but still unwilling to see America lose or quit. Defying convention, practice, and the customary deference and gentility of the retired general officer corps, Jack Keane got involved. General Keane had served with the 101st Airborne Division in Vietnam, ending his career as Army vice chief of staff in 2003 as the Iraq campaign began. Tall, direct, and forceful, Keane stayed close to the war effort, serving as an adviser to key military and civilian senior leaders in Washington, including Rumsfeld and Cheney. He traveled to Iraq and Afghanistan as an unofficial consultant to the commanders there. He also cultivated long-standing ties with Washington insiders like former secretary of state Henry Kissinger and Senator Hillary R. Clinton. The general used his considerable influence to get in front of President Bush at an Oval Office meeting on December 11, 2006. Keane's message was simple: Time is running out. Send more troops now. Tell them to protect the Iraqi people. Hold off on transition until the situation settles. It's our last and only chance to win this thing.

Along with the surge, Keane advocated one other step. When another participant raised the matter of a new commander in Iraq, the retired four-star emphatically recommended David Petraeus. Casey had been there long enough — maybe too long. All agreed that it was time for him to move on. The current crop of four-stars, who opposed the surge to a man, had their own candidates: Chiarelli, Dempsey, Mattis, Vines, maybe even Pete Schoomaker. That last one would have been an inspired choice indeed. And none of them wanted a surge. They could sort it out with what they had, that and the Iraqis, who had to win it anyway.

But Petraeus? He wasn't a four-star (yet). He hadn't been a corps commander in country. He wasn't some SOF manhunter. Indeed, he was running the Army's staff college out in Kansas. Bush pointed out that some of Petraeus's fellow generals didn't want him to take over. The word around the Army, which had reached the White House, was that Petraeus was all about Petraeus, and he had been politicking pretty openly for this promotion. Keane told Bush to ignore all of that inside baseball. The retired four-star had known his energetic protégé well for fifteen years and was sure that, unlike the other candidates, Petraeus would back the surge all the way. Keane told Bush to pick the former MNSTC-I commander. Bush took it in.

On January 10, 2007, Bush announced his decision to his fellow citizens. "America will change our strategy," he said. "I've committed more than 20,000 additional American troops to Iraq." He defined victory in Iraq as creating "a functioning democracy that polices its territory, upholds the rule of law, respects fundamental human liberties, and answers to its people." It was the president's most significant strategic decision since the initial choice to invade. Once more, this time by a definite accountable action, the president doubled down.

Not mentioned in the speech was the man Bush had chosen to make it all work, the new commander of Multi-National Force–Iraq, General David Howell Petraeus. The most uniquely talented and openly ambitious U.S. field commander since Douglas MacArthur surely had his work cut out for him. Those in uniform who knew him — and thanks to his extremely high profile, most did, or thought they did — knew exactly what to expect when Petraeus and his substantial entourage reached Baghdad. It would be just like the arrival of Field Marshal Bernard Law Montgomery at U.S. First Army headquarters on a particularly bleak day during the Battle of the Bulge in December of 1944: "like Christ come to cleanse the temple."

11

MALIK DAOUD

He talks about the big picture and command problems and knowledge of terrain but all that has nothing to do with it — it's this other thing that slips along just under the surface.

— ANTON MYRER, *ONCE AN EAGLE*

THEY CALLED HIM Malik Daoud, King David, and he liked it. "I don't know where the King David thing actually came from," Petraeus said later, "but you had to play the role a little bit." He was willing to try, and the American civilian leaders, President Bush and Secretary Gates, were willing to let him. Petraeus had personal goals beyond Baghdad, beyond the Army — not king, but high enough, perhaps the White House. In early 2007, that could wait. First, the Army's newest four-star had to deliver a victory. It called to mind President Lincoln's response when the bombastic Major General Joe Hooker suggested the country needed a triumphant general as a dictator. "What I ask of you," wrote Lincoln, "is military success and I will risk the dictatorship." Come the hour, come the man. All eyes turned to Petraeus.

Petraeus's acolytes arrived in Baghdad in February of 2007 armed with the holy writ: Field Manual 3-24, *Counterinsurgency,* came right from the font, written by the master and his team during the general's interlude commanding the Army's Combined Arms Center at Fort Leavenworth, Kansas, from 2005 to 2007. Leavenworth writes doctrine, the Army's how-to manuals. During his tenure on the Kansas prairie,

David Petraeus ensured that his views on counterinsurgency were recorded. Now they would be applied.

FM 3-24 was an idiosyncratic document, very much reflecting its guiding mind. Most Army manuals read like telephone books, only less interesting. There have been exceptions, most recently the 1976, 1982, and 1993 versions of Field Manual 100-5, *Operations.* All of those got attention in and out of the uniformed ranks because they marked key shifts in military thinking, notably a post-Vietnam reemphasis on conventional warfare in Europe (1976), an embrace of the air-land battle of maneuver that paced Desert Storm (1982), and a recognition of "operations other than war," small wars, and counterinsurgencies (1993). But this one went well beyond even those controversial predecessors. It prominently included nine, odd Zen-like counterinsurgency axioms:

> Sometimes, the more you protect your force, the less secure you may be.
> Sometimes, the more force is used, the less effective it is.
> The more successful the counterinsurgency, the less force can be used and the more risk must be accepted.
> Sometimes doing nothing is the best reaction.
> Some of the best weapons for counterinsurgents don't shoot.
> The host nation doing something tolerably is normally better than us doing it well.
> If a tactic works this week, it might not work next week; if it works in this province, it might not work in the next.
> Tactical success guarantees nothing.
> Many important decisions are not made by generals.

Clever, thought-provoking, and paradoxical, they played very well in the halls of academia and among the informed public. The prestigious University of Chicago Press issued the manual as a book, complete with an introduction by Sarah Sewall, a Harvard professor who played a major role in helping Petraeus complete the effort. Compared to the standard, terse military jargon and acronyms, this stuff sang a pleasing intellectual tune. It's hard to say what the hell it meant to a platoon sergeant or a lieutenant trading bullets west of Samarra or ducking EFPs on the fringes of Sadr City. Then again, the U.S. Army rarely reads its own doctrine, preferring to rely on leadership, organization, training, technology, and the guts and initiative of young men and women. Americans are not much for instruction manuals. They like to get going, to try things. FM 3-24 mattered only if this guy Petraeus issued orders that applied its principles.

He did, and fast. In his first days in Baghdad, in February 2007, the new commander exerted his well-known energy to chart the course for all of Multi-National Force–Iraq. He was everywhere, out daily and nightly, flying, driving, walking, engaging in pushup contests (he almost always won), and spreading the word with a zeal akin to the prophets of old. Petraeus talked up FM 3-24 but understood that young sergeants facing a grinding patrol schedule didn't have time to pore through a 282-page field manual. So practical orders followed. Looking at those orders, you'd almost think he was following the popular wedding tradition: something old, something new, something borrowed, and something blue.

Something old referred to wholesale retention of the primary George Casey strategy. Al-Qaeda out, Sunni in, Iraqis increasingly in the lead remained the approach. But rather than continuing to hand over areas — especially in Baghdad — Petraeus wanted to beat down the insurgency first, clear before hold, then build. Whereas Casey's method cleared with MNF-I and Iraqis but held with the latter, Petraeus intended to commit U.S. surge units in small outposts to hold ground, side by side with Iraqi soldiers and police. It was a tactic that had long been in use in many regions by many units, among them Bill Simril and 2-11 Cavalry back in 2005 in north Babil Province. The extra forces helped hold key ground.

Something new meant the Petraeus style, open and active in telling the command's story to all who would listen and aggressive in pushing the narrative even on those who didn't think they cared. Most Army officers and NCOs look at the press and other busybody civilian visitors much as city dwellers regard pigeons — noisy, obnoxious, often underfoot, and too numerous. You deal with them, but by no means would you feed them. With his Princeton doctorate, French-speaking wife, sharp wit, and endless desire to network, Petraeus saw the inquiring journalists, visiting academics, and members of Congress not as dirty, interloping pests but as kindred souls. He fed them plenty. They flocked to him, and he schooled many. Like docile carrier pigeons, they conveyed his messages far and wide. It all bought time and deflected criticism as casualties mounted during the spring and early summer of 2007. Dave Petraeus fought it out in the media with the same vigor that he brought to the struggle for the Sunni triangle. Probably more.

Something borrowed meant twenty thousand troops, the famous surge, five brigade combat teams of soldiers plus two more Marine battalions. Petraeus would hold them for a year or so, and he knew where he wanted to put them. Two would go into Baghdad city proper. Three were destined for the Sunni Arab rural districts around the city, areas that captured al-Qaeda documents called "the Baghdad belts." Those dusty little hamlets hosted the safe houses and garages that hid suicide cars and their drivers. If you could stop all, or even some, of those high-casualty bombings, you'd remove the principal excuse for Shia-on-Sunni death-squad raids. The extra American forces also allowed the notorious Iraqi special police to be pulled out battalion by battalion, reorganized, and retrained, cleaning out the Shiite militiamen. Once the sectarian stuff was tamped down, the Americans might restart the transition with more confidence that the Iraqis could hold.

Something blue meant old Army blue, harking back to the rough wool shirts of the frontier regulars. In the long, sad series of U.S. Army campaigns against the Indians, Brigadier General George Crook had found a way to win. He successfully recruited and trained Apache scouts to chase and fight renegade tribes. "Nothing breaks them up like turning their own people against them," Crook remarked in 1886. That opportunity beckoned in Iraq too. Few in Baghdad expected such a development, although to his credit, George Casey had the first inklings of it back in September 2005 when the Abu Mahal tribe asked for help battling al-Qaeda way out near the Syrian border. As Petraeus took command in Iraq, another officer, his era's George Crook—although not a general—had already solved the problem. As Petraeus himself noted, in counterinsurgency, many important decisions are not made by generals.

In the summer of 2006, the Marines came very close to giving up on Anbar Province. On August 17, 2006, Colonel Peter H. Devlin, the First Marine Expeditionary Force intelligence chief, put it in stark terms: "The social and political situation has deteriorated to point that MNF [Multi-National Forces] and ISF [Iraqi Security Forces] are no longer capable of militarily defeating the insurgency in al-Anbar." Emboldened by an active al-Qaeda vanguard, discouraged by the Shia-dominated Maliki government in Baghdad—those damnable Persian-loving *sheruggis*—and frustrated by daily clashes with the American occupiers and their Iraqi collaborators, over a million Sunni Arabs had seemingly collapsed into perpetual revolt. Asked by MNF-I and the MNC-I headquarters to project when the province might come under Iraqi government control, the Marines gave no date. They just didn't see it happening.

Fallujah remained prostrate. The enemy never recovered from the November 2004 MNF-I offensive. But farther west along the Euphrates River, each town stayed actively opposed to the Americans and the Baghdad regime. The most dangerous by far was the provincial capital of Ramadi, once the home of more than four hundred thousand Sunni Arabs, and contested daily. The U.S. had long held outposts in the city, but the enemy never stopped fighting. Ramadi had been one of Zarqawi's favorite support zones, possibly the future capital of his al-Qaeda emirate.

Since 2003, three successive U.S. Army brigades attached to the Marines had been bloodied holding parts of the city. Marine battalions and SOF fought as part of these BCTs. The First Brigade, First Infantry Division of Fort Riley, Kansas, lost fifty soldiers. The Second Brigade, Second Infantry Division out of Korea went toe to toe with AQI and Sunni rejectionists and sustained the highest death toll of any BCT in country: ninety-eight killed in action from 2004 to 2005. The Second Brigade, Twenty-Eighth Infantry Division of the Pennsylvania Army National Guard followed and endured eighty-three killed. The commander of that division, Colonel John Gronski, told the generals that it would take three U.S. BCTs working simultaneously to secure Ramadi—one for the city itself, one for the surrounding rural area, and a third to hold the Habbaniyah area between Ramadi and Fallujah. Lacking enough American manpower, Gronski pulled out from some of Ramadi's central urban checkpoints. He couldn't keep them resupplied without constantly losing soldiers to IEDs and gunfire. More troops made military sense, but they weren't available. In the months before the surge, well . . . forget it. The next U.S. Army brigade would have to go in and do its best.

In June of 2006, the First BCT, First Armored Division arrived from Tal Afar, where they had continued counterinsurgency operations begun by H. R. McMaster's Third Armored Cavalry Regiment in 2005.

Colonel Sean MacFarland received orders from the Marines that came directly from General George Casey. "Fix Ramadi but don't do a Fallujah," the MNF-I commander said. With a BCT plus one Marine and one Army infantry battalion attached, five maneuver battalions total, MacFarland couldn't have run a Fallujah-scale operation even if he'd wanted to.

Starting on June 18, MacFarland's BCT and the experienced First Brigade, First Iraqi Army Division entered central and then southern Ramadi. Three battalions from the new, Anbar-recruited, Marine-trained Seventh Iraq Army Division also joined the operation. Another Iraqi battalion, sent south from Mosul, arrived, crumbled, and quit. Losses mounted, with five tanks destroyed, four Bradleys gutted, and twenty-four Americans dead in the first month. Artillery, tank main gun rounds, and aerial bombing all contributed to the fray, tearing apart blocks of the already battered city. It looked like the same-old, same-old in Ramadi. But MacFarland wasn't willing to settle for that. He was way out west, on his own under an overstretched, skeptical Marine command, trying to quell the enemy in a place that had never knuckled under. The colonel could keep pounding the screw. Or he could seek a screwdriver.

MacFarland's answer was not more tanks, more bombs, or more infantry. The solution came from the George Crook playbook, and that frontier general, after all, had been as much a cavalryman as Sean MacFarland. But the answer in the Arizona Territory and in Ramadi wasn't the U.S. Cavalry. Rather, it bubbled up from the fertile imagination of a pugnacious thirty-two-year-old captain named Travis Patriquin.

It would be easy to underestimate Travis Patriquin. In the words of Lieutenant Colonel Pete Lee, the brigade's executive officer, "He had been injured, he'd put on a few pounds, he had long hair and a mustache, he pretty much chain-smoked." The captain had not attended West Point or commanded a rifle company. He was commissioned through Officer Candidate School. In the First BCT, First Armored Division, Patriquin initially served in the S-3 section as an assistant operations officer. Older than most captains, with a bum knee, Patriquin seemed an unlikely candidate to take over a rifle company.

Under a different colonel, Patriquin might well have remained where he was, stuck in a windowless cubbyhole, cranking out PowerPoint slides and editing operation orders. But along with his quiet demeanor, Sean MacFarland had a rebellious streak, a wry sense of humor, and a fondness for people with unusual talents. When he looked at Patriquin, he saw more than met the eye. The captain had spent his enlisted years in the 1990s in signals intelligence with the Special Forces. He learned Arabic and practiced it on extended deployments to Jordan and Kuwait. In 2001, although assigned to the Eighty-Second Airborne Division recuperating from a knee injury that cut short his stint in Ranger school, Patriquin had talked his way into accompanying the Fifth Special Forces Group into Afghanistan. During Operation Anaconda, First Lieutenant Patriquin and his team of electronic eavesdroppers went in on the initial air assault and fought for three days on the harsh east slope of the Shah-i-Kot Valley. Patriquin's long dark hair and mustache were not accidents. He judged they made him look more Iraqi. MacFarland agreed, and he pulled Patriquin from the S-3 section. The colonel designated the Arabic-speaking captain as the BCT's S-9, the civil affairs officer. That slot usually went to a school-trained major, not a relatively junior in rank (though older in age) infantry captain. Most BCT commanders wouldn't entrust that position to someone so lacking in official credentials, but MacFarland knew he had the right man. He told Patriquin that his job would be more than traditional civil affairs; he would be aiming at outreach to key Iraqi leaders, working on governance and the like. MacFarland provided him with no staff, no armored truck, no radio — he figured, rightly, that a guy like Patriquin knew how to get things like that when he needed them.

Patriquin responded to MacFarland's trust. The captain read voraciously, devouring every work he could find on the Arabs and their culture. He knew and quoted both T. E. Lawrence's *Seven Pillars of Wisdom* and, in passable Arabic, the Koran. Like Lawrence, Patriquin believed that the ageless Arab tribes held the key to controlling Anbar Province. He wasn't the first American to think so. Paul Bremer, George Casey, and a succession of Marine commanders certainly agreed. But they'd stumbled over two obstacles. First, the tribes represented old Iraq, pre-Islamic Araby, not the 2006 world of a parliament and ministries. Second, and worse, the tribes were knee-deep in the insurgency. A Venn diagram that showed Anbar Sunni tribes and Sunni rejectionist militants would overlap almost entirely. With few exceptions, these tribes were the enemy. If Travis Patriquin and his BCT approached the tribal sheikhs, they would be well outside the box.

Given the alternative — another year wasted, another hundred dead Americans, and Ramadi in turmoil — MacFarland let Patriquin give it a shot. The first opening came northwest of the city in late June. Sheikh Sattar Abu Risha led the tribe that owned lands north and west of the U.S. Camp Ramadi, fronting the Euphrates River. A dapper, slight man in flowing white robes trimmed in gold, Sattar had enjoyed — or perhaps the word is *endured* — a long history with the American occupiers. Usually, the sheikh supervised roadwork contracts, hustling day laborers to fill in IED holes or clear brush from culverts before the rainy season. He sometimes sold information to the Task Force. Along with his rudimentary construction enterprise, Sattar operated what he termed an import-export firm. In fact, he ran an extensive smuggling network, moving cigarettes, fuel, sheep, cars, weapons, and ammunition to and from Jordan. To keep the sheikh's monopoly intact, his men waylaid trucks passing through Abu Risha territory and demanded — and got — appropriate tribute. At times, Sattar worked with and for al-Qaeda, as did his brother sheikhs. As they say in the Mafia, it wasn't personal, just business. Those activities earned him a short detention by U.S. SOF in 2004. His Abu Risha forefathers had been doing such things for centuries. It all allowed Sattar to smoke Cuban cigars, race good Arabian horses, and slip off to Dubai to drink and entertain himself. In usual sheikh fashion, he provided for his tribe: employing the men, aiding the sick, and resolving petty disputes for the families. Now it was time for some more roadwork. Captain Travis Patriquin was there with a couple of lieutenant colonels to grease the sheikh.

Sheikh Sattar had met a lot of Americans. Buzzcut, lean, and bluntly direct, the U.S. officers all seemed alike. But not this one — this energetic captain looked, and talked, like an Iraqi. Most Americans came with paperwork and ten-point agendas. Patriquin arrived with smokes and small talk, discussing everything but business. The lieutenant colonels with him went along with it, fidgeting some, but letting it play out. Patriquin talked. But, more important, he listened and understood. The men discussed hometowns, politics, religion, horses, cars, and children. After an hour or so, Sattar got around to asking Patriquin what the new BCT intended for Ramadi. In the elliptical Arab way, the captain answered by directing a query back to his host.

"What idea do you have that will make us succeed?" asked Patriquin.

"The thing we have to do is build a new military force," replied the sheikh. "Let me use my own fighters," Sattar said. He elaborated: The Abu Risha tribesmen wanted to protect their families, not clean out ditches. The sheikh could furnish young men to fill the ranks of the police. The Ministry of Interior had authorized 3,386 police for Ramadi and its surrounding rural districts, but only 420 had been recruited. If 100 showed up for duty at the three isolated, forlorn downtown stations, it was a very good day. Sattar promised to fill those ranks with his men and the men of his fellow sheikhs. He asked only that the initial recruiting occur at his residence and that the first police station be established on Abu Risha turf.

A more experienced staff officer than Travis Patriquin might have equivocated. But with the two lieutenant colonels nodding assent, Patriquin agreed. Lieutenant Colonel Tony Deane, who commanded the First Battalion, Thirty-Fifth Armor that oversaw the Abu Risha area, jumped in to assure the sheikh it would indeed happen. They set the first assembly for July 4. Almost a hundred Abu Risha males showed up. The Americans didn't ask too many questions about what those young men had done in the past.

Patriquin, Deane, MacFarland, and their soldiers and Marines moved ahead, ignoring the usual rules. To get around the Ministry of Interior requirement that all Iraqi police be assigned by the Baghdad headquarters after training in the capital, the BCT designated the new Abu Risha facility a temporary substation. As the new IPs weren't on the official payroll, MacFarland sought funds from the U.S. intelligence community. That bunch knew Sattar well, and they paid. Recruiting continued; more than 200 signed on in late July and 395 in early August. Those two drives happened near the Marines' Camp Blue Diamond downtown and included men from tribes other than Abu Risha's.

What changed? Why did Sheikh Sattar make his offer? It came down to business, sheikh matters, the freedom to live the life of a tribal Arab. Squeezed hard by the Task Force, Zarqawi's AQI successor Abu Ayyub al Masri used the hard edge he'd honed in the Afghan camps in 1999 under the original Osama bin Laden cadres. AQI needed money, recruits, and supplies. The Anbar tribes had always helped AQI. Now, driven by desperation, al Masri's people didn't ask. They demanded. They took. When individual tribal sheikhs objected, the elders lost their heads. Families were attacked. Houses were demolished and cars burned. The AQI men began to impose Wahhabi discipline — no gambling on horses, no drinking alcohol, and no smoking. The AQI leaders had crossed the line at last. The Persian-influenced *sheruggis* in Baghdad were far away from Anbar Province. The Americans were right there, and they had little interest in what sheikhs did with their tribes. Forced to choose between the AQI boot on their necks and the U.S. military, Sattar and others decided to try the Americans.

Abu Ayyub al Masri saw the danger. He turned his suicide bombers on the new tribal IP stations around the outskirts of Ramadi. On August 21, a dump truck loaded with explosives charged toward the new Abu Risha IP substation north of Ramadi. The alert tribal fighters killed the driver and slammed the gate shut, but the truck detonated in an immense gout of fire, blackening the walls. Eleven Iraqi police died, most of them drawn from the Abu Ali Jassim tribe. Several American soldiers were wounded. In the past, that would have emptied the station. This time, the tribesmen ran up a burned, torn Iraqi flag and manned the walls. Later, as if nothing had happened, they carried out their afternoon foot patrols. They weren't leaving. This was their land.

That same day, AQI tried a more personal approach. To punctuate the damage done at the IP facility, assassins nabbed and killed Sheikh Khaled Arak Ehtami of the Abu Ali Jassim tribe. They slew his son and nephew too. The AQI gunmen dropped leaflets warning of a similar fate for any other Sunni leaders who supplied IPs or worked with Sheikh Sattar and the Americans. Then AQI overplayed its hand: al Masri's men desecrated and dumped the three bodies far from Abu Ali Jassim land. It took five days to find them. Sheikh Sattar spent that time rabble-rousing, telling his fellow tribal leaders that they could expect more of this unless they stood together. This time, the Americans would help. Sattar pointed to the new IP posts as proof.

Enraged by the AQI counterstrikes, Sattar asked Patriquin for a meeting with Sean MacFarland. Until now, the BCT commander had let things putter along with Patriquin and Tony Deane of 1-35 Armor. The

Marine generals and their staff assumed Sheikh Sattar was a small-time operator out for himself. The Marines had dealt for years with much bigger guns, richer, more prominent sheikhs in exile in Jordan. This local IP recruitment helped, but it did not promise much beyond a transitory effect. The Marines had tried this stuff before. It had never really panned out. To their credit, the Marine commanders didn't stop the Americans' new outreach efforts in Ramadi. But they weren't convinced it would do any good.

When Sean MacFarland entered Sheikh Sattar's compound on September 9, he realized something significant was happening. Resplendent in his gold-trimmed white robes, Sheikh Sattar presided over a full house: forty sheikhs and deputies representing seventeen of the twenty-one Ramadi tribes. Sattar swept his arm over the crowd and, with a flourish, informed the American colonel that these tribal leaders formed the Sahwa al Anbar. Patriquin whispered to his commander that the literal translation of *sahwa* was "awakening," but used in this context, the word connoted a rude awakening, a wake-up call. Oppressed by AQI, the Sunni sheikhs were coming in late, but not too late. MacFarland immediately saw the implications of the Sahwa. Within days, the Marine generals, Pete Chiarelli at MNC-I, and even George Casey at MNF-I, knew. In a visit to Ramadi on September 25, Casey stated the obvious: they had to "keep Sattar alive." Preoccupied with growing violence in Baghdad, fencing with the testy Maliki, and dealing with spreading pessimism in Washington, neither Casey nor Chiarelli spared many more thoughts about these busy Sunni tribesmen out near Ramadi.

Sean MacFarland continued IP recruitment and got the Marine senior leadership to see the benefits. Patriquin and Sattar smoked a lot of cigarettes, drank a lot of tea, and ate a lot of rice and mutton as they traveled from sheikh to sheikh. The Arabic-speaking captain continued to win friends and influence people. He even took the time to create an eighteen-slide briefing featuring stick figures and school-primer language to explain to one and all why it made sense to engage the tribal sheikhs. In essence, the Sunni Arabs knew their own. Empowered, they would reliably find and clean out the terrorists. Patriquin called his little pitch "How to Win the War in Al Anbar." He gave it to everyone he met, paying particular attention to the Marines, the SOF, the CIA, and State Department representatives. The file made it to Fort Leavenworth, where David Petraeus, already preparing for his tenure in Iraq, noted on his copy "this makes eminent sense."

Of course, the Iraqis weren't interested in PowerPoint slides, so Patriquin gave the sheikhs something more practical: satellite phones. The captain preprogrammed his own number into the devices and told the sheikhs to call anytime, especially in an emergency. On November 25, 2006, one did.

"We're under attack," said Sheikh Jassim al Suwadawi Abu Soda. "I've got my men here, we're establishing defensive positions and we're going to fight these guys off. The fight is going on now. We need American help." Patriquin did not hesitate. Although he was just a staff officer and could not formally commit to helping on his own authority, he knew his colonel's intent. He promised U.S. intervention.

It was just after 1:00 p.m. on November 25, a Saturday, the first day of the Iraqi workweek. The Abu Soda tribe, a small one with fifty fighters at most, held the Sufiyah peninsula, about two miles wide, on the south bank of the Euphrates just outside urban Ramadi. For years, insurgent mortar men had used Sufiyah to rain rounds onto Camp Corregidor in eastern Ramadi. Once Sheikh Jassim joined the Sahwa, that stopped. Jassim got some help from the nearby Abu Mahal tribe, an eastern branch of the same group that in 2005 had formed the Desert Protectors out near Syria.

Seeking a symbolic target, Abu Ayyub al Masri personally chose the Abu Soda tribe to be converted back or liquidated. AQI messengers told Jassim to quit cooperating with the Americans and permit mortar teams to reenter the Sufiyah peninsula. Having lost a brother and three tribal fighters to AQI in September, Sheikh Jassim chose instead to trust his new satellite phone and the Americans at Camp Corregidor. He defied AQI.

So at noon that November Saturday, al Masri had sent in his people, about seventy-five in total. They rode pickup trucks, not horses, but their tactics were those of Bedouin raiders from the Prophet Muhammad's era. They meant to clear each house, killing men, women, and children and burning anything that would ignite. The attackers wore masks, but this was not a suicide attack. They didn't expect to lose.

At Camp Ramadi, Patriquin's report got immediate attention. Sean

MacFarland was out of the country on leave, but Lieutenant Colonel Chuck Ferry, the First Battalion, Ninth Infantry commander at Camp Corregidor, stepped up to organize a company-strength quick-reaction force with tanks and Bradley fighting vehicles. Other officers and NCOs at brigade headquarters began coordinating for air support and artillery fires. At the First Marine Expeditionary Force headquarters, officers scrambled F/A-18 fighters to intervene. The SOF shifted an armed Predator. Iraqi army units prepared to join the effort.

The fighting in Sufiyah intensified. An AQI mortar shell killed two of Jassim's men, leaving him with fifteen effectives at his riverside compound. Among the low adobe houses of his people, pillars of smoke rose from burning blankets and straw. Gunshots sounded from the mud huts: executions in progress. Seven Abu Mahal gunmen sprinted in from the east, shooting as they advanced. Their fire slowed the AQI movement. The tribesmen knew the village and fields well, and they circled back behind the AQI fighters and picked them off. Sheikh Jassim kept calling Travis Patriquin, who was watching the firefight from the command post at Camp Ramadi. The screen there showed the picture from the brigade's Shadow drone above Sufiyah. "Hang on. We're coming," Patriquin assured Jassim.

Ferry's column had four miles to go. It might as well have been forty. The soldiers left Camp Corregidor in midafternoon but immediately had their hands full clearing IEDs and shoving aside a tangle of thick palm trunks blocking the passage to Sufiyah. The AQI people had prepared well, blocking the obvious relief route. Trigger teams and RPG ambushers kept the Americans shooting and getting out of their vehicles to clear out holes and knots on the narrow, overgrown road. It took hours to go a hundred yards, shooting all the way.

At four o'clock, with ninety minutes of daylight left, the Sufiyah firefight centered on the sheikh's walled compound right near the river. The Shadow feed showed AQI men trying to use hut walls and palm trees to work up close to Jassim's house. A fire burned in the compound, straw ablaze in the animal pens. Marine F/A-18 Hornets circled overhead, unable to discriminate friend from foe. It was all just a gaggle of Iraqis in civilian clothes shooting at one another.

Talking on the satellite phone to the trapped sheikh, Travis Patriquin had an inspiration. "Have your men take out white rags and wave them over [their] heads!" he directed in Arabic. "Any pieces of cloth, wave them over your heads!" Jassim complied. Within a few minutes, the Shadow camera showed little figures in and around the walled farm all waving rags. Now oriented to the friendly forces, the Marine jet fighters swooped down, afterburners roaring. The Marine aviators did not shoot or drop anything — people were still way too intermixed for that — but as usual, the arrival of aggressive U.S. airpower panicked the enemy.

Chuck Ferry's approaching Americans, guided by the same overhead images, cleared the big 155 mm howitzers at Camp Ramadi to open fire. The artillerymen pounded open fields near Jassim's compound, pumping out almost a hundred high explosive and smoke rounds. The shock effects of the jets and howitzer barrage broke the AQI attack. As darkness fell, the shaken enemy pulled back, carrying their dead and wounded. They clambered into their trucks. Three of those pickups blundered south, smacking into Ferry's lead tank. The M-1 punched them to bits with 120 mm main gun rounds. There wasn't much left of them or their AQI passengers.

Four other trucks departed to the east. The drone camera showed that each dragged a corpse behind it, just as in *The Iliad,* Achilles had pulled dead Hector behind his chariot below the ramparts of Troy. Al-Qaeda wanted to make a vicious propaganda point even if the assault failed. But the jihadis hadn't reckoned on American overhead surveillance. The pulled bodies tagged the four trucks like beacons. Cleared hot, the F/A-18 Hornets swung low in the dark. At 6:15, 20 mm cannon fire blew apart three of the trucks. The SOF Predator then unleashed a Hellfire to knock out the fourth. The shattered trucks burned for an hour afterward.

Around ten thirty, Chuck Ferry's reaction force finally arrived at the Jassim farm and linked up. "It was a very surreal scene," he remembered, "night time, smoke, houses on fire." All around him, men with AKs walked or stood. They were friends. Just to be sure, Ferry told Jassim to have the men tie white engineer tape to their upper arms.

The Americans began evacuating casualties. Patriquin was with the column, aboard a Bradley full of ammunition, AK-47s, and radio batteries. Additional supplies came from the river, thanks to Marine watercraft. The U.S. soldiers assured Sheikh Jassim they would stick around. By daybreak, a rudimentary combat outpost was in place. Sufiyah had been cleared for the last time. With the Abu Soda tribe in the lead, it would be held.

The Abu Soda and Abu Mahal tribes lost six fighters and eleven civilians. Their AQI attackers suffered at least forty-five dead, perhaps as many as sixty. The Americans honored their promise to Sheikh Sattar and Sheikh Jassim. The other sheikhs took notice. In the prescient words of historian and Marine Vietnam veteran Francis J. "Bing" West, the Sunni Arab sheikhs of Ramadi now made common cause with "the strongest tribe."

General David Petraeus took command on February 10, 2007. Within days, he was at Camp Ramadi, his first troop visit. "I'm so happy I could kiss you guys," he exclaimed. The MNF-I commander immediately saw the promise of the Sahwa. The CIA had written about it in the fall, and President Bush had even mentioned it in his January 10, 2007, announcement of the surge. Petraeus saw it was for real. He wanted to spread it to the other Sunni Arabs, starting in Baghdad. The American troop surge would pass. So too would the Americans. But the Sunni Arab surge had legs. It epitomized al-Qaeda out, Sunni in, and Iraqis increasingly in the lead.

Captain Travis Patriquin did not live to meet Petraeus. On December 6, 2006, less than two weeks after the Sufiyah clash, a random IED explosion in Ramadi killed Patriquin and two other Americans. Sheikh Sattar attended the memorial ceremony at Camp Ramadi, as did many other sheikhs, Iraqi police, and Iraqi soldiers. Sattar called Patriquin Hisham (the Pulverizer) Abu Risha, and the IPs named a substation for him in Ramadi. Sattar himself was killed by an AQI bombing on September 18, 2007. Though Patriquin and Sattar were no more, the great work they began together continued.

The Sunni Awakening expanded rapidly. As the Americans fought their way into various Baghdad neighborhoods and into the city's rural belts, Sunni Arabs came forward. Whereas in Anbar, the Marines had enlisted the tribal men as police, in other areas, the Sunni took different titles, such as security contractors, neighborhood watch, and concerned local citizens. Ever conscious of marketing, Petraeus and his inner circle settled on a more inspirational name. With the approval of Prime Minister Nouri al-Maliki, the Sunni became the Sons of Iraq.

None dared call them what they really were: a Sunni militia. For the first time since the 2003 invasion, the Sunni Arabs matched the other two major political players in country. Before Petraeus took over, Senator (later Vice President) Joseph Biden had proposed a federal Iraq, formally split among Sunni Arabs, Shia Arabs, and Kurds. Many objected to the Biden formula, though it had been employed by the Ottoman Turks and the British in their day. In the spring of 2007, without an announcement, the Americans accepted a version of it.

The Sons of Iraq completed the Iraqi factional lineup. Each of the three Iraqi groups now featured political wings and de facto authorized militias separate from the formal Iraqi military and police. The Kurdish Democratic Party and the Patriotic Union of Kurdistan both owned heavily armed *peshmerga* units. The Shia Sadrists ran candidates under the Office of the Martyr Sadr and fielded Jaysh al-Mahdi elements in Shiite regions. Shia supporters of the Supreme Council for the Islamic Revolution in Iraq kept their Badr Corps, forever salting selected members into the Ministries of Defense and Interior. Now the Sunni hosted the Sahwa, an Anbar-based political movement with an American-equipped militia. Each community balanced the others. Combined with the troop surge in Baghdad, the Sunni Awakening effectively ended the sectarian bloodshed by the summer of 2007. It split the Sunni resistance, and they stayed fragmented during the remainder of the U.S. campaign. It was not a victory, not by any of the criteria the optimistic Americans set for themselves back in 2003, seemingly in another lifetime. But it was something like progress.

David Petraeus grasped the Sahwa and never looked back. Just as there had been fierce combat in Ramadi to establish the tribal sheikhs as U.S. partners, so it played out across the country. It took months to get the five surge BCTs into country. The U.S. Army extended all of its units in theater out to a grueling fifteen months. It was the only way to square the circle of delivering the surge with not enough soldiers. President Bush and Congress authorized more strength for the Army and the Marine Corps, but those numbers would not ripple into the force until well into 2008. Petraeus and MNF-I had to fight it out in 2007 with what they had, bumped up by five U.S. BCTs and greatly bolstered by the fortuitous Sunni Awakening. Everyone braced for higher casualties. Those arrived on their own sobering schedule.

In and around Baghdad, the reinforced First Cavalry Division cleared neighborhoods. At one point, the division controlled ten BCT equivalents. A new Baghdad Operations Command formed in February with absolute authority over all Iraqi forces and full backing from Nouri al-Maliki. The Iraqis contributed four army and two police divisions, some forty thousand other Iraqi police, and almost fifty thousand Sons of Iraq. Behind the U.S. and Iraqi fighting forces, to hold neighborhoods and rural villages, came the Iraqi army, the Iraqi police, and, in the Sunni areas, the Sons of Iraq. Americans joined the Iraqis in living and patrolling from dozens of platoon and company combat outposts. The cost was high, with 492 Americans killed in and around Baghdad, more than half of the U.S. dead in Iraq in 2007. By comparison, the heavy fighting of 2004–05 cost 169 First Cavalry Division lives. Spring broke hard in Baghdad.

The war escalated outside the Iraqi capital too. Up north, the Twenty-Fifth Infantry Division battled in Tikrit, Samarra, Kirkuk, and Mosul. Fighting in Baqubah proved especially tough. To the south, the Third Infantry Division returned to Iraq earlier than scheduled, coming in to supervise the clearance of the southern belts of Baghdad, down to the north Babil hot zones once contested by 2-11 Cavalry. The British continued to fight and talk in and around Basrah, killing the irreconcilables and making "arrangements" when they could. SOF kept raiding, tearing up the enemy networks, killing and capturing leaders.

To America, to Iraq, and to the wider world, the words and actions of one energetic American four-star general told the story. Dave Petraeus consistently drove the narrative for this campaign. His vision and will dominated, just as General Jack Keane and others had predicted. Petraeus became the active, pressing, endlessly engaged face of the war effort, a role impossible to imagine for the thoughtful, selfless George Casey, let alone the likes of Ric Sanchez or Tommy Franks.

A common military adage designates certain particularly tough soldiers — men like Nate Sassaman and Mike Steele — as being in the category of "break glass in case of war." They are the ones you want with you under fire, although they are not always agreeable personalities.

Dave Petraeus belonged in a very special category: "break glass in case of an unpopular, misunderstood war." Sometimes, to run a faltering war effort, you need a politically savvy promoter. Invest this guy with a need to succeed, and he'd sell what was left of the war. In 2007, he did.

Petraeus was the man to do it. He had a compelling personality and an interesting history. Unlike most soldiers with similar experience, he publicly trumpeted his many achievements, and he encouraged interested parties to learn more. A small, slight man, Petraeus radiated purpose and intensity. He greatly resembled the famed British field marshal Bernard Law Montgomery and shared with that estimable figure a single-minded devotion to his profession and a burning desire to rise above his peers. Petraeus graduated high in his class at West Point, chose the infantry, and from then on, he aimed solely for number one. He graduated first in his Ranger class and, like Dwight Eisenhower, first in his Command and General Staff College course. In between, he commanded a rifle company, served as a battalion S-3 as a captain when all his peers were majors, earned a doctorate from Princeton, and taught in the highly regarded social science department at West Point. As commander of the Third Battalion, 187th Infantry in the 101st Airborne Division, Petraeus took a bullet in the lung during a live-fire exercise and was evacuated with his superior Jack Keane at his side. Six hours of surgery followed. A lesser man would have been medically retired. Incredibly tough and superbly fit, Petraeus was back in the field with his men within weeks. As a brigade commander with the Eighty-Second Airborne Division, he fractured his pelvis in a sport parachute mishap. Again, he bounced back incredibly fast and tough as ever. He led the 101st Airborne Division from 2003 to 2004 in northern Iraq, achieving some success in that treacherous region. His work in MNSTC-I gave Iraq its military and police. At Leavenworth, he wrote the book on how to fight the war. Now he was in charge of fighting it.

It was a great military biography, a Petraeus-approved narrative for the ages. And yet, there was another thread in there, one that the smitten journalists missed or soft-pedaled. Soldiers noticed, though. They knew his reputation inside the Army as a self-promoter, and so when orders went forth to the combat outposts, junior soldiers wondered about his real motivations. Service or self? With Petraeus, you never knew for sure, but you often suspected the latter, and it meant trouble. Even among a tremendously competitive officer corps, Petraeus struck other soldiers as inordinately intent on climbing the ladder. He had more connections than ten of his peers, and he wasn't shy about using them. The litany was amazing. Petraeus had married the daughter of the three-star West Point superintendent. He brought his general father-in-law to Fort Stewart for a company ceremony and made sure to invite his entire chain of command. He served as either an aide or an executive officer for his division commander, the Supreme Allied Commander–Europe, the chief of staff of the Army, and the chairman of the Joint Chiefs of Staff. When shot on the range at Fort Campbell, he was pulled out by Brigadier General Jack Keane and operated on by surgeon Bill Frist, later a U.S. senator and Senate majority leader. As a brigade commander, he sought coverage by Tom Clancy and became the central actor featured in that author's book *Airborne.* In Iraq in 2003, when embedded reporters matched up with units, Petraeus asked for and got Rick Atkinson of the *Washington Post.* Petraeus kept track of all of these mentors, well-wishers, and contacts and made certain they kept track of him.

Now Petraeus was at the summit, the man in the arena. All the networking on earth would not cover him if the war didn't turn around. But it did turn, and it turned on his watch — not a win, but no longer a descent into chaos. Others pointed to the contributions of Third Corps commander Ray Odierno, the SOF, the Iraqi Sahwa, and the unrelenting bravery of thousands of Americans, Coalition troops, and Iraqis, all key factors to be sure, especially that last one. But even the most vehement anti-Petraeus factions in the U.S. military well understood the hard reality. A commander is responsible for everything his unit does or fails to do. In Iraq, a good share of the credit belonged fully to the man who owned the whole lash-up, David Petraeus.

He cemented his achievement in Iraq at the seven-month mark, when he returned to Washington, DC, to testify before the U.S. Congress. It was the performance of a lifetime, and he pulled it off brilliantly. On September 10, 2007, hours before the general and his partner, the taciturn yet consistently effective Ambassador Ryan Crocker, reported to the Cannon Office Building to begin their testimony, a full-page advertisement ran in the *New York Times.* "General Petraeus or General Betray Us?" read the big black title, a product of the sardonic minds of the antiwar advocacy group MoveOn.org. When the general arrived on schedule and proved to be calm, measured, and tireless in politely answering pointed questions, and then repeated the performance in the U.S. Senate the next day, the citizenry of the United States saw the solid soldier they'd been looking for. The ad backfired, and Petraeus carried the day.

In the last quarter century, three American soldiers have made valid runs at the heritage of Dwight Eisenhower, the last general to hang up his uniform and move into the White House. Colin Powell certainly might have done it in the 1990s had he so wanted. Wesley Clark wanted it very much, and even won the Oklahoma primary in 2004, but America didn't want him. In 2007, David Petraeus looked like the one to follow in Ike's footsteps. The presidency hung just beyond his grasp.

Success in Iraq also shimmered just out of reach. The casualty and hostile-attack rates went down in the fall of 2007, never again to rise to their previous heights, at least during the remaining years of the American campaign. But the fighting never stopped either. It lingered, a third of the previous rate, but that was no comfort to those who fell, killed or wounded, or to their families. Al-Qaeda in Iraq, unrepentant Sunni rejectionists, surly Sadrists and Iranian handlers all kept their pieces on the board. As long as the occupiers remained, there would be attacks. As long as Iraq was Iraq, violence remained part of the picture.

Determined to secure some return for all of this pain, the Bush administration labored to formalize a lasting agreement with Iraq, the final and necessary piece of the Casey long-term counterinsurgency strategy. As 2007 became 2008, polling indicated strongly that the next U.S. president would be a Democrat. Senator Barack Obama of Illinois made it very clear during the Petraeus testimony and in his campaign that if elected, he planned to pull out of Iraq. The Bush people, led by the president, hoped to nail down an enduring U.S./Iraqi relationship, a lasting foothold for America in a strategic region. The thinking included a division's worth of troops on the ground.

Prime Minister Nouri al-Maliki might not have been the most decisive leader, but he could bargain with the gusto of any brass merchant in a Baghdad souk. Maliki embraced the Sunni Awakening and even lashed out at Sadr and his JAM when Basrah erupted in March of 2008. The Iraqi-led Operation Charge of the Knights demonstrated all the finesse of a train wreck, but when the parts stopped sliding around, Maliki's Iraqi soldiers held the city. The embarrassed British, having already made their sometimes odious bargains with the area's militias, stood aside and counted the days until their final withdrawal. They would retire to their bases outside the cities until mid-2009. The rest of the allies folded their tents and departed during 2008 and early 2009. America alone remained.

The Bush team did not dare to draw up a formal treaty, dreading the ratification fight in the Democrat-controlled Senate. They settled for a "Strategic Framework Agreement," signed by Ambassador Ryan Crocker and Nouri al-Maliki on November 17, 2008. It was a very nicely phrased instrument that featured no mutual defense provisions and referred to a "long-term relationship," largely undefined and likely indefinable. The more detailed accompanying document offered a significantly more telling title: "Agreement Between the United States and the Republic of Iraq on the Withdrawal of U.S. Forces from Iraq and the Organization of Their Activities During Their Temporary Presence in Iraq." This one had teeth, requiring a full U.S. withdrawal by December 31, 2011, and an interim pullback of American units from "cities, villages, and localities" by June 30, 2009. The accord included provisions for discussing a follow-on security relationship, but anyone looking for a commitment by either party to a long-term American troop presence, as in Korea, would not find it. For the United States, this deal was all about leaving.

Petraeus was commanding the U.S. Central Command by the time Ryan Crocker signed those papers in Baghdad. General Ray Odierno commanded MNF-I. In Washington, President-Elect Barack Obama readied himself and his new administration to take office on January 20, 2009. The Iraq campaign ground on with three more years ahead. But you could see the finish line. The hope for victory was long gone. Salvage became the order of the day.

12

REQUIEM ON THE TIGRIS

EVELYN MULWRAY: When was the last time?
JAKE GITTES: Why?
EVELYN MULWRAY: It's an innocent question.
JAKE GITTES: In Chinatown.
EVELYN MULWRAY: What were you doing there?
JAKE GITTES: Working for the District Attorney.
EVELYN MULWRAY: Doing what?
JAKE GITTES: As little as possible.

— ROBERT TOWNE, *CHINATOWN*

FADHIL WAS AN evil place, the oldest neighborhood in Baghdad. Maybe an irregular square mile and a half in area, Fadhil squatted a few blocks east of the languid Tigris River that bisected the Iraqi capital into its west and east sides. Fadhil's narrow, twisting, uneven dirt streets stretched ten feet wide at most, flanked by tightly packed, ancient two- and three-story buildings that leaned toward each other like old drunks trying to stay on their feet. Fadhil reeked of hookah smoke and rotting garbage; its dank alleys were haunted by shades. This was the Baghdad of Ali Baba and the Forty Thieves, furtive Aladdin creeping through the darkness, djinn and devils wafting like smoke along the gutters. It felt all wrong, a cursed place akin to Germany's Hürtgen Forest in 1944 or Vietnam's Hue Citadel in 1968, a meat grinder with its maw ever gaping for more. The American soldiers knew it. So did the Iraqi soldiers. They all hated it.

They'd gone in there before, always at a high cost. In 2007, during the surge, U.S. soldiers probed and fenced in the dark warrens of Fadhil. Every time, they paid, with ten killed in total and more than thirty wounded in countless minor scrapes and four separate major encounters, including an all-day smashup in April and a vicious three-hour firefight in July. One of the wounded was an American brigade commander, and the enemy very nearly nailed the First Cavalry Division commanding general in that firefight too. Their partners in the Eleventh Iraqi Army Division suffered as well, with twelve killed and more than forty wounded. The insurgents in Fadhil downed a Blackwater contract helicopter in January and shot up a State Department convoy in May. Sunni every one, the fifty thousand denizens of cramped, hostile Fadhil boldly decorated their homes with eight-pointed Baath Party stars. They delighted in mounting forays into nearby Sadr City to slaughter sleeping Shiite families.

Several U.S. Army commanders estimated it would take an entire brigade combat team to clear Fadhil, a mini-Fallujah with the same rotten American butcher's bill. Many Iraqi civilians would die, too, especially when the U.S. used artillery and airpower to break into all of those tightly packed old tenements. Instead of doing that, in late 2007, the Americans went with the Sahwa model pioneered in Ramadi. In classical mythology, the labyrinth hosts a monster at its center. In Fadhil, that monster was Adel al-Mashadani. Now he was on the American payroll. The deal ended the shootouts and bought time. But no deal lasts forever.

Adel al-Mashadani styled himself the Don of Fadhil, with no apologies to Mario Puzo, Francis Ford Coppola, or Tony Soprano. He was a charismatic, imposing figure, big and beaming, with his head shaved bald and a wide black mustache. He favored leather jackets and always surrounded himself with armed retainers as he strolled through the cubbyholes of Fadhil almost daily, checking on his people. Lieutenant Colonel Craig Collier, who commanded the Third Squadron, Eighty-Ninth Cavalry in Fadhil in 2008, remembered how "old women would come and kiss his hand, thanking him for stuff."

Much of the largess Mashadani showered on his people was delivered by the Americans in the form of school renovations, street-corner electrical generators to light up the old buildings, and make-work street cleanups to keep the youth busy. Nearly two hundred young men carried weapons as Sons of Iraq, paid by the Americans but reporting directly to Mashadani. His militia scoured out al-Qaeda cells, taking their lives and their weapons. Mashadani knew the foe well because he used to serve them. One key AQI leader owed Mashadani a cut of a reward equal to $600,000 for attacking some Iraqi government officials. The Don of Fadhil didn't get his money, but he got his revenge. By early 2008, the local AQI was liquidated. The U.S. troops appreciated Mashadani's bloody handiwork and made no complaint when he kept the arms and ammunition that had been seized in the score-settling.

Rumors surfaced through Iraqi sources. Informants brought news of numerous shop owners strong-armed to pay protection money, with Mashadani's entire take running up to $160,000 monthly. Stories circulated of husbands kidnapped, their wives taken by Mashadani for his pleasure. Both U.S. and Iraqi intelligence sources tied Mashadani to IED incidents. Some suggested he still did some side jobs for his old AQI contacts. A few Fadhil men who tried to stand up to the Don found themselves strung up to the walls of a subterranean dungeon in western Fadhil known as the Gonch and subjected to the usual Baathist motivational techniques involving electrical cords and power drills. In the midst of it all, the Fadhil Sons of Iraq kept vigil, smiling at passing American patrols. Cowed Iraqi army units stayed out. In Mashadani's view, the U.S. contributed their money, so they received their due under his protection racket.

This cozy arrangement changed on January 1, 2009, a consequence of the Bush administration's security agreement with Maliki. In accord with articles 4 and 22, all detentions after that date required an Iraqi warrant. Although the Americans could (and did) still capture Iraqis they deemed threats, missions worked best when done with Iraqi partners. While many Americans worried about the warrant drill, those documents proved amazingly easy to get and could even be procured after Americans had picked up a miscreant. Unlike a domestic U.S. warrant, the Iraqi version lasted pretty much forever and had no limits regarding location either. A warrant issued in Basrah could be executed a year later in Baghdad. Keeping a close eye on the target lists, Iraqi police maintained solid contacts with favorable judges. Sometimes warrants for Shiite subjects took more elbow grease, but with Maliki in office, Sunni could be indicted pretty easily. Throughout 2009, the Iraqi Baghdad Operations Command and the First Cavalry Division worked together well on getting and using warrants.

Yet the Maliki government had its own views. As part of the security agreement, the government took over the payment of the Sons of Iraq and transferred the militias to Iraqi command. The prime minister fully embraced the program and rode Sahwa support to a substantial victory for his political coalition in the January 2009 Iraqi provincial elections. In accord with this effort, in Baghdad, the Sons of Iraq were told to report to the local Iraqi army and police commanders. In Fadhil, that meant the Forty-Third Brigade of the Eleventh Iraq Army Division, a unit that had never dealt with Mashadani. Their American partners now had to make this work. Complicating matters, in December of 2008, an Iraqi judge had indicted Adel al-Mashadani for murder, extortion, and working with al-Qaeda. The evidence against him would easily have stood up in any U.S. court. It was compelling.

The outgoing American unit let it be, unwilling to go into the dreaded Fadhil buzz saw. "We let everybody know you're going to have a hornet's nest on your hands," Lieutenant Colonel Craig Collier said he'd warned his higher commanders. "You're screwing with something that doesn't need to be screwed with." Come 2009, to back up their Iraqi teammates, the new U.S. commanders chose to screw with it.

Decisions like that brief very nicely up at MNF-I, MNC-I, and division headquarters. It fell to Captain Frank Rodriguez and his Troop A, Fifth Squadron, Seventy-Third Cavalry (Airborne) and his fellow Eighty-Second Airborne paratroopers to pull it off. A former Air Force combat controller commissioned through ROTC at the University of North Carolina, the man the Iraqis called Naqeeb Raad (Captain Rod) was a bit older and wiser than his airborne peers. But he was as tough and shrewd as they came, and his airborne cavalry troopers knew how to shoot and how to move mounted and dismounted. Even so, Rodriguez didn't relish a knife fight in the tight, dusty Fadhil alleyways.

The captain had been thinking hard about the problem. He was among those who'd eagerly read the Petraeus-authored FM 3-24. Rodriguez taught his men from the manual, using it extensively in their training. He often quoted the list of cryptic paradoxes. One really appealed to him: *Sometimes, the more force is used, the less effective it is.* Rodriguez and his paratroopers knew how to unleash a Fallujah. Could they take out Mashadani — or, even better, get the Iraqis to do it — without destroying Fadhil?

Rodriguez, his squadron commander, and the brigade combat team came up with an approach they called Operation Shwey Shwey (meaning "slowly, slowly" in Arabic). In essence, they wanted to take a few months to infuse Lieutenant Colonel Ali Mahmoud's First Battalion, Forty-Third Iraq Army Brigade into Fadhil, to supplant Mashadani's mafia with legitimate authority. They would keep meeting with Mashadani and his men and get the Sons of Iraq to join the American-Iraqi patrols. They anticipated emplacing Iraqi army checkpoints in Fadhil alongside those manned by the Sons of Iraq. They planned to assist the Iraqi military in civic action programs instead of funneling funds through Mashadani. Finally, when the Iraqi soldiers were ready, they would raid several suspected weapons cache sites. Mashadani would either cooperate or, at some point, fight. Either way, the Americans planned to be present in force to quash any violent reaction if and when the Iraqis got definitive orders to move on Mashadani.

Operation Shwey Shwey accomplished something else that turned out to be crucial. Each long walk through Fadhil, each inspection of a little storefront or invitation to share tea inside a cramped apartment, stripped away the dark mysteries of the neighborhood. The American and Iraqi squads spent hours and hours on foot, learning all the nooks and crannies of the district. Over the weeks, they compiled a house-by-house, shop-by-shop, block-by-block census of Fadhil. They placed checkpoints wisely and positioned an Iraqi company in the southeast corner of Zaibeda Square, a key vantage point that dominated the major entry road into Fadhil. The district ceased to be a no man's land and became familiar turf. When it came time to go in hard, Rodriguez and his men — and the Iraqi soldiers too — would know their way. It happened little by little, *shwey, shwey.*

The Shwey Shwey approach got Mashadani's attention. In his meetings with Rodriguez, he stayed cordial, but he dropped some telling hints. "Do not believe what you hear," said the Fadhil strongman, "because you do not want things to go back to the way they were." As the weeks went on, and the squeezing persisted, the Don began to warn openly of upcoming trouble as the Iraqi army checkpoints proliferated and the American-Iraqi foot patrols continued. After an Iraqi army raid removed an AQI operative from the center of Fadhil on March 15, Mashadani again reminded the U.S. captain of the bloody past. The Don smiled and bluntly stated, "You cannot control Fadhil for one day." When Rodriguez replied evenly that in his view, the Iraqi army already controlled Fadhil, Mashadani's deputy Qassim spoke up: "What if Naqeeb Raad makes a mistake?"

Two weeks later, that question still hung there unanswered. About one o'clock in the afternoon on Saturday, March 28, on the outskirts of Fadhil, a team from the Ministry of Interior's emergency response unit blocked Mashadani's SUV and apprehended him without a shot fired. In the militia leader's truck, the Iraqi unit found eight hundred PKC machine-gun bullets and six thousand AK-47 rounds. The Ministry of Interior troops sped away with their prize. Behind them, Fadhil erupted.

The Iraqi police's lightning raid surprised the American officers, who'd received no advance notice of it. As a result, it found the U.S. elements badly out of position to contain the ferocious reaction in Fadhil. Evidently, the Iraqi army also found out only after it happened, as did the Baghdad Operations Center and the First Cavalry Division. Within a half hour of Mashadani's arrest, the Second Company, First Battalion, Forty-Third Iraq Army Brigade reported itself surrounded and under heavy small-arms fire at its Zaibeda Square outpost. The Americans had no soldiers there. East of Fadhil with his paratroopers at Joint Security Site Bab-al-Sheikh, Frank Rodriguez watched as the Iraqi police manned the walls and dispatched reaction-force trucks. He assembled Troop A's First and Third Platoons, leaving the rest of the unit to guard the police post. A radio call confirmed the worst. His paratroopers strung out on other missions, Rodriguez's squadron commander might be able to get Troop C into the area by nightfall. Reinforcing Iraqi army units were spinning up, but none would be there for hours. The American attack helicopters couldn't fly due to blowing dust. Maybe some jet fighters might be available. The squadron was working on it. Meanwhile, Lieutenant Colonel Mahmoud of the First Battalion, Forty-Third Iraq Army Brigade was calling for immediate help. Right now, that amounted to Frank Rodriguez and his two airborne cavalry platoons. All of those hours of walking through Fadhil were about to pay off.

By about three o'clock, Rodriguez and his Third Platoon reached the outskirts of Fadhil a quarter mile from the surrounded Iraqi rifle company. The Iraqi Fourth Company commander met the American troop commander and told him that an attempted linkup had already failed. "Too much fire," he said. Rodriguez sized up the situation. The Iraqis thought they could add five of their own armored Humvees to the American thrust. As the leaders worked out details through interpreters, Third Platoon was already shooting at the enemy. The former Sons of Iraq, now back to being insurgents, kept up a steady volume of gunfire. "Rounds could literally be seen skipping off the ground around us and off the vehicles in the blocking positions," Rodriguez wrote later.

Within twenty minutes, First Sergeant Robert Cobb arrived with First Platoon, bringing the U.S. strength up to ten up-armored Humvees, about fifty Americans in all. Each truck mounted a machine gun in its open turret. Cavalry troops are small units by design, built for reconnaissance, not street fights. But they know how to fire and maneuver, and Rodriguez and Cobb had trained their paratroopers well on how to do so. If they moved fast and hard, they could catch a large number of the Mashadani men in Zaibeda Square. The opponents had not seen their leader's arrest coming, so they probably hadn't buried IEDs yet. Thus far, nobody had reported mortars or RPGs. Without Mashadani's leadership, Qassim and the other second-tier guys had no particular claim on the loyalty of the militiamen. One good bloody nose might break this enemy. And nobody knew Fadhil better than Rodriguez and his people.

Rodriguez told the Iraqis to lead. He'd follow with Third Platoon, and Cobb would provide covering fire with First Platoon. Two F-16 jets came over to help get the relief column going north. The jets made one low pass, dipping out of the clouds with afterburners echoing like metallic thunder off the buildings. As the jets streaked across Zaibeda Square, the Iraqi and American trucks rolled north.

The enemy let up a little when the F-16 pair roared over but picked up again as soon as it became obvious the jets weren't dropping or shooting anything. Rodriguez said later that "the volume of fire became so intense that clouds of dust and rounds enveloped the vehicles." Enemy gunmen fired from rooftops and upper windows, from doorways and street corners. The place was crawling with bad guys, every one with a barking AK. Fortunately, they didn't aim, merely pointed and yanked the trigger. That might work in Hollywood, but in real combat, it causes automatic weapons to skew up and to the right, so rounds soared uselessly over the heads of the Humvee gunners. The American machine gunners in the turrets of the moving trucks shot back, doing what they could as they bounced north. But the real work was being done by the machine guns atop the first sergeant's stationary trucks back at the launch point. Those paratroopers shot low, the way they'd been trained. They aimed and hit. Together, all of the U.S. gunners suppressed the enemy enough for the Iraqi platoon to get to the Second Company outpost. The Iraqi trucks stopped, doors opening before the tires stopped moving. Ducking a sheet of AK rounds, Iraq soldiers piled out and sprinted for the outpost entrance.

Airborne cavalrymen went in with them. Rodriguez followed his men inside. The captain had to dodge and shoot as he did so, with hostile rounds hitting near the front entrance. Next door, a building was burning furiously, belching oily smoke. It made it hard to see inside the Iraqi-held structure. Nearly out of ammunition, the Second Company soldiers were glad indeed to see that the cavalry had arrived. They assembled to depart with the armored trucks.

There was a problem, though. One Iraqi officer told Rodriguez that he could not find seven of his men and presumed they'd been captured. A quick head count of Second Company tracked with the bad news. Rodriguez lacked the manpower to hold the square against a hundred militiamen. With no idea of where the missing seven Iraqi soldiers could be — and with fifty thousand Fadhil residents, thousands of apartments to sort through, and night coming soon — the captain convinced the Iraqi commander to pull back. Having burned through a lot of ammunition, the relief column had to get out and resupply. They intended to come back.

With Second Company's men aboard the Humvees, the American-Iraqi column fought their way south. When they reached the relative safety of the start point, the men got out and began reloading ammunition. As they did, most had a few minutes to look at their scarred trucks: sand-painted metal sides pitted by bullet hits, bulletproof-window blocks spider-webbed by impacts, and multiple tires flattened. Luckily, and typically, the enemy's marksmanship did not match their enthusiasm. In the cases where it did, the armor did its job.

Rodriguez expected to go back in as soon as the Apache attack helicopters came on station. The blowing dust was abating as the sun set. In congested Fadhil, only a few dim bulbs glowed in upper windows as the shroud of darkness descended. The thousands of civilians in there stayed put, paralyzed, waiting to see how it all played out. Rodriguez and his paratroopers reloaded their rifle magazines and kept watch. Any time now, he expected an order to head north.

Instead, his squadron commander told him to hold his position. All around the perimeter of Fadhil, Iraqi and U.S. units arrived, one after another, including Troop C to the west, additional battalions from the Eleventh Iraq Army Division to the south, and T-72 tanks from the Ninth Iraq Army Armored Division to the east and north. A Black Hawk helicopter clattered overhead, dumping hastily printed leaflets. The fluttering handbills announced that Mashadani had been lawfully arrested for murder and collusion with al-Qaeda and that his men must put down their arms. The assembling American and Iraqi forces left no doubt what would come next, beginning at dawn. A few gunshots continued on the edges of Fadhil as the evening wore on. But clearly, after the wild late afternoon at Zaibeda Square, the air was leaking out of Mashadani's militia movement.

Rodriguez was right there to the south with Troop A, preparing to return to Zaibeda Square with his Iraqi army partners. Things were trending differently, though. As Rodriguez listened, he heard Mashadani's deputy Qassim on the cell phone with the Iraqi battalion commander. The interpreters caught and relayed snatches as Lieutenant Colonel Mahmoud spoke. Qassim had his hands on the seven missing Iraqi soldiers, and he and his former Sons of Iraq wanted to negotiate. In the tradition of Ulysses S. Grant, Mahmoud gave Qassim a hard answer: Deliver the prisoners. Lay down your arms. If you don't, face annihilation. While Qassim thought it over, the Iraqi soldiers announced they were ready to return to the Second Company outpost to the north. The Americans of Troop A mounted up. It was time.

With attack helicopters overhead in the dark sky, Rodriguez and Mahmoud returned with their men to Zaibeda Square, rolling slowly north under sporadic fire. Within an hour, just past ten o'clock, the Iraqi army found all seven prisoners just outside the Second Company outpost. Evidently, Qassim had set them free. The Americans and Iraqis spread out and held the square all night. Just before sunup, thousands of Iraqi soldiers, a dozen sand-yellow T-72 tanks, and the paratroopers of Troops A and C moved in. Some gunmen shot at the advancing soldiers. Well-aimed return fire quelled those erratic, isolated attempts to fight back. Resistance had dissipated by midmorning. Iraqi soldiers led a house-by-house clearance of Fadhil that took a week to finish. Every militiaman was disarmed, with some twenty detained briefly. Amnesty was eventually granted. Numerous weapon caches were removed from basements and storage rooms. Store owners stopped fearing for their lives. Women held against their will went home to their families. The Gonch torture chamber went out of business for good.

Four of Mahmoud's soldiers were killed and four wounded, all in the Second Company. Mashadani's militia suffered twenty killed in action and some fifty wounded, all in the afternoon firefight around Zaibeda Square. Despite the massive hostile fusillade, the Americans didn't suffer a single casualty, defying the law of averages and perhaps ballistics too. It turned out that, precisely as Mashadani had warned, Rodriguez could not hold Fadhil for a day. But after that one brief spasm, he held it for good, side by side with the Iraqi army.

Frank Rodriguez, Robert Cobb, and Troop A found plenty of combat in Fadhil. Their work reflected the nature of U.S. operations in Iraq in those final years, smaller American units backing up more numerous Iraqi forces carrying out orders from the Baghdad government. Iraqi soldiers took the lead, with the Americans providing air support, helicopter cover, overhead surveillance, key intelligence leads, medical assistance, civic action funds, and, most important, combat units to stiffen their partners' resolve. In the words of article 4 of the U.S./Iraqi security agreement, the Americans had both the authority and the duty to aid their Iraqi teammates by "training, equipping, supporting, supplying," all while retaining "the right to legitimate self defense." In soldier-speak, that meant the U.S. would advise and assist the Iraqis in their fight.

Without Frank Rodriguez and Troop A, the Iraqis might have retaken Fadhil in their own sloppy way over a few days, with a lot more death and destruction.

By 2009, when you rolled with the Iraqis, you could find fights if you wanted them. But you had to look harder than you did in the old days. They didn't come to you as often as at the height of the war in 2006 and 2007. In all of 2009 in Baghdad, attacks averaged fewer than ten a day; there had been nearly sixty every twenty-four hours in the bad old days. The First Cavalry Division suffered thirty-eight dead, and they killed or captured 1,602 insurgents. Each of those American deaths hurt badly, but by comparison with the nearly five hundred lost in and around the Iraqi capital in 2006 and 2007, it showed just how much the violence had tapered off.

In quelling the insurgency, the American troop surge played its part, especially in Baghdad and its surrounding belt villages. The last reinforcing units departed by early 2009. The U.S. Army was able to revert from the grueling fifteen-month deployments to twelve-month stints. In 2011, deployments dropped to nine months, for the first time since 2003. Even at nine months, Army unit deployments still exceeded the standard Marine Corps, Navy, Air Force, and SOF tour lengths. But the shortened stretches, coupled with troop reductions, certainly eased the pressure on America's soldiers.

Although the troop surge made the news in America, in country, the Sunni Awakening delivered the real and lasting difference in the rate of attrition. Renegades like Mashadani were exceptions. The Sons of Iraq proved overwhelmingly loyal. Nearly a hundred thousand strong, half of that number in and near Baghdad, the Sahwa movement allowed the Sunni to carry weapons lawfully and get paid, effectively removing much of the incentive for the "honorable resistance." It was by far the most successful and widespread jobs program in Iraq. The various reconstruction and civic action projects had patched up the roads, oil pipelines, power plants, schools, and hospitals, employing hundreds of locals at times. The Sahwa, however, paid tens of thousands of Sunni Arabs to kill each other, not Americans. Cynical it might seem, but you couldn't argue with the results. The Sons of Iraq fielded some six times as many Sunni with firearms as the highest estimate of enemy strength. It showed the potential depth and resiliency of the Sunni insurgency.

As George Crook learned with the Apaches, better to pay some than fight all.

Iraq being Iraq, enemies persisted. Somebody was always trying to take a whack at the infidels. Shards of al-Qaeda in Iraq still struggled to mount a car bombing or two every other month. They pulled off three major multisite suicide car attacks in Baghdad, in August, October, and December of 2009, all horrific, yet equal in total to a week's work back in 2006. The Americans stayed after AQI. On April 18, 2010, they nailed Abu Ayyub al Masri up near Tikrit, not far from where they'd yanked Saddam out of his hole back in December 2003. The manhunters didn't let up until the end; neither did the opposition.

Not all Sunni rejectionists reconciled under the Sahwa program. Die-hards hung in there. Gunmen of the 1920 Revolutionary Brigade and Jaysh al-Rashadin (Army of the Rightly Guided) made attacks near Baghdad, and the Jaysh Rijal al-Tariq al-Naqshabandi (Army of the Men of the Naqshabandi [mystical Islamist] Order) fought up north near Mosul and Kirkuk. These groups kept laying IEDs and shooting up American convoys, although at a much reduced rate. With the Americans committed to departure by the end of 2011, few Sunni Arab fighters wanted to be the last to die tangling with the aggressive Americans. Whether insurgents or Sahwa, the majority of Sunni Arabs bided their time, jockeying for their place in Iraq after America's time.

On the Shiite side, the Sadrists officially remained quiescent after their March 2008 humiliation in Basrah, Moqtada himself supposedly studying Shia Islam in Iran. The JAM's Iranian-backed affiliates and competitors stayed active. A new version of JAM, the Promised Day Brigade, took occasional cracks at U.S. targets. They were not very capable. The Sadrist breakaway faction Asib al Haq (League of the Righteous) maintained a tenuous cease-fire through most of 2009, broken by occasional rocket strikes on U.S. bases in Baghdad or the downtown Green Zone. Baghdad Khataib Hezbollah (Party of God Brigade) still placed EFP-shaped charges and fired their own rockets. They perfected a truck-borne contraption called an IRAM (improvised rocket-assisted munition), a calliope of heavy rockets on a flatbed covered with a tarpaulin. An IRAM truck could drive up near a U.S. compound, park a hundred yards outside the wall, yank off the tarp, then shower the base with a dozen high-explosive projectiles in a few seconds. After some ugly surprises in Baghdad during 2008, effective Iraqi and U.S. patrolling and raids preempted any incidents in the capital throughout 2009. But like sharks circling in the ocean just beyond the sight of a scuba diver, even if you didn't see them, they still saw you. The Hezbollah IRAM gunners watched and waited for the Americans to drop their guard. A few times in 2010 and 2011, that happened.

It was all wrapping up in slow motion. Four weeks before Frank Rodriguez concluded America's business with Adel al-Mashadani, the new U.S. president formally announced the end of the Iraq campaign. President Barack H. Obama ran on a platform that emphasized hope and change. He focused on matters at home. Like most prominent Democrats, Obama considered the Iraq campaign a bloody botch. In U.S. domestic partisan politics, after initial enthusiasm in March and April of 2003, the Democrats judged Iraq to be the bad war, brought on by (best case) George Bush's arrogance or (worst case) outright dishonesty emanating from the White House. Afghanistan, however, was the good war, neglected under Bush, outsourced to overmatched allies like Britain, and overdue for attention. Deliberate and nuanced in contrast to his often impulsive, inarticulate predecessor, Obama decided to defer to the timeline already set by Bush and Maliki. The American departure wouldn't be precipitous. But it would happen on schedule.

President Obama announced the Iraq pullout at Camp Lejeune, North Carolina, on February 27, 2009, then cemented the policy during a surprise trip to Baghdad on April 9, 2009. In addition to emphasizing the final American withdrawal in December 2011, the new president added an interim step: "Let me say this as plainly as I can: by August 31, 2010, our combat mission in Iraq will end." The U.S. troop strength at that point would be no more than fifty thousand, less than half of where it stood in 2009, and all would be advising and assisting. Lawyers and press flacks in the White House comforted themselves that such a mission didn't constitute combat, which would have been news to Americans in uniform in Iraq. In any case, Obama was shifting to Afghanistan. Iraq was over.

Behind the scenes, certain of Obama's people made noises about keeping a substantial U.S. follow-on force in country after 2011, an open-ended troop commitment akin to Korea. That neatly matched counterinsurgency doctrine. But it didn't match American domestic sentiment.

Soldiers and airmen heard what they wanted to hear, but they should have taken the president at his word.

Any follow-on U.S. force level depended on negotiating with Prime Minister Nouri al-Maliki, and he took his usual stance: he would cooperate only as far as it suited his purposes. Maliki had no intention of keeping any substantial American troop presence in his country. The Kurds desired a strong American presence, and so did the Sunni Arabs. Riding the Shiite majority wave, with enough Kurds and Sunni on his side to keep things semi-calm, Maliki was well past the need for U.S. soldiers in his country. He believed his own press releases.

One of the stranger provisions of article 24 of the security agreement directed that "all American combat forces shall withdraw from Iraqi cities, villages, and localities" by June 30, 2009. To Maliki, that meant the U.S. units would move to their bases and sit there as Iraqi army and police took over the entire war. His own Iraqi generals did not want that, knowing that, like Lieutenant Colonel Mahmoud in Fadhil, they needed a lot of reliable U.S. help. Ever legalistic, the Americans saw some wiggle room. They had to carry out their support role under article 4. What if such units were not defined as combat forces? Clearly, President Obama opened the door to that interpretation at Camp Lejeune. Could it be sold to Maliki?

The new U.S. ambassador to Iraq, Christopher Hill, inherited that task. He'd made his name trying to negotiate with the duplicitous, intractable North Koreans. Unlike the outgoing Ryan Crocker, Hill thought he could take the sole lead in managing Maliki. The military role was ending. General Ray Odierno at MNF-I, a team player all the way, was rankled by Hill's overt insistence on taking charge. The general preferred the easy partnership the military had built with the embassy under Crocker. But he also knew his limits. Odierno had served during the rocky Bremer-Sanchez period and refused to get sucked into any similarly dysfunctional arrangement. The general set aside his ego and tried to follow the new ambassador's lead.

As Odierno expected, Hill got nowhere with Maliki beyond basic pleasantries. The two civilians discussed diplomatic visits and generalities but no hard matters, certainly not the true scope of the pending out-of-the-cities deadline. After a few sessions enjoying the circuitous stylings of Nouri al-Maliki, Hill probably longed for the old days dealing with Pyongyang. At least you knew for sure they would stiff you. With the new ambassador neatly boxed in by the dour Iraqi prime minister, it fell to MNF-I to sort out what the getting-out-of-the-cities thing really entailed.

Having served as both a division and corps commander in Iraq, Odierno chose an interesting, bottom-up solution. The big four-star made it clear that partnering with the Iraqis must continue, in and out of the cities and villages, as well as in the "localities," whatever the hell those were. Odierno then directed Lieutenant General Charles H. Jacoby Jr. and his First Corps headquarters to back off and let the division commanders sort it out. In the spring of 2009, all the division commanders were in-country campaign veterans maintaining strong relationships with their local Iraqi counterparts. Jacoby had never served in Iraq and had been in country only a few weeks when the out-of-the-cities horse-trading began in earnest. He stepped back. Odierno figured each division would make local arrangements, and it would work. Maliki could claim his great "victory" over the occupiers, and article 4 support would continue.

Odierno's idea produced the results he wanted. Up north, the Twenty-Fifth Infantry Division kept key outposts open in Mosul, Kirkuk, and other urban centers. Vehicles carried placards indicating they performed a support role, reinforcing the Iraqi forces. In the northern provinces, the larger, uniquely American MRAP (mine-resistant ambush-protected) armored trucks stayed out of the cities. Only Humvees, which the Iraqis also used, went in and out, lowering the U.S. profile. The Americans in Kirkuk and Mosul trod lightly, cajoling the Kurds and Arabs daily to keep them happy along what all euphemistically referred to as the "disputed internal boundaries" that traced the edges of semiautonomous Kurdistan. Out west, the Marines continued their successful embrace of the Sahwa, moving in and out of cities and villages as necessary with full support of the Anbar Sunni and without any attention from the *sheruggis* in distant Baghdad. Down south, the deadline coincided with the British departure, and they had long ago left the inner cities. The Americans slid into the old British bases without much fuss. Maliki and his cronies chortled over the departure of Abu Naji (Father of Those Left Behind), the same British who'd once ruled Iraq under the League of Nations mandate after 1918. Maliki's circle then realized they had just evicted the forces of the country training their small navy, whose sailors and patrol boats guarded the offshore oil platforms that powered the entire Iraqi economy. Maliki backtracked and quietly got the British Royal Navy trainers to return for a while.

A quiet approach also worked in Baghdad, the acid test for "out of the cities." In the capital, all of the U.S. brigades had areas of responsibility that extended from central Baghdad out to the rural belts, like slices of a pizza. Accordingly, the BCTs shifted the bulk of their power to the countryside over the spring. Fourteen combat outposts remained in the city to ensure article 4 support to the forces of the Baghdad Operations Command. The First Cavalry Division directed all of its urban-based American units to execute on a reverse cycle commencing on July 1. For about a month, the U.S. units in the city joined their Iraqi partners by night, and together they patrolled, raided, and resupplied. To the average Baghdad citizen, the Americans were gone. Nobody ever called them on the subterfuge. Given the 120-degree summer heat, nights worked out better operationally too.

Nouri al-Maliki celebrated the June 30 deadline as a great day of independence for all Iraqis. Television stations ran countdown clocks as if awaiting the new year. In Baghdad, there were fireworks and a Saddam-scale parade. Maliki designated the day a national holiday. He proclaimed: "The national united government succeeded in putting down the sectarian war that was threatening the unity and the sovereignty of Iraq." He never mentioned the United States, by whose sufferance his regime existed. Obama had been right. It was time to go.

Maliki needed one more thing from the Americans. On March 7, 2010, after several delays, Iraq voted for a new parliament to replace the one chosen in December of 2005. The three primary candidates were truly the usual suspects: Maliki, al-Jafaari, and Allawi. Backed by a substantial number of Sahwa figures disenchanted with Maliki, Allawi's list won in a squeaker, securing ninety-one seats to Maliki's eighty-nine. The 2005 Iraqi constitution designated the leading party to form the government, and Allawi set to work doing just that. American soldiers saw this as a good outcome. Allawi was by far the most level-headed of the three. As for Maliki . . . well, he would not be missed.

Maliki did not go quietly—far from it. He demanded and arranged recounts in Shia areas sure to favor his prospects. The tenacious prime minister also argued that because Allawi's list did not win a majority of the 325-seat assembly, he had no right to form a government. Maliki began courting smaller parties in an attempt to cobble together a majority coalition and retain his office. He also announced that as long as the electoral outcome remained unsettled, he planned to continue as acting prime minister.

At MNF-I, Ray Odierno looked forward to the return of Allawi, especially with that leader's substantial Sunni backing. The old Casey strategy of "Sunni in" remained a key MNF-I idea, emphasized by the Sahwa. Ambassador Chris Hill, however, wanted to build on his relationship with Maliki, such as it was. He advocated backing Maliki and cajoling Allawi to accept some ill-defined role running a new strategic council. The Kurd Jalal Talabani would continue as president of Iraq, and the Sunni could count on subsidiary posts. In other words, the same-old, same-old. Hill convinced Vice President Joe Biden to back the plan. Preoccupied with domestic policy and Afghanistan, Obama concurred. It took until November to convince Allawi to knuckle under. America cast its lot with Nouri al-Maliki.

He repaid the trust with his customary ingratitude. By 2013, he had moved to arrest his Sunni vice president, snubbed the Kurds, and was working to modify the Iraqi constitution so he could stay in office in perpetuity. The following year, Maliki reaped the whirlwind when the Sunni north and west rose up against his Shia-dominated ruling coalition. Defiant and stubborn, Maliki blustered in Baghdad, seeking help from Iran and the U.S. to hold his fractious state together. Whatever Nouri al-Maliki ran, it sure wasn't a democracy. That was already quite evident in 2010.

It did not matter anymore; Maliki's missteps were no longer of much interest to the United States. The Americans focused on finishing up their mission. Hill left. Odierno did too. All the Coalition allies were long gone, as were the Marines. Effective September 1, 2010, U.S. soldiers drew down to fifty thousand in six advice and assistance brigade combat teams. Sixty-six more Americans died before the campaign officially ended, on December 18, 2011. The last soldier killed, Specialist David E. Hickman of the Second Battalion, 325th Airborne Infantry, died in an

IED strike in Baghdad on November 14, 2011. Few Americans noted his passing.

No division of American soldiers remained behind to honor Hickman's sacrifice, though the military had expected to keep such a force in country right up to the end. Army planners had even designated the Third Infantry Division to carry out the enduring mission. The U.S. generals should have listened more closely to Obama and Maliki. It was not to be.

Many of the generals and admirals thought that Iraq's strategic value, bordering Turkey, Syria, Jordan, Saudi Arabia, and Iran and sitting on huge oil reserves, supported keeping a significant U.S. force of soldiers and airmen in the country. Any vacuum left by the U.S. would certainly be filled by Iran, already present in country, playing footsie with Maliki, arming the Sadrists, and courting all the other Shia Arabs now running Iraq. To the senior leaders in the Pentagon, Iraq in 2011 looked like Korea in 1953, an unpopular, mishandled war that nevertheless positioned America in a pivotal location. President Eisenhower, a soldier and strategist to his bones, knew how to read a map. Newly elected on a wave of discontent over the frustrating Asian land war, Ike had held his nose and stayed on the Korean peninsula. He refused entreaties to shift forces to the failing French campaign in Indochina. But Barack Obama was not Eisenhower. His strategic education was just beginning. As he'd promised in his election campaign, Obama trained his sights on crumbling Afghanistan and unreliable, nuclear-armed Pakistan. The new president was willing to leave a token presence in Iraq, but nothing more.

So one American military element remained. To keep the security relationship going, the Americans employed a 4,318-strong contractor force superintended by a 157-man military Office of Security Cooperation–Iraq (OSC-I) under a U.S. Army three-star general. This "son of MNSTC-I" ran ten training and equipping entities at Irbil, Kirkuk, Tikrit, Basrah, Umm Qasr, and five Baghdad-area sites. Fielding plans for Iraq included M-1 tanks and F-16 fighter jets as well as other weapons, ammunition, and repair parts. The OSC-I resembled similar establishments in Egypt and Saudi Arabia. Along with the U.S. embassy, OSC-I represented all that remained of the once-massive American commitment to Iraq.

Even this relatively small contingent had its issues. The 2008 U.S./Iraqi security agreement ended with no follow-on document, another ball dropped in the haste to get out. Maliki didn't much care, nor did his ministers. It all rendered movement difficult and left the contractors at the mercy of Iraqi law, such as it was. America intended to provide arms and training. The Iraqis wanted to receive them. Still, somehow, and characteristically for Mesopotamia, the fumbling continued. In the words of Lieutenant General Robert J. Caslen, OSC-I chief: "And this drives to one of the biggest issues we have and that is, what is our mission, what is the organization for this mission, what are the authorities, and what is our doctrine?" No answers came back. Caslen and his small headquarters were left to figure it out on their own.

Nobody in Washington wanted to look at Iraq anymore. It was in the rearview mirror. Bush's war of choice, the bad one, the foul-up, was over. As the partisan critics had long predicted, it had not turned out well, with 4,486 Americans dead, 218 Coalition fatalities, at least 103,775 Iraqis lost, and a suspicious authoritarian regime running Baghdad under strong Iranian influence. President Barack Obama and his administration turned fully to Afghanistan, the campaign they believed to be the good war. As they soon discovered, the only good wars are the ones that are over.

PART III

NEMESIS

THE AFGHAN CAMPAIGN

APRIL 2003 TO DECEMBER 2014

Hello, I must be going.

— CAPTAIN SPAULDING IN GEORGE S. KAUFMAN, *ANIMAL CRACKERS*

13

UNDONE

> When you're wounded and left on Afghanistan's plains,
> And the women come out to cut up what remains,
> Jest roll to your rifle and blow out your brains
> An' go to your Gawd like a soldier.
>
> — RUDYARD KIPLING, "THE YOUNG BRITISH SOLDIER"

BRITISH SOLDIERS KNEW Afghanistan only too well. The heart of the British army, especially the infantry, has long been the regiment, the battalion of six hundred or so who live, train, fight, and, when necessary, die together. Cynical observers often refer to these units as tribes, and indeed they are, each with its own customs, totems, uniform variations, and traditions. Their tactics follow common British army doctrine, but the style, the nuances of the application, is uniquely regimental. Each regiment carries two flags, the Queen's Color and the Regimental Color, emblazoned with the battle honors of the command. Soldiers knew what made their regiments famous. In pubs, on training grounds, and under fire, they stood up for those traditions.

By 2006, most British regiments represented amalgamations of numerous older units. The Royal Gurkha Rifles (RGR), for example, recruited tough hill men from Nepal and integrated Gurkhali-speaking British officers with the Nepalese officers and NCOs. The RGR combined the lineage and traditions of the Second King Edward VII's Own Gurkha Rifles (the Sirmoor Rifles), the Sixth Queen Elizabeth's Own

Gurkha Rifles, the Seventh Duke of Edinburgh's Own Gurkha Rifles, and the Tenth Princess Mary's Own Gurkha Rifles. The RGR's battle honors ran from the Battle of Assaye in India in 1803, where the regiment served with Arthur Wellesley, the future Duke of Wellington, and on through numerous colonial conflicts, both world wars, clashes tied to the postwar British imperial withdrawal, and the 1982 Falklands War. The Royal Gurkha Rifles saluted Charles, the Prince of Wales, as its honorary colonel in chief, and every Gurkha officer and "other rank" (enlisted soldier) knew the brigade motto: *Kaphar hunnu bhanda marnu ramro* ("Better to die than to live a coward"). Their curved kukri knives gave abundant evidence of the well-documented Gurkha willingness to fight in close. Six of the eighty-five battle honors stood out: Afghanistan, 1878–80; Kabul, 1879; Kandahar, 1880; and Tirah, the Punjab Frontier, and Afghanistan, 1919. On July 1, 2006, a platoon of Gurkhas lived out the regimental heritage in the hardest possible way. Under Major Dan Rex, the Gurkhas inherited the defense of the district center in Now Zad, Helmand Province, Afghanistan. It soon became evident that the local Pashtuns had their own traditions and plenty of firearms too.

At the time, the basic British tactic in Helmand Province was to establish platoon houses and secure key population centers. Once that was completed, the British intended to revitalize major construction projects, building on work previously done by the U.S. Agency for International Development to dam the Helmand River for hydropower and irrigation. The Kajaki Dam project dated back to 1951. Like most good ideas in Afghanistan, it languished half finished, yet another casualty of constant civil strife.

Getting Kajaki going again necessitated planting a solid Coalition footprint in Now Zad, a mud-brick settlement of ten thousand citizens serving as the district government seat. The brown adobe town stood about four thousand feet above sea level on the arid outer fringes of the Registan Desert. Now Zad lay west of the meandering, muddy Helmand River and north of Highway 1, the Ring Road segment that connected Herat two hundred miles to the northwest and Kandahar sixty miles to the east. With nearly six thousand troops in country and British general Sir David J. Richards in command of the International Security Assistance Force, the British were serious about choking off the resurgent

Taliban movement stirring among the Pashtuns of Helmand Province. That depended on holding Now Zad. And keeping Now Zad depended on Dan Rex and the Gurkhas.

Two Royal Air Force Chinooks brought them into town. The big twin-rotor helicopters landed in the late morning on July 1, 2006, in the flat, dusty open space just north of the walled Now Zad district center. The helicopters blew up huge yellow dust swirls in the 110-degree heat. Small muscular men in British army yellow-and-khaki-mottled desert camouflage, the heavily laden Gurkhas trotted down the dirt-streaked Chinook ramps. The roar of the helicopter jet turbines and the chop of the big rotor blades drowned out all other sound. After quickly clearing the rotor disks, the Gurkhas proceeded to the lip of the open area and knelt down, backs to the aircraft, shielding their helmeted faces with gloved hands. Then the Chinooks took off.

As the helicopter noise abated, two cracks sounded, then two more — Kalashnikovs, firing at the Gurkhas. The rounds went way high, even above the walls of the district center. The Gurkhas stood slowly, shrugged under their bulging packs, leveled their SA-80 5.56 mm rifles, and plodded deliberately to and then through the open gate of the Now Zad government compound. Gurkhas do not scare easily, if at all.

The district compound was a small, rectangular city block, fifteen hundred feet by five hundred feet. An adobe brick wall ten feet high and two feet wide ran all around the facility, pierced by multiple openings, each hung with a gray metal gate. Inside the block stood sixteen separate buildings, each fronted by a square courtyard rimmed with four-foot-high packed mud walls. Eight small structures clustered to the west, lined up like eggs in a Styrofoam container. Those served as living quarters for the Afghan national police (ANP) station. The biggest building, two leaning, brown-brick stories in the eastern quarter of the site, furnished offices and meeting rooms for the district governor. That notable had not been seen lately. Major Dan Rex used the governor's office as his headquarters. He focused his defenses right there.

The commander positioned Gurkhas on the corner rooftops and at the gates. The Gurkhas began to work immediately, using mud bricks and sandbags to build up bunkers they called, in accord with British squaddie parlance, sangers. Each featured a high false position and, below it, a well-protected actual post with an L7A2 general purpose machine gun, the same wicked, reliable, hard-punching 7.62 mm belt-fed weapon the U.S. called the M-240. The Gurkhas manned the northern sanger. An attached Afghan national army (ANA) platoon held the one to the south.

The men quickly established a sustainable routine. One section (the British equivalent of a U.S. rifle squad, seven soldiers led by a corporal) manned the north sanger and the wall posts. A second section acted as a quick-reaction force and patrolled on foot in the town. A third section trained the Afghans, did fatigue details to maintain the mud-brick buildings, cleaned weapons, ate, performed personal hygiene, and rested. That section often had to kit up and reinforce either of the first two sections if something happened. In Now Zad, something happened every day and most nights.

In theory, Rex and his men were to patrol Now Zad, securing the population, setting up for the arrival of a provincial reconstruction team to carry out civic action projects. Shortly after his men had settled into the district center, Rex got a few sections out and about in the bazaar, the local marketplace. Only a few shops did any business. The locals had little to say. Gurkhas reported back that they'd seen beat-up old Toyota pickups and sagging Opel cars, all packed completely with men, women, children, rugs, chairs, and bags of clothing. Many civilians evidently didn't care to be "reconstructed." They were leaving. The Gurkhas also reported other vehicles entering town, all with young men aboard. That couldn't be good.

For two days, an uneasy quiet held around the district center. Gurkha riflemen in the sangers reported robed Pashtun men watching from roofs and doorways. The British called these enemy scouts dickers and would learn over time to pay close attention to them. But in Now Zad, they didn't know that yet.

Around 11:00 a.m. on July 3, another hot, dry day, the Gurkhas in the northern sanger reported that the shopkeepers in the market were shuttering their stores. Young men in twos and threes moved among the buildings on the far side of the open helicopter landing zone. The British couldn't see any weapons. From the southern sanger, the ANA announced that they saw men popping up and down on the bare mound of what they called ANP Hill, because the police sometimes put a team up there. Not today — the Afghan national police were not out. Alerted by the reports, Rex moved up to take a look. He had been in enough fights already to recognize what was brewing. "Soon, they were all around us," he said. "It was clear they were preparing an attack."

The Taliban initiated with an RPG, shooting six hundred feet across the open landing zone right into the Gurkha north sanger. The defenders heard the opening hard bang of the launch rocket, then saw the fat grenade zip their way and explode above them in the dummy position. Five hostiles stood up just across the alley and charged the sanger, AKs chattering wildly and, as it turned out, uselessly. The Gurkha machine gunners ripped off a burst of twelve or so, then another, right out of the book. All five Taliban pitched over, stopped in their tracks by the big slugs.

To the south, the ANA soldiers had a nasty surprise when two Taliban emerged from a hole in the base of a fronting house. The enemy tossed hand grenades at the ANA sanger. Both blew harmlessly in the narrow dirt street, but the blasts alarmed the Afghan troops. Other Taliban up on ANP Hill fired AK-47s at the Afghan soldiers. Most of their rounds never even hit the dummy bunker. Shooting high came easy with an AK, and the Taliban did it consistently. The ANA soldiers did too. Both sides furiously traded bullets. The steady Gurkha quick-reaction section shifted to the south and fired a series of calm, well-aimed single shots from their SA-80 rifles, killing both the close attackers. The Afghan soldiers took down a few more up on ANP Hill. Within twenty minutes, two British AH-64 Apache attack helicopters arrived overhead, and the enemy broke off the contact.

Rex and his Gurkha soldiers took stock. They'd clearly won the daylight fracas, knocking out the Taliban without any friendly casualties. There were, however, some worrying signs. "They used cover well and they moved about very fast," Rex noted. He went on: "It's obvious they were testing us out. They were examining our arcs of fire, our fire times, how soon before air support would arrive." The Gurkha defenders of the little British outpost knew well the many tales of encircled detachments and what became of them.

Dan Rex borrowed pages from John Lennon, Northern Ireland, Basrah, and many colonial scrapes: He decided to give peace a chance. The next morning, working through translators, the British major offered the Taliban a truce. The enemy accepted the deal. They respected the British as fellow warriors. As one opponent later put it, "Fighting the British seems like unfinished business for us." The Pashtuns remembered well the stories of the tough, disciplined khaki-clad regiments standing fast in their hard squares. The British were not bad soldiers, a Taliban fighter said. "They are not cowards. They do not cry, or shout 'Oh my God' in the front lines as the Americans do." The Gurkhas watched a few Taliban drag the dead bodies off ANP Hill.

Major Rex hoped the truce might hold. It did not. Instead, as the major guessed they might, the enemy kept probing. The Taliban hewed to the Maoist dictum of *enemy halts, we harass.* After nightfall on July 5, the Taliban again shot up the north sanger. Rex called in an American jet fighter. The plane dropped a JDAM right on the enemy machine gunners across the helicopter landing zone. That ended the night's skirmishing.

Checked, the enemy changed tactics again. Taliban marksmen, probably not trained snipers but definitely able riflemen, squirreled deep into neighboring structures. When they shot, you couldn't see a muzzle flash at night or a dust puff by day. They began trying to pick off anybody who ventured into the open. Most shots missed, but two Gurkhas suffered gunshot wounds.

The enemy also brought up three separate mortar teams. The 82 mm rounds plonked here and there and did some useful work keeping the Afghan national police off their namesake hill. One of the hostile teams showed some real skill, putting the fat brown bomblets right into the open spots in the district center compound. The Taliban mortar men fired the classic bracket, one over, one short, and then, in the words of Captain Jackie Allen, "You just knew the third one was coming straight down the chimney." Over a nerve-racking week, roving U.S. jets nailed two mortar teams. A Gurkha machine gunner got the third.

Having sized up their foe, the Taliban next tried a major assault. Just after dawn on July 12, an ANP foot patrol ran into an advancing group of enemy gunmen. The two elements exchanged AK fire, and one policeman fell wounded. Their attack prematurely unmasked, the Taliban opened up on the district center from across the landing zone. They used RPGs, PKM machine guns, and plenty of AK fire. It seemed like every fronting window and door spouted bullets. The loud RPGs blew up one after another, leaving blackened stars across the adobe-brick face.

American A-10 Warthog jets and British Apache attack helicopters rolled in to smash up the Taliban-held buildings. They killed and wounded some of the opposition, enough to matter. Hostile fire slackened. The enemy briefly backed away.

When the air support departed, the Taliban advanced again. Having judged the British tactics carefully, the enemy took advantage of the gap in air cover. From the north, west, and east, three groups, each of twenty-plus Taliban, slipped down alleys and along building sides, closing in on the Gurkhas and their Afghan soldier and police comrades. Gurkha Corporal Kailash Kebang stood up under fire, took aim, and slammed an 84 mm bazooka projectile into a Taliban machine-gun team firing from a window. The Swedish-designed Carl Gustav rocket, "Charlie G" to the British soldiers, detonated with a hot burst. The enemy gunmen kept it up. Under the suppression provided by the continuous enemy fire, Taliban fighters kept creeping forward on their bellies. "The shooting was just too much, and so close," Kebang said.

Kebang fired a machine gun at the approaching Taliban, stitching across six low crawlers. The heavy volume of enemy fire started to unnerve his men. "The boys told me 'We are all going to die!'" Kebang said. "I calmed them, giving them tasks to do." Those tasks mostly entailed shooting advancing hostiles. A busy soldier forgets his jitters, even in tight situations. Kebang kept his people very busy.

The corporal then pulled the pin on two fragmentation grenades and flung them one by one down out of the sanger. Each plopped into the tight dirt street. No Gurkha had used that weapon in almost five decades. After a few seconds, the things detonated in quick succession, spraying hot metal into the recoiling Taliban men. Kebang's grenades did the job. Bloodied, leaving one dead man behind, the closest enemy backed off.

Meanwhile, the A-10s and Apaches returned. They methodically bombed and strafed the house fronts north of the open landing zones and ANP Hill to the south. That broke the Taliban assault. Although they got near the main north sanger, they never closed on the compound proper.

If he'd had more men, Rex would have permanently placed an outpost on ANP Hill and also secured the ravaged row of abandoned structures north of the landing zone. But he was hard-pressed even to hold the district center and mount an occasional local patrol. The Gurkhas reported no women or children to be seen anymore, only men furtively moving between buildings — or shooting.

The ANP still slipped out on patrol more often than not. One day, they brought back a prisoner. Rex happened onto the interrogation of the captive. The Now Zad chief of the Afghan National Directorate of Security (the NDS, the Afghan CIA analogue), told the British major that the prisoner commanded a company-strength Taliban group. The NDS officer had been nicknamed "Hazmat" by the Gurkhas. He lived down to his reputation, discharging a pistol near the detainee's head to encourage discussion. Rex and his soldiers got control of the prisoner and evacuated him to the British headquarters for what Rex called "more orthodox" questioning. Things had definitely gotten primal in Now Zad, but British discipline held firm.

Shredded by air in daylight, the enemy made another try by night. The foe figured that the Apaches and A-10s couldn't intervene as effectively in the dark. In this, they were partly right, as the poor visibility allowed them to get closer and thus make it harder to discriminate targets. But the British army has long emphasized night fighting. The Taliban were about to learn the hard way that the stationary, prepared force has a huge advantage in darkness. The extensive British use of night sights and image-intensifying goggles made their edge all the greater.

Just before midnight on July 13, the Taliban again opened fire from the north. Rocket-propelled grenades thudded into the governor's building and peppered the main northern sanger and its many small mates now dotting the wall tops. The sandbags and mud bricks swallowed the blast and searing fragments. Green tracers crossed overhead, high as usual. A few AK shooters closed to within a hundred yards, and one bold enemy foursome tried again to get up to the north sanger. They didn't make it. The A-10 Warthogs came in low and again bombed the ruined strip of northern buildings. The darkness didn't confuse them at all. Anyone not in the district center was fair game.

The enemy took a few days and nights to reassess. Try as they might, they couldn't get into the Gurkhas' position. Crossing the final hundred yards stopped them every time. Had they been able to coordinate their mortars, their RPGs, and their machine guns, shot low, had men creep up out of a hole, and focused on just one sanger, they might have done it. But that kind of skill eluded the Taliban. They liked to fight, but in their usual haphazard style. Having tried most things, the Taliban again did some digging. For hours over the next two nights, Gurkhas reported the crunch and ping of shovels. The defenders got their hand grenades ready.

Before midnight on July 16, the Taliban tried again. They started by hitting both main sangers and the little ones too, a timed barrage of RPGs that lit the night with molten white bursts. Then enemy gunmen rose like ghosts from carefully scraped ditches and chiseled holes within twenty yards of the ANA-manned south sanger. A Gurkha reaction section moved immediately to back up the desperate Afghan soldiers. The gunfire skipped and banged off the hard points on the dummy bunker. Green AK tracers arced over the helmeted defenders. The Taliban was almost at the south gate, the closest they had gotten so far.

The Gurkhas would have none of it. Their SA-80 rifles spat 5.56 mm bullets, muzzles flashing yellow in the dark. One alert Gurkha shimmied into a latrine with a bag of hand grenades. A vent gave him a perfect chute aligned right over the enemy below milling in the black, dusty street. Slowly, methodically, the Gurkha pulled pins and dropped grenades. He unloaded twenty-one in all. Frantic Taliban fighters tried to throw back their own grenades and ended up inflicting what the sarcastic British referred to as "own goals." Cut to ribbons by an unseen tormentor, the Taliban pulled back. Two British Apaches sped them on their way.

The next morning, the British finally brought in reinforcements: a platoon of the Royal Regiment of Fusiliers, a mortar section, two more machine-gun elements, and a sniper team. The Chinooks landed and left without any hostile reaction. The enemy was too beat up. The reinforcements allowed the British to put Afghan police and a British platoon on ANP Hill, along with a .50-caliber heavy machine gun. A final attack on that position failed on July 22.

The Gurkhas departed that day, having held their platoon house at the cost of five wounded. In killing an estimated hundred Taliban,

Dan Rex's men expended 170,000 rifle cartridges and thirty thousand machine-gun rounds. They counted twenty-eight distinct shooting incidents in twenty-two days, including forty-eight separate RPG attacks and forty mortar-round impacts. It would have been appropriate to record that as the finish. But it wasn't — far from it.

The Royal Fusiliers reinforced to company strength. In a hundred and fifty days, they sustained 107 separate attacks. British troops did not secure the population. Indeed, they did not hold anything beyond the ground on which they stood. Things got so bad in Now Zad, Musa Qala, Sangin, and surrounding towns that by early 2007, the British found themselves reevaluating the entire platoon-house concept. They arranged a de facto truce, pulling back south of Sangin and relying on occasional forays by mobile columns to patrol the lands ceded to the Taliban. As former member of Parliament Paddy Ashdown put it, "We do not have enough troops, aid, or international will to make Afghanistan much different from what it has been for the last 1,000 years." He was overly optimistic. It wasn't much different from the vicious reaction that greeted the Macedonian phalanxes of Alexander the Great in 330 B.C. Except now, the Pashtuns had Kalashnikovs.

British military skill and Gurkha bravery could not reverse the facts on the ground. Following its pasting in 2001 and 2002, the Taliban came back strongly in the desert south and mountainous east. Try as they might, the British, Canadians, Dutch, Danes, and other willing contingents could not clear, let alone hold, the Pashtun south. Nor could the single American division in country clear or hold in the Pashtun east. The slowly forming Afghan security forces didn't contribute near enough men, with 36,000 soldiers and 49,700 police on duty about the same time Dan Rex was fighting for his life in Now Zad. It was easy to blame the Pakistanis for not cleaning out the border sanctuaries, but in truth, most of the Taliban were homegrown. Absent anything preventing them, they reseeded like weeds. The conditions that spawned the Taliban in the 1990s remained in effect, compounded by just enough infidel occupiers to make it sporting but not enough to stop it.

In July of 2006, the United States was consumed by Iraq, with a thousand attacks a week, more than seventy dead Americans monthly, and 140,000 troops on the ground. By comparison, Afghanistan endured a hundred attacks a week, fewer than a hundred Americans dead in all of 2006, and about twenty thousand Americans on the ground. Whatever was going on in Afghanistan, it was not as bad as Iraq, not yet. The trends, though, all ran in the wrong direction.

Since 2002, the U.S. had done just enough in Afghanistan to keep things moving. NATO officially assumed the lead but never really sent the troops needed. Until 2006, the NATO-led International Security Assistance Force (ISAF) functioned mostly in benign areas and avoided contact. Meanwhile, a U.S. division operated in the east, along the Pakistan border, chasing al-Qaeda ghosts and the increasingly active Taliban. The U.S. troops joked that ISAF stood for "I Suck at Fighting," "I Saw Americans Fighting," or "I Sunbathe at FOBs [forward operating bases]." There was a NATO chain of command focused on some kind of jackleg, underresourced peacekeeping venture and a U.S. chain trying to cobble together a counterinsurgency, counterterrorism, or just plain counterinstability effort. It was a recipe for floundering. Until ISAF and the U.S. commands unified, it always would be.

The United States kept its own fairly useful two-tier command structure of 2002. Until 2007, successive American three-star commanders handled strategic and operational matters, including coordinating with ISAF. Below that level, a single U.S. Army division, sometimes reinforced by Marine expeditionary units, ran the tactical fight along the porous Pakistan border. As the authors of an astute book on the subject put it, each arriving commander "suggested that the situation he had inherited was dismal; implied that this was because his predecessor had had the wrong resources or strategy; and asserted that he now had the resources, strategy, and leadership to deliver a decisive year." The decisive year never came. The Taliban, however, kept chipping away, growing like a resurgent cancer in the sullen Pashtun villages.

In 2002 and 2003, Lieutenant General Dan K. McNeill of the Eighteenth Airborne Corps emphasized trying to develop and extend Afghan governance into the provinces. While his infantry battalions pursued Taliban cells and helped in the chase of al-Qaeda elements, McNeill and his paratroopers partnered with the U.S. embassy to activate the first provincial reconstruction teams (PRTs), interagency groups of State Department officers, soldiers, and contractors to serve as a funnel for civic action programs. The first three went into quiet Bamyan, home of the pro-government Shiite Hazara; Tajik Kunduz up north; and restless Pashtun Gardez in the southeast. President Hamid Karzai helped place the initial set. More followed, many sponsored by ISAF partner countries.

The PRTs had three effects. First, they dovetailed nicely with the underfunded ISAF's country-development schemes, beginning the necessary integration of the two chains of command; only America had the money to do something useful with governance and economic development. Second, the PRTs began to stretch the U.S. forces across the country, again moving toward a day when the American scope extended to more than the eastern border regions. Finally, the PRTs provided a single central authority for civic action plans and execution. The idea was so good that General George Casey in Iraq adopted it there too, with the full blessing of Ambassador Zalmay Khalilzad, who had helped get the PRTs going in Afghanistan.

Of course, PRTs might make sense of the American-NATO governance efforts, but in Afghanistan, by definition, the entire idea faced a steep uphill climb. Afghan villages traditionally ran their own affairs, each valley a law unto itself. Afghanistan never enjoyed much of a central government. Even the brutal Soviets could not impose one. The British had tried once, from 1839 to 1842. The resultant debacle saw failure, a massacre of some sixteen thousand British soldiers and civilians, and the last survivors of the Forty-Fourth Regiment of Foot going down hard on a barren hilltop west of Jalalabad, cut to pieces by Ghilzai Pashtuns swarming out of a January snowstorm. Only one military man, surgeon William Brydon, rode out of the bloody retreat. His skull gaped open, half cleaved by an Afghan sword, no thanks for the many locals he'd cured. Thus it went. Afghanistan defied governance.

Hamid Karzai and his clique threw their own unique handfuls of sand in the gears. The U.S., NATO, the United Nations, and various nongovernmental organizations had little choice but to deliver governance training, economic aid, and material assistance through the Karzai government. That meant filtering every dollar through the most corrupt sieve on earth. Every level took a cut. As long as the bribes arrived on time, the officials in Helmand and Kandahar looked away as the farmers raised acres of poppies to produce opium, which, when refined, became heroin. Ahmed Wali Karzai, the president's half brother and council chairman of Kandahar Province, earned quite a healthy income from the same poppy producers he was supposed to be suppressing. Yet even as the Afghans skimmed and smuggled opium, U.S. officers were told they had to "strengthen" the existing elected and appointed officials by relying on them. In the words of McNeill's chief of staff, Brigadier General Stan McChrystal, "The strategy to help build Afghan institutions was well conceived, but the West's effort was poorly informed, organized, and executed." You could only do so much with Afghan institutions.

After McNeill and, briefly, John Vines, Lieutenant General Dave Barno took over. He recognized that "NATO and the other countries contributed less than the sum of their parts." Accordingly, he reorganized the U.S. headquarters for the long term as Combined Forces Command–Afghanistan. Those countries willing to fight, train Afghans, or work on development would be included to the extent of their national caveats, the long list of what each country would and would not do. Barno emphasized teamwork with Ambassador Zalmay Khalilzad at the U.S. embassy. Together, they crafted "a more classic counterinsurgency strategy."

Barno designed a campaign based on five pillars: defeat terrorism and deny sanctuary; enable the Afghan security structure; sustain area ownership (by U.S. units in place); enable reconstruction and good governance; and engage regional states. In all of this, Barno tried to protect the Afghan people. He planned to clear with U.S. and willing NATO forces, hold with newly formed Afghan units, and build with PRT-focused civil assistance. It was a superb formula, right out of the Casey/Petraeus model, but Barno lacked the forces—American, NATO, and Afghan—to make it work. Iraq still held priority. That became obvious when the Department of State reassigned the gifted Afghan American ambassador Zalmay Khalilzad to Baghdad.

Barno departed just ahead of Khalilzad in mid-2005, mission not accomplished and impossible to accomplish, despite a sound approach. Barno's backfill, Lieutenant General Karl W. Eikenberry, went in a different direction. A China specialist who'd never commanded a brigade or a division, Eikenberry headed the early efforts to train a new Afghan national army from 2002 to 2003. He stated that he did not find the Barno/Khalilzad campaign outline "elegant" and told a British deputy, "This is ridiculous. That is like a Soviet Five-Year Plan. We won't have any of that." Eikenberry had clear direction from Secretary Donald Rumsfeld in Washington and General John Abizaid at U.S. Central Command: Turn this thing over to NATO.

Long experienced in working with the Department of State, Eikenberry limited what he considered unproductive contacts and coordination with the U.S. embassy. Building on his own experience and the clearly identified need to turn ground over to someone who could hold it, the new commander put his emphasis on building more Afghan forces. He continued an idea, originally advanced by Dave Barno, to bring Afghan national police training under U.S. military authority. Various foreign and contractor efforts in that realm had not borne much fruit. To oversee the formation of the Afghan army, police, and the tiny air force, Eikenberry established the Combined Security Transition Command–Afghanistan, consciously emulating the Dave Petraeus example of MNSTC-I in Iraq. Although recruitment, training, and equipping increased, the need for Afghan forces far outpaced the output.

Meanwhile, at U.S. urging and under British leadership, ISAF moved into the lead countrywide. British general Sir David Richards became ISAF's first four-star commander in May of 2006. A veteran of the Troubles in Northern Ireland and the intervention in Sierra Leone, Richards described the situation in the south as "close to anarchy." His solution was "establishing bases rather than chasing militants." So Major Dan Rex and his Gurkhas set up their platoon house in Now Zad, and numerous other British units did likewise in Helmand Province, with Canadians and Dutch following suit in neighboring Kandahar Province. It did not work. The resultant British/Taliban modus vivendi in northern Helmand starkly demonstrated the limits of British-led ISAF operations.

With the United States fully engaged in the Iraq surge, NATO couldn't count on more U.S. troops. They settled for a renewed American commitment, emphasized by the appointment of a U.S. four-star commander. Cerebral, sometimes touchy, and long past his years in the infantry, Eikenberry lacked the operational experience to step up. But the U.S. had a general with all the right credentials, a real fighter. In February of 2007, about the same time Petraeus assumed command in Baghdad, General Dan K. McNeill returned to Kabul and took over ISAF. With McNeill's ascension, the separate ISAF and U.S. chains of command were unified at last. Resources, especially infantry battalions, remained far short of the amount needed. But the command organization finally settled into a stable pattern. Most of it remained intact for the rest of the campaign.

Along with commanding U.S. Forces–Afghanistan in his ISAF role, McNeill directed the operations of five regional commands (RCs), division-size NATO contingents. Germany led RC-North, headquartered in Mazar-e-Sharif. The Germans, Norwegians, Swedes, and others supervised the Tajik and Uzbek north, a relatively quiet area very much under the thumb of Fahim Khan and the Northern Alliance (Jamiat-e Islami, the Party of Islam). Italy ran RC-West out of Herat, another inactive region. With Spanish and Albanian elements, RC-West showed the strong influence of Tajik strongman Ismail Khan and, too often, the Iranians. In Kabul, the Turks commanded RC-Capital. Directly supervised by the Karzai regime, that small area, too, stayed calm. The Taliban staged strikes into the north, west, and Kabul. But with a few exceptions, they didn't dwell there. The hot zones were the Pashtun home territories. The U.S., French, and Poles fought in RC-East, a huge area that ran down the mountainous, ill-marked, dangerous Pakistan border. A British/Canadian/Dutch headquarters, with the Australians, Danes, Romanians, Bulgarians, Estonians, and others, operated in RC-South, which included the original Taliban heartland of Kandahar and Helmand Provinces. Both RC-East and RC-South lacked forces, with some individual six-hundred-man battalions trying to secure land equivalent in size and topography to Vermont or New Hampshire.

While the basic structure made sense, McNeill wasn't impressed by the situation he had inherited. He knew Afghanistan and understood the difficult nature of the fighting. Privately, he told his confidants that he "was particularly dismayed by the British effort." He went on: "They had made a mess of things in Helmand, their tactics were wrong, and the deal that London had cut at Musa Qala had failed." The new U.S. secretary of defense, Robert Gates, sent about five thousand more troops: another brigade combat team, and some key combat support forces, including more trainers and advisers for the new ANA and ANP units. Given the ongoing Iraq surge, McNeill would have to make do with that.

Unable to control much terrain, McNeill unleashed his forces to hunt

Taliban. This proved to be a rather uneven effort. The Americans spread out and went at it. So did the Canadians and the British. Most NATO members, however, refused to detain insurgents, passing them immediately to the Afghan partner units or, in certain cases, to the Americans. Some countries, like Germany up north, Italy out west, and the Turks in Kabul, avoided offensive combat operations. When confronted by the Taliban, these NATO forces, and many others like-minded, resorted to airstrikes to avoid closing with the enemy and thereby limit friendly casualties. They preferred to use U.S. aircraft too. Bitter Pashtun villagers coined a slogan: "The government robs us, the Taliban beat us, and ISAF bombs us." Corrupt he might well be, but President Hamid Karzai regularly and loudly decried civilian casualties.

When General David McKiernan replaced McNeill in June of 2008, he ratcheted back the airstrikes. By ISAF estimates, shared and coordinated with the Afghan authorities, 2007 saw about four hundred Afghans killed by NATO munitions, more than a thousand slain by the Taliban, and nearly two hundred dead by undetermined causes. Bad as the toll was, it hardly equaled the bloodbath of Karzai's ravings. Indeed, given rugged terrain, foul weather, isolated platoons surrounded by opponents in civilian dress, and extensive use of helicopters, jet fighter-bombers, and field artillery, the number seemed incredibly low. Yet because the Afghans readily queued up for compensation money, the figures were reliable, and might have been a little on the high side. Even so, as the new commander, Dave McKiernan tried to accommodate the feisty Afghan president.

Soldiers often groused about overly restrictive rules of engagement (ROE) supposedly written by overly sissified military lawyers, but they missed a key fact. The actual rules of engagement are written by operations officers, checked by lawyers, and approved by commanders. The ROE remained rather permissive throughout the war, with little change since 2001. The real screws got applied in tactical directives and daily unit orders. McKiernan's adjustment came via the latter. It was the first acknowledgment of the constant Karzai carping about civilian casualties. More would follow.

Tall, gray-haired, and self-effacing, Dave McKiernan gained his commission through ROTC at William and Mary. An armor officer, nicknamed by some "the Quiet Commander," McKiernan had effectively run the opening 2003 Army/Marine invasion of Iraq. This one was different. McKiernan pretty quickly figured out that to arrest the downward slide in Afghanistan, he needed more troops. Airstrikes went only so far. With an Afghan election coming up in August 2009 and the long lead times to get U.S. combat power halfway around the world into the remote reaches of the Hindu Kush, McKiernan wanted to get things flowing soon. He rapidly determined that the NATO allies had given about all they could stand to and that too many seemed to consider the war to be "like summer camp." The heavy lifting, as usual, fell to the Americans.

McKiernan's intelligence officers painted a grim picture. Attacks were running at over a hundred a week even in the winter, traditionally a period of reduced enemy activity. By summer, hostile acts spiked up to four hundred for a couple of weeks, and they consistently topped three hundred. Afghan forces finally put more than a hundred thousand men in the field, with about fifty thousand NATO troops, just under half American. To secure thirty million Afghans in a mountainous country the size of Texas, that wasn't nearly enough, even given that most of the bad guys infested the southern and eastern Pashtun homelands along the Pakistan border.

With Iraq settling down, some forces became available for Afghanistan. The Bush administration sent more troops, bringing the U.S. contingent to more than thirty thousand. In October 2008 McKiernan requested four more BCTs. He got one. The Bush people studied the ISAF commander's full proposal of thirty thousand more troops, found it valid, sent what they sent, then left the rest of the request open. It would fall to the next president to decide.

President Barack Obama clearly stated his position in the 2008 campaign: America must wrap it up in Iraq and commit to Afghanistan. "Afghanistan was the war that had to be won," Obama argued. The 9/11 attacks came out of a failed Afghan state, a country that bordered one known and one aspiring Muslim nuclear power. Like any new president, Obama entered his office with many priorities. Shifting the U.S. out of Iraq and into Afghanistan ranked high on that list. But as all would soon learn, Obama did few things quickly. He much preferred to think it through, discuss it, judge options, and then, long after, decide.

Logistics limited what the president could consider. Most supplies for landlocked Afghanistan rumbled slowly up the crumbling, insecure highways of America's favorite frenemy, Pakistan. Other commodities came through Russia and the Central Asian countries north of Afghanistan, a very long, convoluted, and expensive network. U.S. Air Force and commercial airlifters brought in a few key equipment items and all of the troops. But as Afghanistan was without a seaport, sending anything into the country necessitated a careful assessment of the long, difficult sustainment requirements. Not surprisingly, the Americans avoided deploying main battle tanks and self-propelled artillery, sticking to lighter forces; the Marines eventually sent a single company of fourteen M-1 Abrams tanks. The British and Canadians also used some armored forces. By and large, though, the heaviest combat vehicles were the various types of MRAP armored trucks. Fuel, ammunition, food, spare parts — every bit of it had to crawl up the tortuous roads from the port of Karachi, Pakistan. Even if Obama gave Dave McKiernan every additional soldier he requested, it would take many months to get them there and a huge effort to keep them supplied.

When Vice President Elect Joe Biden traveled to Kabul for the mandatory meeting with Karzai, he saw McKiernan. The new vice president asked for the commander's view. "We're not losing," McKiernan said, "but to get off the fence where we're actually winning we need these additional troops." Biden assured him he would get more forces. How many, and when, were left hanging.

A people person by nature, Biden won election after election in Delaware the old-fashioned way, by shaking hands and making small talk. In this trip, as in others, he took time to meet with junior soldiers. He got some interesting answers when he asked them to describe what their mission was. One said, "Basically, we're trying to rebuild this country." Another offered, "We're trying to get al-Qaeda." The most common answer rang true: "I don't know."

Biden, like Obama and the rest of his new team, had to deal with that reality. The mission remained obscure because the enemy remained ill-defined. The U.S. had steadily backed into a counterinsurgency in Afghanistan and gotten snarled up in firefights with the Pashtun Taliban, who were never a threat to the American homeland. Al-Qaeda wasn't in Afghanistan in any appreciable strength, and what was there couldn't mount international operations, not anymore. Karzai amounted to a weak, crooked partner, "the mayor of Kabul," living in a bubble partly of America's making. He looked too much like the Afghan version of Nouri al-Maliki, a petty, paranoid American-installed authoritarian leader-for-life with delusions of grandeur. NATO provided legitimacy but also bogged down substantial American assets backing up secondary efforts. Pakistan next door was untrustworthy at best. The constraints on any significant U.S. action piled up.

Various smart guys up in the stands offered Barack Obama's administration new strategic visions of counterterrorism, counterinsurgency, Afghanization, cross-border sanctuary strikes, you name it. At U.S. Central Command, General Dave Petraeus, fresh from his MNF-I stint, wondered about a concerted push for Pashtun reconciliation, an Afghan Sahwa equivalent. McKiernan thought that would be highly unlikely, and time proved him right. Pashtuns are not Arabs. They prefer to fight, each other if need be, but infidels if available. Retribution and revenge undergird the Pashtunwali, their tribal code. So none of the brilliant alternatives were possible, not this far down the rabbit hole. The United States was tied and trussed like Gulliver laid out among the Lilliputians. The usual range of options loomed: Reinforce. Stand pat. Get out.

McKiernan clearly advocated for more troops. So did the Joint Chiefs of Staff, Petraeus, and Secretary of Defense Gates, who stayed on from the Bush term, mainly to oversee the continuing war. The ultimate numbers floated out there: 100,000 Americans; 150,000, like Iraq; maybe even 200,000. You could make the case that Afghanistan was bigger and harder than Iraq. Many said so. Biden, frustrated by the unwillingness of most to discuss the mission, argued for shifting from bashing the far-flung Taliban to concentrated counterterrorism, targeting al-Qaeda in both Afghanistan and Pakistan. He didn't have much support. Amazingly, especially given Obama's comfort with yanking the plug in Iraq, no senior voice advocated pulling out of Afghanistan. The campaign rhetoric about the necessary war, the Democrats' consistent position from late 2003 onward that Afghanistan was worth America's effort and Iraq wasn't, and the strong advice from the military to fight it out . . . well, it all added up. More U.S. troops were going to Afghanistan.

The president bought some time right away. On March 29, he approved and announced the deployment of twenty-one thousand troops, the other three requested BCTs, and four thousand trainers and advisers to increase and improve the Afghan army and police. His speech supposedly described a new strategy, but it really just announced more resources for the standard U.S. statement of purpose. Barack Obama said it better than his predecessor, but he reiterated the damn near bottomless Bush mission, now adopted as his own:

> I want the American people to understand that we have a clear and focused goal: to disrupt, dismantle and defeat al Qaeda in Pakistan and Afghanistan, and to prevent their return to either country in the future. That's the goal that must be achieved. That is a cause that could not be more just. And to the terrorists who oppose us, my message is the same: We will defeat you.

In later days, some near the president objected that the military boxed in the new commander in chief, offering him nothing but narrow variations — small, medium, and large — on the same theme, a force increase. But once Obama embraced the long-announced Bush mission of unscrewing Afghanistan enough to prevent any al-Qaeda comeback, and maybe even doing likewise in Pakistan — well, there probably weren't enough troops in the American military, or even the Chinese military, for that one. To his credit, Obama did not consider it all a done deal. He was determined to do a full strategic review before sending any more Americans into the campaign. And even if nobody else was thinking about pulling out, he was.

General Jack Keane had played a major role in the Iraq surge, and in the spring of 2009, that gave him significant credibility with the new administration. A native New Yorker, Keane had long advised Hillary Clinton, and when the former First Lady and senator became the secretary of state, Keane went to see her. He very much agreed with mounting a full-scale counterinsurgency in Afghanistan. And the man who'd once spoken up for installing Dave Petraeus to command the Iraq surge had some views on this campaign too.

"The strategy in Afghanistan is wrong," he began. "And not only that, but the leadership is wrong." Keane knew the strategy and troop levels would get fixed, and trusted the president's ongoing review to do so. He knew Clinton had a big role in those deliberations. But McKiernan was the wrong guy to run the war. Keane thought him too conventional, too cautious, not good in telling the story of the war through the news media. "He's got to come out of there," Keane said. To replace him, Keane offered Lieutenant General Stanley A. McChrystal, a standout as the top SOF field commander and, at the time, the director of the Joint Staff. McChrystal understood counterinsurgency thoroughly. With Dave Petraeus at USCENTCOM and McChrystal in country, Afghanistan could be turned around pretty quickly. Iraq had shown the way.

As in the Iraq surge, Keane's counsel coincided with what others thought. McKiernan seemed to be lacking something, and the president, the secretary of defense, and their key advisers couldn't quite put their fingers on it. It's clear in retrospect. McKiernan was Casey, not Petraeus. And after the experience of Iraq in 2007, to lead a surge in one of these unpalatable counterinsurgencies, America's elected and appointed leaders preferred the Petraeus type, or thought they did. Admiral Mike Mullen, chairman of the Joint Chiefs of Staff, put it dramatically: "I cannot live when I know I have a better answer when kids are dying every day." Secretary Bob Gates agreed, as did Obama. McKiernan was removed, strictly business, not personal. So they all said.

In going with Stanley A. McChrystal, Gates, Mullen, and the president chose one of the most impressive officers in the U.S. Army. Tall, whipcord-thin, intense, and as well-read and articulate as they come, McChrystal was the son of a major general. He was not an aide or throne-room type but a thoroughly grounded field soldier. He commanded Airborne, Ranger, and Special Forces, America's most elite military organizations. As smart as Petraeus and even more physically fit, McChrystal had commanded the major SOF campaign and done it right. He had developed manhunting to an art and gone on to spread the gospel to the conventional military as well. Now he would go to Afghanistan to rescue the good war.

The mission was assumed to be counterinsurgency. For the first time in years, there would be enough troops — exact numbers as yet undefined — to do the job. Gates told McChrystal that as the ISAF commander, he would help define the resource demands. Already, Afghanistan was proceeding differently from the Bush surge into Iraq. A new commander had what he thought was his mission. But he didn't yet know what resources he'd have to accomplish it. Nor did he know if the mission would hold firm. Obama kept his options open.

14

THE GOOD WAR

When enough men had died on a hill,
their comrades began to show a grim fondness for it.

— T. R. FEHRENBACH

THE RIFLE COMPANY in Barg-e Matal on July 12, 2009, was there for a reason. Not often can a soldier say that he stands his ground by the direct order of his four-star commander, but those in Company A, First Battalion, Thirty-Second Infantry could do so. General Stanley McChrystal, the forceful new commander of the International Security Assistance Force and U.S. Forces–Afghanistan, directed that Barg-e Matal in Nuristan Province be seized and held by a combined American and Afghan force. In issuing the order, McChrystal responded to a personal plea from Afghan president Hamid Karzai. Facing a very close election on August 20, Karzai wanted to show strength. To do so, he chose a small village out in the most distant and isolated of Afghanistan's thirty-four provinces.

If Americans knew anything of Nuristan, they could thank Rudyard Kipling as well as director John Huston, whose popular 1975 movie *The Man Who Would Be King* was an adaptation of Kipling's 1888 story. Set in the heyday of the imperial raj, two former British army sergeants trek across the forbidding Hindu Kush to reach a fabled mountain fastness known as Kafiristan (Land of the Kafirs; that is, those who don't follow Islam), the traditional Afghan name for Nuristan Province, where the ethnically distinct locals finally converted to Islam about 120 years ago. On arrival, joined by a Gurkha rifleman, the adventurers find redheaded descendants of Alexander the Great's Macedonians, a trove of priceless riches, and an opportunity to set themselves up as rulers. One sergeant convinces the gullible Nuristanis that he is the long-lost son of Alexander the Great. When the fraud collapses, scores are settled in blood. Only one of the sergeants escapes, barely alive after being crucified by the vengeful Nuristanis. The Kipling story and the film were both enjoyed by U.S. soldiers bound for Afghanistan. Sharp viewers recalled that the three soldiers briefly backed a short, quick-tempered, balding local warlord with a trimmed graying beard who styled himself "Ootah the Terrible." He looked a lot like Hamid Karzai. With that unpleasant similarity noted, the Americans prepared to go into remote Nuristan.

Thus the orders came down, ISAF to corps (there was one, at last), corps to division, division to brigade, brigade to battalion, and then battalion to the company, all at the speed of e-mail: *Retake Barg-e Matal from the Taliban. Hold it until relieved.* Carl von Clausewitz carried out such tasks himself during the Napoleonic Wars. He wrote that war was "a continuation of political intercourse, carried on by other means." Clausewitz got it right. Among the "other means" manning Company A, references to intercourse abounded in later days whenever they talked about their time in Barg-e Matal.

They went in by CH-47 Chinook, as usual. The sun was just beginning to creep above the forbidding eastern ridges. Glancing through the small helicopter portholes or looking back over the open ramp guarded by a crew chief with a machine gun and an anchor line, one could see a lot of green. Everybody said Nuristan had water, trees, and very few people. The word was passed: thirty seconds. Looking out the right side, a couple of the men saw the village, a tiny island of structures in a vast green carpet. A river gleamed, dividing the town in two. The Chinooks descended to the western side, the lesser portion, flaring over the open edge of a tall cornfield. The two big helicopters settled — no fire, despite all of the awful predictions. The riflemen ran out, knelt, and formed a ragged line against a waist-high mud wall to the east. Green cornstalks waved as the helicopter blades churned the air. With a spiraling gust of dirt and sticks, the Chinooks were off.

Crack. That was an AK-47.

Crack. One of the NCOs pointed to the south, toward a hill. Almost as if summoned by the gesture, an AH-64 attack helicopter fluttered down and unleashed a Hellfire. The missile streaked into the green hillside, blossoming light gray. A few seconds later, a dull thud echoed. The riflemen later heard that the aviators had spotted and nailed an RPG gunner and his partner. That done, the soldiers stood up, shifted their M-4 carbines into the front ready mode, muzzles a bit below level, and shook out into a loose column. They were heading toward the ANP station a few huts to the east.

Most people would think that, at 9,100 feet above sea level, Barg-e Matal stood at a high elevation. In fact, the settlement formed the bottom of the bowl, neatly defined by four hulking 10,000-foot massifs. At the center of the village, the ice-cold, neck-deep, south-flowing Karigal River, thirty feet across, met the equally chilly, knee-deep, west-flowing Mahalgal River, more a mountain brook, six feet wide at most. Some five hundred farm families raised animals, tilled small vegetable plots, and grew corn for feed in and around the hundred or so flat-roofed, scrubby one-room mud huts and wooden-slatted sheep and cattle corrals. The soldiers saw only a few men outside. Women and children huddled in gaping doorways or peeked from windows. They definitely looked different from Pashtuns, with lighter complexions. Many of the men had bright red beards, some natural, most dyed. Nobody seemed happy to see the Americans.

A second pair of Chinooks landed out in the cornfield, bringing in the rest of Captain Mike Harrison's Company A. Harrison led a mixed force made up of his own Third Platoon, the battalion's scout platoon (including snipers), a battalion mortar section, a company of 80 ANA soldiers, and a platoon of 40 Afghan border police. It totaled about 90 Americans and 120 Afghans. The U.S. soldiers planned to go in, reclaim the place, and stay for about a week. The Afghans would remain to hold the town.

The rest of Lieutenant Colonel Mark O'Donnell's 1-32 Infantry had its hands full to the south, fronting twenty-five miles of mountainous Pakistani border. O'Donnell spared what he could. Ideally, he'd have sent more, all he had, to climb up and hold that high ground around Barg-e Matal. But he was told by the Third Brigade Combat Team, Tenth

Mountain Division that his primary mission remained blocking cross-border infiltrators. This thing in Nuristan was a secondary task. Harrison's understrength company was it. The scout/snipers and mortars added some needed punch. And these Afghan troops were reputed to be okay.

Along with mortar bombs and extra belts of machine-gun ammunition, Harrison's soldiers carried another burden into Barg-e Matal. Six days earlier, Stan McChrystal had issued a new tactical directive. He'd been listening, evidently with interest, to Hamid Karzai. The language left little doubt as to what the new ISAF commander thought:

> Our strategic goal is to defeat the insurgency threatening the stability of Afghanistan. Like any insurgency, there is a struggle for the support and will of the population. Gaining and maintaining that support must be our overriding operational imperative — and the ultimate objective of every action we take.
>
> We must fight the insurgents, and will use the tools at our disposal to both defeat the enemy and protect our forces. But we will not win based on the number of Taliban we kill, but instead on our ability to separate insurgents from the center of gravity — the people. That means we must respect and protect the population from coercion and violence — and operate in a manner which will win their support.
>
> This is different from conventional combat, and how we operate will determine the outcome more than traditional measures, like capture of terrain or attrition of enemy forces. We must avoid the trap of winning tactical victories — but suffering strategic defeats — by causing civilian casualties or excessive damage and thus alienating the people.
>
> While this is also a legal and a moral issue, it is an overarching operational issue — clear-eyed recognition that loss of popular support will be decisive to either side in this struggle. The Taliban cannot militarily defeat us — but we can defeat ourselves.

That last line turned out to be only too true, but not in the way the ISAF commander thought. If you'd ever wondered what handwringing looked like on paper, this was it. McChrystal believed it with all his heart. Maybe all of those years with the Rangers and Special Forces convinced him that you could be purely precise with armed force and kill only those who merited killing. In any event, the general believed his conventional forces must become significantly more discriminating. He spoke of the value of "courageous restraint," the willingness to take ISAF casualties to protect the Afghan people. To tough Pashtun tribesmen, that smacked not of discipline or kindness, but weakness. It wasn't clear that any significant portion of the Afghan population would ever embrace thousands of infidel foreigners. The locals often hated the Taliban, but those insurgents were natives, and ISAF troops were not. As for "strategic defeats," McChrystal mistook Karzai's daily bleatings for the views of Afghan villagers. Many of the average Pashtuns proved to be made of sterner stuff and accepted that in a war, innocent people sometimes got killed. Afghans would never love ISAF, but they might well fear and respect the occupiers. Now, with this kind of guidance, even that was unlikely.

Having described his overall philosophy, the ISAF commander then got down to specifics, no doubt helped immensely by clever military lawyers and earnest public-affairs flacks who had never heard a shot fired outside of a rifle range. Well, it kept their bright minds clear, if nothing else. "The use of air-to-ground munitions and indirect fire [artillery and mortars] against residential compounds," the directive stated, "is only authorized under very limited and prescribed conditions." That amounted to hardly ever. So you could not fire on buildings occupied by people in civilian clothes — which was to say, pretty much every building in country — unless someone inside shot at you first. Structures with civilians in or around them, even if the Taliban hung around there as well, were off-limits unless someone in there fired first.

The list went on. "Any entry into an Afghan house should always be accomplished by Afghan National Security Forces [army and police], with support of local authorities." This required U.S. and other ISAF units to operate side by side with the Afghans, a good step forward, once the locals put enough units in the field. Left unsaid was what happened when Afghan forces shied away or outright refused when told to go into the houses of their fellow countrymen. Also unclear was how to gain surprise on a raid when you have to notify the local village elders. If Americans entered an Afghan house, they could expect a lot of scrutiny and second-guessing from senior officers. Raiding by conventional battalions got very, very difficult. For the moment, the Task Force operators remained an exception. But they, too, would eventually be reined in.

For line battalions, this aspect of the ISAF tactical directive discouraged detentions. Most ISAF countries already turned prisoners over to the Afghans. The Americans took their own captives, but far below the rate seen in Iraq. In that other country, the U.S. had held as many as twenty-six thousand at the height of the surge. The Afghan detainee population never got much above four thousand, even after the ISAF troop increases of 2009 and 2010. Prisoners were almost always given to the Afghans. Whether they stayed confined or went home depended on bribes, whims, and the Afghan way, which tended toward fines and apologies, not incarceration. Less than a tenth of those in Afghan prisons ever got sentenced. A favorite way to avoid conviction involved assuring the judge that the defendant had killed infidels only and so didn't deserve punishment. That was often a winning plea. Most detainees stayed in Afghan hands a few months and then walked free. You'd see them again on the battlefield. Disgusted American soldiers called it "catch and release."

One additional stricture applied. "No ISAF forces will enter or fire upon, or fire into any mosque or any religious or historical site except in self-defense." For an enemy that used mosques as meeting sites, command posts, and arms caches, this was welcome news. A lot of new mosques, religious facilities, and historical sites were quickly designated by shrewd Taliban seeking safe houses. ISAF willingly accepted this charade.

Stan McChrystal left in one big exception to these restrictions. To his credit, he'd been right there on the ground accompanying enough raids with the SOF to realize that armed forces in hostile lands must retain the ability to fight to the death. "This directive," he emphasized, "does not prevent commanders from protecting the lives of their men and women as a matter of self-defense where it is determined no other options are available to effectively counter the threat." In other words, under the most dire circumstances, a commander could use all his weapons. But afterward, if there was an afterward, he could expect a lot of high-level attention and pointed queries.

ISAF compounded the negative impact of this tactical directive by issuing most of it as a press release, in accord with McChrystal's rather expansive view of open communications with the Afghans and the wider world. Even the master Petraeus had never gone this far. Some tactical details stayed classified, but it didn't take a war-college graduate to fill in the blanks. The Taliban greatly appreciated the insights and immediately adjusted their tactics. For Karzai and his Kabul circle, running up to an election, it created expectations that ISAF operations would cease to kill any Afghan civilians at all. Even McChrystal wouldn't try to promise that, but the tactical directive sure leaned that way, and written words take on a life of their own.

Stan McChrystal did more than just issue the tactical directive. Americans and other ISAF countries often paid compensatory monies to families of civilians who'd died during operations. The desperately poor Afghans were quick to arrive at the claims tables. As every dead person in country had worn civilian dress, more than a few dollars likely went to Taliban families. The Pashtunwali emphasized resolving feuds with blood money, and if the infidels went along, so much the better. The ISAF commander now added his voice to the effort. Up north in the German sector, a U.S. Air Force F-15E Strike Eagle engaged and destroyed two Taliban fuel-truck bombs. The tankers went up in a fireball, but the attack also killed at least 15 civilians, although tribal chiefs later claimed 179 dead, among them 69 known Taliban, with forty-nine fire-blackened weapons found onsite. In the first report, long before any investigation by the Germans in RC-North, McChrystal personally apologized to Karzai. The U.S. general then went on Afghan television and announced, in a sober speech, that he was determined to make amends to the Afghan citizenry. The chastened Germans gave five thousand dollars for each supposed victim, a higher amount than usual. They paid off the Taliban households too. It set a pattern. Allegations begat a public ISAF apology. A public ISAF apology begat more money. And then the various ISAF and subordinate headquarters operations officers generated more orders limiting more aspects of the use of force. A week or two later, another event would occur, and the cycle started again, with Karzai himself merrily turning the crank. The Taliban probably found it all amusing, as the ignorant occupiers essentially provided their foes' death benefits.

Determined to curb the errant Afghan deaths, the July 6 ISAF tactical directive effectively curtailed the offensive, preemptive capacity of NATO forces, especially the Americans, who naturally responded quickly to the bidding of a U.S. four-star general. In the document, the general encouraged a "cultural shift," one that leaders "clearly communicated and continually reinforced." His subordinate commanders spread the word. One lieutenant general told the leaders of an incoming unit, "If there is any chance of creating civilian casualties or if you don't know whether you will create civilian casualties, if you can withdraw from that situation without firing, then you must do so." Staffs at each echelon set up working groups to address civilian losses. These officers generated a blizzard of documents that tried to specify when a soldier could drop a bomb, fire a howitzer, or even shoot a rifle. Even at his most zealous, Lieutenant General Pete Chiarelli in Iraq never went this far. But then again, he and his troops were dealing with a thousand dead civilians a month in Baghdad alone. The civilian casualty rate in 2009 Afghanistan would not even have registered in Iraq at the height of the insurgency.

The results of all of this pretty much tracked with McChrystal's hopes. Civilian casualties dropped. As feared, friendly casualties increased, although how many were due to "courageous restraint" and how many reflected increased troop levels proved hard to determine. The biggest change involved the willingness of ISAF, especially U.S., units to engage the enemy. McChrystal didn't mean to do it, but his carefully phrased proscriptions ceded a lot of the initiative to the Taliban.

In fact, this was the inevitable result of the new ISAF tactical directive. All military units work hard to prevent blue-on-blue fratricide — the horrible friendly-fire mistakes that suck the morale out of combat units that clash with each other. Americans use computer trackers, rehearsals, and careful fire-control measures, and they train hard with heavy weapons on live-fire exercises, all to prevent blue-on-blue. But even in combat at its most chaotic, with Clausewitzian friction running wild, disciplined military organizations have a good prospect of avoiding each other. Random civilians, however, intermixed with civilian-clothed enemies determined to be among them, cannot help but get in the line of fire. The Geneva Conventions accept this reality and place any blame on the side that willingly shields itself inside noncombatant populations. Trying to avoid all civilian casualties leads inexorably to no casualties. Admiral Sir John "Sandy" Woodward, British commander in the 1982 Falklands War, put it succinctly: "The essentially bureaucratic peacetime mind will, for the sake of avoiding a single Blue-on-Blue, cause Blue-on-Red [hitting the enemy] to cease!" That mentality was too much alive in ISAF's Kabul corridors during that high summer of 2009.

So units like Captain Mike Harrison's Company A in Barg-e Matal now had to take the first hit. Then, if there weren't too many civilians around, they had to hope that when they called for artillery or air support, the distant command posts agreed that things were bad enough to risk later investigation, censure, and relief. A game of twenty questions is not the ideal response from higher up when you're under fire. It might have comforted Harrison to know that Stan McChrystal had been on the ground around Barg-e Matal a few years before, walking with a U.S. SOF team on one of those surgical strikes so beloved of pundits. Harrison and his riflemen were going to do their surgery with rifles, machine guns, mortars, and, if the higher headquarters saw fit, artillery and airpower.

In Barg-e Matal, the Afghan border police moved into the abandoned ANP station. The Americans secured the empty surrounding huts. Their accompanying ANA soldiers hoisted an Afghan flag. In the station, the enemy had left behind 82 mm mortar rounds, PKM machine-gun belts, RPG rockets, and cases of AK-47 cartridges. On a table, the Taliban left a stack of voter-registration cards for the upcoming election. The Afghan border police heard from one very forthcoming local man that the cards provided a means to intimidate the townsfolk. The Taliban had gone door-to-door and made it clear that voting would be fatal.

With things quiet since the morning landings, a combined Afghan and American foot patrol moved through the western side of the village. The Afghans removed Taliban flags and banners. Nobody saw any women and children. That usually meant trouble coming.

At one barnyard, a few Nuristani elders offered information. They spoke in broken Pashto rather than their distinctive native tongue. They'd counted two hundred Taliban, all up in the hills, watching. A

U.S. technical team overheard cryptic Taliban handheld radio chatter. Something was up. A year ago, that would have been enough to bring in the A-10s and Apaches and clear off those frowning heights. But in the age of courageous restraint, the Americans and Afghans could only exchange anxious glances and wait.

At around 6:30, about forty-five minutes before dusk, the watching Taliban opened fire. All four hills erupted in what Corporal Jeff Sutton termed "a 360 degree ambush." He saw "rounds whizzing in front of me." He and his squad mates looked up and shot back. They had their choice of targets, as the Taliban gunmen moved confidently among the boulders dotting all four hillsides. The bad guys had the high ground, and that wasn't good. RPGs crossed over the American positions. Sutton reported, "They were everywhere. I hit my trigger as fast as my hand would allow."

An RPG "sailed down the valley and smacked a building." As Sutton watched, three men fell. One looked bad and had to be dragged inside a doorway. Sutton didn't have time to dwell on it. His platoon leader, First Lieutenant Jacob Miraldi, zigzagged up, shouting. "His leg was drenched in blood," Sutton said. "He looked like he was hit pretty bad, but it didn't seem to faze him." Miraldi told Sutton and two other men to move to the opposite side of the one-story building. They did. One of the soldiers carried a 40 mm grenade launcher and popped one fat little projectile after another to burst on the southern hillside.

The American mortar section fired round after round, trying to clean off the hills. The long-barreled M-240 machine guns cranked off one belt after another, all aimed up at the hills. The Taliban didn't let up. Miraldi said, "It was raining bullets." Staff Sergeant Robert Ridge, a scout platoon squad leader with multiple deployments to Iraq and Afghanistan, thought it was "the most intense amount of enemy fire I have ever experienced." But Jeff Sutton summarized it best: "At this point we were completely pinned down without air support." The A-10s at Bagram were thirty minutes away. Under the new tactical directive, they didn't scramble until Captain Mike Harrison declared "troops in contact," the extreme-circumstances exception. In the meantime, from the base of the funnel, the Americans and their Afghan mates shot upward at the flitting figures.

In Afghanistan's Helmand Province on July 22, 2007, a section of British Royal Marines waits to move out. Thick vegetation and a network of canals restrict movement and visibility. The British had the lead in the south during the tough years from 2006 to 2010.

Inside a rooftop bunker in Helmand Province, a British soldier of the Royal Gurkha Rifles keeps watch as night falls. Gurkhas defended the platoon house in Now Zad, Afghanistan, against repeated Taliban attacks in the summer of 2006.

Unless otherwise noted, photographs are reproduced courtesy of the United States Department of Defense.

The British long gone from Now Zad, a U.S. Marine patrol enters the town from ANP Hill in 2013.

A U.S. Air Force A-10C Warthog drops flares after a low pass. The hot-burning flares ward off heat-seeking missiles. Warthogs provided superb close air support in both Afghanistan and Iraq.

Mortar men of the First Battalion, Thirty-Second Infantry, fire an 81 mm round from their position in Barg-e-Matal, Afghanistan, in the summer of 2009. The mortar can fire a projectile almost four miles. When it hits, hot metal fragments from the explosion can kill out to about fifty yards.

A soldier aims an M-249 squad automatic weapon from his position at Barg-e-Matal, Afghanistan, in the summer of 2009. The Taliban held much of the high ground most of the time.

American riflemen move out along a rocky trail overlooking an Afghan village in the Hindu Kush Mountains in the spring of 2012. Holding the high ground was a major tactical advantage.

Brigadier General Lawrence D. Nicholson (right), commander of the Second Marine Expeditionary Brigade, escorts Admiral Michael G. Mullen, chairman of the Joint Chiefs of Staff, through a market in Nawa, Helmand Province, on December 17, 2009. Nicholson led the major operation to clear Marjah in February 2010.

A Marine checks to his rear on foot patrol in Marjah, Afghanistan.

General David Petraeus, commander of the International Security Assistance Force and former commander of the 101st Airborne Division (he wears its combat patch on his right sleeve), meets with Colonel Arthur Kandarian at a U.S. forward operating base near Kandahar, Afghanistan, on July 9, 2010. Kandarian's Second Brigade Combat Team, 101st Airborne Division, was just beginning its grim campaign to clear the Arghandab River Valley.

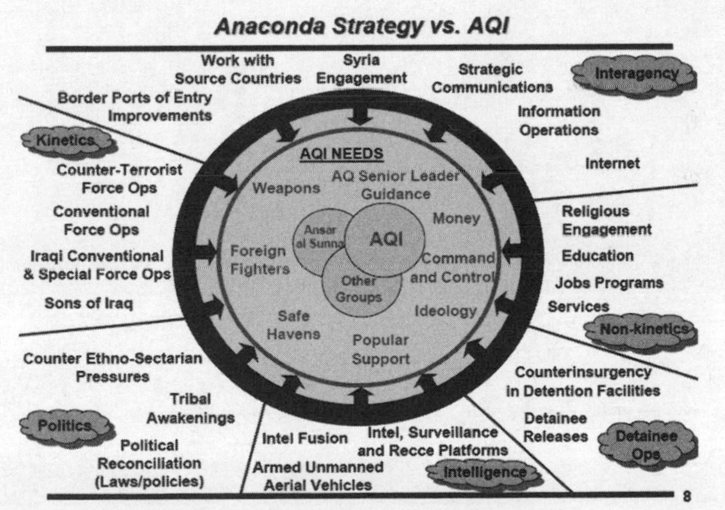

General Petraeus developed this busy chart to explain how he wanted to squeeze the enemy in Iraq in 2007–8. He made a similar version for Afghanistan in 2010–11. He named it Anaconda, after the Civil War strategy developed by General Winfield Scott in 1861. Unlike Scott's vision, the Petraeus version never came together in either Iraq or Afghanistan.

An F-16C Viper takes off from Bagram Airfield, Afghanistan, on February 11, 2014. F-16s have been the workhorses of the air campaign in both Afghanistan and Iraq. Under the wings are a pair of JDAM (joint direct attack munition) smart bombs. These guided bombs were the weapons of choice in theater.

Afghan national army soldiers patrol in the Arghandab River Valley in the winter of 2010–11. They all carry U.S. weapons, and seem rather nonchalant, never a wise posture in that embattled district. During the warmer months, dense vegetation would limit visibility to a few feet around each soldier.

Two soldiers check the brush near an unidentified combat outpost in the Panjwai District, Afghanistan. The plants grew tall and thick in the summer, and had to be cut back to keep the COP secure.

Lieutenant Colonel Steve Miller, commander of the Third Battalion, Twenty-First Infantry, before his deployment to Panjwai in 2011. By this point in the war, Miller and most of his leaders had been on multiple hard-combat deployments. His eyes show it.

A Stryker wheeled combat vehicle bursts through a wall west of Kandahar in 2013.

Seen through the troop ramp of a Stryker, a U.S. Army rifle squad with a mine detector prepares for a foot patrol in the Panjwai District in 2012. The man with the baseball cap and scarf is the Afghan translator, a hired civilian but very much part of the team.

Aboard a UH-60L Black Hawk helicopter, a door gunner mans an M-240 flexible machine gun as the aircraft lifts off from Kandahar Airfield in 2012.

Searching from his window position, Corporal Jeff Sutton spotted a target. He grabbed a soldier with a grenade launcher. In the steady roar of fire, he got right up in the other man's face. It was getting dim out there in the dusk.

"When I tell you to get up," Sutton yelled, "I'm going to point at a spot in the mountainside behind us. I want you to shoot at it, okay?"

"Okay, Corporal," the grenadier shouted back, nodding. The man loaded the first egg-size round into the breech. At his side, Sutton watched closely. Dust from a mortar hit drifted across the darkening, scrubby hillside. He waited. Then he saw the Taliban fighter move. An AK muzzle flashed. "Now!" he commanded. The soldier stood and pumped out two 40 mm grenades. They blew on the hillside, hot little stars in the gloom. The Taliban guy was gone.

One motivated group of insurgents tried to breast the cornfield to get up near the ANP station. In all the noise and crossing fire, they slipped through the dense stalks, some of which reached eight feet tall. Prone on rooftops, a small but useful height advantage, members of a scout machine-gun team saw the threat. The hostiles closed to a hundred yards, but no farther. Bursts of ten to twelve hot 7.62 mm bullets shredded the corn and bowled down the Taliban too. The ripped-up corn plants smoked in the waning light.

About dark, two A-10s arrived. They each took one dry run to pinpoint the enemy. The Taliban ignored the jets and kept banging away, guessing it all was an empty show of force. They guessed wrong. The first aircraft swung around for business. The lead A-10 looked like an iron cross hanging in the dusky sky, gliding slowly down. Its blunt nose dipped. On the southwest hill, a string of high gray dust devils spurted up. Then came a noise like the world's loudest chainsaw. A few hundred 30 mm explosive shells, each the size of an old-school glass soda bottle, slammed into the hillside. The Warthog pitched up. Then the other one repeated the drill. That hill done, they moved methodically to scour the other three.

The airpower ended the enemy attack. Staff Sergeant Eric Lindstrom was dead, killed by one of the opening RPG shots. Lieutenant Jake Miraldi and another man were wounded by the same enemy round. An ANA soldier lost a hand. As usual, the Taliban gunmen shot way too high, scattering rounds everywhere, scaring the hell out of the Americans but not harming many—not this time. Their aim would improve over time.

The enemy backed off some over the next few days. Each side traded shots. The Americans got better established, with the scouts west of the river watching the cornfield and landing zone and the rest of the Americans and Afghans on the east side, just across a rocky ford, snuggled into a few of the more substantial adobe structures. Captain Mike Harrison let his battalion commander know this wasn't going to wrap up in a week. Accordingly, Lieutenant Colonel Mark O'Donnell and the experienced Command Sergeant Major James Carabello came up with a rotation schedule to transition platoons through the Barg-e Matal hotbox. A battalion tactical command post was set up in the town to run the fight and provide needed continuity. The team included the battalion S-2, Captain Raymond Kaplan.

Kaplan figured out a lot about the enemy. The overall commander was Abdul Rahman, and he also had men across the border in Pakistan. There were about a hundred and fifty fighters. His technical specialists, mortar men and a few snipers, came from the Pakistan villages. Local Nuristanis supplied the AK-carrying rank and file. Rahman well understood the new ISAF methods. The Taliban made camp five miles down the river at a place Kaplan nicknamed Pop Rock. Women and children lived with them, providing human guarantees against American air attacks or artillery barrages. When the Taliban used their little walkie-talkie radios, they made certain to keep some females and toddlers nearby. The Americans listened but could not act. Each day, just before dawn, Kaplan watched the hostiles hop into vans and pickup trucks looted from the ANP. A few wives and offspring also climbed aboard. The impromptu caravan then motored north and dropped off the commuters on the two southern hills overlooking Barg-e Matal.

To attack a Taliban van, Ray Kaplan would have needed to ensure the enemy showed weaponry (they didn't) and had no women or children with them (they did), and then kept continuous surveillance on the vehicle until an aircraft arrived, often an hour later. Even if he had a drone overhead that was not otherwise tasked, Kaplan also needed two independent confirmations of hostile intent, such as an obvious AK or a radio intercept announcing a coming attack. And all of that only worked in the absence of women and children. As the enemy fully understood these litigious games, they played along with alacrity.

An example suffices to demonstrate the raging frustration caused by this procedural morass. One particular day, Ray Kaplan was pretty certain he'd tagged Abdul Rahman himself, the big cheese, getting into a pickup truck at Pop Rock. He called for a nearby armed Predator drone to engage. Under the new tactical directive, that request went all the way to RC-East, divisional level. That day, it went fairly fast, with approval in only twenty-three minutes. But Kaplan lost sight of the truck for a ten-minute stretch due to some intervening trees. The mission was canceled. The intelligence officer determined that he spent about 70 percent of his energy trying to talk his higher headquarters into allowing any kind of distant engagement. "There were echelons of staff sitting above me," he said, "like owls in trees." And this all happened in Barg-e Matal, a personal priority of General Stan McChrystal and President Hamid Karzai, a location where Americans took mortar, RPG, and small-arms fire almost every day at sunset.

Even face-to-face encounters had moments when separating friend from foe got very tricky. A foot patrol from Lieutenant Miraldi's First Platoon, moving through the east side of town in pursuit of a Taliban shooter, halted to allow the passage of women and children. Two of the "women" lifted their head-to-toe burqas, produced AK-47s, and clipped two American soldiers. The Americans killed the disguised opponents. The Taliban, of course, put out the word that the infidels had slaughtered a pair of women. Guerrilla warfare thrives on such sordid incidents. No tactical directive could fix that.

Stymied in mounting preemptive strikes, the Americans used what they owned. Their hard-working mortar men fired an average of sixty-six rounds every day, pummeling the surrounding crests. Not cleared for high-explosive artillery missions unless already in direct-fire contact with the enemy, the smart artillery forward observers used red phosphorus smoke to blanket the hillsides. When the enemy did shoot, the Americans brought in air support, which sometimes arrived in time to do some damage. The Taliban never again mounted an attack on the scale of the one on July 12. But sniping, mortar fire, and chance encounters kept up for the Americans' entire sixty-eight-day stay in Barg-e Matal. Mao would approve fully: *Enemy halts, we harass.*

Cleaning the cornfield proved especially painful. On July 24, a hidden gunman killed Specialist Justin Coleman from a few feet away. Seven days later, another ambush in the tall stalks killed Specialist Alexander Miller. The enemy made sure to fire all the AK bullets just below Miller's Kevlar body armor. The Americans got the shooters and, after Miller's death, erased the cornfield too. The daily resupply Chinooks kicked up a lot more dust after that. Nine of those flights came out with holes in their hulls, but the enemy never knocked one down. It wasn't from lack of trying.

After losing Miller, the Americans decided to deal with the Taliban camp at Pop Rock. Unable to bomb it or shell it, they used infantry. After dark, Harrison sent Jake Miraldi's First Platoon out on a mission that resembled some western front trench raid in 1916. The platoon got about two hundred yards down the west side of the Karigal River when Sergeant Sam Alibrando at point halted the column. He heard a person snoring. The sleeper was part of a four-man Taliban outpost, all of them conked out. Evidently they were not expecting guests.

The Americans looked through their night-vision goggles. Laser aiming devices put a bright dot on each sleeper. After a whispered countdown, the Americans squeezed their triggers together. Alibrando remembered what followed: "We opened fire on them, and despite the moonless night, they returned fire." The M-4 rounds hit the enemy but didn't kill them. One of the AK slugs clipped Alibrando's neck. "I got up and charged, firing at them, killing one of the lead guys." His squad finished the others. Then the Taliban on the southeast hill opened up, flinging green PKM tracer rounds and even an RPG downslope at the source of the shooting. Four Americans were hit by fragments off the rocket. The platoon pulled back. Pop Rock remained a Taliban sanctuary.

A night mission on August 17 turned out much better. Sergeant First Class Kevin Devine led First Platoon out of Barg-e Matal, seeking to get atop the northeastern hill. The five-hour climb wasn't easy, but they found the hill unoccupied. From that night onward, the Americans held it. Spotters and snipers found the perch very useful. The mortars now had eyes on the high ground, and their rounds began to do even more damage.

American ownership of the hilltop position took most of the fight out of the enemy. Maybe a hundred adults, a fifth of the prebattle populace, remained in the village on voting day, August 20. They dutifully cast their ballots. The Americans stayed for almost another month, departing on Chinooks on September 18 after sixty-eight days in Barg-e Matal. The ANA and ANP remained to protect the town. The next summer, Barg-e Matal came under assault again, changing hands from government to Taliban and then back in what became an annual summer ritual. But with some U.S. SOF backup and American airpower, the Afghans hung on, good enough by their lights.

The mission cost the lives of three American soldiers; thirty-nine were wounded. Affiliated Afghan troops lost seven men, and nine were wounded. Although hampered by the new ISAF tactical directive, there had been so much direct-fire contact that a lot of firepower still got used. U.S. forces dropped 5,940 mortar rounds, called in 1,135 155 mm artillery rounds, and employed 40 separate close-air-support sorties. The First Battalion, Thirty-Second Infantry believed they killed 372 Taliban, but nobody could be sure of that tally. It far exceeded any projected number of local enemies. Barg-e Matal held. For those tracking at ISAF headquarters in Kabul, not a single noncombatant died, although every dead Taliban wore civilian clothing.

What of the election? In that, Hamid Karzai had to be pretty happy. The hundred or so voters in Barg-e Matal on August 20 generated four thousand votes. Every one of them favored the serving Afghan president. During a series of recounts, the number inflated to twelve thousand, then went as high as twenty-five thousand. Those sorry numbers spoke for themselves.

The August 20 voting elected the Afghan president as well as members of provincial councils. The presidency was the real prize. The person in that office held a lot of power in post-Taliban Afghanistan and enjoyed broad authority to appoint officials in all ministries, provinces, and districts. Governors, for example, took office by appointment, not election. The two houses of the assembly retained the power to approve funding and remove ministers, but in general, they deferred to the president.

The constitution allowed a single person to serve two five-year presidential terms. Elected in 2005 to his first official term, Hamid Karzai very much wanted to win his second. He had been the de facto leader of the country since the demise of the Taliban.

To win the election outright, a candidate needed 50 percent of the vote. If no one reached that threshold, then a runoff would pit the top two vote getters against each other. Forty-four candidates registered to compete. Of these, only two — Hamid Karzai and Abdullah Abdullah — looked like serious contenders. A physician of mixed Tajik/Pashtun descent, Abdullah had served as foreign minister from 2001 to 2005. He railed against Karzai's corruption. Most pre-election polling showed Abdullah losing but possibly gaining enough votes to compel a runoff. If that came to pass, Abdullah's chances improved immensely. Anti-Karzai sentiment might well coalesce around a single champion.

Some senior American diplomats favored Ashraf Ghani Ahmadzai, a Westernized, articulate English speaker with a doctorate from Columbia University and a résumé that included teaching at the University of California at Berkeley, a stint with the World Bank, and a position as the minister of finance from 2002 to 2004. He was a more honest, less cynical Afghan version of the ubiquitous Ahmed Chalabi. Like that Iraqi worthy, as long as Ghani projected himself as a cosmopolitan intellectual, he had no prayer of gaining the backing of average Afghan citizens. He earned less than 3 percent of the vote. Ghani retained some influence in Kabul but not as much as his many American admirers credited him with. No Western-educated white knight was going to rescue this election.

On August 20, less than half the eligible Afghan citizens cast ballots. Turnout was low in the Taliban-influenced south and east and higher in the capital, the north, and the west, tracking well with ISAF reports of enemy activities. The early counts confirmed what most outside observers expected. Karzai won the plurality, but he did not top the necessary 50 percent. Violence, allegations of fraud, and intimidation affected many polling sites. Various United Nations and independent watchdog entities fretted over it all. It wasn't as clean as any of the Iraq votes, but it also wasn't all that bad for one of the five poorest countries on earth with an 85 percent adult-illiteracy rate and an active insurgency. Yet because of all the complaints, it took a month to sort out the results. Karzai claimed 54 percent of the votes, and he believed it, even if few others did. It took until November before Abdullah Abdullah gave up and gave in. Every indicator suggested Karzai's people manipulated the counts to ensure he did not face a runoff. Because the senior American officials stayed warily aloof, and because some in the background accused Karzai of cheating, the Afghan president grew even more volatile in word and deed, increasingly distrustful of those he saw as meddling outsiders. In Kabul, the ISAF commander alone retained some positive influence. It all formed a very sobering backdrop as McChrystal's team got going in country and Obama's administration tried to chart future strategy in Washington.

Stan McChrystal did not let the festering election aftermath distract him from the larger purpose. Obama might take a while to set the overall direction, but McChrystal moved quickly. The tactical directive ensured an immediate change in the character of operations, including a renewed relationship with Hamid Karzai. Meanwhile, the new ISAF commander sorted out his command structure, his strategy, and his input to the ongoing assessment in Washington.

To carry out his evolving campaign plan, McChrystal brought in two key subordinates to lead two new commands. Both were fellow members of McChrystal's West Point class of 1976. To run the new ISAF Joint Command (IJC), the long-needed corps headquarters, McChrystal tapped Lieutenant General David "Rod" Rodriguez, a tall, plain-speaking paratrooper of the John Vines and Dan McNeill school. Rodriguez had served as a brigadier general in the Fourth Infantry Division in Iraq in 2003, then commanded a division-size force in northern Iraq from 2005 to 2006. He'd led the Eighty-Second Airborne Division in Afghanistan and commanded RC-East from 2007 to 2008. He'd been in as many fights as any American general and more than most. Rodriguez wasn't a Petraeus type. Never an aide or great man's horse holder, he didn't court the press. He would spend his time out among the brigades and battalions, Mr. Inside while McChrystal played Mr. Outside. It suited the tough former West Point football player.

To command the new NATO Training Mission–Afghanistan (NTM-A), building on the old two-star Combined Security Transition Command–Afghanistan, McChrystal secured Lieutenant General William B. Caldwell IV. Son and namesake of a three-star general, Bill Caldwell consciously emulated the Petraeus example, having followed the man himself as executive officer to the chairman of the Joint Chiefs of Staff and commander at the Combined Arms Center at Fort Leavenworth. A former White House Fellow, Caldwell had also worked as the military assistant to the deputy secretary of defense early in the Iraq campaign. He actively sought media attention and believed strongly in crafting and pushing a Petraeus-caliber narrative; whether he could do it as well as the master remained to be seen. Caldwell had served in Panama, Desert Storm, Haiti, and Iraq, but always as a higher-level staff officer, not as a commander. The NTM-A position would be his first combat command. Now, Caldwell tried to do in Afghanistan what Petraeus had done in Iraq with MNSTC-I in 2004 and 2005. He had to speed up the growth of the Afghan national army and police, numbering about 173,000 in July of 2009. McChrystal demanded 400,000 by 2011, sooner if possible.

The new U.S. ambassador in Kabul was not McChrystal's choice, although he, Rodriguez, Caldwell, and the other U.S. generals knew the man well. Karl Eikenberry returned to Afghanistan less than three years after he left. Now he wore a suit. The Obama inner circle thought putting a former general into the embassy would ensure closer cooperation between ISAF and the diplomatic effort in country. Senior to McChrystal, Rodriguez, and Caldwell by three years at West Point, Eikenberry had already done their jobs in Afghanistan too. Unlike the other three, although he'd seen plenty of infantry duty up through lieutenant colonel, Eikenberry didn't continue through the traditional upper command channels but rather was a China expert prior to his time in Afghanistan. The new ambassador tended to speak, not listen. He did not work well with the embassy staff in his past Afghan assignments and thought little of Karzai, who loathed him. Eikenberry also did not favor robust counterinsurgency because he believed Karzai's regime was corrupt and erratic and thus would be unable to hold what ISAF cleared. When asked about Eikenberry back in the spring, Dave McKiernan had strongly recommended choosing another person. Having a retired general, especially this one, grade his successor might be problematic. But nobody cared what McKiernan thought, then or now. He was old news. Eikenberry became the ambassador.

Realizing that Eikenberry had enough to do in Kabul and wanting to link together the entire Afghanistan-Pakistan diplomatic effort, Obama added one other key man to the mix: Richard C. Holbrooke. The veteran foreign service officer had spent much of the 1960s in Vietnam, and he didn't forget that formative experience. The Clinton administration made him the lead negotiator for Balkans affairs, and he'd engineered the 1995 Dayton Accords that ended the bloody Bosnia civil war. Holbrooke also played a key role in resolving the U.S./NATO–Kosovo intervention in 1999, and then served as the American ambassador at the United Nations. President Obama and Secretary of State Hillary Clinton, who knew the energetic Holbrooke well, thought he just might be able to arrange talks with the Taliban, Pakistan, and Karzai that could wrap up the entire mess. Vice President Joe Biden called him "the most egotistical bastard I've ever met," which was saying something. But he also stated that "he's maybe the right guy for the job."

So the new players came together. President Hamid Karzai fought for his political life, looking over his shoulder for an American sign of affection, ever watchful for a knife in the back. Stan McChrystal's ISAF, strengthened by IJC and NTM-A, pushed ahead to institute a counterinsurgency campaign. Ambassador Eikenberry watched it bubble but held his counsel. And Dick Holbrooke began casting lines in the South Asian pond, hoping for Taliban nibbles. Everybody provided his recommendations for President Barack Obama's strategic review. The days, weeks, and months ticked by.

In the absence of decisions, Stan McChrystal thought through what he wanted to accomplish. To ensure a fresh look, he assembled a large number of military, diplomatic, academic, and intelligence service intellectuals. The slate included people like Colonel Kevin Owens, who'd previously commanded the 173rd Airborne Brigade in country; noted Army strategist Colonel Chris Kolenda; Stephen Biddle, who taught at the Army War College; renowned defense analyst Anthony Cordesman; Iraq veteran and political adviser Catherine Dale; former Ranger and counterinsurgency thinker Andrew Exum; and Professors Fred and Kim Kagan, both of whom had worked closely with General Jack Keane in the run-up to the Iraq surge. McChrystal turned them loose with three questions: Can we do this mission? How would we do it? What will it take to do it?

Their answers reinforced McChrystal's inclinations, already expressed in the tactical directive. He packaged it up and sent it to Washington as his input to President Obama's ongoing review. McChrystal focused on strategy: "Protecting the people means shielding them from *all* [emphasis in the original] threats." He meant the insurgency, cross-border threats from Pakistan, drug dealing, and corruption, a very wide mandate. The general thought the mission could be done.

McChrystal's method involved controlling key population centers. There would be no more platoons sent to eat bullets at far-flung hamlets like Barg-e Matal. ISAF planned to go where most Afghans lived and cede the sparsely populated periphery to SOF, drones, and whoever wanted it. The north, west, and capital were already relatively secure. So ISAF must clear and hold the Kandahar-to-Kabul corridor, the core of the country that united the two largest cities and tied together southern Pashtuns with northern Tajiks. To do so, in the winter of 2009–10, ISAF would start in the Taliban heartland west of Kandahar, in Helmand, where the first of ten thousand Marines had begun to arrive as directed in Obama's March 2009 deployment decision. Next, during the rest of 2010, ISAF planned to clean up Kandahar and its environs. Finally, in 2011, they would press on into selected areas in the east, clearing the Taliban pockets in Logar and Wardak Provinces, the ones that menaced the Kandahar-Kabul population axis. American drone strikes would deal with Pakistan-based terrorist-cell leadership, and, of course, McChrystal's former SOF colleagues had to keep going after the enemy network kingpins inside Afghanistan, too. Finally, the embassy would pressure Karzai to clean up his act. That's how it would be done.

Resources would be the hard part. The math suggested six hundred thousand soldiers to protect thirty million Afghans, the favored 1-to-50 ratio. Bill Caldwell at NTM-A and the Afghans would supply a lot of those as quickly as they could. With four hundred thousand Afghans eventually on side, plus forty thousand ISAF partner troops and the sixty-thousand-plus U.S. forces already present or en route, McChrystal's planners wanted eighty-five thousand more Americans. That strength, an Iraq surge-force level, would let ISAF clear Kandahar and start on the east in 2010. The general knew it wasn't going to happen. But Stan McChrystal thought he could do it sequentially—Helmand, then Kandahar, then the east—with forty thousand additional Americans, four more BCTs, and combat-support enablers like helicopters, intelligence teams, and engineers. He kept those numbers out of the assessment proper. But they were floating around in ISAF, U.S. Central Command, and the Pentagon by August. The ISAF commander concluded his submission in no uncertain terms: "Failure to provide adequate resources also risks a longer conflict, greater casualties, higher overall costs, and ultimately, a critical loss of political support. Any of these risks, in turn, are likely to result in mission failure." So there it was: Send the troops or lose.

Dave Petraeus put the debate out in the open. Not him directly—he was much too polished to stumble into that near ambush. But his shining example of dominating the press coverage in 2007 Iraq spurred Stan McChrystal to try a similar approach. He had the moves, but he lacked the rhythm. His long life in the shadows hadn't featured much of a press program, preferring to say nothing, so McChrystal was a novice at this game. In his well-meaning attempts at media outreach, he established a reputation for candor. But he also made it clear that he didn't understand how to do business inside Washington.

Someone in the Pentagon leaked McChrystal's assessment, which the *Washington Post* printed on September 21 with a few minor redactions. That kind of thing happens in the capital. McChrystal had nothing to do with the revelation. But to those in and near the White House, who, like most political people, did such things habitually when it helped their cause, the disclosure looked like a way to pressure the president to send more troops now.

On October 1, McChrystal added fuel to the fire. During a speech to the International Institute for Strategic Studies in London, the ISAF general casually shared his thoughts that counterterrorism in Afghanistan wasn't enough. The war demanded a full-scale counterinsurgency to protect the Afghan people. That put McChrystal publicly opposite the U.S. vice president while policy was still being developed. Simultaneously, a prerecorded interview with the news show *60 Minutes* featured McChrystal's admission that he'd spoken to the president only once during his first few months of command. Together, the two comments, widely reported, implied that the vice president was wrong on strategy, and the president detached from the entire war effort.

Admiral Mike Mullen called McChrystal. "Obviously, you need to be careful," Mullen told him. McChrystal would have been even more chagrined had he heard Barack Obama's comment. "We've got to stop this," the president had said. "This is not helping." When the president met the general the next day in Copenhagen, Denmark, the press (wrongly) categorized the previously scheduled session as a dressing-down. It wasn't. But it looked like one. Some of those around the president didn't mind that image.

McChrystal pulled back his press availabilities. He had to lie low for a few months and let the president's review run its course. "Still, in retrospect," McChrystal later wrote, "I never felt entirely the same after the leak of the strategic assessment and the unexpected storm raised by the London talk." Around the Army, the vast majority who favored the subdued Casey/McKiernan way of dealing with journalists sympathized with the ISAF commander. Playing with the press was like playing with fire. Sooner or later, you would get burned. There was only one Petraeus. McChrystal went back to running the war and waited for Obama to decide. And waited.

The delay wasn't just due to lack of information or conflicting recommendations. It also offered Obama both his first and, likely, final chance to get the campaign right. He had watched Bush go into Afghanistan, overstay, and then lose interest and try to let others carry the ball. He had seen what he considered the Iraq blunder — getting neck-deep into an ill-advised counterinsurgency — now being reversed at his direction. Obama didn't think the Iraq surge did anything more than stabilize the terminally ill patient before the family went home. It was a salvage ploy at best. Now the new president was being asked to repeat the Bush strategy in Afghanistan, to surge. But in the end, did it matter? The patient looked likely to die anyway.

By mid-2009, the top civilians and senior military officers in the decision chain regarded Afghanistan through the prism of post-surge Iraq. But where Obama and his team looked at Iraq and saw Vietnam 1971, a bad war righted just enough to crawl to an end, the generals and admirals thought they glimpsed Korea 1953, a botched war redeemed to a stalemate and, with a long-term U.S. commitment, perhaps even a sort of win. A troop surge in Afghanistan would jump-start the counterinsurgency campaign. Success in Afghanistan, however, depended on follow-through, on America staying for decades. Obama had not made such a commitment in Iraq, although in the summer of 2009, despite ample contrary evidence, high-ranking military leaders remained hopeful and assumed it would be so. They certainly figured on a long-term U.S. effort in Afghanistan. If unwilling to stay, America might as well skip the surge and pull out. Go long or go home — so went the thinking. In field command posts and Pentagon meeting rooms, generals talked to one another about it. But in public, when asked, none raised the crucial strategic matter of a lengthy U.S. commitment. They talked about the immediate issue. They supported McChrystal. Send the troops. Surge. Maybe the jump-start would work.

Maybe so. The new administration liked to think that way. After all, Obama and his people truly believed that they would do it better, smarter, than the previous bunch. They were taking the time, thinking it through, putting all the right people in place. And yet, as they convened and deliberated, wrote papers and gave presentations, deconstructed briefing slides and exchanged views, in the backs of many minds, maybe including that of the president, a small, spiny thought pricked: Were they repeating Bush's mistake?

Obama took in all the inputs. He looked at the proposals for more BCTs and where they would go. He reviewed the counterterrorism options in both Afghanistan and Pakistan and considered a shift to emphasize such a strategy, strongly urged by the vice president. He heard both Eikenberry and Holbrooke denigrate Karzai, with the former going so far as to call the Afghan "not an adequate strategic partner." Yet that's who America had, very much Our Man in Kabul.

President Obama finally made the call on November 29. In an unprecedented move, recalling Abraham Lincoln's personal letters to his Civil War commanders, the president himself dictated the multipage strategic guidance. He chose his words precisely, like the constitutional law professor he was: "to deny safe haven to al-Qaeda . . . to deny the Taliban the ability to overthrow the Afghan government . . . building sufficient Afghan capacity to secure and govern . . . an extended surge of 18 to 24 months . . . not fully resourced counterinsurgency or nation building." Obama added one other crucial phrase, requiring "the United States to begin reducing its forces by July 2011." In other words, McChrystal would get his shot. But it would be that, a single surge, a last chance to get it right. Then, unless Hamid Karzai straightened up, Dick Holbrooke worked a diplomatic miracle with the Taliban, or Pakistan cleaned its border sanctuaries, America would wrap it up, like in Iraq.

Numbers followed. McChrystal got 33,000 Americans, less than he'd asked for. NATO agreed to send 10,000 more troops. So that helped, even though those troops came with the anticipated strings attached. It would take until mid-2010 to see any of the reinforcements in theater. The U.S. strength would grow to nearly 100,000, although it would then be reduced back to 68,000, the already approved pre-surge number. The Afghans had to build more forces, but the 400,000 goal wasn't codified. It would be less.

The split nature of the decision preserved options for Obama. If it worked, we would stay. If it didn't, we would depart. Yet even as the orders went out, they undermined the one thing, maybe the only thing, that might have deflated the Taliban at this point. The enemy feared a long-term U.S. commitment of troops on the ground, airpower overhead, and sustainment for the Kabul regime. If the United States agreed to keep them in place for decades, as in post-1953 Korea, as few as ten thousand Americans might have cracked the Taliban will to fight. But nobody in Washington wanted to talk about U.S. troops on the ground for decades. Far from it. Rather, the president proclaimed the July 2011 deadline to start drawing down. In their hiding places in Qetta and Miranshah across the Pakistan border, in their hovels in north Helmand and their mud huts west of Kandahar, the Taliban heard that message loud and clear. The U.S. was leaving. The Taliban fighters were not. They were willing to take that bargain, suffer a few more years of attrition—*enemy advances, we retreat*—then finish off Karzai's faction in their own good time. *Enemy retreats, we pursue.*

President Obama announced his decision at West Point on December 1, 2009. He'd selected an appropriate venue. Like West Pointer Jake Miraldi at Barg-e Matal, these young men and women in dress gray would soon enough lead the soldiers who would enact the president's policy "by other means," as Clausewitz said. Not all of them would make it home. Everybody in Eisenhower Hall that night knew it.

15

TALIBAN HEARTLAND

> BRITISH LIEUTENANT COLONEL: Your Marines seem to have exceeded the ops plan.
>
> U.S. STATE DEPARTMENT OFFICER: Well, of course. They're Marines.
>
> — OVERHEARD IN HELMAND PROVINCE

THE MARINES ASKED for Helmand. United States Marine Corps forces had been in Afghanistan from the earliest days, with a brigade operating south of Kandahar in December 2001 and expeditionary units going in and out over the next seven years. With Iraq drawing down and the Marine effort in Anbar largely completed thanks to the Sunni Awakening, the commandant himself, General James Conway, sought a greater role in Afghanistan. Almost as soon as General Dave McKiernan requested more U.S. troops, in late 2008, Conway offered Marines. He requested one concession. The Marines wanted their own distinct area, like Anbar Province in Iraq.

Kandahar was the obvious place. But the Canadian brigade operated there. They knew the area, had paid for it with a lot of lives, and they wanted to keep it, at least in 2008 when Conway made his offer. Within three years, Canada reassessed and gladly handed it over to the Americans. But they weren't there yet.

The British were. Having faced a buzz saw at Now Zad, Sangin, and Musa Qala in 2006, the British very much wanted help. Getting 10,672

Marines into north Helmand sounded great to the hard-pressed UK brigade. It would more than double the ISAF troop strength in the desert province. So the Marines went to Helmand.

Ruddy, pugnacious Brigadier General Lawrence D. Nicholson commanded the Second Marine Expeditionary Brigade going into Helmand. An ROTC graduate of The Citadel, a state military college in South Carolina, Larry Nicholson looked, talked, and led exactly like the Marine infantry officer he was. He had a lot of troop time, highlighted by command of Marines in the Balkans and Iraq. Nicholson had also been a United Nations observer in the Palestinian territories and had attended the NATO Defense College in Rome. As was becoming more typical of general officers in the age of Petraeus, Nicholson thought about his public image and cultivated it a bit more than most. But that was a secondary consideration. His mission and Marines came first.

Nicholson assessed the area and quickly identified Marjah as a problem, probably the biggest one in Helmand. The farming community of eighty thousand, seventeen miles southwest of the provincial capital of Lashkar Gah, served as a Taliban base camp. Gunmen walked openly there. Buildings held stocks of arms and ammunition. Some houses served as IED factories. The Taliban dunned the locals, who harvested the vast surrounding poppy fields and refined some of the opium right in town. Drug money came easy in Marjah. The poppies grew well, a heritage of an irrigation-canal system tied to that moribund 1950s U.S. Agency for International Development Helmand River project. The Kajaki Dam remained unfinished, but the irrigation grid near Marjah watered a hell of a fine poppy crop. Taliban bomb planters had wired up all the trails, booby-trapped the little cement-slab footbridges over the drainage ditches, and planted traditional land mines along major dirt roads. British soldiers stayed out. Afghan soldiers did too. British aircraft even avoided overflying Marjah.

"If Marjah is the worst place you've got, let me go there first," Nicholson said to McKiernan in early 2009. The four-star said no. He didn't want a big, bloody battle during the August 20 Afghan election. Both men remembered well what had happened in Fallujah, Iraq, in 2004. And Marjah smelled a lot like Fallujah in the bad old days. It offered a Taliban-liberated space full of cranky Pashtuns who would immediately scream to fellow southern Pashtun Hamid Karzai in Kabul. "It's going to be too political," McKiernan warned. "It's going to be too big of a fight." It all conjured up the admonition of Marine General Jim Mattis outside Fallujah, by way of Napoleon Bonaparte: if you're going to go after an objective like Marjah, you need to take it. It wouldn't do to get started and then back away. In the spring of 2009, the ISAF commander wasn't up for it.

Now Stan McChrystal ran ISAF. He very much wanted to take Marjah, clear it, hold it, build it, and, in a new addition to the Bush-era formula, transfer it to the Afghans. At first, he wondered what good his strongest new contingent, the Marines, could do way out in Helmand. Then he saw the opportunity. Marjah was enemy-infested, isolated, and achievable. What worked there could also work in Kandahar and, later, points east. Following President Obama's December 1, 2009, announcement of the troop surge, McChrystal recognized "there was an appetite for an operation with rapid, observable impact." That became Marjah.

McChrystal pointedly followed his own orders. The assault wave included three Afghan battalions from the Third Brigade, 205th ANA Corps, as well as the Third ANA Commando Battalion (Afghan Rangers), and police elements. With the large Afghan contribution, the operation became known as Moshtarak (Dari for "together"), which had also been the name of two failed 2006 attempts to secure Baghdad and so was not exactly a good omen. This Operation Moshtarak would be different; committed to making it work, Stan McChrystal sought approval for it from the local elders at a public gathering in Lashkar Gah. He personally briefed President Karzai, asked for consent, and got it. He even arranged what he called a "government in a box," a group ready to take over as soon as Marjah came under ISAF/Afghan control. Karzai appointed a new governor, picked out a new supporting cast, and made noises about going to Marjah himself after the fighting ended. ISAF promised funding for projects and services. ISAF also drew up a detailed media coverage plan. The world would see what happened in Marjah.

All of the preparations went on in full view of the Taliban. They watched as Marine and Afghan companies tightened the noose around Marjah, securing a line of platoon positions fronting the town. Leaflet drops announced the pending operation, and although literacy was uncommon among rural Pashtuns, word got around. Beat-up cars and trucks full of people and property clogged the roads out of town. Nearly 2,700 Marjah residents showed up in Lashkar Gah seeking shelter. A good number of Taliban left too. But hundreds stayed. The usual reports surfaced naming Arabs, Chechens, and Pakistanis among the holdouts, although, of course, almost all of the actual enemy fighters encountered turned out to be locals. The Taliban might have been being brave, or stupid, or arrogant, or perhaps they were just unwilling to accept that this time, ISAF meant it. Marine observation posts reported, drones filmed, and rumors circulated. Every night, shadowy figures put more bad things in the ground around Marjah.

Meanwhile, at corps level, Lieutenant General Rod Rodriguez lined up the forces. The detailed planning and mission execution fell to British major general Nick Carter, commanding RC-South. A peppy, pushy fellow, portrayed as dashing by the London press, Carter proudly styled himself "the Petraeus of the British Army" and fully embraced the July 2009 ISAF tactical directive. His operations orders included colorful language telling subordinate commanders to "discombobulate" the opposition and place the enemy "on the horns of a dilemma," all while incurring "zero" civilian casualties. "Our young men and women refrain from using lethal force," one of Carter's staff officers said, "even at risk to themselves, in order to prevent possible harm to civilians." Other than brief helicopter flights to a few quiet forward positions, Carter supervised RC-South from his large, well-appointed command post at the Kandahar Airfield. So be it. Day-to-day leadership in Marjah fell to Brigadier General Larry Nicholson. He was up to it, and had no intention of running things from some distant computer terminal.

Under Nick Carter's willing thumb, Stan McChrystal's micromanagement of lethal fires greatly constrained the entire undertaking. In their extensive preliminary reconnaissance, Marines identified dozens of hostile locations, all good targets, but, of course, all thoroughly interleaved with civilian homes and activities. Denied the usual preparatory airstrikes and artillery fires, Larry Nicholson and his Marines chose to clear Marjah from the inside out. They would employ helicopters to hop over the IED-encrusted outer layers surrounding the town and land in the hole of the Taliban defensive doughnut. The Marine Corps had pioneered vertical envelopment in the Korean War. In Marjah, they would draw on these long-established close ties between Marine aviators and infantry. Although McChrystal, ISAF, and RC-South surely telegraphed a lot of the punch, they had not disclosed its timing or exact nature. Prevented from using shells and bombs, Marine riflemen and Afghan *askari* (soldiers) would drop out of the night sky. Then they'd do the dirty work in the old style, house to house.

Marine Aircraft Group 40 dumped infrared parachute flares from circling KC-130 transports. A person on the ground would see nothing with the naked eye. Through their night-vision goggles, Marine aviators followed a string of flares that lit the way into the grid of canals and mud huts like streetlamps floating in the still winter air. The helicopters went straight into Marjah proper, delivering Company A and Company B of the First Battalion, Sixth Marines at 3:00 a.m. sharp on February 13, 2010. Each of the three-engine CH-53E Super Stallions inserted a platoon or so of Marines and Afghan soldiers. The men expected hot landing zones, given the estimated four hundred Taliban in town. Instead, the packed mud was cold, the dust minimal, and the starry night black and quiet. The temperature in the high desert hung in the upper thirties, damn cold for southwestern Afghanistan, even in mid-February. Dogs barked, but not an electric light or oil lamp shone. Marine riflemen and Afghan *askari* shifted their heavy loads, an average of seventy pounds of Kevlar body armor, water, and ammunition for the Americans, less for the thinner, slighter Afghans. The Marines aligned their night-vision goggles on the front of their helmets, turning the dark world of Marjah luminous green. Unopposed, they moved out.

Up north, somebody was shooting at the British Eleventh Brigade landings at Nad Ali. Marjah remained quiet, but the Taliban a few miles away tried to fight back. As he plodded forward, Lance Corporal Matthew W. Hunter glanced up and saw it. "I remember seeing those AA [antiaircraft] guns go off — that was the sign," he said. "It clicked that this was real." They weren't really air-defense weapons, just Russian-made DShK 12.7 mm machine guns, the heavier cousin of the U.S. .50-caliber. But Hunter and his brother Marines got the Taliban's message: *Enemy advances, we retreat.* But when it suited them, they shot on the way out.

To the southeast, Company C of 1/6 Marines didn't enter by helicopter. They planned to come overland, across the grain of the north–south pattern of canals and footpaths. They knew better than to use the little cement bridges or the hard-packed trails. Their own patrols and constant surveillance identified so many known and likely IED locations that the Marines guessed it might take all day to walk a mile into town to link up with Company A. But to resupply their forces in Marjah, the Marines needed a ground route cleared. Company C got the task.

For weeks, the company had been holding a series of outposts, overwatching Marjah, getting ready. The enemy moved openly in front of them, and their foes' constant digging probably wasn't preparation for planting petunias. There had been clashes and losses, two Marines killed and three badly wounded by IEDs in a vicious late-January encounter. Marine first lieutenant Aaron McLean watched four unarmed Pashtun males with black flags taunt them from a mile or so away. The enemy knew that if they didn't show weapons, the Americans couldn't engage. "I didn't call air to destroy that damn house," McLean explained later. "I was too sick about our losses to argue rules of engagement." The Marines watched and waited.

Now, at last, in the predawn darkness as the helicopters clattered away, the men of Company C acted. They walked slowly away from Outpost Husker, looking up through their night-vision goggles to see the marking lights of the departing helicopters heading back north out of Marjah, toward the long runway twenty-five miles away at Camp Leatherneck. The Marines shook out into a long, loose column. A point squad led the way. Four-man fire teams moved out two hundred yards to either flank. A trail team drifted back, turning around every ten steps or so like human gun turrets, keeping an eye out for unwelcome followers. You had to put out such teams or the men would gradually drift together into a long, single file, boot after boot, step after step, nobody watching anything but the back of the man in front of him. That's when the Taliban usually opened fire. Good infantry units spread out and kept their security up. Sergeants saw to it. But it slowed progress.

So did the ground itself. On the photo map, the mile did not look too bad. There were only a few abandoned huts, easily skirted. The silent, black poppy fields stretched all around them, with little irrigation ditches every ten yards or so, bigger channels every thirty, and waist-deep waterways every hundred. Each watercourse had a knee-knocker wall rimming it. On the big canals, the mud wall rose up almost to your neck. You had to lean forward, grope for vegetation handholds, crawl up one side, splash across, then clamber out, with helping hands offered and accepted. The water was still, oily, and ice-cold. To avoid the IEDs, the Marines and their Afghan partners slowly staggered right across row after row of these water obstacles. The drier stretches meant walking through fields of waving poppies, wet, mud-streaked boot tops scraping through the premier opium-production fields of Afghanistan. It beat tripping off IEDs.

You had to walk carefully. The column stopped and started every few minutes. Each time it halted, the men took a knee or leaned up to an irrigation mud wall and propped their rifles over the top, scanning, waiting to pick up and move ahead. Using night-vision goggles meant looking at the route through a field of view about the size of a cardboard toilet-paper tube. Peripheral vision didn't exist. That combined with heavy loads of gear affected your balance on the already uneven dirt. Marines slipped and slid on the clammy mud. The clatter of stumbling men marked almost every barrier crossed. Sometimes you heard some sloshing in the canals. People kept the cursing to themselves. Nobody talked except in whispers, to pass orders or send a radio report. The less heavily burdened Afghans, more at home in the rural night, moved confidently, heads up, pacing the slower Marines. Stop and start. On it went. Nobody hit an IED, though. Doing it the hard way was paying off.

By 6:30 a.m., gray light grew strong enough that the Marines could stow their night-vision goggles. As a rooster crowed, two shots cracked — Kalashnikovs, of course. Some of the Taliban must have stayed behind to face the attack. The Marines and Afghan soldiers dropped behind the canal walls. Nothing followed. The Marines slowly rose, like tired old men, dirt-stained, wet from the water, wary. They again started forward.

Another shot sounded. That was an American rifle. Word came back. The point man had shot an angry dog outside a farmhouse. Somebody seemed to be inside. The Marines asked for an Afghan fire team to come up and enter the house — the tactical directive *über alles*, of course. Everybody took a knee, facing "outboard," as the Marines put it.

A series of AK shots hammered from straight ahead, from a canal bank, it appeared. The Marines and Afghans returned fire with rifles and machine guns. The Marines aimed, the Afghans less so, but better than usual. Of course, nobody really had a clear look at the bad guys. The field manual says that when you take shots like this, you should return fire at "known or suspected enemy positions." Experienced riflemen learn to judge based on noise, muzzle flashes, puffs of dust, and just plain instinct. Those worked. When the Marines ceased fire after a few minutes, no more AK rounds came back. After about twenty minutes, a passing drone puttered overhead and sent pictures of three dead insurgents. When a search team found them, one had a handheld radio. All were armed.

The commander of 1/6 Marines, Lieutenant Colonel Calvin Worth, linked up with Company C about this time. His Company B and their Afghan teammates had just cleared the Koru Chareh bazaar in north-central Marjah. The happy *askari* ran up a big Afghan flag. Now Worth and his security team got with Company C to see how things were going on the outskirts of town.

Worth and his translator approached Jawad Wardak, the owner of the dead dog. Wardak's single-story adobe house had dried poppies on the ground in front of it. Five young men, all unarmed, all delivering the standard Pashtun stink-eye, huddled behind Wardak. "I'm very sorry about your dog," Worth offered. "Hopefully we haven't done any damage to your home."

"It's no problem," Wardak replied. He was saying whatever it took to get these Americans to move on. While Worth conversed, men of Company C found a ten-foot mud wall just south of Wardak's home. The wall hid seventy explosive devices. Wardak shrugged — they were news to him. Sure.

As Marine engineers rigged up demolition charges to destroy the munitions, the American and Afghan infantrymen saddled up and moved on. Just to the west, they came across another waterway. More sure-footed in daylight, the Marines and the Afghan soldiers slid one by one down the sandy bank wall. Men kept their weapons up as they waded across the waist-deep cold water in the canal. Something didn't look right about a house to the northwest. It had a dirt-wall corral out front. The point men thought they saw movement, but maybe not.

Maybe so. *Crack. Crack.* AK-47 bullets whipped overhead, followed by the rhythmic chugging of a PKM machine gun. The Marines scrambled up the canal bank and returned fire. Rifles and machine guns joined in. Worth immediately called for his mortars to engage. They did. Two, four, then six dirty-gray bomblets burst on the house. Nobody asked about civilians. The mortar hits finished off the enemy. The point squad, with Afghan soldiers in tow, found four dead Taliban in the house.

Those would turn out to be all of the day's contacts, but naturally, nobody knew that then. So the column kept clearing structures, splashing through canals, and tripping over mud piles. Marine engineers behind them, tracing a parallel route, cleared the major trail into Marjah. They detonated IEDs and land mines by the dozen. It took until four thirty, almost sunset, before Company C linked up with Company A in the front courtyard of a smashed-up building near the shot-up Loy Chareh bazaar in the southeast corner of town. Marjah belonged to the Marines.

The inside-out ploy worked superbly. Company C's coincident cross-country attack also surprised the Taliban. Cal Worth's 1/6 Marines and their affiliated Afghans from the Third Brigade, 205th ANA Corps set up eleven checkpoints in key neighborhoods. While the British cleared out Nad Ali to the north, 3/6 Marines operated northwest of Marjah, and the Third ANA Commando Battalion and their U.S. Special Forces advisers worked in southern Marjah. Taliban spokesman Qari Yousef Ahmadi put out a press release avowing that they wanted to limit civilian casualties, so their tactics would emphasize "hit and run." Thus far it had been mostly the latter. *Enemy advances, we retreat.*

The tactical directive reared its ugly head on February 14 when 3/6 Marines attempted to hit an enemy-held building in north Marjah. Hung up by hostile machine-gun fire, two men wounded already as they tried to close on the structure, the Marines called for a guided 227 mm (9-inch) projectile from the high-mobility artillery rocket system (HIMARS). The rocket resembled a JDAM bomb in that it followed GPS coordinates in order to strike within ten yards of the aim point. This time, the resultant shot hit on target, but either the spot was wrong or, more likely, the Taliban had shielded themselves among civilians. The Afghans immediately reported ten dead civilians, all women and children, of course. The Marines thought eleven had died. ISAF raised it to twelve.

In Kabul, General Stan McChrystal immediately suspended all use of HIMARS fires. The regular routine ensued: Karzai complaints, ISAF apology, compensation money paid, more screws cranked on the fighting units. Thereafter, Major General Nick Carter at RC-South decided that he had to approve any air or artillery fires. He did not approve many. For the Marines, who had traditionally relied on supporting fires of this sort, it was very awkward. Routine requests for fire degenerated into lengthy radio debates. Young riflemen paid the price.

Brigadier General Larry Nicholson and Sergeant Major Ernest Hoopi almost got a taste themselves. At midday on February 16, forward with their men as usual, the Marine leaders took a few moments for a symbolic gesture. They accompanied the new governor, Haji Abdul Zahir, into the Koru Chareh bazaar. Already in town to take charge of liberated Marjah, Zahir represented the government-in-a-box promised by McChrystal. Zahir knew about boxes. A returned expatriate, he'd spent four years in a German jail for stabbing his stepson. Now he strode through the market greeting his new constituents. Most extended their hands, curious and polite.

One elder was less than happy. "We are all Taliban here," he growled. "You represent a corrupt and murderous government. I'll give you a chance. But if you betray me, I'll kill you and your entire family." Zahir listened, unperturbed. He'd heard worse.

Somebody else chose that instant to deliver another opinion. A burst of AK fire erupted from a canal bank to the north. The Marines took cover, and the rifle platoon pulling security for the governor immediately returned fire. Sergeant Major Hoopi did likewise, moving to position Marines. One of them turned to him.

"Dad, I got it." It was his son Sean, a lance corporal in Company L, 3/6 Marines. Sean Hoopi and his fellow Marines quickly cleaned up the episode. Nicholson, trailed by reporters, stood tall. He noted that such things happened. It was all part of clearing Marjah.

The few remaining enemy infiltrators bided their time. At dusk two nights later, as a platoon of 1/6 Marines prepared their night patrols in the courtyard of their outpost fronting the Loy Chareh bazaar, a single gunshot sounded. The enemy aimed at the rooftop bunker, threading the needle to strike a Marine inside. The round caught Lance Corporal Kielin T. Dunn right in the face. Four Marines pulled the stricken Dunn out of the lookout post. They got him downstairs and carried him out on an Air Force medevac helicopter. He didn't make it.

His comrades returned fire in volume. Up on the roof, the gunners traversed the big .50-caliber machine gun, ripping bullets across the unoccupied buildings on the north end of the marketplace. The Marines called in 81 mm mortar rounds too. They even got permission to bring in an A-10 Warthog, an unusual concession from the RC-South headquarters. The A-10 unleashed its deadly 30 mm Gatling gun. The heavy shells tore the roofs off several mud huts. It probably did no good. The Marines never found a Taliban body, partly because nobody was exactly sure where the fatal bullet had originated from.

Taking Marjah and holding it that first week cost the lives of eight U.S. Marines and six Afghan soldiers and wounded nearly thirty. Three of the dead were Marine engineers, indicative of the congested belts of mines and IEDs lacing the fringes of Marjah. A hundred or so Taliban died, and the rest evaporated, submerging into the population from whence they'd come to wait for their moment. The Afghan *askari* fought, not exactly the way the Americans did, but well enough. The new Marjah district governor, Haji Abdul Zahir, was on the job, a start toward something that might evolve into governance. Satisfied, Stan McChrystal invited President Karzai to visit as he'd promised.

On March 7, the Afghan president arrived in Marjah with McChrystal at his side. It was Karzai's show. An overflowing crowd of several hundred greeted Karzai in a mosque for a two-hour *shura,* a traditional Afghan meeting of the minds. The president experienced a reception similar to Zahir's inaugural market walk. Giving evidence of how he got his job and why he kept it, the Afghan chief executive waded right in. He knew how to work a room. After a brief speech, he took questions and comments.

Most locals gave him respect. Some most vociferously did not. Karzai didn't help his case by appearing alongside Abdul Rahman Jan, the predatory pedophile former police chief, and former governor and drug-smuggling kingpin Sher Mohammed Akhundzada, both known confidants of the Afghan president. When Karzai hailed the new civil authority, Haji Zahir earned a round of applause, mainly for not being the other two thugs. For a man who got scant credit from most Americans and little affection from most of his countrymen, Karzai showed moral courage in making the trip, the first by any senior Afghan not wearing a uniform and carrying a weapon.

The Marjah operation buoyed McChrystal's spirits. He hailed it as a "litmus test for the validity" of the strategy. "On display was our ability to conduct effective counterinsurgency in Afghanistan," he said. McChrystal also applauded the performance of ANA and ANP forces and the "Afghan government's commitment and ability to bring legitimate governance to a skeptical population." He didn't oversell it — few buyers existed at this point in the war. But a win was a win. Kandahar would be next.

Success in Marjah, even a qualified success, still left one critical question in the mind of the ISAF commander. As McChrystal put it, "The action offered the chance to determine whether it was possible and appropriate to sharply limit the use of our overwhelming advantage in lethal fires." Possible, certainly — that much was obvious. But was it appropriate?

You could never tell what the Pashtun villagers really thought. Afghans did not have much experience with polling. The farmers and herders often told canvassers what they thought the questioners wanted to hear rather than their actual beliefs. Yet postoperation polls in Marjah didn't comfort those who subscribed to Stan McChrystal's thinking. Despite all the "courageous restraint," 96 percent believed "a lot of civilians" were killed in Marjah. Only 3 percent figured that changing the conduct of ISAF units would prevent Taliban recruitment. As a final kick in the teeth, 67 percent thought that NATO and the Kabul regime would not defeat the Taliban. These numbers didn't deviate much, if at all, from findings in earlier years. If anything, they trended more negatively. Put bluntly, the objects of ISAF's affection were not feeling the love.

ISAF's men and women, especially those in fighting battalions, also had strong reservations, to put it mildly. This courageous-restraint business clearly rankled all but a few true believers or outright sycophants. Companies and platoons lost people trying to adhere to this policy. In the east, a U.S. attack triggered a horrific ambush in the Ganjigal Valley on September 8, 2009, costing the lives of five Americans and eight Afghans. The defense of Combat Outpost Keating on October 3, 2009, resulted in another eight U.S. dead and twenty-two wounded. Both of these brutal firefights underscored serious concerns with limiting artillery fires and air support. In each case, the freedom to engage known or suspected targets before the enemy closed in tight would have greatly limited friendly casualties and might have reversed the circumstances that brought on these desperate engagements. McChrystal's solution, and that of his subordinates in RC-East, amounted to pulling back to larger populated areas and thus avoiding putting small units out in isolated positions. If only they could get the enemy to go along with this plan. Guerrillas by their nature look for little, separate elements. Missions and friction, complicated by awful terrain and unpredictable weather, ensured that platoons and companies kept ending up out on their own. Without ready firepower on call, the results promised to be ugly.

One ISAF NCO raised the question directly with the man himself. Staff Sergeant Israel Arroyo of the First Battalion, Twelfth Infantry, slogging through the poppy fields, vineyards, and drainage ditches in hotly contested Zharay District west of Kandahar, finally had enough. On February 27, 2010, he sent an e-mail directly to McChrystal. Arroyo stated that he understood what he called the ISAF "restraint tactic," but "it is telling the men that they should not shoot even if they are threatened with death." Having seen comrades fall already — seven killed and dozens wounded from the battalion in the last two, supposedly quiet winter months — Arroyo said that he and his squad "don't want to lose any more." He invited McChrystal to come without a large security team and go on foot patrol with Arroyo's men. McChrystal accepted the invitation. Within two days, the general arrived at platoon Combat Outpost JFM in Zharay, ready to patrol.

McChrystal joined up with Arroyo's squad. After a short briefing, the men strapped on their body armor, buckled their helmets, and leveled their carbines. Off they went, with the four-star in the column. It wasn't the usual faked-up great-man stroll but a real mission. The general watched, listened, and learned. And he sweated. He had been out before with SOF raids. But this one was different. It went on and on, and it took something out of the men even in the temperate late-winter air. In the blazing heat of Afghan summer, it would be incredibly draining.

For hours, the men worked their way through claustrophobic grape fields. The Pashtun farmers in Zharay didn't use trellises. Instead, to support the vines, they built row after row of head-high mud walls. Every field became an intricate waffle iron, a packed dirt and foliage maze with but a few entrances and exits. The Taliban knew it and aimed to catch the Americans in the narrow corridors between the mud walls.

They put IEDs on the obvious turn points and exits, and then, out of spite, put them at random spots too. The ground played you false every time. It was like Chutes and Ladders with live bullets. "They could see eighteen inches to their left and right, and rarely more than fifty feet to their front or rear," McChrystal later said. "Above, only a slice of sky." Nothing happened that day. It usually didn't. But when it did . . .

Afterward, the general sat and talked with the soldiers. Like Karzai visiting the citizens of Marjah, Stan McChrystal had both the physical courage to go out on the ground and the moral courage to try to explain himself to the skeptical riflemen. They still didn't buy the new tactical directive, but they respected the general for coming to their outpost and walking with them. Arroyo thanked the general in person and then followed up and thanked him again by e-mail, a sentiment echoed in another e-mail sent to him by Corporal Michael Ingram, one of the squad's aggressive fire-team leaders.

On April 17, a Taliban IED killed Mike Ingram.

A lesser man that Stan McChrystal would have sent a sympathy note and avoided a wrenching session with Arroyo and his soldiers. But Stan McChrystal knew he had to see these riflemen again. One of the hardest things for any commander is losing soldiers, and this loss hurt more than most. McChrystal took it hard. His second meeting with the squad, on April 28, became tense and emotional. The general listened more than he talked.

The men mentioned that their battalion had overinterpreted and amplified the ISAF tactical directive, making it even more restrictive. One 1-12 Infantry order directed squads and platoons to "patrol only in areas that you are reasonably certain that you will not have to defend yourself with lethal force." Which IED-dotted grape field would that be? Which hateful Pashtun village? Given such strictures, one soldier asked McChrystal what they were doing in Zharay. Another emphasized that he wanted to use preemptive fires. Better them than us, the private said. McChrystal warned the young men of the moral burden that comes with killing — even when the deaths are justified — and ended quietly with "don't get cynical." The troops listened, but their body language showed they did not, and likely could not, agree. The meeting ended in sadness.

To the general's credit, these events did make him think. With this strategy of tactics, reducing the counterinsurgency to prevent civilian casualties, McChrystal had chosen an exceptionally difficult course. Men like Kielin Dunn and Michael Ingram paid for that decision with their lives. Because he truly cared for his people, Stan McChrystal paid in his way too, carrying a horrific weight that only those at his level truly understood. Staying steady under pressure, strong in sorrow, defines a commander. Though he definitely second-guessed himself, McChrystal did not modify the ISAF tactical directive. By McChrystal's reckoning, it worked in Marjah. Kandahar was next. Maybe success there might make Ingram's death count for something.

The poignancy of Stan McChrystal's gathering with Israel Arroyo's riflemen featured one jarring aspect. Among those in the small room at the platoon compound was Michael Hastings, a reporter for *Rolling Stone*. It didn't seem right to have a journalist present for such a difficult personal discussion. But there he was.

Hastings was there with the general's full agreement. The journalist had come warmly recommended by McChrystal's personal press assistant Duncan Boothby, himself a well-connected civilian sent over by the ISAF commander's media-conscious West Point classmate and NTM-A commander, Lieutenant General Bill Caldwell. Well, McChrystal didn't know much about dealing with journalists, as his misadventures the previous fall had shown. You didn't worry about such things in the SOF arena. McChrystal trusted Caldwell, though, and learned to value Boothby's instincts.

Boothby saw a lot of good in Michael Hastings. The writer's brother had served as an infantry officer in the Tenth Mountain Division. Hastings understood the military, having been embedded with line units in both Iraq and Afghanistan. His fiancée had died in a Baghdad ambush, a personal tragedy that sharpened Hastings's fervent interest in the war. Perhaps emboldened by the media avalanche that accompanied, and by and large ably reported, the Marjah mission, he thought an in-depth article about the ISAF commander made sense. In addition, the idea of granting access to a writer from *Rolling Stone* appealed to McChrystal's creative, unorthodox streak. Maybe he would make the cover of *Rolling Stone*, an unusual place for a general. Well, it was an unusual war.

Given what later transpired, many questioned the wisdom of granting continuous close contact to the representative of a publication better known for covering the likes of Snoop Dogg, Lady Gaga, and Bono. But McChrystal read widely and recalled that *Rolling Stone* also regularly published excellent political and cultural commentary. Hunter S. Thompson, Joe Klein, and William Greider had all covered elections and domestic politics for the magazine. Tom Wolfe's colorful history of the early space program, *The Right Stuff*, started as a series in *Rolling Stone*. Michael Hastings seemed to be part of that tradition.

Hastings followed the ISAF commander in April and May of 2010, even accompanying him on a trip to Paris and Berlin. McChrystal and his personal aides didn't get the usual fifteen-day rest-and-recreation leave given to every U.S. soldier who'd been in country for a year, so they tried to relax a little between meetings in Europe. McChrystal's wife, Annie, flew over and joined him in Paris to mark their thirty-third wedding anniversary. She had seen her husband fewer than thirty days a year since 2003. In their off-hours, such as they were, McChrystal and his people drank and talked freely. Volcanic dust from an eruption in Iceland delayed air travel across Europe, prolonging the trip for a few days. Hastings took a lot of notes.

Most journalists understand that with full access to any senior figure comes the opportunity to hear loose talk. The military tells commanders that the press covers what's said under several categories. *On the record* means that if you say it, they print it. *On background* allows a reporter to quote you as "a senior military official" or "a senior commander." *Off the record* means they don't print it word for word, don't attribute it at all, but they gather the information. But the experts always warn you: Treat it all as on the record. Watch what you say.

Duncan Boothby never quite pinned down the rules for Michael Hastings. Stan McChrystal and his team thought their personal off-duty banter to be off the record. They figured that any quotations used had to come from the various official meetings that Hastings observed, including events like the blunt discussion at Combat Outpost JFM. Hastings decided otherwise. He personally liked McChrystal, found him impressive, and wanted to let him and his men speak for themselves. Most journalists would not have run with many of the quotations. But some would, and Hastings was one of them. In World War II, such warts-and-all reporting explained how General Joseph Stilwell became known as "Vinegar Joe" and why Dwight Eisenhower quailed almost every time George Patton spoke to reporters. The tolerance was higher back then with the world at war. And neither Stilwell nor Patton had offered thoughts about his commander in chief.

Hastings's article "The Runaway General" appeared in the June 22 edition of *Rolling Stone*. Hastings characterized McChrystal as "a snake-eating rebel." The writer offered a compelling description of the foot patrol and both meetings with Staff Sergeant Israel Arroyo's rifle squad, gave good background on Stan McChrystal's long service and ascetic personal habits, and commented on the general's well-meaning attempts to limit civilian casualties. Nobody paid attention to those parts.

Instead, the trenchant, inflammatory pull quotes made the news. McChrystal said that Ambassador Karl Eikenberry "betrayed" him and "covers his flank for the history books." Ambassador Dick Holbrooke was "a wounded animal." National security adviser and former Marine four-star James Jones was a "clown" and "stuck in 1985." President Hamid Karzai was "locked up in his palace the past year." Senators John Kerry and John McCain were "not very helpful." When Vice President Joe Biden's name came up, McChrystal asked, "Who's that?" An assistant replied, "Biden? Did you say: bite me?" Finally, the general called President Obama's 2009 strategic review "painful," a process that left him "selling an unsellable position." He thought that the president was "uncomfortable and intimidated" dealing with the military leadership. Some words came from the ISAF commander's own mouth. Most came from his staff. But the damage was done.

The published profile cold-cocked Stan McChrystal. "How in the world could that story have been a problem?" he thought at the time. The ISAF commander expected a gonzo portrait, sure—it was *Rolling Stone*, after all—depicting him out on patrol, consulting with NATO officials, dealing with Hamid Karzai, drawing up plans for combat operations, and so on. He was appalled to see word-for-word transcriptions of long-forgotten asides and trash talk. Stan McChrystal knew the old warning: Play with media fire, get burned. Hastings had charred McChrystal to a cinder.

In retrospect, Stan McChrystal faltered because he tried to be something he wasn't. The success of Dave Petraeus inspired many, including McChrystal, to try it themselves. Yet the Petraeus touch with the press reflected a lifetime of calculated preparation. You couldn't just show up after decades in the company of rough-hewn riflemen, let alone years in the shadows, and ride the tiger that was the contemporary American news media, not without a hell of a lot of homework, a generous scoop of good luck, and a nose for the media near ambush. Brains, bravery, devotion to duty, and love of soldiers weren't enough. This explained why most soldiers, especially the senior ones, the George Casey and Dave McKiernan types, dealt with reporters only when they had to, said little, and moved on. You never really won with the press. If you stuck to your talking points, you didn't make much news and broke even. But if you started yapping, well . . . Perhaps McChrystal should have read more closely the admonition offered by that other *Rolling Stone* contributor Tom Wolfe:

> Every career military officer, and especially every junior officer, knew that when it came to publicity, there was only one way to play it: with a salute stapled to your forehead. To let yourself be turned into a *personality,* to become *colorful,* to be portrayed as an egotist or rake-hell, was only asking for grief, as many people, including General George Patton, had learned.

The ISAF commander immediately apologized, but that hardly quelled the ongoing media uproar in Washington. He well understood that this was the end. When summoned back to Washington by the president, McChrystal offered his resignation. Obama accepted it. First McKiernan, now McChrystal — Afghanistan was chewing through generals, all right.

With McChrystal gone and the Afghan campaign in a critical state as the first surge brigades arrived, only one man would do: General David H. Petraeus, architect of the 2007 Iraq turnaround, commander of U.S. Central Command, counterinsurgency guru, the first military man since Colin Powell whose name meant something to the average American. On the early afternoon of June 23, the president walked into the sunshine of the White House Rose Garden and introduced Petraeus as McChrystal's successor. "He is setting an extraordinary example of service and patriotism by assuming this difficult post," said Obama. Resplendent in his dress greens bearing numerous bright badges and row after row of ribbons, standing erect and pensive next to the president, Petraeus looked like a coiled spring, born ready for this moment.

Certain commentators referred to his new position as a demotion from overall theater command to running the Afghan war. They missed the larger point. With the good war hanging in the balance, America knew exactly whom to call. Like the song said, our nation turned its lonely eyes to him.

16

MALIK DAOUD AGAIN

Brilliant: they would still say he was brilliant,
inventive, tireless. No topographic wrinkle had not
been examined, no eventuality had not been explored.
Nothing ever changes, he thought bitterly.

— ANTON MYRER, *ONCE AN EAGLE*

THE NORMANDY HEDGEROWS chewed up U.S. paratroopers in the summer of 1944. Over and over again, the men of the 502nd Parachute Infantry Regiment moved out in the early-morning light, the dripping green trees just above their helmeted heads as they pushed slowly forward at a crouch, rifles ready. Tall dirt walls topped by trees and bushes loomed on either side. You couldn't drag a dogcart down the narrow pathway. Maybe this morning, it would be different.

It never was.

A few hundred steps down the dirt trail, after the tired, dirty riflemen got well spread out, one of the lead men tripped a mine. It happened so regularly each day you could almost set your watch by it. The sharp blast might be an S-mine, the size of a coffee can, spewing up a mortar bomb that burst at waist level, the dreaded bouncing Betty, ripping open a soldier's entrails, shredding his genitals, and slicing both femoral arteries in a welter of blood and gristle. Men rarely survived that. But sometimes, a smaller, sharper, muffled explosion signaled the nasty little Schu-42, which could slice off a booted foot as neatly as a butcher knife.

The Germans used a wooden box for the Schu mine. The U.S. Army–issue magnetic mine detectors never sniffed it. Whatever the cause, now a paratrooper lay sprawled in the hedgerow lane. His fellows hesitated a second. Hidden German defenders did not.

Instantly came the loud, urgent, metallic snort of the MG-42 machine gun. Men immediately scattered to either side of the dirt wall, dropping prone, cheeks pressed into the warm, dark, firm French soil. But the Germans had trained their gunners well. They shot low, skipping the 7.92 mm bullets right along the surface of the trail, tearing holes in prostrate men. You'd hear the wet smacks as the bullets hit. The Americans fired back, banging away blindly at the next line of low shrubs at the base of the berm, more hoping than aiming.

Then came the German mortars. You could hear the rounds slide into the tubes, like the noises made by plumbers slipping smaller steel pipes into larger ones, smooth sounds, tight fits. Next came the *bloop* of the rounds firing, arcing up, tipping over, then — *bam, bam, bam,* one after another, pounding the men on the path. Hot fragments shredded legs, arms, and torsos. The groaning rose like a chorus.

It went that way day after day, in what one veteran called "the Gethsemane of the hedgerows." The 502nd Parachute Infantry Regiment never ran out of courage that summer in Normandy. Half of the regiment fell dead or wounded. The colonel lost his mind. Tough, aggressive, ingenious paratroopers figured out how to crack the earthen walls with demolitions and use the black of night to creep right up to the Germans. They used bullets, bayonets, and grenades, plus hundreds of bombs and thousands of artillery shells. Eventually, the Germans ran out of hedgerows and *landsers*.

The deadly checkerboard thickets of the 1944 Normandy *bocage* had nothing on the lethal agricultural puzzle box that comprised the Arghandab River Valley of Afghanistan in the hot summer of 2010. As outside Carentan and Saint-Lô, each Arghandab field featured a defining packed high dirt wall several feet thick. At the corner of every plot stood a *qalat,* a small cluster of one to four one-story tan adobe huts enclosed by a dried mud wall. But whereas the interior of the Norman rectangles were usually plowed furrows or fallow pastures with nothing growing much above ankle height, many Arghandab fields hosted rows of chin-high dirt berms festooned with grapevines, and the walls rarely stood more than two feet apart. About a third of the Afghan land instead featured long, orderly lines of gnarled fifteen-foot pomegranate trees. Some farmers also grew corn to feed their cattle, sheep, and goats. Hot summer sun and centuries of irrigation from the Arghandab River made for fulsome crops. In June of 2010, the greenery ran riot. So did the Taliban.

Into this verdant hellish tangle came the Second Brigade Combat Team, 101st Airborne Division, the current incarnation of the 502nd Parachute Infantry Regiment that had fought and bled in Normandy. Colonel Arthur A. Kandarian knew well the heritage of the 502nd, and he embraced it. A forceful, intense commander, commissioned through ROTC at Washington and Lee University, Art Kandarian had earned the Silver Star with the Rangers in Iraq during the 2003 invasion and commanded a cavalry squadron in the troubled Diyala Province in 2005–06 Iraq. He had never served in Afghanistan, but he knew all about the 1944 campaign in the Normandy hedgerows. These Arghandab grape rows looked all too similar.

When Kandarian thought about the 502nd's 1944 experience, he found it hard to understand how the *bocage* terrain — and its skillful German defense — could have come as such a hideous surprise to the paratroopers. All the aerial photographs, French Maquis (Resistance) reporting, and lengthy terrain studies must have somehow missed the significance of the endless succession of small, ridged farm fields, each a natural killing ground available to an entrenched foe. In fact, the existence of these killing fields had been well known for two millennia. In his account of the Roman offensive in Gaul from 58 to 50 B.C., Julius Caesar noted that the *bocage* country presented "a fortification like a wall through which it was not only impossible to enter but even to penetrate with the eye." Add in Schu mines, MG-42s, mortars, and an occasional Panther tank, and it all proved horrendous. Yet the Americans found out about this meat grinder only when they entered it.

Art Kandarian knew better in his war. American and Canadian riflemen had been operating in the Arghandab for years. The Taliban had no Panther tanks, nor did they possess a tenth of the Germans' discipline and tactical skill. But the Pashtun opponents owned plenty of IEDs, PKM machine guns, and mortars, and in addition, they had the support of the populace. Norman villagers had welcomed and aided their American liberators. The unhappy people of the Arghandab neither trusted nor helped the ISAF infidels. They sided with their Taliban cousins, the devil they knew. Secure among fifty-five thousand villagers in the various hamlets dotted up and down the lush river valley, the Taliban had long considered the Arghandab their favorite base area for attacks into nearby Kandahar city.

Until 2010, ISAF had never sent in enough forces to change that calculus. One Canadian battalion, and then a single U.S. battalion, struggled in this hostile medium. When General Stan McChrystal proposed the Afghan surge with a focus on Kandahar, he wanted to commit an entire brigade combat team to clear the Arghandab. Once President Obama agreed to the surge, the orders went to Art Kandarian's Second BCT, 101st Airborne Division.

Kandarian's BCT epitomized the experience of most surge units. Originally tagged for an Iraq deployment, Kandarian's air assault soldiers represented the physical manifestation of Obama's strategic shift to Afghanistan. As McChrystal directed, the surge allowed ISAF to build on initial success in Helmand Province and move in strength to secure the Kandahar population. McChrystal departed as Kandarian's soldiers arrived. But the plan remained intact.

Mindful of the 1944 Normandy experience, careful in his predeployment reconnaissance, poring through all available information, Art Kandarian and his team sought every possible edge to meet the challenge of the IED-encrusted Arghandab grape rows and *qalats*. They appealed to the U.S. Army authorities for some key equipment. Not all of it arrived quickly, and none of it came in the numbers needed. But enough of it showed up to make a difference.

In a perfect world, once President Obama announced the Afghan surge, the institutional U.S. Army should have looked at the calendar, checked a map, and force-issued the right sets of gear. In the real world, it took Colonel Art Kandarian and Command Sergeant Major Alonzo Smith twisting the arm of every general and senior sergeant major up to and including General George Casey, the Army chief of staff, and Kenneth Preston, sergeant major of the Army. They also involved General Dave Petraeus, the new ISAF commander and former 101st Airborne Division commander, and Command Sergeant Major Marvin Hill, the new ISAF senior NCO who had served in the same capacity in the 101st Airborne Division in Iraq from 2003 to 2004. Petraeus and Hill ensured that their staff followed up on all such requests but definitely took a personal interest in anything bubbling up from the Screaming Eagles.

As air assault infantry, the Second BCT trained with helicopters regularly and planned to fly into action when they could in Afghanistan. But however they got there, once on the ground, when not on foot, the brigade needed armored trucks. Art Kandarian wanted the latest version, the sturdy mine-resistant ambush-protected (MRAP) vehicles. Thanks to the personal intervention of Secretary of Defense Robert Gates, he got them.

To be more specific, the Second BCT sought the sixteen-ton MRAP–all-terrain vehicle (M-ATV), a lighter version of the greater family of armored trucks. M-ATVs were not as top-heavy as the larger, older kinds of MRAPs, which had an unfortunate tendency to roll over when hit or when bouncing along the edges of rough, narrow roads. The broad-beam, stable, sand-colored M-ATVs provided mobile firepower with their turret-mounted M-2HB .50-caliber heavy machine guns; they had powerful radios and computers for command and control and were sheltered transportation for soldiers and critical supplies. The extra protection came because, like all MRAPs, the M-ATVs incorporated a V-shaped hull that deflected IED blasts away from the cabin crew. The South Africans had developed this key anti-mine feature after years of contested mounted patrols in Namibia during the apartheid era. The loathsome white-minority government eventually ended, but the South African arms industry kept what it had learned and continued to manufacture the world's best armored trucks for defeating mines. Urged by Defense Secretary Gates, BAE (British Aerospace), General Dynamics, Navistar, Oshkosh, and other companies adopted licensed versions of the South African V-hull. All MRAPs, including the M-ATV, benefited from it. Flat-bottomed up-armored Humvees, Stryker wheeled armored personnel carriers, and even hulking M-1 Abrams tanks lacked this critical design advantage. On the IED-infested dirt roads of the Arghandab, Kandarian's soldiers very much needed those V-shaped hulls.

Kandarian also sought conventional minefield-breaching equipment not much used since the initial offensive into Iraq back in 2003. His Second BCT logisticians scrounged through various storage depots in

Kuwait and found six trailer-mounted mine-clearing line charge (MICLIC) systems parked in a row. Nobody had asked about them for years. The MICLIC launcher shot a rocket out a hundred yards. The projectile dragged a thick rope that snaked back to the trailer. After the thing landed, it looked like a meandering electrical cable. But the cord was actually packed with C-4 plastic-explosive filler. When blown, a MICLIC sympathetically detonated all the mines in the vicinity, scouring a swath eight yards wide out to the full range of a hundred yards. That made a lane wide enough for an M-ATV. A smaller version, the anti-personnel obstacle-breaching system (APOBS), came in a backpack. Its line charge created a lane one yard across and forty-five yards long, good for a foot patrol. Art Kandarian and his engineers thought these munitions, built to breach standard military minefields, promised good results against Taliban IED belts. Such brute-force measures beat the usual drill. In the typical practice, soldiers — too often exposed riflemen — spent hours probing painstakingly for hidden booby traps and placing chunks of plastic explosives on each located item. That was when all went well. When it didn't, soldiers guessed wrong and went up in an awful flash and bang. The MICLICs and APOBS offered a much better approach, albeit one unkind to local structures, crops, and livestock, let alone curious onlookers. Laboring under the intermittent supervision of the risk-averse Major General Nick Carter at RC-South, Kandarian wouldn't ask permission to employ these powerful weapons. He'd just use them.

Other capabilities helped in the counter-IED fight. The Joint IED Defeat Organization (JIEDDO), reporting directly to Secretary of Defense Bob Gates and headed by an energetic former U.S. Army division commander, delivered some very useful new items. The Vallon Minehound used both metal detection and hard-object sensing to find buried items. Had Minehounds been available in 1944, they could have located even wooden Schu mines. The Minehound promised to find all types of Taliban IEDs. Other mine detectors, including the Gizmo, Gadget, and Goldie series, also contributed. None were foolproof — all gave a lot of false readings. But they beat tiptoeing and hope.

In addition, JIEDDO used its high-level access to overrule objections from the U.S. Army military police leadership and deploy a number of contracted off-leash IED-detector dogs. The MPs believed in a well-established, step-by-step training regimen that produced outstanding soldier-and-dog teams, but there were never enough of them. JIEDDO and the Marines had developed a much shorter course that had proved good enough in Helmand. Art Kandarian asked for and got his share of the off-leash detector dogs. Eventually, some MP working dogs also showed up. By then, of course, the JIEDDO contract dogs had long proven their worth.

These qualitative improvements depended on enough soldiers to use them. Art Kandarian did the math. To have a realistic chance to control the sprawling, Taliban-held districts west of Kandahar, Zharay and parts of Maiwand and Panjwai as well as Arghandab, Second BCT needed at least twelve terrain-holding battalions. Kandarian owned three maneuver battalions: the First and Second Battalions, 502nd Infantry (each made up of 665 soldiers), and the First Squadron, Seventy-Fifth Cavalry (305 soldiers). Two more maneuver battalions were to be attached from neighboring U.S. brigades already in country. To get a sixth U.S. battalion, Kandarian reconfigured his 307-man First Battalion, 320th Field Artillery to fight as infantry. Another battery from another BCT would provide 155 mm howitzer fire when needed. It wasn't ideal, but the Second BCT trusted its gunners to fight as riflemen.

The other six battalions came from the Afghan army. Thanks to the consistent churn of new units generated by Lieutenant General Bill Caldwell's NATO Training Mission–Afghanistan, all came manned, equipped, and trained. External critics complained about various deficiencies, including desertions, shortages of heavy weapons, and lack of skilled sergeants. Those problems were very real, but by any measure, the ANA brought enough to contribute. Three of the battalions — *kandaks*, in the Afghan vernacular — had fought earlier in Helmand during the Marjah operation in February and March. Each ANA *kandak* paired with a U.S. Army counterpart. These thousands of Afghan troops guaranteed a far different complexion to the 2010 campaign west of Kandahar, an ANA surge to parallel the American reinforcement.

Once Second BCT and their attached Afghans cleared the districts west of Kandahar city, the ANA soldiers planned to hold it. Afghan national police would follow. As for the build phase, nobody knew when a degree of local government might arrive, a Kandahar equivalent of that overhyped government-in-a-box promised — and indeed overpromised — in Marjah. Local strongman Ahmed Wali Karzai looked unlikely to support such developments unless they included significant lining for his deep pockets. For Kandarian and his soldiers, clearing and holding promised to be hard enough.

Weapons and dogs helped, converted cannoneers and newly arrived Afghans added necessary combat power, but the Second BCT integrated all of them because Kandarian and Smith insisted on training their soldiers for the hot, draining, close-quarters war they faced in the perilous greenery and cramped *qalats* west of Kandahar. Given the prevailing McChrystal rubric regarding prevention of civilian casualties and the even more constricting attitude of Nick Carter at RC-South, some colonels and sergeants major might have assiduously schooled their soldiers in the niceties of well-digging, courageous restraint, and avoiding trouble. That wasn't going to work for Second BCT, bound for a region teeming with all kinds of trouble. The air assault soldiers prepared accordingly. The brigade emphasized physical fitness, marksmanship, first aid, and land navigation. Soldiers ran endless live-fire battle drills that included the use of demolitions to blow openings through earthen walls. They also practiced using mortars, artillery, attack helicopters, and airstrikes.

In their tactical training, the BCT soldiers took to heart some shrewd advice from retired Master Sergeant Rob Pittman, an Asymmetric Warfare Group contractor teaching Special Forces methods to the conventional forces. After studying the grape rows and fruit-tree orchards, Pittman figured out that by-the-book fire and maneuver seemed likely to get hung up on the adversary's well-crafted IED clusters and physical obstacles. Closing on the hidden Pashtun guerrillas would be as tough as approaching those German machine gunners in the Norman hedgerows of 1944. There would always be a dried-mud barrier or a waist-deep irrigation canal between the enemy and the Americans. Pittman advocated not the classic close-and-finish fight but rather a conscious "attack by fire" from two small units. He told the first soldiers engaged to return frontal fires to pin the enemy then quickly and boldly move a fire team or squad to get around to the enemy flank, about ninety degrees offset, if possible. Pittman argued for always seeking the harder path, not taking the obvious, inviting trail, which was invariably filled with IEDs. The maneuver team needed to plan on belly-rolling across the top of skin-scraping, clothes-tearing grapevine walls, and plunging through the grasping, spiky lower branches of the pomegranate trees, getting only as close as necessary to see and hit their antagonists. Closing to ideal positions usually meant stepping on death. A second- or third-best position would do. Once in place, the two American units could then deliver a high volume of well-aimed shots to kill the foe, who did not maneuver all that well, especially under fire. "If you can get them with interlocking fires," the former SF NCO said, "they won't know whether to shit or go blind." So taught, Second BCT small units in contact maneuvered rapidly, and not in the ways the Taliban anticipated.

Such tactics put a premium on officer and NCO leadership. In congested terrain, the chain of command had to be out front and hands-on, not sitting in some air-conditioned bunker watching a computer screen. Kandarian and Smith set the example themselves and encouraged their key leaders to do the same. They expected this from the senior people, not just the lieutenants and staff sergeants. It took a toll, but it also ensured that the BCT continued to press the fight when the Taliban inflicted losses.

The Battery B lieutenant wanted out. So did some of his men. They'd gone hunting in the Arghandab. In their view, they had been through more than enough. They didn't believe they had much left to give.

Two days of hard walking, scrambling over grapevine berms, and sliding into slimy canals, all of it in the draining heat, had used them up. The men hit no IEDs but had checked every foot as they slowly advanced. They took the difficult route, across the grain of the grape rows. The constricting body armor, sloshing water, dead-weight ammunition, and other stuff wore them down. Every piece of the seventy pounds of combat gear, great and small, weighed on them, dulled their reactions, and sapped their strength. They moved in a daze. When they halted, they did not want to get back up.

The seven-hundred-yard march on July 11 proved particularly galling. Occasional potshots made the artillerymen jump for cover, craning their necks over the grape rows and under the shrubs in search of their tormentors. They never saw the enemy, but they heard the distant muzzle cracks and then saw the bullets zipping overhead. Opportunities for the kind of bold flanking maneuver that Rob Pittman had suggested back in training never arose. Instead, hour after hour, the men plodded on. They sweated so much behind their ballistic sunglasses that the wide black lenses became streaked with dried salt, a crazy-quilt curtain gradually swirling into a hot, dim gray-green tunnel.

In the early afternoon, the numbed column finally reached an abandoned, damaged *qalat* with a few single-story flat-roofed hovels and the usual hard-packed mud-wall courtyard. The light brown adobe compound smelled of animal waste and moldy straw. It was a key location, overlooking a road junction that dominated an important canal crossing and connected the towns of Jelawur to the northwest and Khosrow Sofla to the south. All the water for cultivation, as well as all the foot and vehicular traffic crossing through the western Arghandab, passed through this spot. The 1-320 Field Artillery intended to control the valuable junction point, but first they had to see if it measured up to its advertising. Hence, Battery B's patrol to check it out; they would spend the night and then go back to Jelawur to prepare for a more permanent stay next time. But this reconnaissance in force came first. Twenty exhausted men, five of them near collapse from the unforgiving heat, amounted to a rather tenuous initial American footprint.

The heat exhaustion cases needed treatment. With his defenses set in the baked-mud enclosure, the platoon leader called for a medevac helicopter. Within minutes, the HH-60 came from nearby Kandahar Airfield. As the Black Hawk hovered to land next to the farmstead, the tree line across the road, conveniently protected by a chest-deep canal, erupted with gunfire. The helicopter aborted, climbing away to circle. The tired artillerymen opened fire. Their machine guns and rifles did not quiet the Taliban. Elated to see the HH-60 peel off, the antagonists continued to fire away.

Artillerymen excel at bringing in supporting arms. After some radio calls, two AH-64D Apache attack helicopters showed up. Directed by the small U.S. platoon, the lead Apache tilted downward and unleashed a brace of 30 mm cannon shells that punched holes through the trees. Minutes later, the other Apache followed suit. That did it for the opposition. The medevac got in and safely pulled out the stricken soldiers.

As the medevac bird angled up and away, the lieutenant contacted the 1-320 Field Artillery command post. He requested immediate heliborne extraction. He didn't want to walk out due to the heat and nearby hostiles, not to mention the numerous suspected IED belts. As for the planned overnight stay, the lieutenant thought his bone-weary fifteen men lacked the combat power, not to mention the willpower, to hold if the Taliban swarmed them after sunset.

The platoon leader's call came at an awkward time. To the west, a Battery A platoon traded bullets in a sharp clash; they took three wounded. That platoon held its position, killed some Taliban, evacuated its casualties, and made no noises about withdrawal. True to the Second BCT lead-from-the-front ethos, the acting brigade sergeant major was on the ground with the men of Battery A, joining in the firefight. This lieutenant in Battery B, though, sounded shook up. And the battalion very much wanted to keep that vital ground.

Certain commanders would have pulled out the shrunken, bedraggled platoon. Other battalions might have issued a harsh radio order to stay put and called it done. This battalion took a more positive course. True to their air assault heritage, 1-320 Field Artillery reinforced by helicopter, air-assaulting Captain Elijah Ward, the Battery B commander, and a second platoon to strengthen the makeshift patrol base. They brought in more water, greatly needed in the horrific heat. In addition, Lieutenant Colonel David Flynn himself, accompanied by his security team, flew in with the fresh troops. Army tradition says that the commander goes to the point where he can make a difference. Flynn judged that Battery A had their fight in hand — not so Battery B.

The son of a U.S. Army sergeant major, an ROTC graduate from Clemson University, Flynn knew the area. In 2004, when he was a major, his howitzer battalion worked out of Kandahar Airfield in the provisional infantry role. The Arghandab had been a backwater then, a pleasant green diversion from the red dust of the Registan Desert. Now the area sustained at least two hundred insurgents, according to the intelligence staffers back at RC-South. But they were using 2006 estimates. The ANA partners put the number at about five hundred and believed the Arghandab had been under Taliban control since late 2008. Flynn agreed with the Afghans. Leaving fifteen shaken men out all night in the middle of those predators risked a terminal result.

So Flynn went there in person to look in his soldiers' eyes and judge their tone of voice. While Captain Ward and the new platoon, buttressed by the battalion commander's team, established security on the compound perimeter, Flynn got with the lieutenant and his NCOs. General

George Patton wrote from long experience: "There are more tired division commanders than there are tired divisions. Tired officers are always pessimists." The lieutenant was tired. His sergeants were not. They wanted to continue the mission.

A later official document put it clinically, in no-nonsense U.S. Army prose. The battalion commander "immediately isolated the fainthearted, eradicated dissent, and eliminated the defeatist attitude." Flynn "empowered junior non-commissioned officers" and "charged the previously dejected leaders to hold the position." Dave Flynn was not a screamer, nor a rah-rah chest-beater. But he understood soldiers. He made clear what he expected. Then he did something else, the act that cemented whatever words passed in the hot, stinking *qalat* courtyard that late July afternoon.

After dark, with the two Battery B platoons tied in for the night, Flynn and his eight security soldiers slipped out of the compound and walked a hundred and fifty yards to the northeast, to the canal and tree line where the Taliban had engaged the medevac that afternoon. Flynn went very slowly, mine detectors waving across the dark ground. Over several hours, the short column picked its uneven, tortuous way. It took a while to snake across the watercourse. The soldiers set up a covering position.

At first light, overwatched by Flynn and his men, Battery B walked out and returned to their launch point seven hundred yards to the northwest at the combat outpost in Jelawur village. Captain Elijah Ward led his gunners on the deliberate movement. "The moment crystallized for me," he said, "the toughness of the enemy, the challenges of the climate and terrain, but also the confidence that our soldiers were prepared and that the enemy could be beaten on his own ground." Lieutenant Colonel Dave Flynn brought up the rear, covering the Battery B elements until they reached the relative safety of the fortified Jelawur outpost. Every man knew he'd be back.

As Ward's Battery B methodically pushed north, Battery A to the west also repositioned. They had a rougher time. Sensing a buried IED, the point man halted the movement. When Staff Sergeant Kyle Malin slipped forward to take a look, he stepped on a booby trap, set it off, and fell, severely wounded. The blast hit three others. During the subsequent air medevac, another IED blew under Private First Class Corey

Kent, badly hurting him and less seriously injuring two more. The Taliban chose to watch and learn. *Enemy advances, we retreat . . .* for now.

When Lieutenant Colonel Dave Flynn returned two weeks later, he brought friends—lots of them. Pashtun Lieutenant Colonel Mangal Andar and his First Kandak, First Brigade, 205th ANA Corps brought more than four hundred Afghan *askari* to the fight. Mangal came from eastern Afghanistan, near Ghazni. Most of his Third Company hailed from Jalalabad in the eastern Pashtun region. The *kandak* also included Tajiks, Hazara, and Uzbeks, a typical mix for an ANA organization. By design, the Afghan Ministry of Defense and the NTM-A advisers mixed the various ethnic and provincial peoples. Orders were issued in Dari, but much day-to-day business went on in Pashto. Mangal and many of his leaders, as well as a lot of his riflemen, could talk to the villagers in the Arghandab.

Guerrillas normally pull back in the face of force, but the Taliban in the Arghandab dared not abandon the crucial road/canal junction between Jelawur and Khosrow Sofla. The enemy interpreted the U.S. advance and pullout of July 11 and 12 as a close call, more of the customary ineffectual infidel activity, the old in-and-out they'd seen dozens of times in the past. The Americans, and the Canadians before them, cleared but seldom held. Even the Soviets never controlled this area, although they'd swept through it now and then and bombed it repeatedly. The Taliban-dominated Ghilzai clans—Mullah Omar's people—oppressed the Alkozai tribal elders who traditionally ran the Arghandab. Along with confiscating buildings for arms and insurgent rest sites, the Taliban extracted money, food, labor, and information at will, Mafia-style. It was a nice setup for the enemy, and one they'd fight to keep.

Before dawn on July 30, the Americans and Afghans attacked. While Battery A operated to the west and Battery B held the key outposts in and around Jelawur, Headquarters Battery—reorganized and fighting in two provisional rifle platoons—and the Afghan Third and Fourth Companies infiltrated south under cover of the last hours of darkness. They planned to retake the critical compound, build it into a combined U.S./Afghan outpost, and hold it against all comers. Flynn, Mangal, and their men expected trouble. They found it in spades.

The American First Platoon and a platoon from the Afghan Third Company led, crawling over grapevine berms and slithering between tangled fruit trees, taking the long, hard way. It was slow but avoided the damn IEDs on the paths. Sergeant First Class Kyle Lyon, by training an artillery meteorological survey team leader, was acting as the U.S. platoon leader. Aware that the enemy had no doubt wired the objective *qalat* with IEDs, aimed to block the obvious northern entrance, Lyon took the combined American/Afghan column all the way around to the south. They spread out along a low wall beneath a grove of waving pomegranate trees as the dawn broke. The adobe farmstead stood before them.

Behind Lyon's soldiers, the U.S. Second Platoon and another Afghan element from Third Company moved across their own set of grape rows and orchards. They intended to establish an overwatching position on the west side of the farm buildings, providing that ninety-degree offset to ensure the interlocking fires taught by Rob Pittman in so many training exercises. As the sky brightened, the Americans and Afghans crossed the canal near the target compound. One squad from Second Platoon cleared a path to bring up armored trucks. Two big M-ATVs trundled in their wake. Packed full of ammunition and water, one vehicle mounted a .50-caliber machine gun and the other had a Mark-19 40 mm automatic grenade launcher. Along with fire support and powerful radios, the M-ATVs allowed for casualty evacuation if that became necessary. It soon did.

As the light got better, an inquisitive ANA *askar* found the top of a small IED poking up on the side of the road. When Specialist Michael Stansberry moved forward to investigate, he stepped on a pressure plate. The ground erupted. Sergeant Kyle Stout raced to his aid but tripped another IED as he did so. Both men were badly wounded, ripped up and bleeding. Eight others were cut up and stunned as well. Unseen Taliban began firing AK-47s and PKM machine guns.

Meanwhile, on the other side of the *qalat,* Sergeant First Class Kyle Lyon and his people began to take fire from mud-brick structures two hundred yards to the south. Staying low under the fusillade of AK rounds, most, thankfully, zinging high, Lyon scrambled over the wall. He grabbed an American with an M-203 grenade launcher and pulled him a few yards down into a better position. His hand on the soldier's shoulder, Lyon pointed to the source of enemy resistance, and told the grenadier to mark it with smoke once the Apache attack helicopters clattered in overhead. When the helicopters appeared ten minutes later, Lyon nodded. The M-203 man began to plop red smoke rounds onto the enemy-held farm huts.

The two Apaches took turns. The first shot a Hellfire guided missile, and the second followed up with 30 mm cannon shells. Then the first aircraft wheeled around and administered another dose of explosive 30 mm rounds. The attack aviators did not use rockets. Though deadly, the Hydra-70s, upgrades of the 2.75-inch models used in Vietnam, had a tendency to go wide at times. The Americans and ANA *askari* were too close for such blunt instruments.

The Apache gun runs got the Taliban to back off for a few hours. Lyon and his platoon got into the compound, as did Second Platoon. The Vallon detectors came in very handy. The artillerymen had to work around booby traps buried in every doorway and even a few squirreled into the courtyard walls. Engineers and explosive-ordnance teams dealt with those by using C-4 plastic explosives. The Afghans joined their American counterparts. Lieutenant Colonel Dave Flynn and his tactical command post came too. The two armored trucks, turret weapons at the ready, positioned themselves on the trail just north of the dirt corral barriers. A succession of booms marked the steady demolition work of the engineers.

Just after noon, as the hot July sun reached its zenith, the insurgents to the south cranked up again. This time, they added RPGs to the mix, lofting the small rockets toward the farmhouses. Taliban machine gunners raked the western side of the *qalat*. The Americans ducked down as the enemy slugs punched into the packed-dirt wall. Not all got down fast enough. Rob Pittman fell, hit hard by one well-placed 7.62 mm bullet. The retired master sergeant was trying to identify the hostile PKM team when the Taliban got him.

"Medic!" Even in the raucous din, that shout got attention. Around the yard, despite the pelting AK fire, men looked up.

From their aid post near the battalion forward command post, Private First Class José Rosario and Specialist Cameron Fontenot responded immediately. Bulky first-aid bags banging on their backs, they zigzagged through the hundred yards of open courtyard to reach Pittman. AK rounds dug dirt around them as they sprinted.

Two captains from the battalion tactical command post ran with them. While the medics bent over Pittman, the officers provided security. So did Kyle Lyon's cannoneers, steadily popping M-4 bullets toward the well-hidden enemy. Taliban shots hit all around Rosario and Fontenot, but the duo stayed focused. Lyon, meanwhile, had a medevac helicopter inbound. Getting that thing on the ground wouldn't be easy.

The gun trucks began engaging, the heavy .50-caliber rounds chewing into the enemy-held mud-brick buildings to the south. The Mark-19 gunner chugged 40 mm grenades into the base of the bushy trees rimming the hovels. All the other men in First Platoon increased their volume of fire. The HH-60, red crosses on white squares marking each door, crabbed over the embattled farmstead. Instead of choosing the open spot inside the *qalat,* the helicopter slipped a hundred yards out to the northwest. There was no choice. Pittman had to be moved into that maelstrom.

The medics and Captain Patrick McGuigan, joined by another soldier, hefted Pittman's litter out into the fire-swept landing zone. As the HH-60 nosed up to settle on the open dirt, AK rounds punched into the aircraft. Brown dust rose like a gritty, ragged pillow. Rosario and Fontenot stopped the litter and folded themselves over the bleeding SF NCO, protecting him from the flying dust and sticks. The second the chopper touched ground, the men were up, bearing their comrade to the aircraft. As soon as the stricken NCO was in the cabin, while the onboard medics were still strapping in the patient, the aviators pulled pitch. An angry RPG trailing gray smoke just missed the tail boom as the HH-60 soared away.

No sooner did the medevac helicopter depart than another was called. During the firefight, Staff Sergeant Jason Hamilton had stepped on a wooden crush panel, initiating an IED with two 82 mm mortar rounds. The explosion blew Hamilton into a grapevine wall and bowled over two more men. Again, it took a lot of U.S. and ANA gunfire to clear the decks for the HH-60 to get in and evacuate Hamilton. He survived.

By sunset, both sides had run out of gas. The day's toll for Flynn's artillerymen was three dead and twelve wounded, with similar casualties among the ANA. But 1-320 Field Artillery and First Kandak held the vital ground. It had a new name: Combat Outpost Stout, after the sergeant killed in action. He, Pittman, and Stansberry paid for this real estate in the hardest possible way.

Although at least fifteen of them had been killed, probably many more, the Taliban needed to counterattack. They saw engineers and vehicles inside the farmstead and, notably, watched the tired, dusty Americans unload dozens of the metal mesh-and-fabric Hesco baskets. Filled with dirt, Hescos became the outer wall of a combat outpost (COP). Building a COP meant holding terrain. In a day or two, it would be impossible for the enemy to stop the construction. So the clarion call went out to all the Ghilzai households: Attack COP Stout. Insurgents rallied.

The next day, as enemy gunmen watched and planned, others probed. All day on July 31, COP Stout took fire. Enemy marksmen shot at exposed heads and pinged rounds off the armored trucks. A dismounted local security patrol suffered one wounded from an RPG. You could feel something bad building up. In and around the *qalat,* the Americans and Afghans stayed alert. Hesco baskets were filled and placed. COP Stout grew more sturdy by the hour.

One Taliban team used a Soviet-era AGS-17 30 mm grenade launcher to pummel the inner courtyard and disrupt construction work. Similar to the bigger U.S. Mark-19, the intermittent series of Plamya (Flamer) bomblets wounded four ANA soldiers. Sergeant First Class Kyle Lyon volunteered to go out under the incoming barrage to look at the shallow craters dimpling the packed dirt. By inspection, one can see how the round came in and thus figure out the direction of the shooter. Lyon ran out into the open area and had a look. "Probably wasn't the smartest thing I've ever done," he said later. It did give the Americans a good idea of where the AGS-17 might be. Steered by the estimate, Apaches came in with a pair of Hellfires and more 30 mm cannon strafing. They raised hell in that suspect *qalat* to the south. The Plamya didn't fire again.

August 1 dawned hot in every sense. Colonel Art Kandarian joined Dave Flynn and Mangal Andar that morning. This day would show what the Taliban had in mind.

As the sun rose, the Americans and ANA took the offensive, pushing Sergeant First Class Kyle Lyon and the First Platoon plus ten Afghan *askari* out to the pair of treacherous structures and their surrounding courtyard two hundred yards to the south. Over the last two days, they had pounded that site with Apache fires, Mark-19 grenades, .50-caliber bullets, and every manner of U.S. and ANA small arms. But enough was enough. It was time to erase the problem altogether.

With Lyon's men marched an engineer squad laden with demolitions: blocks of C-4, Bangalore torpedo segments, and rolls of explosive detonation cord. The column crossed several files of vine-encrusted grape berms, getting about halfway to its goal before enemy AK shots sounded. Once they'd scrambled into the cover provided by a long adobe wall that ran right to the target structures, engineer Staff Sergeant Christopher Young and his men unpacked an APOBS. Insurgent fire increased, some of the bullets now skipping into the dirt here and there. The intent engineers kept working. They aligned the little rocket to parallel the sheltering barrier. Then they set it off. It darted out, its trailing cord looping behind. Once the line charge came to rest, Young ignited that too. The ground shook with the detonation. A ragged curtain of brown dust spouted up — and two buried IEDs went off. Lyon's platoon moved up, machine guns firing at the enemy-held buildings while the rest of the artillerists pushed up along the low mud partition, arrowing toward the enemy.

Young put down another APOBS. Again, the small projectile arced out, the line flopped, and, once more, with a sharp roar, the tan dirt cloud arose. Another IED went off. Their way made safe, Lyon's men kept moving, heads low, carbines level.

Alarmed by the relentless approach of Lyon's platoon, the Taliban defenders opened up with all they had. Men with AK-47s stood up to fire. The insurgents added a PKM machine gun and then resorted to the previous day's big gun, the AGS-17 30 mm Plamya grenade launcher. But the U.S. engineers and Lyon's men were huddled under the lee of the *qalat*'s projecting farmyard wall. All the hostile fire sailed high. From their well-chosen positions back near the first APOBS, Lyon's machine gunners kept up their deadly bursts, firing and pausing, attracting the enemy's full attention and pushing the foes' heads down. Meanwhile, crawling up and down their shielding barrier, Young and his engineers unpacked their wares. The men tamped down more than five hundred pounds of demolitions, equivalent to an air-delivered JDAM bomb but placed precisely where it could do the most damage to two sides of the enemy-held farmyard and both of its buildings. Young's people made sure to set the explosives close, but not too close. The Taliban were going for a ride.

"Fire in the hole! Fire in the hole! Fire in the hole!" The U.S. and ANA soldiers ducked down. Oblivious, the enemy did not.

The resultant thunderous explosion sucked the air out of chests and heaved up a massive cascade of dark brown dirt, adobe shards, tan grit, fine dust, sticks, wood chips, and pieces of enemy positions and men. Lyon and his people charged in before the mushroom cloud subsided. The Taliban did not survive the encounter.

While Lyon and Young and their elements cleared out some other hostile positions just to the south, another U.S. NCO dealt with another Taliban threat. On the western rim of COP Stout, Staff Sergeant Benjamin Tivao of Second Platoon, with ten ANA attached, saw armed men moving in buildings two hundred yards to his west. The Taliban commenced shooting, a lot of volume, but most bullets going well high. Tivao's men took shelter, peeking up to see their foe.

A technical team reported that the foe's leader had identified Tivao's exact position. The opposition chatter accurately described the mixed military-civilian garb of two local national interpreters who had been standing near Tivao before the firefight. When Tivao and his Americans saw four insurgents moving their way just past the one-story adobe structures, Tivao directed the U.S. and ANA soldiers to concentrate on the foursome. They ducked behind the farm's corral berm.

Two AH-64D Apaches came on station as if summoned by telepathy. They were part of a series of attack-aviation relays arranged by Colonel Kandarian. Tivao got on the right radio net and talked the Apaches onto the hidden hostile quartet. The aviators easily found the targets. A Hellfire streaked out and blew in the front of the enemy farmhouse. Then 30 mm cannon shells ripped across the yard, all hot white flashes and brown dust geysers. Taliban shooting stopped. The technical team said the enemy had sustained four dead and five wounded.

The two actions broke the Taliban attempt to run off the American and Afghan soldiers at COP Stout. There were five more U.S. wounded, including Headquarters Battery First Sergeant Jonathan Brown, setting the example in the expected Second BCT style, and the gutsy medic Private First Class Jose Rosario. Two battalion staff captains were hit. One of Art Kandarian's security team took some RPG fragments. The ANA also had some wounded, but no dead. COP Stout lived up to its name.

Establishing COP Stout started the retaking of the Arghandab River Valley. Over the next two months, 1-320 Field Artillery and First Kandak shifted from the three COPs they'd inherited to seventeen combined American-Afghan outposts. The Alkozai tribe reemerged as the dominant bunch in the valley, and they refused to knuckle under to the Taliban Ghilzai gunmen. Indeed, enraged Alkozai elders told Dave Flynn and Mangal Andar where to attack. The two lieutenant colonels followed this advice.

In an incident that garnered some fairly hysterical media interest as an allegedly inhuman scorched-earth measure, on October 6 and 7, 2010, Flynn's men called in B-1 bombers and HIMARS rockets — 49,200 pounds of munitions — to attack known IED storage sites in the empty hamlet of Tarak Kaloche and the mostly empty village of Khosrow Sofla. Follow-up American-Afghan ground operations confirmed the widespread storage of IED materials, much of which went up thanks to the aerial pounding. When the media heard of the bombardment, charges arose that the Americans razed the towns. With seven Americans dead and seventy-six more wounded, ten of them missing a leg or a foot and four with both legs gone, Dave Flynn agreed with the Alkozai elders: "There's no space for the Taliban to return to this district during the spring." Significantly, ISAF did not overreact to the media criticism of 1-320 Field Artillery's hard war. Indeed, General Dave Petraeus personally pinned on Sergeant First Class Kyle Lyon's Silver Star, and Lieutenant General Rod Rodriguez did the same for Lieutenant Colonel Dave Flynn.

Clearly, times had changed. General Stan McChrystal's restrictions were eased, and not much missed. It took killing to clear and hold the Arghandab, and Art Kandarian and his Second BCT did not shy away from slaughtering the Taliban. In one ninety-day stretch west of Kandahar city, Kandarian's soldiers employed 2,035 155 mm artillery shells, 2,952 mortar rounds, 60 HIMARS guided rockets, 266 aerial bombs, 19 Hellfire missiles, and more than 50 30 mm strafing runs by A-10

Warthogs and Apache helicopters. The Americans counted hundreds of hostile dead. The Afghan villagers farmed in peace for the first time in years — in some cases, in decades.

Success did not come cheaply. Sixty-five Americans died, two-thirds of them officers and NCOs, well out of proportion to their numbers in the BCT but exactly in accord with Art Kandarian's insistence and personal example of combat leadership. The wounded totaled 426, more than 50 missing one or more limbs. The soldiers' retaking the area west of Kandahar impressed even the commander of RC-South, who pivoted and adjusted adroitly to the new tougher tactics. "This is the greatest achievement of my command," said Major General Nick Carter. Indeed it was.

Securing the Arghandab directly reflected the impact of General David H. Petraeus. As in Iraq in February of 2007, he had inherited an ongoing campaign. In the other country, he ably carried out George Casey's strategy of al-Qaeda out, Sunni in, and Iraqis increasingly in the lead, taking full advantage of the U.S. troop surge and the Sunni Awakening movement. Moreover, Petraeus explained it far better than Casey to one and all. It still led to a U.S. withdrawal, but at least we left with our heads up and, to a degree, on our own terms.

When Petraeus got to Afghanistan in July of 2010, the strategy was less solid: send a troop surge to clear the Pashtun south, continue to increase Afghan forces, address the shaky (and shady) Afghan governance, and do something about the Taliban's Pakistani sanctuaries. In July of 2011, President Obama would review the bidding and determine the future of the U.S. commitment. So the new ISAF commander had a year to show enough progress to keep the war effort going. Staying did not guarantee victory any time soon. But leaving . . . well, that finished it in more ways than one. The Taliban waited patiently.

Petraeus was not patient. He got started quickly at the tactical level. From his previous role in U.S. Central Command, he'd heard of the awful smashups at COP Keating and in Ganjigal and watched Marjah unfold and Kandahar begin, and he knew Stan McChrystal had gone too far to curb ISAF firepower. The new general didn't fully rescind Stan McChrystal's insistence on reducing civilian casualties, but he unscrewed the pressure a few turns, especially for units in tough areas like Second BCT. Yet backing off did not mean ending the onerous oversight. President Hamid Karzai still complained, and so did people in Washington, London, and elsewhere, especially in journalist circles. Determining which particular military-age male was truly the enemy remained very, very difficult.

Even though Art Kandarian fought it out with all his weaponry, his soldiers continued to abide by ISAF orders that required each operation to answer fourteen questions before, during, and after execution.

1. Did the mission include ANA or ANP?
2. Were any civilians killed or wounded?
3. Was the local civil leadership notified ahead of time?
4. Did they agree?
5. Were bombs dropped, attack helicopters used, artillery delivered, or mortars fired?
6. Did the enemy present an imminent threat? (Fussy military lawyers loved that one.)
7. Were any houses entered without invitation?
8. Did the Afghan forces do the home entries?
9. Did any element enter a mosque?
10. Were any civilians searched?
11. Did the unit search the objective to locate enemy and civilian casualties and material damage?
12. Were there any unobserved fires?
13. Had any escalation-of-force incidents (warning shots to ward off curious or confused locals) occurred?
14. Were the Afghan or Western media informed?

Pessimists decried the seemingly endless, niggling top-down scrutiny. Optimists cheered that the question count stayed below twenty. But higher headquarters could always add more. This is how things went in the ninth year of a war of attrition.

A war of attrition, in essence, is what Petraeus commanded. All the heady talk of counterinsurgency theory and supposed progress in Iraq aside, the strategy amounted to keeping America and NATO committed. The operational campaign plan protected the major population centers, taking risk in the other areas. The tactical methods boiled down to looking for bad guys and killing or capturing same, increasingly side by side with Afghan partners.

Petraeus came with his Iraq perspective. He referred to the other country constantly, as did so many who had served there. But at his level, it grated some. Many, a bit unfairly, hoped to see Malik Daoud say the magic words and fix Afghanistan, à la Iraq. Instead, they saw that the great man possessed only one template. And this theater didn't match it very well.

The general remained courtly, polished, and media-savvy. He charmed visitors with a very busy PowerPoint slide titled "Anaconda Strategy vs. Insurgents in Afghanistan." The idea was to encircle and squeeze the opposition, like an anaconda constricts its prey. As in the 2002 Afghan operation — which seemed very long ago and very far away — the name intentionally echoed the 1861 American Civil War strategy of Union major general Winfield Scott: to slowly press the Confederacy from multiple angles with a sea blockade and overland campaigns.

The Petraeus slide teemed with buzzwords and memorable images. In the middle of the constrictor ring squatted the Taliban, the Haqqani network, and "other groups" supposedly in need of safe havens, weapons, money, foreign fighters, an ideology (as if they didn't have one already), popular support, command and control, and senior-leader guidance. Outside, seven axes assailed the foe: intelligence ops (fusion and analyses of all types of collection, the white whale of every MI and G-2 type), detainee ops (including "rehabilitation of detainees," unlikely as that seemed), politics (unscrewing the Karzai regime and seeking a "Pashtun Awakening"), info ops ("credible voices" and "strategic communications," perhaps incompatible and hard to pin down in anyone other than Petraeus himself), international ops (influencing Pakistan and "source countries," which was to say, other Islamic frenemies like Saudi Arabia), nonkinetic ops (working on governance and economic development, a bottomless pit in an illiterate, impoverished country), and, finally, kinetic ops (killing the enemy). Petraeus briefed it all smoothly to many opinion makers. Those with good memories recognized it as very nearly a word-for-word lift of an MNF-Iraq slide labeled "Anaconda Strategy vs. AQI." More than a few eyebrows went up: same-old, same-old. Well, what did they expect? Hire the Iraq guy; get the Iraq solution.

The Petraeus Anaconda Strategy included everything, all of it thrown at the wall to see what stuck. "Does he really think it's our mission to do all of this?" asked one White House insider. "It's laughable." Maybe so.

Outsiders might disparage the Petraeus anaconda approach, but it was all he had. The ISAF commander knew that most of the heavy lifting rested on him and those in country. But as the slide showed, others could help.

Some hoped ambassador at large Dick Holbrooke could broker a deal with Mullah Omar or one of the other Taliban honchos. That didn't happen. When Holbrooke died after a sudden, acute medical calamity on December 13, 2010, the prospects for a grand settlement died with him. The Taliban had never wanted a deal. They wanted to win. Holbrooke's final verdict resonated: "It can't work."

The other part of Holbrooke's remit involved Pakistan. He got nowhere with them either. Despite steady, low-level cross-border military coordination, Afghan liaison officers, and Holbrooke's many meetings with senior Pakistani generals, that unhappy neighboring state kept its own counsel and pursued its own ends. Its intelligence operatives supported the Taliban — better a weak Afghanistan than a resurgent Pashtun nation straddling Pakistan's western border to complicate the never-ending threat from India to its east. If the Americans stayed, Pakistan would enjoy control of the U.S. supply conduit, earning money and exerting pressure as needed. If the Americans left, as Islamabad leaders suspected they would, then Pakistan's keeping good relations with the Taliban ensured its influence in Afghanistan when Karzai's regime succumbed.

Senior Americans tried to secure better cooperation. The weak civil administration proved moribund and unresponsive. Inside Pakistan, in or out of formal power, the army held sway. So military-to-military avenues were opened and maintained. Admiral Mike Mullen, the chairman of the Joint Chiefs of Staff, developed a strong relationship with General Ashfaq Parvez Kayani, the senior military man in Islamabad. Dave Petraeus met regularly with Kayani, a fellow graduate of the infantry school at Fort Benning and the Command and General Staff College at Fort Leavenworth. Speaking impeccable English, Kayani invariably came across as urbane and polite. He also invariably never delivered anything. The boundary remained wide open.

Unable to secure reliable, large-scale Pakistani cooperation, the Americans resorted to drone strikes and some SOF missions by the CIA and the Task Force, culminating in the May 2, 2011, raid that killed Osama bin Laden. Pakistan cooperated most of the time, offering occasional public complaints while privately facilitating operations. It all kept the Taliban and the remnants of al-Qaeda leadership nervous. But it offered a half-measure at best. This condition showed no evidence of improving.

Inside Afghanistan, Petraeus moved to do something about corruption. He worked closely with Ambassador Karl Eikenberry. The two got on better than many had expected. They found common cause in pressuring President Hamid Karzai to clean up his act. Confronted in his office by the two senior Americans, the Afghan president blamed all the illicit money on foreign funds and international intriguers, especially the United States. He exploded: "I have three main enemies [the U.S., other international entities, and the Taliban], and if I had to choose today, I'd choose the Taliban!" Apparently, many of his fellow Pashtuns agreed. The president's own half brother, the wily Ahmed Wali Karzai, continued to run his lucrative rackets in and around Kandahar even as Dave Flynn's Americans and Mangal Andar's Afghans fought and died to secure the Arghandab.

Petraeus's ISAF and Eikenberry's U.S. embassy worked together to go after such monetary diversions. If nothing else, they'd try to ensure that U.S. dollars did not knowingly go into corrupt coffers. Petraeus brought in one of his favorite officers, Brigadier General H. R. McMaster. Charged to put a public spotlight on Afghan malfeasance, McMaster assembled an international team of intelligence analysts, lawmen, and investigators, including agents from the U.S. Federal Bureau of Investigation. He called it Task Force Shafafiyat (*shafafiyat* is Dari for "transparency"). McMaster generated his usual share of publicity, as was both his wont and his duty. His team uncovered $150 million in Kabul Bank assets diverted to villas in Dubai; outright theft of medical supplies by the ANA surgeon general; and double-dealings by segments of the small

Afghan air force. The evidence wasn't all that convincing, much of it rumors repeated through different channels. It made a splash, though, which is what the ISAF commander wanted. Some Afghan officers and officials were fired, and some court cases began, but the entire undertaking generated more light than heat. In Kabul, Karzai continued to blame the cash-flush foreign hand. Outside the country, observers highlighted the rampant corruption, and many used it to argue for an end to ISAF. Mistrust increased on all sides. In some ways, it was the worst possible outcome.

Task Force Shafafiyat's very public examination of certain shortcomings in building Afghan security forces tarnished the otherwise superb work of Lieutenant General Bill Caldwell's NTM-A. During Caldwell's tenure, the ANA nearly doubled in size (going from 95,000 to 179,610); the ANP added many (from 95,000 to 143,800); and the little Afghan air force (about 5,000) was established. Mandatory literacy education provided a popular recruitment incentive and increased performance. Marksmanship, leader training, and upgraded weapons and equipment all helped the cause. These forces promised to hold what ISAF and the Afghans cleared in 2010. The ANA and ANP were not perfect, and in many cases not all that good. But they were better than their Taliban antagonists, and they'd fight. That was good enough. All in all, it marked a dramatic achievement.

Yet again, as with some of H. R. McMaster's more lurid press releases from Task Force Shafafiyat, Bill Caldwell's insistence on a heavy dose of "strategic communications" nearly undermined the whole edifice. NTM-A created an outsize press organ constantly pumping out good news, always looking for the silver lining. Smart alecks nicknamed the effort "Operation Fourth Star," the implication being that Caldwell was promoting himself too much. But that was untrue and unfair. Having inherited a steady-state, small-scale enterprise, Caldwell ramped it up fast and hard. Telling the command's story to the media was part of the job. So he thought.

Unfortunately, the well-meaning NTM-A commander ran afoul of Michael Hastings, fresh from nailing Caldwell's West Point classmate Stan McChrystal. The reporter wrote a feature story for the February 23, 2011, issue of *Rolling Stone* titled "Another Runaway General: Army Deploys Psy-Ops on U.S. Senators." Disgruntled subordinates had kvetched about preparing for visits by U.S. legislators, darkly hinting about the use of secretive mind-control measures, whatever those might be. The charge sounded ludicrous, and it was. While Hastings might have had most of his notes straight on McChrystal, he greatly overshot the target with Caldwell. Still, the media uproar and subsequent Department of Defense investigation hardly helped matters in the busy ISAF headquarters. The investigation eventually exonerated Caldwell. But Dave Petraeus didn't appreciate the distraction. Those who couldn't handle the press needed to stay off the screen. The ISAF commander made certain that the next officer chosen to command NTM-A was the kind who sought no press profile.

Caldwell's West Point classmate Lieutenant General Rod Rodriguez at ISAF Joint Command diligently avoided publicity. A fighter to his core, Rodriguez just worked. The surge had given him a lot of work to do. While Petraeus handled strategic aspects, Rod Rodriguez completed Stan McChrystal's theater vision. He placed all the surge Marines in Helmand Province, creating a new Marine-led RC-Southwest. Along with providing more police for the villages, Caldwell's NTM-A and the Afghans generated a new 215th ANA Corps. Tweaking the British in Helmand, the Afghans chose the nickname "Maiwand" for the new organization. It commemorated an 1880 battle in which twenty-five thousand Pashtun tribesmen had defeated twenty-five hundred British soldiers. At Kandahar Airfield, the Tenth Mountain Division took over the reduced RC-South from the British. This became the final ISAF structure: three American divisions in the contested areas and three NATO-led formations in the much quieter west, north, and capital. It completed the transition that had begun in 2007 when General Dan McNeill took command. The next major hand-over would be to the Afghan national army.

Rodriguez and Petraeus tried one other initiative. Seeking a Pashtun version of Iraq's Sunni Awakening, Petraeus pressed Hamid Karzai to allow village militias in the most exposed areas. The locals didn't like the rapacious Kabul regime, but they had no real love for the strict, demanding Taliban members who took money, crops, and buildings either. Pashtun families just wanted to be left alone and would gladly bear arms to make that happen. The two U.S. generals proposed putting Special Forces teams in villages to recruit and train Afghan local police (ALP), a euphemism to avoid the poisoned term *militia.* Karzai, of course, saw himself and his cronies as the true voices of Pashtun Afghans. Reluctantly, he approved a few ALP sites. The Afghan president imposed a very slow buildup of the village protectors and demanded direct supervision by his own Ministry of Interior. The Americans agreed and got started. In a tribute to the value of the ALP, Taliban prisoners and communications intercepts soon indicated that the enemy considered these forces their number-one target. They stood squarely between the guerrillas and the rural population. But a few thousand ALP could not spur a Pashtun reconciliation.

In the main, however, Rodriguez and Petraeus and their forces, hard men like Art Kandarian and Alonzo Smith, did what soldiers did best. Petraeus exhorted all to "get our teeth into the insurgents and don't let go." He went on: "When the extremists fight, make them pay. Seek out and eliminate those who threaten the population."

At about the same time Dave Flynn's 1-320 Field Artillery stood toe to toe with the Arghandab Taliban, Petraeus told a journalist that he estimated that his SOF and conventional battalions had together killed or captured more than 350 enemy leaders and about 2,400 lower-ranking insurgents over the summer. ISAF casualties peaked in 2010, with 499 Americans killed and 212 other ISAF soldiers dead, reflective of the scale of surge operations. Intelligence began to circulate suggesting that the Taliban wanted to back off, to wait out the Americans. *Enemy advances, we retreat.* Pushed back into northern Helmand, rolled away from the gates of Kandahar city, beaten down in parts of the east, the Taliban absolutely felt the pressure. Bloodied and demoralized they might well be. But they were still out there.

ISAF certainly knew how to clear. And the ANA definitely showed sufficient skill and will to hold. But the rest of it — the build part, the hand-over to viable Afghan authorities, the other great notions emblazoned on Petraeus's Anaconda chart — was not in evidence. Absent some kind of real "Pashtuns in" power-sharing bargain, even an imperfect one, Hamid Karzai's bunch just couldn't wrap it up. For its part, America refused to stay for decades. The Taliban, however, wasn't going anywhere. At such an impasse, killing became a measure of progress, perhaps the only measure. It wasn't enough.

Educated soldiers saw the flaw. Attrition by itself reflects a lack of strategy. It's what happens when armies grind against each other because they don't know what else to do. Wars of attrition conjure up images of Abraham Lincoln's terrible arithmetic of the 1864 Petersburg siege, of the blood-soaked western front of 1917, and of the futile chase of furtive Vietcong guerrillas in 1967. As a White House staffer commented, "The model had become clear, hold, hold, hold, hold and hold. Hold for years. There was no build, no transfer." Stalemate. Game over.

On June 22, 2011, President Obama announced the end of the surge. He ordered ten thousand troops out by the year's end, and the other twenty-three thousand gone by July of 2012. He went on: The entire ISAF mission would conclude on December 31, 2014. After remarking on the Iraq drawdown well under way that summer and referring to the recent demise of Osama bin Laden, the president stated, "These long wars will come to a responsible end." The door remained open to a residual force. But what had transpired in Iraq made that prospect unlikely. The Office of Security Cooperation–Afghanistan might well be it — if that.

Whatever it would be, Petraeus wouldn't be there to execute it. He departed Kabul on July 18, 2011, disappointed at not being elevated to chairman of the Joint Chiefs of Staff but honored to be chosen as the next CIA director. In his stead, Marine general John R. Allen commanded ISAF and U.S. Forces–Afghanistan. Allen would oversee the withdrawal.

In the uniformed ranks, heads shook; some sadly, others knowingly. The great Malik Daoud had labored mightily, moved the pieces around the board, said the right things . . . and brought forth less than met the eye. This time, the sparks hadn't flown up. Something lacked. Something distracted. Maybe all the months on deployment had finally worn down even this most fit of U.S. soldiers. In any event, he moved on. Behind him, the unwon war continued.

17

ATTRITION

> Thus the portrait of McNamara in those years at his desk, on planes, in Saigon, poring over page after page of data, each platoon, each squad, studying all those statistics. All lies.
>
> — DAVID HALBERSTAM

THE FRENCH CALLED it *le cafard.* Out in the barren, windswept deserts of southern Algeria, atop the unforgiving outcroppings of the Atlas Mountains, men of la Légion Étrangère — the famous and sometimes infamous French Foreign Legion — grew bored keeping watch over the endless vistas of sun, rocks, and sand. At small garrisons, the real equivalents of lonely Fort Zinderneuf in the novel and many film versions of *Beau Geste,* hardened men drilled, labored, stood guard, and marched out on long, fruitless foot patrols, seeking Bedouin tribesmen who lingered forever just out of rifle range. The grueling routine wore men down. A few tried to walk away and ended up feeding the vultures. Others turned on one another, or themselves. Many drank to excess, guzzling the cheap red military-issue *pinard* wine, "the supreme consoler" for legionnaires. Most found rough amusements. One such involved shooting cockroaches (*les cafards*), no easy feat with an unwieldy long rifle.

Similar things happened on other empty frontiers where small contingents watched and waited for vicious, lurking foes. On the American Great Plains among the frontier regulars, suicides, homicides, and desertion marred the calm at many little fifty-man outposts. Out on the endless grassy steppes of Asian Russia, more than a few bored czarist soldiers passed the time shooting pistols at each other in a blacked-out cabin, a lethal game known as Can You Hear Me, Ivan? You shot at the voice, and hit all too often. Some of the czar's men put a single bullet in a revolver and tried their hand at that grim national pastime, Russian roulette. But as only they can, the French chose the exact right name, *le mot juste*, for the dark phenomenon: *le cafard*, the cockroach, the blues, going bug-fuck.

As autumn became winter in 2011 and then as 2011 became 2012, *le cafard* crept through the back corners of Afghanistan. Soldiers crawled into plastic port-a-johns and blew out their brains. Marines filmed themselves urinating on Taliban corpses and posted the videos on the Internet. Denied the diversions of marijuana, hard drugs, liquor, and beer, rear-echelon types experimented with painkillers. Sometimes, alcohol was found, used, and abused. Young men huffed compressed air, a particularly risky high. The military chain of command dealt with all of this swiftly, often preemptively. Yet still it happened, a bit more with every passing month.

All of these were isolated incidents, nothing widespread, nothing akin to the much more prevalent drug abuse, fragging of leaders, and racial strife seen in Vietnam after 1970. There have always been such events in any war. Yet when a conflict drags on with not much chance of victory, morale decays. Here was the unhappy cost of President Obama's decision to leave, but not right away. In Iraq and now Afghanistan, the thoughtful, deliberate U.S. president thoughtfully and deliberately condemned Americans in uniform to years of deadly, pointless counterinsurgency patrols sure to end in a wholesale pullout. It bred *le cafard* as surely as putrefaction bred its loathsome namesake.

Le cafard touched the Panjwai District west of Kandahar in the fall of 2011 and the winter that followed. The Taliban lurked on the fringes of the long-contested rural area, south of the Arghandab River and north of the red rocks of the Registan Desert. There was just enough fighting to keep things very tense, yet nothing near the brutal slugfests seen in the Arghandab in 2010. Out in the COPs, sergeants and officers kept their soldiers busy on patrol, hunting the enemy. Leadership by example, firm discipline, and hard work banished *le cafard.* Yet in a few dark corners, things went very wrong.

Combat Outpost Palace defined the middle of nowhere. The COP stood just south of the Arghandab River, a mile north of the only paved east-west road in Panjwai, surrounded by a broad plain covered with poppy fields and grape rows. An Afghan national army rifle platoon held a similar small facility just to the west. Three layers of dirt-filled Hesco containers formed a thick retaining wall nine feet tall that framed a three-hundred-by-four-hundred-yard rectangle of limestone gravel. Each corner of COP Palace featured a twenty-foot guard tower capped by a small sandbagged bunker. The compound boasted a small vehicle parking area, a platoon command post, ten metal "container housing units" — militarized versions of double-wide house trailers — a similar container rigged as a shower trailer, a small kitchen, a laundry tent, and a recreation tent. The Canadians built it and the Americans of Third Platoon, Company C, Third Battalion, Twenty-First Infantry took it over in the summer of 2011. Michael "Beau" Geste wouldn't have felt out of place there.

Company C, including Third Platoon, suffered a rough summer. At the outset, most of the NCOs and the company commander had prior combat experience while the lieutenants and privates did not. Panjwai evened that up soon enough. In an especially bad three days from August 25 to 28, Taliban IEDs killed two soldiers and wounded four others. A buried bomb in a dirt trail flipped over an M-ATV, crushing Private First Class Brandon S. Mullins. On a foot patrol, an IED exploded, felling three. Hostile gunfire from hidden ambushers then struck the already wounded Specialist Douglas J. Green. He didn't survive. Gunfire sounded most days outside COP Palace.

The Third Battalion, Twenty-First Infantry Regiment considered COP Palace a quieter area than most of Panjwai. Things got progressively worse as you went west, into the triangle defined by the river to the north and the desert to the south, the area the Canadians had labeled, after its shape on a map, the Horn of Panjwai. The small area teemed with IEDs and hidden gunmen. Company A lost two men in July in a single major attack, then two more in two separate incidents in mid-September. The Horn accounted for thirty-seven separate IED attacks in July, August, and September, as well as more than eighty buried devices found and cleared. The key western hamlets of Mushan with a thousand people and Doab with fifteen hundred bowed to the Taliban. Americans and their Afghan partners wanted to clear and hold both villages. With most of the action out west in the Horn, the battalion's leadership focused there. After the August attacks, the battalion left Company C to itself.

Company C's commander, in turn, left COP Palace to Third Platoon. He operated out of a larger base to the south and rarely got out to Palace. The Third Platoon's riflemen kept up their missions. Contacts lessened over time as the Taliban shifted to address the American/Afghan military threat to Mushan and Doab. But one young soldier just didn't get the hang of the rifle platoon's routine. In an excellent unit, wiser leaders found places for such a man, maybe in front of a computer in the command post or running biometric scans of local farmers. An average outfit left a troubled soldier behind at home station to be administratively discharged, not hard to do in a volunteer army. Problem privates got out pretty quickly. But those in a less capable unit, especially one far from more experienced leaders, might apply other, more cruel solutions. Young men under extreme conditions resort to such things.

Private Danny Chen, a nineteen-year-old from New York City, deployed late into country, months after the rest of the outfit. It's never easy to be the new guy, and this one had his issues. Sometimes confused, physically frail, the tall, thin son of first-generation Chinese immigrants struggled as an infantryman. He consistently misplaced or forgot key equipment, including his Kevlar helmet. Judged unable to go out on patrol, Chen was relegated to guard duty in the lookout towers, light menial labor, and cleanup details. With the others working long days and nights out in the deadly poppy fields and grape rows, Chen drew the ire of his sergeants and his platoon leader. His fellow junior enlisted men joined in the pressure. Over the course of six weeks, it got out of hand, way out of hand, to the point where it included physical abuse. Remedial calisthenics went well beyond building strength and endurance. Along with extended rounds of pushups and sit-ups in the blazing sun, Chen found himself low-crawling across the hot gravel, knocked about, and subject to racial slurs. Unable to "motivate" the soldier, the NCOs wanted to get rid of him, to evacuate him to Kandahar and then back to Fort Wainwright, Alaska, to be discharged. Chen acted first. Late on the morning of October 3, 2011, while standing watch in a tower, Chen shot himself in the head with his M-4 carbine.

The subsequent investigation switched on a long-overdue flood of light, sending the cockroaches running for cover. Right out of *le cafard* in the Foreign Legion, COP Palace gave up its sordid secrets. The lieutenant had known his men, including his platoon sergeant, possessed and drank alcohol at the outpost. He'd chosen to look the other way, and had done the same as Chen suffered. The officer also allowed some dangerous horseplay involving the detonation of a live hand grenade. Five NCOs and two senior specialists were found complicit in abusing the dead soldier. Four of the eight eventually faced formal courts-martial; all eight were punished by judicial or nonjudicial proceedings. The lieutenant and one other left the service as part of their formal arrangements. None of it brought back Danny Chen.

The chain of command, however, went beyond investigation and judicial matters. Lieutenant Colonel Steve Miller of the Third Battalion, Twenty-First Infantry relieved the absentee commander of Company C, who claimed he'd had no knowledge of Chen's travails or any other lapses of discipline at COP Palace. The new captain took charge and reorganized, bringing in steady NCOs and a new lieutenant. The revamped Third Platoon focused on its mission. The outpost got a lot more checking, as did the other isolated elements in Panjwai. COP Palace stayed quiet for the rest of the winter.

West Panjwai, the Horn proper, continued to fester. Mushan and Doab villages teetered, hosting the Taliban by night and the ANA and Americans by day. Company A of 1-5 Infantry, attached to Steve Miller's battalion, occupied COP Mushan and then pushed west, establishing COP Lion in Doab itself. Afghan soldiers of the Second Kandak, First Brigade, 205th ANA Corps joined the Americans in these posts. The enemy fought back, sowing IEDs in each hamlet and on every dirt road. On November 13, a buried bomb smashed up a Stryker armored vehicle, killing a platoon sergeant and a specialist. Three days later, another IED penetrated a Stryker hull. Two more soldiers died. In the colder months, the Taliban often fell back. But in Panjwai, the enemy fought through the winter. The opposition intended to stay in the Horn, one of their final redoubts west of Kandahar city.

Between COP Palace and COP Lion, American Special Forces worked to hold the ground gained. At COP Belambai, an SF A-team, reinforced by a squad of conventional infantry, prepared to recruit, train, and advise Afghan local police. This village militia offered a way for the Pashtun people of Panjwai to protect their families. Many belonged to the same Alkozai tribe that supported U.S./ANA operations north of the Arghandab River as well. The Taliban hated the ALP above all other Afghan forces, as these armed locals greatly constrained its freedom of movement.

Just after 3:00 a.m. on March 11, 2012, Staff Sergeant Robert Bales slipped out of COP Belambai. The NCO wore Afghan robes, including a cowl that hid his head, over his U.S. Army uniform. Bales walked with the aid of an AN/PVS-14 night-vision device, its single light-intensifying tube projecting like a snout. He carried an M-4 carbine with an attached 40 mm grenade launcher, an M-9 pistol, and a large knife. Soldiers didn't simply leave a U.S. compound alone at night, especially in a dangerous area like Panjwai. But Bales did.

Those who later watched surveillance footage saw him slowly move out of view to the north. Gunshots sounded, not an unusual noise in this district. Then, after an hour of silence, more single shots rang through the night, this time from south of the COP. All of this activity alerted the Special Forces team, who immediately awoke all at COP Belambai. As the soldiers scrambled to man the outpost's security positions, a head count found Bales missing. And then he wasn't.

A hooded figure shambled slowly to the front gate. Slowly, uncertainly, the shrouded apparition placed his weapons on the ground, then stood up and raised his hands. Bales admitted, "I did it."

By sunrise, Panjwai resounded with wailing and outrage. Bales had slain four and wounded six in the first settlement. Then he killed twelve others in the southern hamlet. The victims included four men, four women, and ten children. Bales had attempted to burn the bodies. The first Afghan national police on the scene got the same story from the survivors: One American shot them all. Later, hysterical Afghan media accounts, fanned by President Hamid Karzai, speculated that more than a dozen drunken Americans had killed the villagers. That overwrought rumor mill gradually wound down. The evidence all pointed squarely at Staff Sergeant Robert Bales.

Nobody knew why he did it. A high-school graduate with three years of college credit, the thirty-eight-year-old had served ten years in the Army, including three tough Iraq deployments. He'd suffered a foot injury on the first and trauma from a vehicle rollover on the third. His family life wasn't always happy. He'd been passed over for promotion. He'd been drinking alcohol with others at COP Belambai the night of the murders — *le cafard* again, the dry rot even among the elite SF, among tough veteran sergeants. Clearly something in him had snapped, but even Bales wasn't sure what.

The Taliban leaders chortled with glee. Bales's horrific acts underscored every Pashtun fear of the infidels, every mistrust of the occupiers. For years, the enemy had defended the Horn of Panjwai from battalion after battalion of resolute Canadians, killing 157 of them. Now, for the few years the Americans remained in country, the insurgents fully intended to battle and best them. The Third Battalion, Twenty-First Infantry Regiment was strung out all the way into the embattled Horn, near the end of a tough yearlong deployment, one of the last U.S. Army units to serve that long stretch. How much fight could these Americans have left? *Enemy halts, we harass.*

Before they left in July of 2011, the Canadians asked Lieutenant Colonel Steve Miller to actually sign for Panjwai District. Miller framed the document. He hung it in his headquarters, down the hall from the radio room, past the row of solemn photographs of the battalion's eleven dead, Danny Chen among them. Even before the mayhem in Belambai, it had been a long deployment.

Tall and sturdy with crewcut blond hair, a piercing gaze, and a Hollywood-issue command voice, Miller benefited from all the right preparation to command the Third Battalion, Twenty-First Infantry. He'd commanded an air assault rifle company and a long-range surveillance detachment in Korea, then saw combat in Iraq as a major with a cavalry squadron. He'd also served as an exchange officer with the British army, a valuable opportunity to learn from America's most battle-wise ally.

Calm and direct, Miller led by example. He'd been in a lot of firefights. And he kicked himself for what happened at COP Palace, a place he hadn't gone enough, preoccupied as he was with actions out west.

At this point in the war, in the wake of the ugly suicide of Danny Chen, a more nervous chain of command might have thrown Miller under the bus. His brigade commander tried to avoid dealing with the entire sad affair. But the RC-South commander, Major General Jim Huggins of the Eighty-Second Airborne Division, backed Miller "all the way," as they say in the Airborne community. A veteran of Panama in 1989 and Desert Storm in 1991, Huggins also served earlier in Afghanistan and twice in Iraq, the second time as a one-star during the surge. He followed the hard examples of Dan McNeill, J. R. Vines, and Rod Rodriguez, a fighter from the start. Huggins saw the same trait in Steve Miller. A Japanese American, Jim Huggins well understood the potential damage of any incident with racist overtones. But he also rapidly determined that Miller was committed to fixing his unit's cohesion and continuing to try to clear and hold Panjwai. Jim Huggins gave him the chance to stay at it. Miller repaid that trust.

Two weeks after the Belambai massacre, the battalion's attached Company A/1-5 Infantry and Fourth Company, Second Kandak, First Brigade, 205th ANA Corps pushed once more into Mushan village. This time, they planned to find a good, abandoned *qalat*—there were several—to set up an ANA outpost in the center of the cluster of one-story adobe structures. The soldiers' coming and going just made it easy for the Taliban to squeeze the people. With but a few weeks left in country, the Americans wanted to install their Afghan partners in central Mushan once and for all. That required a lot of dirty work.

Beginning at dawn on March 26, the American and Afghan rifle companies each committed a platoon as well as a company commander. Company A added a U.S. Air Force explosive-ordnance disposal (EOD) team of three NCOs. The Afghan Fourth Company brought along its own EOD team of four. The troops approached on foot from COP Mushan to the north of the town itself. Expecting trouble, Steve Miller and his security team moved to Mushan to follow the all-day effort.

The battalion commander and his team rolled west aboard two eight-wheeled Strykers. These green armored personnel carriers ran superbly on paved roads, pretty well on dirt trails, and decently on relatively level hard ground. Each held up to a squad of soldiers, sheltered behind armor capable of shrugging off small arms and most shell fragments. The majority of Strykers mounted a .50-caliber machine gun in a small automated turret, although there were also mortar carriers, ambulances, and specialized reconnaissance models with thermal sights. To ward off RPGs, Strykers bolted on a wide cage of metal slats that jutted out on all four sides. The rows of thin bars served to pre-detonate any approaching shaped-charge warhead. The Third Battalion, Twenty-First Infantry Regiment also used a number of V-hull variants, which fared better against IEDs. The flat-bottomed types suffered more damage. Miller's battalion employed both kinds, but clearly, the men preferred the safer V-hulls.

The morning didn't start well. Not far from Mushan, an Afghan civil order police patrol struck an IED out on the single paved road. One Afghan tripped an IED, which took off both of his legs. A U.S. Army medevac helicopter picked up the stricken Afghan. Miller's two vehicles moved carefully past the IED site. Scattered debris marked the blackened roadside depression, a common sight in the Horn of Panjwai.

At about 11:15, the battalion commander stopped his pair of Strykers north of Mushan. The ramps went down and the men got out, leveling their carbines. Having formed up his security element, Steve Miller walked into the north edge of the village. Captain Mike Nolan and First Lieutenant Mohammed Zacharias, both small in stature but full of energy, met the commander. All three men took in the scene. "We're just getting started," Nolan said. "Not many people on the streets," replied Miller. The soldiers knew what that portended.

The American and Afghan riflemen provided security, with mixed elements one compound forward, one back, and a block over on each flank. Those men crawled over mud barriers, slid around trees, and slogged through fetid sewage rivulets, places where the Taliban wouldn't likely put IEDs. The Afghans wore their bright green camouflage uniforms, and you could see them easily from a distance. Americans used the duller green mottled pattern commonly issued by 2012; they blended in much better. The two rifle platoons spread out as the day wore on.

One of the Afghan EOD NCOs, Sergeant Abdul Hamid, a squat young man with a fringe of beard and a broad forehead visible beneath his tilted-back Kevlar helmet, led the search of Mushan. Trained by the meticulous German advisers at the ANA EOD school up north at Mazar-e-Sharif, Hamid and his three men slowly, carefully probed their way forward. One Afghan next to Hamid used the Vallon device, about the size of a small broom. He swept it back and forth to find any buried items. The drill went very slowly.

While this tableau unfolded on their major dirt road, all of ten feet wide, the people of Mushan stayed inside. As the soldiers passed, they saw glum women, children, and old men — no young ones — in windows and doors, watching, silent, staying out of the entire episode. That night, when the Taliban males came back, the long-suffering villagers could say honestly that they'd had nothing to do with the infidels and the Afghan lackeys. It's how people stayed alive in a place like Mushan.

At each empty house, Hamid and his team stopped. The Vallon man waved his detector slowly along the ground, paying particular attention to the doorways. Then he ran the stick up the doorjamb, across the lintel, and down the other side. Sometimes the enemy liked to rig grenades at eye level, a nasty surprise indeed. Once safely through the doorway, the Afghans made a quick scan of the dim, low hovel. They found nothing but trash and old straw. The first dozen structures, some on each side of the path, all seemed clean. Picking along like this, it took them hours to go the equivalent of a city block along the meandering lane and its fronting mud huts.

About a third of the way down what passed for Mushan's main avenue, the column came to the town's mosque. By far the largest adobe structure in Mushan, nearly a story and a half tall to allow a large gathering space, the mosque sported some actual glass windows, a real wooden door, and a fifteen-foot rusty iron lattice minaret topped by speakers for the call to prayer. Three preteens, two girls and a boy, stood quietly in the mosque's courtyard. They waved at the Afghan soldiers. They stared at the Americans.

The gregarious Lieutenant Zacharias asked them where the men of the village had gone. The trio pointed three hundred yards to the southeast, toward a small hill covered with rocks and scraggly brush. Steve Miller's interpreter stepped in close to the lieutenant colonel: "Taliban Hill." The Americans knew it well. They had gone round and round there in previous skirmishes. Miller looked up. Above them, the flat gray clouds hung low, a heavy overcast, a gloomy day without shadows. "No more helicopters today," the battalion commander said to Captain Mike Nolan.

Certain units might have called it a day right there, with aerial medevac and Apaches off-station, having found no IEDs, few locals to meet and greet, and no likely facility to take over as a prospective ANA outpost. But Miller trusted his people. He knew Zacharias and Nolan wanted to press on. With the Strykers a few hundred yards to the north and COP Mushan itself a mile away, the risk appeared worth taking. Of course, that's how most really bad days in Panjwai started. But Steve Miller knew his business. They could handle whatever percolated out there on Taliban Hill.

Fifty yards south of the mosque, Hamid stopped abruptly. He stood in the narrow street between two of the larger *qalats* in the center of the village. Those both seemed likely spots for an ANA platoon position, and both looked to be long empty. Hamid squatted down, then pointed up at the three U.S. Air Force sergeants. The Afghan then slowly swept his arm down to the dusty opening leading into the eastern courtyard. The EOD men nodded. A thin black wire, exposed for a foot or so, ran toward the back wall and into a small, low-ceilinged mud-brick stable.

"Command wire," said one of the Air Force sergeants. That meant the enemy triggered it with a live observer. As none were evident, it could be disposed of rather simply with some demolitions. Evidently the Taliban, too, had figured out the two best locations for an ANA COP. As usual, they prepared something to welcome the Afghan soldiers.

Like most good EOD technicians, Hamid kept track of his finds and clearances. He indicated that this was his thirty-fifth. The Afghan found the bomb itself, marked it, and then crept catlike past it, sliding to the left. He pointed at the dirt yard, just inside the door. A dull wooden board peeked through a wide patch of disturbed dirt. "Pressure plate," said one of the American EOD men. Those types were victim-operated, which meant they went off when you stepped on them. The Taliban liked to mix varieties to cross up men like Hamid. This ANA sergeant, however, was way ahead of his adversaries.

While the U.S. and ANA riflemen kept all-round security, Hamid dipped into a team member's rucksack. He extracted detonating cord and two C-4 plastic-explosive blocks. He carefully placed the charges, ran the cord, inserted thin silver blasting caps, and then checked his work. Satisfied, he backed into the street, pushing against the thick adobe outer wall. With a minimum of orders, the experienced soldiers got the word and settled low, sheltering in the lee of various mud-brick partitions. Almost in unison, Afghans and Americans lowered their helmeted heads. In English, Hamid announced: "Fire in the hole! Fire in the hole! Fire in the hole!"

The twin explosions thundered, belching up two tall pillars of brown dirt and bits and pieces of both IEDs too. As soon as the debris cloud dissipated, Hamid crept forward with some plastic storage bags. He wanted fragments of the Taliban hardware to bring back to the base. He'd see if the analysts could figure out who made the weapons and how they worked. Exploiting the site for intelligence — it came right out of the book. Hamid notched numbers 36 and 37.

Across the dirt pathway, in the opening to the other *qalat,* one of the Air Force EOD men found another pair of buried bombs. One used a metal pressure plate. The other relied on a piece of buried tire with a crush switch. In that version, when a soldier stepped on it, nothing happened right away. Then, when he lifted his boot, the trigger snapped back, closing a circuit and, the enemy hoped, taking out both the soldier who'd hit it and those behind him. The Americans showed the clever, brutal items to Hamid. The Afghan EOD sergeant turned to his partners.

Crack. Crack. Crack.

The Americans ducked. Those were AK-47s. A few heard the zip of nearby enemy rounds. Not good. About ten more single shots sounded. Of course, they came from on or near Taliban Hill.

Afghans armed with M-16s and Americans with M-4s immediately returned fire. Then a heavier sound ripped the air — an M-240 machine gun, sweeping across the crest of Taliban Hill. The machine gun always got the enemy's attention. The American gunner worked through two twelve-round bursts, peppering the high ground where the Taliban AK gunmen likely hid. The U.S. soldiers knew it only too well.

"Medic!" Uh-oh.

Steve Miller looked up. South of the two compounds with their embedded IEDs, a four-foot mud wall ran down each side of the ten-foot-wide street. About fifty yards south, a T-intersection loomed. The left turn led due east to Taliban Hill. Both Americans and Afghans huddled against the dirt berms. The wounded soldier lay sprawled near the T-intersection.

Nolan and Zacharias moved forward at a crouch. Their sergeants and lieutenants had matters in hand. Those two machine-gun bursts had clearly done some good. No more AK-47 bullets came from Taliban Hill. Fleeting forms ran off both the north and south slopes of the low rise, seeking shelter at the base of an angular mud-brick grape-drying hut to the south and among some mud walls to the north. Maybe two went each way. Maybe there were more. You couldn't tell. In movies, both sides see each other clearly. Not so in reality when bullets fly. Men get behind cover and try hard to stay there. What you see when you glance up might be a flash of motion . . . or nothing at all.

With the Taliban run off the crest of Taliban Hill, U.S. and ANA soldiers pushed up the left leg of the T, worming their way closer to the enemy. The AKs kept banging away in ones and twos. Some hostile shots came from both sides of the now-empty mound. Americans and Afghans replied in kind, husbanding their ammunition.

As the firefight sputtered on, the medic reached the wounded soldier at the intersection, who turned out to be First Lieutenant Patrick Higginbottom, the Company A fire-support officer. An AK slug had caught the young West Pointer squarely in the hindquarters. Fortunately, the penetration wasn't deep. He'd be okay as long as he got evacuated and cleaned up. The medic packed the wound. Around him, his mates kept shooting.

For more than ninety minutes, from 3:45 until almost 5:30, both sides fired. Neither maneuvered much after the opening moves. Miller and his soldiers knew from previous encounters that the insurgents wanted to draw them into prepared IED belts. Taliban gunmen kept flinging odd rounds in the hope of prying loose their disciplined opponents. The Americans and Afghans anticipated the ploy. For their part, American riflemen used the dirt walls for cover, moved back and forth to get glimpses of the foe, and just kept engaging. In a contest of aimed shots, Miller's people held the advantage.

Miller's men nailed one black-clad foe near the southern grape hut. A machine-gun team thought they knocked over another. Going out to check would have to wait for another day, for fear of the IEDs likely planted between the two sides. The engagement kept going, mostly rifles now, with the powerful U.S. M-240 machine gun being held for an occasional burst. Rather than their all-too-common "spray and pray," individual Afghan soldiers actually tried to find targets and shoot single rounds. Zacharias beamed.

The weather didn't allow for helicopter support, but about 5:15, two U.S. Marine Corps F/A-18D Hornet fighter jets checked in. Miller turned to his Air Force controller. "Bring one down for a low pass to identify targets," he directed.

A few minutes later, maybe a hundred feet up, racing ahead of its sound, a massive gray airframe suddenly materialized. It scorched overhead from west to east, both jet engines blazing like yellow-white stars. A wave of heat washed over the firefight, giving way to the throaty rumble of the jet. Pressing sensations pushed you down to the earth, even if you knew what it was, the U.S. deus ex machina come to deliver death.

As the Marine aviator pulled back up into the low clouds, his voice sounded on the air controller's radio. Matter-of-fact, as if describing holiday traffic on the interstate, the calm Marine confirmed at least three hostiles crouched behind a wall north of Taliban Hill. The insurgents were very close, within two hundred yards. This could get tough. Maybe a strafing run parallel to the Americans might work.

Miller decided to wait a bit. He told Nolan and Zacharias to cease fire. Sometimes merely bringing in a warplane scared off the adversaries. Maybe this was one of those times.

It was.

The intelligence people later determined that the enemy lost two men. The U.S. and ANA suffered one wounded, the lieutenant hit in the first minutes of the engagement. All in all, it made for a good afternoon's work, leading to a better week to follow. The ANA COP went into Mushan village to stay.

At ISAF headquarters, General John Allen didn't know the names of Danny Chen, Robert Bales, or Patrick Higginbottom, at least not at first. A suicide, a homicide, or a soldier wounded in action constituted "significant acts," *sigacts* in military lingo. The staff handled most of it. They boiled it all down, giving only the choice tidbits to the four-star commander. That's the way Allen preferred to get his information: distilled to the essence, with details left for subordinates. Most senior officers worked this way. It certainly suited John Allen well. He consistently sought the critical measurements, the correct bits of data.

A 1976 Annapolis graduate, Allen earned his degree with honors in operational analysis, the use of numerical metrics to determine progress and value. A contemporary of the Class of 1976 West Pointers Stan McChrystal, Ray Odierno, and Bill Caldwell, Allen shunned the spotlight, although he received some press attention simply because he performed so extraordinarily well. As commander of the Second Battalion, Sixth Marines in 1995, he was noticed by author Tom Clancy. "He is always alert. If you watch his eyes, they are always moving, always taking note of details." A gentleman in every sense, Allen treated people of all ranks kindly and listened attentively, both notable traits in a senior general, especially a Marine infantryman. Many senior officers tended to transmit. Allen chose to receive.

Allen's background included the usual Marine assignments up through battalion level. He'd earned awards and recognition time after time, always the best, the first, the most able. But he'd performed some other duties that marked him as a man of unusual potential, demonstrating a strain of Petraeus-style political/military acumen unhampered by Malik Daoud's more open ambition. Allen held three master's degrees in national security studies. He'd served in the prestigious, highly selective ceremonial unit at Marine Barracks at Eighth and I Streets in Washington, and then taught political science at Annapolis. In addition, he'd attended the demanding Defense Intelligence College and graduated first in his class, a real distinction for an infantry officer. Allen's tenure in 2/6 Marines included something far more consequential than meeting Tom Clancy: contingency operations in the Balkans and a brush with combat. After his battalion command, he served as senior aide to the commandant of the Marine Corps.

At that point, John Allen's path diverged from the normal route. An officer like him typically commands a Marine expeditionary unit or a regiment, then earns his star. Instead, Allen commanded the Basic School at Quantico, which all new Marine officers attend. He then went on to become the first Marine appointed as the commandant at the U.S. Naval Academy. After promotion to brigadier general and duty on the Joint Staff in the Pentagon, Allen served as a deputy commander in Anbar Province in 2007 and 2008. He played a key role in accelerating the Sunni Awakening. Promoted to major general and then lieutenant general, Allen became the deputy commander at U.S. Central Command under Dave Petraeus and his successor, Marine general Jim Mattis. When asked who should follow him to command ISAF, Petraeus confidently tagged John Allen. The quiet gentleman got the job.

After Dave McKiernan's removal, Stan McChrystal's untimely demise, and the larger-than-life Dave Petraeus persona, taciturn, thoughtful John Allen looked to be the perfect officer to guide the transition of Afghanistan from an ISAF counterinsurgency campaign to national authority under the Kabul government. And yet, out in the force, among the upper ranks, there were reservations.

Nobody questioned the idea of placing a Marine in charge. Indeed, the generals in the Marine Corps had long overcome the prejudices once expressed by the otherwise open-minded General Creighton Abrams. Asked to accept a four-star Marine deputy in Vietnam in 1968, Abrams replied, "None in the Marine Corps have the full professional qualifications." By the time John Allen took command of ISAF, Marines had performed with distinction on multiple occasions at the four-star level, including as chairman of the Joint Chiefs of Staff, vice chairman of the Joint Chiefs of Staff, and Supreme Allied Commander Europe. Since the 1990s, Marine generals had commanded U.S. Atlantic Command, U.S. Central Command, U.S. Joint Forces Command, and U.S. Strategic Command. Indeed, the founding commander of what became the vital U.S. Central Command was General P. X. Kelley, later the commandant of his service. Schooled from induction to work closely with the larger services, Marines proved able joint commanders.

Allen's issue — for those who thought he had one — involved his lack of relevant command time, especially in combat. Service as a deputy in Anbar, however estimable, didn't match the experience of men who'd commanded divisions and corps in both Afghanistan and Iraq. John Allen had made his name as a staffer and deputy, not as a field commander. But people had leveled that charge at Dwight D. Eisenhower, and he'd turned out pretty well. The grumbling never really went anywhere.

Like Eisenhower, Allen excelled as the commander of a large multinational headquarters. He made time for the various senior NATO officers and, like most of them, preferred to run the war from the command post. Allen cared deeply for his men and women, but didn't mix easily with the troops. You wouldn't find him on foot patrol in Zharay like Stan McChrystal, or out pressing the flesh and doing challenge pushups like Dave Petraeus. John Allen supervised his campaign from Kabul, via e-mail, video-teleconferences (VTCs), and occasional formal gatherings. When he got out and about, he gladly pinned on medals and thanked young people in uniform. He spoke well, and looked every bit the part of the general in command. But mainly, Allen favored conversations with senior officers and formal briefings. He especially liked to look at numbers on slides.

Few matched his nose for statistics. He absorbed numbers and sorted through them quickly, seeing subtle connections that others missed. Ever polite, he also unerringly asked the right questions. "I am going to manage you by slides," Dave Petraeus once told his staff In Iraq. John Allen outdid even Petraeus in that category.

The trust Allen placed in the virtue of these lengthy sessions belied the presence of a key defect. The data always seemed suspect. Many metrics hung on things that defied numbers. How did you measure the annual poppy cultivation? This illegal activity provided spending money for Pashtun subsistence farmers and, incidentally, financed the Taliban. It wasn't like the locals registered their illicit crops. Analysts resorted to guesses based on aerial surveillance and patrol reports. Then they rolled them up. Over time, the numerals gained a certainty not present when the inputs were gathered.

Similar things happened with Afghan national army and police readiness ratings. Military staffs assigned ANA and ANP units to various readiness categories — newly established, developing, effective with assistance, effective with advisers, independent — and created numeric charts of the same. Yet the underlying assessments, excepting a nose count of men and weapons, largely reflected subjective views of how the Afghan leaders performed. Again, over time, the subjective estimates were taken as concrete facts.

The problem with numbers in Afghanistan related to the overall difficulty of combat reporting. All the computers and spreadsheets on earth couldn't change that basic old rule: garbage in, garbage out. People under fire, or those slumped in exhaustion in small malodorous COPs or cursed with balky IT systems, tended to be late and incomplete in filing summaries. Americans reported more accurately than many other ISAF partners. Afghan reporting varied from adequate to ludicrous and rarely met Western standards of timeliness and thoroughness. Initial notices seldom got it right. It took a week before anyone could figure out how badly things had gone at COP Palace, a day or so to sort out the slaughter outside COP Belambai, and it's unlikely John Allen ever heard about that long afternoon in Mushan village. Every day, a selection of various sigacts found their way into the hourlong ISAF morning-update briefing, tagged in small, neat call-out boxes on map slides. If the day before had been slow, an incident like Mushan could bubble up. On a busy day, though, only the big ones made the cut. The Belambai killings certainly did, once it became clear something really terrible had happened.

General Allen's daily briefing brimmed with numbers. The presenting officers read from a preapproved, fully rehearsed script. Although the array of subordinate commanding generals, such as ISAF Joint Command, NTM-A, and the special operators, all followed along on a VTC hookup, this show played to an audience of one. Earnest majors and lieutenant colonels, some not native English speakers from NATO countries, read through the patter. The great man absorbed, assented to proceed, or perhaps asked a question. The usual reply to Allen's unfailingly courteous query became a standing joke around ISAF: "Sir, I'll get back to you." Answers were not in the script. So junior briefers dared not go off the beaten path. The ranks of generals and admirals who assembled daily in the ISAF headquarters typically sat mute, mesmerized by the colorful slides and soothing voices, like cobras swaying to the tune of a fakir's flute. Day in and day out, the performance continued.

One set of numbers got the most attention. Enemy attacks became the metric of metrics, the indicator that above all measured progress for ISAF. But what constituted an attack? Certainly the fracas at Mushan that wounded Pat Higginbottom qualified. But how did the staff count the roadside explosion that took the legs of the Afghan policeman earlier that day? What of the four IEDs found and two blown by Sergeant Hamid and his American partners? Was it all one attack? Two? Three? Six? Counting rules evolved, as detailed and arcane as those developed by the U.S. Internal Revenue Service. For the record, the March 26 attacks around Mushan totaled three: the initial IED on the road, the group of four IEDs in the village, and the firefight.

Brilliant mathematicians and dedicated intelligence officers sliced and diced the attack numbers, but they always showed the same two trends: Enemy action rose in the summer and fell in the winter. You could bank on that, especially in the colder, snowy east and north. Additionally, the overall attack numbers stayed about the same, with a slight decrease as ISAF forces began to reduce in 2011 and 2012. If you thought you saw anything other than a stalemate, you were kidding yourself.

The ISAF bar charts stayed stubbornly consistent. Attacks peaked during the 2010 surge (4,500 per month in the favored Taliban fighting season of June, July, and August) and fell off consistent with the U.S. and NATO troop drawdown in 2011 (4,100 per summer month) and 2012 (down to 4,000). But they never dropped back to the summer 2009 levels (2,600 per summer month), a period when Stan McChrystal assessed the campaign as failing. By cranking down to subcategories, ISAF statistics experts sifted the data ever more finely, claiming to see small but significant trends. One that grabbed General Allen's attention indicated that, over time, more hostile attacks occurred farther away from population centers. Well, maybe . . . whatever that meant. An objective observer assiduously dug through all the figures and charged that ISAF accounting techniques "shows how a carefully rigged portrayal of the trends in EIAs [Enemy Initiated Attacks] can exaggerate progress." You had that sick feeling in your stomach you were looking at the hamlet evaluations from outside Da Nang, circa 1967.

John Allen was an honest and honorable man. As an experienced operations research analyst, he understood that the numbers were not adding up. Directed by President Obama to transition to the Afghans by 2014, ready or not, Allen did his best to play a weak hand well. Despite his stated antipathy for senior Marines, General Creighton Abrams would have certainly sympathized with General John Allen, having endured the same bleak sequence in from 1968 to 1972 as the U.S. commander in Vietnam. It was not a happy precedent.

As the numbers didn't look all that good, the ISAF commander concentrated instead on doing what he could with transition. Allen had inherited a schedule, and that number, December 31, 2014, looked likely to hold up. The north, west, and Kabul all remained fairly quiet, as did parts of the east, south, and southwest. Why not claim credit where the going was better?

The Afghans called it Inteqal, Dari for "transition." Having prevailed in the messy 2009 election, President Hamid Karzai embraced the transition effort fully. As his minister to the Joint Afghan-NATO Inteqal Board, Karzai named his former finance minister and one-time rival Ashraf Ghani Ahmadzai, America's favorite Westernized Afghan. Glib, personable, and independently wealthy — few in ISAF asked questions, but perhaps they should have — Ghani became the authority on which districts and provinces went under Afghan government authority. The shrewd new U.S. ambassador, Ryan Crocker, who knew Ghani well from previous days in Kabul and remembered Allen from Iraq, provided an able, helpful voice alongside the ISAF general.

The Afghans placed districts and provinces into transition in several groups, known as tranches, reflective of the original British proposal that spawned the entire arrangement. The first began under Petraeus on March 22, 2011. The second (November 27, 2011), third (May 13, 2012), and fourth (December 31, 2012) happened under Allen's oversight. By the fourth batch, twenty-three of thirty-four provinces and 87 percent of the Afghan people officially lived under the sway of the Kabul regime. Supposedly, when that occurred, ISAF forces would be greatly restricted in their operations, compelled to adhere to Afghan law, a slippery concept at best. Although they all eventually entered it, not one district ever really emerged from transition. In truth, that awaited the departure of ISAF.

It all made Ashraf Ghani and Hamid Karzai quite happy. In reality, it meant little on the ground. Karzai's writ ran to the outskirts of Kabul. Aside from his Afghan national army and police, plus a few overworked appointed officials, most Afghan citizens got nothing from their government beyond a grasping hand. At least his obnoxious half brother Ahmed Wali Karzai was no more, having been assassinated by a rogue bodyguard the same month John Allen took command of ISAF. Things improved somewhat in Kandahar.

Even as the Inteqal process ran on, Allen devoted much more attention to ISAF teamwork. He adopted NATO's idea of "in together, out together," even as key countries began to shift from combat to supporting roles. Canada did so in 2011. France followed the next year. Other countries did too. It all kept the ISAF commander busy, as both the U.S. and NATO wanted him to develop options for a force to remain after

2014. Allen spent way more time than he liked flying to Brussels, hosting foreign dignitaries, and participating in hours of laborious VTCs with the U.S. Department of Defense, Department of State, and the White House. Although he had no choice but to send the surge troops home in late 2011, Allen successfully kept sixty-eight thousand Americans in country through his tenure. Come late 2013, then early 2014, though, they would begin to depart pretty quickly. NATO would do the same.

As he prepared for American troop numbers to go down, Allen faced some difficult choices. The main contests remained in the southwest, south, and east. In a perfect world, all U.S. combat power belonged there, killing Taliban and securing villages. Yet the German-led north, the Italian-led west, and the Turkish-led capital all demanded valuable American enablers: route-clearance teams, medevac helicopters, overhead surveillance drones, and certain SOF and advisory elements. Keeping the allies in country required U.S. help. It amounted to a tax of about ten thousand men and women, plus highly useful equipment, deducted from what would soon be steadily dwindling American manpower. With sixty-eight thousand U.S. troops, that levy made sense. With forty thousand or fewer, it did not. But agreeing to "in together, out together" ensured that John Allen propped up the allies. So it went, much to the frustration of understrength American elements hungry to own those less active units hanging around up north, out west, and in Kabul.

The final aspect of transition involved that most difficult of Afghan challenges, President Hamid Karzai. Compassionate and decent by nature, Allen found himself ensnared in a seemingly endless cycle of Karzai complaints, ISAF apologies, and then U.S. concessions. Within months, battered by Karzai about the supposed horrors of civilian casualties and night raids, Allen reimposed and then exceeded the McChrystal strictures on firepower. Every allegation, even the most trivial, got a full investigation, often superintended by a general officer. When American and Afghan military police soldiers at the Bagram detention center accidentally burned some Korans marked with detainee codes, Karzai went ballistic. Encouraged, street mobs did likewise in several cities. Allen came hat in hand to ask for forgiveness. And Karzai almost granted it. Then came the Belambai massacre. It was always something.

On John Allen's watch, the U.S. formally signed over the Bagram detention facility, although Americans continued to play a very active role advising the Afghan soldiers there. Allen also agreed to significant limits on ISAF night raids, as well as certain aspects of the Afghan local police program. Given the ongoing resiliency of the Taliban, this made little military sense. But it kept Karzai quiet. And that in turn kept Washington quiet.

On May 2, 2012, a year to the day after Osama bin Laden's death and just prior to a major NATO summit in Chicago, Illinois, President Barack Obama made a dramatic night flight to Kabul to meet with President Hamid Karzai. Together, the two men signed the "Enduring Strategic Agreement Between the United States of America and the Islamic Republic of Afghanistan." Those who read it noted many word-for-word matches with the very similar 2008 document signed in Baghdad, Iraq. As Ambassador Ryan Crocker had negotiated both agreements, that only made sense. It also offered the options that Obama demanded, essentially committing America to nothing and suggesting that Afghanistan would be fortunate indeed to get what Iraq had gotten after 2011. If he'd really wanted to see what came next, Hamid Karzai should have taken a long, hard look at Baghdad.

The man who'd made his name in Baghdad and who'd recommended General John Allen to command ISAF also played a major role in Allen's departure, although not intentionally. Busy running the CIA, David Petraeus got some quite unwelcome news in early November of 2012. In pursuing an Internet stalking complaint, Federal Bureau of Investigation agents uncovered evidence of an extramarital affair involving the CIA director. During his time in Afghanistan, Petraeus had developed a relationship with Paula Broadwell. One of the general's several biographers, Broadwell was a married Army Reserve officer, fellow West Pointer, scholar, and athlete. Apprised of the incriminating information, the CIA director tendered his resignation, which the president accepted. On November 9, 2012, David Petraeus made a public admission regarding the matter. It ended speculation about the general's future in American politics.

The fallout extended past Petraeus. Broadwell apparently suspected another woman, a Tampa socialite, of having an untoward interest in Petraeus. In the subsequent official inquiry, it turned out that General John

Allen kept up a string of e-mail messages with the woman in Tampa, who was also married. The news media pulsed with innuendo linking Allen and the other person to Broadwell, Petraeus, and their activities. Suggestive articles hinted at an inappropriate relationship. A thorough investigation followed. Allen's nomination to become the next Supreme Allied Commander in Europe hung in limbo during most of November, December, and into January of 2013. The ISAF commander tried to keep his mind on his duties, no easy task given the ongoing journalistic feeding frenzy. After months, the results came back, clearing the Marine general of misconduct.

It was too late for John Allen. His reputation unfairly impugned, a gentleman to the last, the general asked the president to allow him to retire rather than take the NATO supreme command. Obama agreed. On February 10, 2013, Marine general Joseph F. Dunford Jr. took command of the International Security Assistance Force and U.S. Forces–Afghanistan. Determined to stay out of the limelight, Dunford had one mission: Wrap it up.

18

GREEN ON BLUE

> We are content with discord. We are content
> with alarms, we are content with blood . . .
> but we will never be content with a master.
>
> — AFGHAN ELDER TO BRITISH ENVOY, 1809

AFGHAN SOLDIERS FOUGHT. Their tactics, at times, left a lot to be desired. They didn't always wear all their body armor. They occasionally left their helmets back at the COP. On static guard posts, it took concerted effort to keep them awake all night. American sergeants observed that, like their Taliban cousins, Afghan *askari* seemed to be solar-powered. It reflected an upbringing in a rural society without electricity. Turn off the sun and they turned off too. But they kept their weapons clean, which was easy when they'd carried simple, robust AK-47s, tougher with their new, more touchy U.S.-issue M-16 rifles. When they made contact, they opened fire and moved to close with the enemy. You could forgive a lot of failings in soldiers as long as they fought.

One thing, though, nagged from the start. Way back in that autumn of 2001, a favorite quip acknowledged that you couldn't buy Afghan loyalty, but you could rent it. Various tribal chiefs switched from the Taliban to the Northern Alliance after a display of U.S. bombing prowess accompanied by the timely delivery of a few cases of cash. Over time, more than a few turned back or, in more typical Afghan fashion, went their own way. The U.S. CIA and Special Forces teams knew from long experience not to get too comfortable with their new friends. Afghans might temporarily work for you, but they always worked for themselves.

The British learned that the hard way during their ill-fated 1839 to 1842 expedition. Determined to block imperial Russian expansion, the British sought to annex Afghanistan as a buffer state, a strong shield for the rich British holdings in India. The Grand Army of the Indus marched north out of Quetta in modern Pakistan. The armed column of sixteen thousand British regulars and British-led Indian troops, as well as some thirty-eight thousand camp followers and servants, sixteen thousand camels, and even the foxhound pack of the Sixteenth Lancers, invaded to install a pro-British regime under Shah Shuja, a Durrani tribe Pashtun just like Hamid Karzai. A British officer and his contemporaries entered Afghanistan with high aspirations, seeking "so much promise of distinction and advancement." The British seized Kandahar in April of 1839, Ghazni in July, and Kabul in August. Shah Shuja took the throne backed by British bayonets. So far, so good—hapless Pashtun tribesmen fell by the hundreds before British musketry.

Over the next two and a half years, things deteriorated. The Afghans, particularly the Pashtuns, refused to submit to the infidel-loving Kabul regime. Spurred by Ghilzai tribe Pashtuns—Mullah Omar's ancestors—a succession of snipings, ambushes, and provocations occurred at various British outposts, rising in number over the months. "One down, t'other come on, is the principle with these vagabonds," observed the senior British envoy. "No sooner have we put down one rebellion than another starts up." Throats were sliced. Supply wagons were burned. Small detachments got cut off and cut up. In far-off London, the government looked at the mounting resistance, casualties, and financial toll. Unwilling to keep pouring money down the drain in Kabul, with better uses for British regiments, London decided to cut its investments in the Hindu Kush.

When Shah Shuja's inner circle got wind of a planned British reduction of cash subsidies and troop withdrawals, they reacted in the normal Afghan manner. Even as they smiled at their British benefactors by day, the senior Afghans arranged clandestine meetings and cut midnight deals with the aggressive Ghilzai chieftains. On November 2, 1841, as temperatures plummeted and snow fell in Kabul, a well-armed, riotous Afghan mob surrounded, penetrated, and burned the British residency. The attackers pillaged the treasury too, taking seventeen thousand pounds in currency. Most alarming of all, hundreds of Kabul rebels overran and looted the military commissary, carrying off or destroying British food, ammunition, and medical stores. Ghilzai men encircled the British cantonment, and even brought up some captured cannon. Essentially, the British were under siege in their Kabul compound. The nearest support was in the British garrison in Jalalabad, ninety miles distant along a narrow, rocky trail that crossed snow-covered mountains swarming with Ghilzai warriors. A brutal winter retreat loomed.

The British commander wasn't up to it. Major General William George Keith Elphinstone enjoyed a reputation as "the most incompetent soldier that was to be found among officers of the requisite rank." Even as his opponents surrounded and picked off his cold and hungry soldiers, Elphinstone dithered. Knowing he lacked the combat power to fight his way out, inert and indecisive by nature, the general put his trust in talk. He hoped to negotiate a safe British departure with the implacable Ghilzai chiefs. Days passed. Casualties mounted. Food supplies dwindled. The winter grew colder, and snow piled up in the key passes that led to Jalalabad. Meanwhile, truce talks dragged on and on, interspersed with sharp enemy forays against the weakening British perimeter. Shah Shuja, of course, publicly backed Elphinstone and privately encouraged the Ghilzais. After weeks of this charade, his trapped countrymen suffering and desperate, the game British political envoy at last agreed to the humiliating Ghilzai terms. During a follow-up meeting, the Afghans seized and killed him. In return for the assent of a dead British official to what amounted to a conditional surrender document, the Ghilzai leaders graciously decided to allow a withdrawal. That finally stirred Elphinstone to action.

On the bright cold morning of January 6, 1842, after issuing orders to withdraw all British garrisons in country, handing over all remaining money, and leaving behind almost all remaining supplies and most of his artillery, Elphinstone's forces pulled out of Kabul. The forty-five hundred troops and nearly twelve thousand civilians, including Indian servants, British families, and terrified Afghan collaborators, trudged off into the snow. Behind them, jubilant Afghans set fire to the former British barracks and ravaged the rear guard. That contingent abandoned two cannon and fifty wounded soldiers as they fled to join the staggering march column. For a week, the thousands endured an "attendant train of horrors—starvation, cold, exhaustion, death." Ghilzai marksmen shot down soldiers and civilians alike, even as their grinning leaders parlayed with the befuddled British commander. Over the days, lacking food, beset by the unforgiving weather and brutal terrain, Elphinstone agreed to place women, children, and wounded under Ghilzai "protection." Vicious Pashtun gunmen and cruel knife wielders promptly slaughtered all of the column's Afghan affiliates and most of the Indian servants and soldiers, keeping the British alive as hostages, presumably to gain ransom money. Elphinstone surrendered personally on January 11, trusting that might mollify the Ghilzais. It did not.

Victorious Ghilzais overran the pathetic remnants of the force on January 13. That afternoon, one cut-up, freezing, wounded man, surgeon William Brydon, rode into Jalalabad on a staggering, bloodied pony. Some of the British captives, too, eventually made it out. But the entire debacle convinced later British commanders that treachery was too much a part of the Afghan culture, that the country could not be pacified or befriended, and that peril mounted once the locals found out you were leaving. In their 1878–80 and 1919 wars, the British went in, did the hard business as quickly as they could, and pulled out. After President Obama's announcement in June of 2011, mindful of the adamantine lessons of their long imperial heritage, British officers raised valid concerns as ISAF began to talk of drawdown and transition. An easy exit could not be blithely assumed.

The Russians, too, recognized and warned of this problem. During their long Soviet-era campaign, from the initial advisory period of 1973–79 through the major intervention of 1979–89 and then in the final reversion to advisory work in 1989–91, the Russians sometimes found their Afghan partners to be duplicitous indeed. As early as March 15, 1979, before the major Soviet incursion, a bloody clash occurred in Herat. Angered that the new socialist government in Kabul had directed the schooling of females, several battalions of the Seventeenth Afghan Division revolted, joining civilians already up in arms. Angry Afghan *askari* turned on their Soviet advisers, killing one major and two civilians. Alarmed, the Russians evacuated their families from Herat under cover of darkness. The resultant reprisals, including aerial bombardment, slaughtered at least three thousand Afghans in and around the city. Sensational tales circulated about the deaths of hundreds of Soviet officers as well as Russian women and children. None of that happened, but it almost did, and the boisterous Pashtun Street enjoyed the resonance of the legend. To this day, you'll find the story in many Western sources. Former Afghan mujahidin repeat it as gospel.

Overblown though accounts of the Herat mutiny might be, the insider threat was all too real. The Russians admitted to two events in addition to Herat, one in Paktia Province in the east and another in Baghlan Province up north. There were more, conveniently wished away as friendly fire or training accidents. Treachery didn't accord with the glorious socialist brotherhood under way in country, so reports were sanitized. To this day in Russia, it remains so. Yet the Russians understood the challenge, and they addressed the danger with some success. Small Soviet adviser teams in particular watched for it. Like their ISAF successors, the Soviet advisers lived with their Afghan counterparts on that era's equivalent of COPs. Already accustomed to living their lives under party surveillance and the heavy hand of the secret police, Soviet Russians exhibited a knack for keeping an eye on their Afghan partners. The Russians preferred positive measures, such as shared training, combined patrols, sporting contests, and common meals. Party meetings also had some effect on teamwork. If nothing else, the Afghans received a daily dose of propaganda, and over time, repetition in the absence of other news exerted influence. That said, in a pinch, if an Afghan acted oddly, immediate arrest, swift evacuation, and even execution also worked just fine. Soviet officers needed no permission to deal peremptorily with a suspicious Afghan *askar*. When the Fortieth Army began its withdrawal, the Soviet leaders increased their vigilance toward their Afghan teammates. With a few exceptions, it worked.

As a result, the Soviet departure in 1988–89 went better than expected, certainly far better than Elphinstone's catastrophic British retreat. There were ambushes and convoy clashes, and some Afghan units changed sides. But in general, the Afghan army stayed loyal, as did the Kabul regime. Of course, both the Soviets and their Afghan proxies expected aid and advice to continue, so the removal of Russian regiments did not signal abandonment. More than three hundred uniformed advisers remained in place. Aid continued to the tune of three hundred million dollars a month, no mean amount in the cash-strapped, wheezing Soviet economy. Some four thousand air-transport sorties continued throughout 1989, as well as six hundred truckloads of supplies weekly. The logistics push delivered spare parts, ammunition, and new weapons, including tanks, jet fighters, and Scud missiles. The Moscow authorities also continued to ship fuel and food. It kept the socialist Afghan forces in the field, and they for their part held off the disorganized mujahidin. When the Soviet Union collapsed in a heap on December 26, 1991, that help came to an abrupt end. The few remaining Russian advisers departed by air before the Kabul regime realized they had been left in the lurch. That was one way to obviate Afghan perfidy: betray them first.

Such measures were not available to the ISAF adviser teams still dotted across the country as troop withdrawals began in earnest after the summer of 2012. Notably inattentive to history, few Americans knew anything of the British defeat in 1842 or even much about the Soviet pullout of 1988–89. But they knew the Durranis, the Ghilzais, and the other Pashtun tribes only too well. Those hardy characters certainly remembered the British and Russians, and what they'd done to them. Mao's guerrilla formula held: *Enemy tires, we attack. Enemy retreats, we pursue.*

Colonel Keith A. Detwiler knew quite well what had happened to the British in 1841–42 and the Russians in 1988–89. In command of the NTM-A detachment in Herat, site of the 1979 rising celebrated among Afghan jihadis then and now, Detwiler understood how bad things could get. The tall, quiet, tough West Point graduate, trained in the infantry, went to learn Russian and become a foreign area officer, one of the handful of highly educated, uniquely experienced soldiers prepared for especially selective military-diplomatic assignments. Karl Eikenberry, retired lieutenant general and former ambassador, exemplified the promise of this program. Keith Detwiler showed as much potential as Eikenberry ever had, probably more. In addition to a combat deployment to Iraq to hunt for weapons of mass destruction (not much luck there), Detwiler had also joined the effort to accelerate building Afghan forces in the early days of NTM-A. Most relevant to understanding the perils of close dealings with Afghans, Detwiler had served as a liaison officer with a Russian airborne brigade in Bosnia and as the U.S. defense attaché in Belarus. The colonel heard plenty of stories of the Soviet war in Afghanistan. The Russians respected the Afghans as warriors—not soldiers in the Western sense, but cunning fighters, wily killers, cut-throats. Given opportunity and motives, the throats they sliced might well be American. In the autumn of 2012, with talk of the U.S. pullout in the air, with units leaving without backfills, you had to be damn careful.

The senior leadership initially labeled the phenomenon "green-on-blue," referring to an attack by the Afghan military or police, green on the computer maps, against ISAF troops, marked in blue. Such attackers might as well have borne the Taliban red, because they struck at the heart of the war effort, the trust between the uniformed Afghans and their ISAF comrades. Counts of such incidents varied. The reporting rules changed over time. As the assaults gained more attention, ISAF senior people began to direct exactly what was and was not a green-on-blue event. Indeed, at one point, the ISAF commander directed that the term be changed from *green on blue* to *insider attack*. The slide titles and press releases changed, but on the COPs, the troops still called it green on blue.

Regardless of which set of numbers you examined, the trends became obvious. By one count, there were two such attacks in 2008, five in 2009, five in 2010, fifteen in 2011, and forty-two in 2012. The 2012 acts killed thirty ISAF personnel and wounded fifty-nine. Some argued that the numbers went up because of the U.S. troop surge, although if that was so, the curve lagged by a year. Almost all green-on-blue engagements involved a single Afghan turning on unarmed or unprepared Americans or ISAF partner troops, typically inside a secure compound. The subsequent response often killed the assailant, who rarely left anything defining his motivation. Although the Taliban immediately took credit for each attack, only a quarter of the incidents could definitively be tied to infiltration or recruitment of the Afghan forces. The presumed causes of the rest were evenly split between some sort of psychiatric condition, a personal dispute, a generic hatred of outsiders, or "unknown." Blaming a uniquely Afghan strain of *le cafard* wouldn't have been out of line either. The precise incentives remained obscure. Without a doubt, however, the biggest jump in attacks came after the June 2011 announcement of the proposed U.S. and consequent NATO withdrawal and transition plans, all aimed at mission completion by December 31, 2014. Evidently, some Afghans decided to express their real feelings before the infidels left.

To add to the murk and mistrust, the Afghans also assaulted one another. So-called green-on-green incidents tracked with what happened to ISAF troops. In 2012, for example, the Afghans sustained at least twenty green-on-green actions — you never knew for sure with their accounting — that resulted in thirty-eight killed and fifteen wounded. Again, the Taliban both encouraged and claimed every attack. Each treasonous strike, green-on-blue and green-on-green, added to the steady attrition caused by IEDs, mortars, rockets, and Kalashnikov bullets. The insidious nature of friends turning on friends greatly weakened resolve in distant ISAF countries, many of which had long since soured on the great Afghan adventure.

A commander like Keith Detwiler had to think about this peril day and night. He and his handful of advisers lived with the Afghans, directly in the line of fire as the threat of green-on-blue attacks increased. In Detwiler's case, he and his teams labored under one other complication: the colonel's organization operated way outside the major American force envelope. In fact, by October of 2012, it was one of the few remaining pieces of the U.S. footprint out in RC-West. The Italians ran the regional command. Their lineup included the Albanians, Salvadorans, Georgians, Hungarians, Lithuanians, Slovenians, Spanish, and Ukrainians. Of course, as the allies demanded and the ISAF/U.S. Forces–Afghanistan commander disposed, the Americans provided an aviation task force and some other key engineer, logistics, and intelligence elements. Small in numbers but tied to vast financial resources and a nationwide network of NTM-A advisers, Keith Detwiler's Regional Support Command West ensured institutional advice and support for ANA and ANP training centers and supply depots in the west. Detwiler and company tried to provide this without getting trapped and killed by their Afghan partners, 99.99 percent of whom were fine and upstanding. But it took only one.

On the morning of October 30, 2012, Keith Detwiler, his engineer officer, his contracting officer, a logistics officer, and a security platoon prepared to take the long ride from the outskirts of Herat City to Camp

Shouz, seventy-five miles to the south. It was one of the more distant stations of his sprawling realm. As the NTM-A commander supporting RC-West, Detwiler supervised a multinational, multiservice roster of 250 uniformed training advisers, logistics mentors, support and security troops assigned to eight separate sites spread across Herat and Farah Provinces. He also looked after an equal number of civilian contractors, very much part of his training and assistance force. In addition, as the NTM-A commander in RC-West, Detwiler coordinated closely with the 240 air advisers at Shindand Air Base, about fifty miles south of Herat on the main road to Shouz. Those people helped the Afghans train their pilots and ground crew, and although they reported formally to the 438th Air Expeditionary Wing in Kabul — the Air Force insisted that airmen work for airmen when doing flight duties — Keith Detwiler made sure they had what they needed. On top of all of this, he oversaw the distribution of NTM-A's considerable U.S. funding, the RC-West share of the ten billion dollars annually provided to pay, clothe, house, train, equip, arm, fuel, and feed the Afghan forces. In that capacity, Detwiler and his busy detachment looked after some seventy million dollars being spent on more than eighty separate ongoing military construction projects. They also checked on the effectiveness of 173 separate service contracts that did everything from overhauling damaged Afghan Humvees to pumping sewage out of police-post latrines. Every aspect of the Afghan national army, air force, and police sustainment in RC-West eventually traced back to Keith Detwiler's advisers and funding. For Detwiler, it meant a lot of time out and about, riding the circuit.

Their task for the day involved assessing the transfer of a barracks and training complex that cost $3,991,635 of U.S. taxpayer funds. The Afghan National Civil Order Police (ANCOP) had assumed authority a few months earlier. Detwiler heard the Afghans had encountered some trouble with operating the camp utilities and pinning down the life-support contracts. Such fumbling often arose when the Afghans took over a facility. It was time to go take a look. On the way, Detwiler planned to check in at two other locations. It broke up the trip and cleaned up other nagging matters.

Getting around wasn't easy. Most of the generals in country hopped aboard helicopters and off they went. But some generals, and most colonels, moved by road. It took time, it raised risks, but you saw a lot more, dirty unpleasant things beneath the notice of those clattering along a thousand feet up. Every movement in Afghanistan, though, constituted a combat patrol. You went out armed and armored.

Detwiler's security platoon, his ticket to ride, came from Company F, First Battalion, 167th Infantry, Alabama Army National Guard. They reminded all of their Civil War heritage as the Fourth Alabama. Hometown cohesion forms the strength of National Guard, and 1-167 Infantry, recruited from in and near Oxford, Alabama, Company F adopted the nickname "Vigilant." They had lived up to it since taking over security duties for NTM-A Regional Support Command–West in August of 2012. Their three platoons of M-ATVs had run more than 650 missions, an average of seven a day for the company. Their commander, no-nonsense Captain Kurt Buchta, emphasized fundamentals of marksmanship, navigation, first aid, and physical fitness. He consistently schooled his soldiers on the reason for the company's moniker. They ran each mission ready to make contact and win. "Vigilance is how we do our job," he said.

Just before 8:00 a.m., four M-ATVs lined up to roll from Camp Stone, right next to the 207th ANA Corps headquarters, logistics site, and training center at Camp Zafar (*zafar* is Dari for "victory"). The wide tan armored trucks each carried four Alabama Guardsmen: a driver, a truck commander, a turret gunner manning the machine gun, and a rifleman in the left rear seat, ready to get out as required. Into the open right rear seat in each M-ATV went Colonel Keith Detwiler or one of his three staffers. Detwiler rode in the second truck with the platoon leader.

For a tall soldier like the colonel, squeezing into the M-ATV took some effort. He had to duck his helmeted head to get through the heavy, hatch-model steel-and-Kevlar layered door. Its bright metal pneumatic piston hissed as the heavy cover clanked shut, producing a sensation akin to being locked in a bank vault. Once he was seated, it took some squirming to get bulky body armor, the carbine, and all the protruding ammunition pouches and the like settled properly in the narrow space between the seat cushion and the leaning frame of the truck commander's position just to the front. Then came the five-strap nylon seat belt, two over the shoulder, two around the waist, and one up from the floor, all united in a hard metal circular lock neatly placed in front of the crotch. To the left, the gunner's lower legs, from knees to boots, grazed

Detwiler's arm as the soldier in the turret flexed and stretched. To the right, about at Detwiler's elbow, a small, thick, triangular window block permitted the colonel to look out if he leaned way down and squinted. He could also see a little through the slanted, equally thick windshield, although that necessitated some leaning to the left to see around the soldier seated in the front right seat. The RPG netting that hung across each window made it seem as if you were looking at the world through a tennis net, albeit one with little metal bolts clipped onto every right-angle intersection. The mesh limited visibility and attracted a lot of leaves and trash, but it pre-detonated RPG-shaped charges and encouraged hand grenades to bounce back toward those who threw them. With all of the protective measures, for a guy in back, trying to look out M-ATV windows was a duty, not a pleasure.

Better situation awareness came via the ubiquitous American IT. Directly in front of Detwiler, bolted to the back of the right front seat, hung a computer display terminal about the size and shape of a hardcover book. The bright silver-gray screen showed a map covered in blue icons, the Blue Force Tracker system. Older, clunky, Atari-era, not Xbox-generation, the device wasn't very elegant. But it did the job. It allowed reliable text messaging via a satellite uplink, valuable when out of line-of-sight radio range, a common condition in Afghanistan. And although it offered elderly graphics, it depicted your buddies. A militarized version of Google Maps, Blue Force Tracker featured live updates of all the pieces on the board, or at least the ones whose transmitters worked. If you saw a red marker — as occasionally you did — some G-2 guy had put it there based on a report of a Taliban ambush. Most worrisome were little outlined cones. Those were IEDs.

The computer ensured a degree of communications outside the truck. In military communications, redundancy matters. With this in mind, each soldier in the truck wore a black plastic headset, allowing normal-tone conversation on an intercom to tie together all five in the M-ATV. By flipping a thumb switch, you could transmit on the various radio nets, including the platoon internal loop, the Camp Stone command net, and the contingency frequencies, such as RC-West command, the different units holding the areas through which the patrol passed, and the vital air medevac channel. A control box let you choose how many of these radio nets to play in your headset. If you made the error of choosing all, the sounds in your head resembled a football stadium during a touchdown drive. Veteran soldiers learned to keep the number limited. Regardless, it meant that every drive added the clamp of the hard headset and the constant chatter of those in the truck and on the radios, most of it business, a good portion of it not. Some days you felt like talking or listening; some days you didn't.

With a radio check to the Camp Stone command post and a text update on the Blue Force Tracker, the four trucks moved off. They bounced out past the U.S. entry checkpoint. Two alert Company F sentries swung up a metal bar. A few minutes down the gravel trail, they rumbled through the larger, outer Afghan checkpoint. A thin *askar* in a green beret and another in an elderly Russian-style helmet waved and raised their yellow metal barrier gate. Ahead was Route 1, the Ring Road, the major Afghan highway. The platoon moved up to the lip of the four-lane asphalt highway — no traffic this morning, which was kind of unexpected for a Tuesday. The M-ATVs turned left, south toward Shouz.

After the trucks had made the turn, they shook out into an open column with about fifty yards between each vehicle. Before Stan McChrystal's time, U.S. and other ISAF trucks used to barrel down the road clearing all comers, employing gunfire if necessary. A few convoys from the more nervous ISAF countries and some of the "great man" motorcades in Kabul still drove in the old, brusque style. But by October of 2012, most ISAF trucks just moved to the right lane and let local traffic pass. The soldiers kept the gunners up in the turrets, watching the traffic and roadsides for signs of trouble. As a rule, on the open highway, M-ATV crews didn't permit Afghan civilian autos and trucks to get between the military trucks.

With a Kevlar helmet, ballistic protective glasses, a headset, body armor, and gloves, the strapped-in soldiers felt as encased as astronauts. As all the M-ATV's external armor made it impossible to roll down the windows, and as the IED threat made it inadvisable, the Army had thoughtfully provided air conditioning for the summer and heat for the winter. Of course, the outside air screaming in through that open turret ensured that the climate-control system always seemed to run a season behind. On this temperate October day, Detwiler's well-swaddled crew left it off.

As the M-ATVs settled into their cruising speed of forty-five miles an hour, the Afghan civilian traffic picked up. On this part of the Ring Road, cargo and fuel trucks predominated. Afghan drivers favored massive Mercedes-Benz stake-and-platform trucks, often piled high with furniture, food crates, or whatever sold in the Herat markets. Even enclosed trailers tended to carry more goods lashed to the top, a very popular method out west, where tunnels were not an issue. The truckers painted their rigs in garish blue, orange, bright crimson, and kelly green, a dramatic rainbow of color. The cabs featured painted eyes, like ships in the Arabian Sea. Intricate landscape and wildlife scenes decorated the trailer panels. As a final touch, most drivers hung a fringe of chains along the front and rear bumpers, and sometimes down the sides too. This aspect caused U.S. soldiers to dub them jingle trucks.

Along with the jingle trucks coming and going, other Afghan civil traffic used Highway 1 that day. Old sedans lurched by with families packed inside, children on laps and boxes roped to the roofs. They passed dirty Toyota pickups full of jostling sheep. A few puttering farm tractors dragged banged-up trailers piled with some kind of grain or hay. The panoply of vehicles all shared the road with the four M-ATVs. Every dozen miles or so, a cluster of one-story mud huts rose out of the hardpan desert. Jingle trucks stopped there for food or to peddle wares. Among every set of hovels stood a few corrugated metal containers, the kind used by Sealand, Hanjin, and Maersk. In Herat Province, they became homes and shops, an unanticipated American contribution to the local economy. Now and then, the discerning observer noticed in the background a line of smaller, rounded-off metal bins daubed with the remains of peeling yellowish paint. Those came from the Soviet army rear services. The Afghans kept it all and put it to use.

As for Taliban activity en route, there was none. The de facto potentate of the west, former mujahidin chief, sometime Karzai administration minister, and all-around strongman Ismail Khan preferred it that way. He enjoyed lucrative arrangements with the Iranians to the west and the Turkmen to the north. Insurgency was bad for business. Although the Soviets had faced stiff opposition in the region in their time, that was not the case with the Italians and their colleagues in RC-West. Incidents arose now and then, a rocketing here, small-arms fire there. But by and large, the west stayed calm. So it went on October 30 as well.

After almost two hours of trundling south, just over halfway to Camp Shouz, the M-ATVs came up on Adraskan Regional Training Center, an ANP site advised by the Italian *carabinieri* (national police) as well as four Americans. The four trucks diverted into the secure compound to give everyone a break, check the M-ATVs, get some fuel, and talk to the Italians. One of those rare attacks had happened there on June 25, when a well-aimed RPG shot from beyond the center's perimeter struck a guard tower, killing one Italian policeman and wounding three. This day, however, everything went fine. After an hour, Detwiler and his men headed out.

Another forty-five minutes south, after waiting for some slow-moving jingle trucks to pass, the M-ATVs turned into Shindand Air Base. Detwiler wanted to check in with the Air Force contingent there. Out over the long runway, a couple hundred feet up in the bright blue sky, an Afghan student pilot in a light gray, high-wing, single-engine Cessna 182 circled lazily. Shindand served as the principal training base of the new Afghan air force. A U.S. logistics task force shared the runway, using it to move supplies to the remote sites of RC-West. Now and then, somebody popped a 107 mm rocket into the vast base. Today wasn't one of those days. The air force squadron had some construction contracts to address. Detwiler's detour to Shindand lasted longer than he'd hoped. But by about one o'clock, the M-ATV column got back on the road.

It took another hour to get to Camp Shouz. The Ring Road got more potholed and beat up here in the hinterlands. Helmand lay one province over, and bad things happened out here: smuggling, kidnapping, and shakedowns. The Taliban didn't operate much here; the predatory local gangs disliked competition.

Just after two o'clock, the four M-ATVs slowed and turned left, to the east side of the highway. With a wave and a smile from the natty Afghan sentinels, the trucks pulled into Camp Shouz. Unlike Adraskan and Shindand, this place belonged wholly to the Afghans. A battalion of the Third ANCOP Brigade, the Afghan analogues of the Italian *carabinieri*, occupied the site. NTM-A funds had paid for the twelve metal-roofed buildings and the tall masonry wall. The ANCOP had been there a few months. Once the trucks got to the parking area near the headquarters building, they halted. Detwiler and his staff officers got out, gear on, weapons in hand. An ANCOP major immediately walked out and saluted.

Behind the four officers trailed four armed U.S. soldiers. Having suffered one of the first, and worst, green-on-blue attacks at the airfield in Kabul on April 27, 2011, with eight adviser airmen and one contractor (a former Army officer) gunned down by a jihadi Afghan air force major, NTM-A ordered use of designated armed, alert guardian angels when dealing with Afghans. Adviser work required close interaction with Afghans, and even with a loaded weapon, you could easily be distracted. By definition, conversing with Afghans took one's full attention and thereby opened a window to a skulking hostile infiltrator. Guardian angels watched for that. They stayed uninvolved with the adviser work, eyeballing the Afghans, aware of the people in the background, ready to intervene if anything went awry. Whatever ceremonial, planning, training, or social events transpired, the guardian angels remained suited up, rounds chambered, ready to shoot. Even in the Ministry of Interior and Ministry of Defense buildings in Kabul, NTM-A people used armed escorts, admittedly carrying pistols rather than rifles in those relatively secure facilities. It wasn't an overreaction. Two ISAF Joint Command advisers had died at the hand of a rogue ANP sergeant inside the Ministry of Interior operations building on February 25, 2012. With all of this in mind, riflemen of Company F filled the close security role at Camp Shouz that October day.

While Keith Detwiler, his other three officers, and the ANCOP major and his staff sat down for tea, the guardian angels stood by, weapons ready. The Afghan battalion commander was out on the road himself, as Detwiler hadn't announced his arrival. If you weren't sure of the Afghan unit — and Detwiler had never met these guys — it made sense to keep your schedule to yourself. So as the deputy, offered the chance to seek help from the Americans, the major explained the problems at the camp. Detwiler took notes as the Afghan unspooled the standard litany of requests: fuel for the generators, parts for the Afghan Humvees, and another gravel dump in the entrance, where rain and constant truck traffic combined to wash out a depression. The major expressed satisfaction with food preparation, water, sewage extraction, and ammunition supply. Discussion ensued on all of these points as the U.S. officers ensured they knew exactly what help to provide and consequently coached the ANCOP officer on what he could do for himself. The Afghan major wrote it all down, nodding gratefully. He then offered to take the Americans around the camp.

Outside the office, the M-ATV crews set up their own security ring around the trucks. When a gunner climbed out of the turret, another man wriggled up. On the ground, riflemen stood, watching the Afghans watch them. It wasn't an aggressive posture, just ready. Most of the ANCOP were out on missions. A few walked by, curious. Some waved.

On the second truck, the driver and sergeant noticed a pronounced list toward the left front tire. The damn thing was going flat. By design, a M-ATV can run on a flat tire. But it made no sense to do so when it could be fixed in this Afghan camp before they set out on the seventy-five-mile return trip. After checking by portable radio with Detwiler's guardian angels, the soldiers learned they had time to change the tire.

A M-ATV tire is big, about the size of a dump-truck tire in America. The M-ATV had a spare attached to the back. While their partners stood guard, the U.S. pair stripped off their helmets, weapons, and body armor. Then they pulled out the tools and got to work. Two other drivers also took off their Kevlar gear and pitched in to help wrestle the spare tire off the back end. This was a drill the men knew all too well.

The tire change took an hour or so. As that went on, Keith Detwiler and his accompanying staff trio followed the major to see the generator farm, the mess hall, the several barracks, the infirmary, and the motor pool. Overall, things appeared to be in good shape. The Afghans did not always take good care of new facilities, but this police unit demonstrated some decent discipline. As the group walked from building to building, the guardian angels kept the truck crews informed.

The ANCOP deputy finished his camp walkabout around four o'clock, just as the U.S. soldiers finished tightening the lug nuts on the new tire. Cleaning up probably demanded another fifteen minutes or so. That would be fine. The ANCOP major, another ANCOP officer, and the American leaders stood in front of the headquarters building, enjoying the sunny late afternoon. The other ANCOP officer experimented with his English phrases, derived from watching DVDs of American movies. He exhibited an impressive command of slang and swearwords.

Meanwhile, the American security men remained vigilant. Along with his teammates, Specialist John Yates kept his finger near the trigger well of his M-4 carbine, as he had since the tire change started. Security was boring, but Yates understood that if something went wrong, it would brew up fast. He'd heard of the green-on-blue ambush on July 22 at Camp Zafar, when a disgruntled Afghan police recruit attacked six U.S. Department of Homeland Security contractors as they smoked and talked after dinner. Three died, two were wounded, and the survivor killed the assailant. They had never seen it coming.

Yates did.

Across the parking area, at the corner of a barracks building, he saw an ANCOP man raising an AK-47 rifle. The black muzzle faced right at him, the parked trucks, and the rest of the Americans and Afghan ANCOP leaders. The AK shooter coughed out one shot. High.

Yates immediately fired back. One crack, then the Afghan was convulsing, wavering in front of Yates's M-4 barrel, the AK up, shooting wild, a sputter of hostile bullets as the gunman tumbled. Yates ignored it, leaned into his sights, and squeezed off a second round, then another. The 5.56 mm rounds did their job, punching three holes in the man. The Afghan got off about half a magazine, but only the first round came close to anyone. In seconds, it was over.

Five ANCOP men, AKs in hand, immediately surrounded the fallen shooter. He was still alive, but maybe not for long. The U.S. platoon medic shouldered his way through the crowd and got to work. A rifleman stood right beside the medic, his M-4 trained on the bleeding attacker's head. Another U.S. soldier deftly kicked aside the fired AK-47, then slowly picked it up, pointed it away from anyone, dropped the half-empty thirty-round magazine, and jacked back the bolt, ejecting the unfired 7.62 mm cartridge still in the chamber.

Yates did not lower his weapon, watchful for other threats. Behind him, Colonel Keith Detwiler and the rest moved slowly to clear the perimeter of the trucks. The ANCOP officers joined them. Another ANCOP officer took charge of the group near the wounded shooter. It all took a while.

As the tumult subsided, the ANCOP major reached out with both hands. He grasped Detwiler's right hand. Through the interpreter, the Afghan deputy spoke: "I am so sorry. I am so sorry."

Thanks to Specialist Yates, Keith Detwiler was able to accept the apology.

The ANCOP gunman turned out to be a Pashtun sergeant from the east, near Ghazni. In the ever-adjusted criteria of ISAF reporting, the incident did not make the official roll-up. That day, ISAF recorded the death of two British soldiers at the hands of an Afghan policeman in Helmand. Yet the message got through. When trained and employed, guardian angels worked. ISAF ordered all subordinate commanders to adopt the proven NTM-A practice.

As a result, 2013 saw a major drop in green-on-blue events, from forty-two the previous year down to nine. The Afghans, too, afflicted by green-on-green attacks, increased their counterintelligence cadre and screened new recruits more thoroughly. It all did some good. Yet it was a hell of a way to work together, with a rifle constantly trained on your counterpart.

The ANCOP major's heartfelt apology contrasted with President Hamid Karzai's silence. Quick to demand apologies from the United States for Afghan civilian casualties — real and alleged — inadvertent burnings of the Koran in country, free-speech Koran torchings during protests in America, U.S. attempts to negotiate with the Taliban, and pretty much anything else that upset him, the Afghan president offered no comments on the green-on-blue attacks. Evidently, he didn't consider it a problem.

For his part, the new ISAF commander soon renewed the accustomed pattern of contrition. On February 13, his third day in command, General Joe Dunford's forces, with Afghan partners side by side, engaged and killed two Taliban chiefs in contested Kunar Province; the opposing leaders sheltered with their families. Karzai objected and arbitrarily ordered his commanders not to call for ISAF airstrikes under any circumstances. Dunford tried to calm him down and expressed "personal condolences." The excitable president settled down, and after a week, quietly backed off his unworkable airstrike ruling. But the tranquility didn't last long. On March 2, when attack helicopters engaged two young men in Uruzgan Province, Dunford immediately stated, "I offer my personal apology and condolences to the families of the boys who were killed. We take full responsibility for this tragedy." In less than a month, Karzai had brought the new American general to heel.

General Joe Dunford walked into a tough job, trying to complete the transition to Afghan authority while extracting his U.S. and NATO forces. Tall, personable, and wry, the St. Michael's College graduate had served in the usual sequence of infantry assignments. After a stint as aide to the commandant of the Marine Corps, he commanded a Marine infantry battalion, then served as executive assistant to the vice chairman of the Joint Chiefs of Staff. During twenty-two tough months in Iraq, Dunford commanded the Fifth Marine Regiment in the initial 2003 invasion, earned a Legion of Merit with V device for valor under fire and the nickname "Fighting Joe," and then became the chief of staff of the First Marine Division in Anbar Province during the Fallujah operations. After selection for his first star, he served on the Joint Staff, then was promoted to his second and eventually third star as the Marine Corps operations chief, followed by command of First Marine Expeditionary Force. He came to ISAF from duty as the assistant commandant of the Marine Corps. With a master's degree in government from Georgetown, another in international relations from Tufts, a lot of time as an aide and general officer in Pentagon circles, and that strong series of infantry commands, Dunford understood how strategy came to be as well as how to execute it.

President Barack Obama had another option. Most of those in country expected to see the return of General Rod Rodriguez, who'd spent two hard years running the corps and the war in country under McChrystal and Petraeus. In a stateside command since August of 2011, he was ready to return to Afghanistan. But for whatever reason, blunt-spoken Rodriguez didn't match what the president wanted. Maybe he was too rough around the edges. Dunford might lack prior Afghan experience, but he checked all the right boxes for the Obama administration. Here was a Marine who'd keep his mouth shut and stay below the radar.

On June 18, 2013, the Afghans officially assumed the security lead across the country. Notably, to mark the occasion, the political figure who went to Kabul turned out to be NATO secretary-general Anders Fogh Rasmussen, not President Barack Obama. Obama had said his piece back in May of 2012 when he signed the vague partnership agreement with Hamid Karzai. "The tide has turned," Obama said. "We broke the Taliban's momentum. We built strong Afghan security forces. We devastated Al Qaeda's leadership." After another hard year, those claims rang hollow.

What wasn't said mattered most. The U.S. made no concrete commitment beyond words. As one "senior administration official" noted in a White House–sponsored press-conference call on June 18: "So the exact shape of our commitment, of our presence beyond 2014, has not been decided. The president is considering a range of options." The U.S. military and the uniformed Afghans heard that as ten thousand Americans. More than a few in the White House thought it meant zero. The compromise position, of course, seemed obvious: Office of Security Cooperation–Afghanistan, just like in Iraq. The drawdown would be gradual, but inexorable.

General Joe Dunford and ISAF, led by the Americans, did their best. So did almost all of the Afghans in uniform. Yet out in the villages of the dry southwest, the hardscrabble south, and the mountainous east, tired U.S. Marines and soldiers pushed through the patrol schedule, packed out their gear, and handed over bases large and small during that long final year. At their side, Afghan officers and sergeants wondered aloud who would help them after the end of 2014 and got few answers from their laconic ISAF partners. In Lashkar Gah, British colonels traded black japes about Op Elphinstone. In Kabul, diplomats speculated on how Karzai had plotted to manipulate the 2014 election and its acrimonious aftermath. And in the unnamed lairs of the CIA and the SOF, the armed drones flew and the raiders went forth, taking enemies off the battlefield, dead or alive.

Behind it all, the old question hovered, rarely asked aloud: Who was the enemy? Al-Qaeda. The Taliban. The green-on-blue turncoats. Karzai. The Pashtuns. The Pakistanis. Everyone. It was past time to go.

EPILOGUE: INFINITE JUSTICE

Bitter as hell on conduct of war.

— LIEUTENANT GENERAL JOSEPH W. STILWELL

IN THE DUSTY fields outside Iskandariyah, the Company C sniper team really did the job. Tipped off by local Iraqi farmers, the Americans had been watching the dirt road for hours. Sure enough, just after dawn on May 20, 2005, three young men cruised up in a battered gray Opel sedan. They stopped just short of an uneven patch of darkened, turned dirt — an old IED hole.

With binoculars, the spotter watched the three get out. The driver walked around the back and opened the trunk. It sprang up, wobbling. Camouflaged in the long grass of a hide site two hundred yards out, the Americans were too far away to hear anything much. Beside the soldier with the binoculars, the sniper adjusted his telescopic sight, barely breathing. He was taking a bead on the Iraqi driver.

As the trio stood behind the car, the first man reached into the trunk and wrestled out a big, earth-toned cylinder. It looked to be an artillery shell, probably a 152 mm, a big one. The Iraqi cradled it in both arms and passed it like a fat baby to his partner. That man accepted the burden, then turned slowly and began walking toward the hole.

The first man then pulled up another heavy artillery projectile. Deliberately, he swayed to face the third man, who stood with arms out, waiting. The driver handed off the thing. The person with the second shell faced toward the hole and walked unsteadily, clearly uncomfortable with the heft of the ninety-seven pounds of metal and explosives.

Both rounds on their way to the chosen spot, the driver then reached into the trunk and pulled up a coil of red rope. Russian detonating cord looked like cherry-colored clothesline, but it consisted of a fast-burning demolition. Wrapped around those 152 mm rounds, then tied to a trigger device — a pressure plate, for example — the red det cord made sure the thing would go up nicely. The sniper soldiers had seen enough.

The American with the rifle held his breath. In the crosshairs of the optical sight, the Iraqi with the red cord bobbed a bit, like a dinghy in prop wash. The sniper's right index finger slowly squeezed the trigger.

Crack!

The Iraqi went down as if poleaxed. That supersonic one-inch 7.62 mm slug hit him in the head or neck — hard to be sure. But he wouldn't be getting back up.

The other two Iraqis dropped their 152 mm shells as if the things were on fire. One began running down the road. The other, clearly confused, went back to the car, opened the door, crawled in, and scrunched down in the back seat, playing possum. Within minutes, a blocking squad from Company C's First Platoon nabbed the runner. Another squad pulled the third man from the back of the Opel.

The squadron commander, Lieutenant Colonel Bill Simril, put out the word to all 2-11 Cavalry troopers: "Yesterday, we sent a strong message to the insurgents trying to operate in our AO [area of operations]. Set up IEDs and we will hunt you down."

For three days, hostile activity ceased along the farm roads north of FOB Kalsu. One good shot made a difference. But on the afternoon of May 23, on another dirt trail not far from where Company C's snipers had nailed the Opel threesome, the insurgents evened it up.

Four Company C up-armored Humvees drove slowly down the slightly elevated road, long grass waving on either side. The thermometer stood at a hundred degrees, the horizon indistinct with heat haze as the turret gunners scanned their sectors, alert for untoward movements. No Iraqi farmers worked in the fields. It was just too hot this early afternoon.

The Mississippi Guardsmen liked being attached to Bill Simril's aggressive squadron. These Cavalry troopers took the fight to the enemy, and that suited the riflemen of Company C. In the counter-IED fight, the squadron had killed or captured twenty hostile bomb emplacers in that warming month of May. Company C had gotten its share, with the recent sniper ambush a particularly proud achievement.

Then the road flashed white and it all went to black, a smashing, grinding, hammering, throaty crash, lightning and thunder at ground level, right here, right now. A massive dirt column vomited up, engulfing the third Humvee completely.

As the brown grit rained down and the savaged air cleared, the armored truck materialized, a ghostly outline wholly engulfed in flames. It was all burning, the turret, the tires, the tan-painted metal, and the Kevlar armor, a licking, roaring furnace. There were four soldiers in there, four Mississippi Guardsmen. But you couldn't see them at all. Tongues of fire barred the way. Heat waves shimmered off the stricken Humvee. Oily black smoke boiled up, swirling around the scene, shrouding the horror.

The men on the other three trucks tried to get near it, to pull somebody out. The fierce heat and shooting flames drove them back. Then the onboard bullets began sparking off, strings of machine-gun rounds going off by tens, individual rifle bullets popping into the air at crazy angles. A grenade cooked off with a dull boom. The JP-8 fuel, formulated to burn rather than explode, fed the towering pyre.

A long, awful hour passed while the fiery twisted mass burned down. Thoroughly stripped of paint and additives, molten raw metal pooled into bright puddles on the dusty road. Charred chunks of tires and curled strips of plastic smoldered, belching up stinking tendrils of hot coal-black smoke.

By the time the heat at last began to subside, the 2-11 Cavalry's quick-reaction platoon had arrived with medics and an armored wrecker truck, all led by Major Kevin Hendricks. They did what they could on the spot. It wasn't enough.

Unable to complete the recovery of the four dead, wary of potential insurgents prowling in the distance, the Americans dragged the mangled truck back to FOB Kalsu. There, Hendricks, Company C Guardsmen, and other 2-11 Cavalry troopers labored for hours to extract the four Americans. Kevin Hendricks said later, "It took me four days to remove the smell from my hands."

• • •

Four lives sacrificed—to what end? Great War veteran Wilfred Owen, before the Germans got him, wrote a verse starkly contrary to the beliefs of the ancient Romans, finding little *dulce et decorum* about dying in the cruel, scaring grip of modern explosives. We dare not ask lightly for our soldiers and their families to endure such unspeakable agonies. Yet we did.

Almost as bad, maybe worse in the long run, what did we inflict on Major Kevin Hendricks? For the rest of his life, he'll be trying to wash away what he saw, what he smelled, what he felt that horrific afternoon near Iskandariyah. No man is hard enough to walk away from that. How many others carry those everlasting stains? We know, but we don't know. We don't want to know.

What do we know? The war resulted in almost 7,000 Americans dead, five times that many wounded, and a much higher number dealing with psychiatric injuries great and small. Our Coalition allies sustained more than 1,300 dead and nearly 6,000 wounded. The Afghan people counted at least 13,000 civilian fatalities, although statistics there are notoriously unreliable. Iraqi dead, more accurately estimated in the face of much higher levels of violence, approached 104,000. In both countries, the lists of dead included a significant number of insurgents. Sorting it all out was just as hard in the morgue as it was on the battlefield.

The war also cost the U.S. a lot of money, almost a trillion dollars since September of 2001, about two-thirds for Iraq, the rest for Afghanistan. Just how much permanent damage this did to our country's economy is hard to determine. War funding certainly elevated the federal government's already burgeoning annual deficits and added a few more unwelcome strata to the accreting mountain of long-term national debt. Both political parties pointed accusing fingers even as the spending continued. By any measure, fighting a protracted war on the opposite side of the world with a volunteer military and a lot of expensive contractors is not cheap.

For all of this pain, what resulted? Pace Barack Obama's May 2012 comments at Bagram Airfield, the al-Qaeda leadership has indeed been devastated. The organization that planned and executed the 1998 African embassy attacks, the USS *Cole* strike, and 9/11 no longer exists. Terror cells take the name and claim the heritage. But classic, centralized al-Qaeda has been crushed. Those who died in New York, Washington, and Pennsylvania on September 11, 2001, have been fully avenged.

Post-Saddam Iraq endures under the authoritarian rule of an unenlightened Shia clique that has long backed the likes of Nouri al-Maliki. Violence from marginalized Sunni Arabs has increased steadily, erupting in widespread antigovernment rebellions in the north and west in mid-2014. The United States maintains formal relations and helps out the Iraqi military, but at arm's length. Iranian influence remains strong. Between the U.S. troop withdrawal in 2011 and whatever evil may come next, the decent interval continues for now. How long Iraq holds together remains to be seen.

Post-Taliban Afghanistan carries on, subject to the whims of the same entrenched self-serving Kabul circle that gave us Hamid Karzai. Out in the southern and eastern villages, the Taliban perseveres, biding its time, awaiting its opening. It may come, or it may not. Nobody can really be sure. America intends to draw down yet stay engaged, as does NATO, to help Afghanistan remain intact. Pakistan agrees at times, as long as that Afghan state stays weak and fractured and is no threat to the Islamabad government. What follows after 2014 may well revert to the tradition of millennia: squabbling tribes, random violence, and an occasional genuflection toward a figurehead in Kabul. And that's the best case.

Out in other places, in remote reaches of Africa and Asia, the CIA, the SOF, and other allies persist in hunting America's terrorist foes. At home, law enforcement entities strive to stay ahead of threats, acting under an expanded mandate and not without objection from citizens unhappy with the consequent intrusions on privacy and personal liberties. President George W. Bush warned that much of the war necessarily proceeded unseen, "covert operations, secret even in success." President Barack Obama kept up the pressure and, indeed, increased it in many ways. As a result, the United States has suffered no major terror attacks since 9/11.

Those things were done, and continue to be done.

What wasn't done?

Here a broad chasm gapes between what the United States accomplished and what it aspired to do in the wake of the 9/11 attack. In the immediate aftermath of the al-Qaeda assault, America wanted victory over Islamist terrorists writ large and suppression of their various hosts and enablers. The citizenry demanded a strong response. The U.S. Congress reflected that sentiment by granting broad authorities to retaliate against and, in time, to preempt al-Qaeda and its collaborators. A clutch of visionaries even advocated remaking the Middle East in America's image: democratic, committed to the equality of women, valuing education, and embracing interaction with the global economy and free societies. Rhetoric soared.

In carrying out this effort, the country faced a choice. America might opt for decisive war through annihilation, as in World War II, or containment through limited conflict and attrition, as in the Cold War. The Bush administration favored the former but lacked a concentrated enemy power to crush. Al-Qaeda operated from Afghanistan but obviously spread well beyond the valleys of the Hindu Kush. Denied a single geographically distinct foe, Bush and his people rightly chose their path with care. Casting too broad a net risked war with multiple Islamist networks tied to a list of actively hosting and clearly tolerant countries, including Iraq, Iran, Libya, Somalia, Sudan, Syria, and Yemen, not to mention Saudi Arabia and Pakistan. Bush's team lacked the stomach to wade into more than a billion Muslims, most of them not actively hostile. Containment had to suffice. Over time, defined as decades, a firm U.S. military stance coupled with economic and cultural penetration promised to erode Islamist roots, as it had once degraded Communist ideology. So went the thinking. The original code name Infinite Justice may have been discarded, but the logic behind it, and the timeline suggested, remains the U.S. strategy to this day.

Bush's war began narrowly, knocking out al-Qaeda and its Taliban backers in Afghanistan. Within weeks of 9/11, the basic goals were fulfilled, not perfectly, not completely, but probably close enough. Had we stopped there and reverted to the long, slow Clinton-era squeeze of terror cells and Islamist supporters, it might have done the job. But after 9/11, in an America beset by rumors and fears, the gambit in Afghanistan, though impressive, did not seem to fill the bill. Other threats existed and demanded action.

Saddam Hussein's Iraq topped that list of known and presumed dangers. Already face to face with U.S. airpower every day, loudly championing terrorist groups, hammering his own people, and suspected of amassing chemical arms, Saddam seemed overdue for decisive American action. In retrospect, Saddam's Iraq appears to have been contained and could conceivably have been boxed in more tightly rather than destroyed outright. But Saddam's survivor's luck ran out. Buoyed by quick success in Afghanistan, the United States chose to act, rapidly overwhelming the Baathist regime. Containment and limited conflict rang up another accomplishment.

Again, as after the fall of Kabul, the swift seizure of Baghdad offered another opportunity to close out the conventional military phase and go back to the slow, steady, daily pressures of global containment of Islamist threats. That moment passed. Instead, satisfaction in brilliant opening-round victories, initial popular acclaim in America, encouragement from many allies, and more than a little pride influenced Bush to try for more, way more. With minimal domestic debate — and, notably, no known military objection — the administration backed into two lengthy, indecisive counterinsurgency campaigns, Afghanistan underresourced, and Iraq overly optimistic. The limited ends of containing Islamist movements and attritting al-Qaeda faded into the background, at least until Sunni Arabs and a resurgent Taliban made it obvious that neither Iraq nor Afghanistan seemed likely to move readily into the orbit of Western democratic republics. By the decade's end, chastened, bloodied, and weary, America returned to containment and limited operations.

We have been to this rodeo before. The same cycle played out early in the Cold War, when President Harry S. Truman, determined to contain Russian Communist influence, intervened on the Korean peninsula in 1950. After sobering early reverses, the brilliant Inchon invasion enveloped and largely destroyed the North Korean divisions, and the initially narrow task of defending South Korea became a great crusade to liberate the Communist north. Massive Chinese intervention compelled Truman to go back to the earlier, limited policy of protecting South Korea, half a loaf, but better than none. Still, the heady promises and then dashed hopes, coupled with the casualties incurred, the very public relief of imperious Douglas MacArthur, and what glum GI wags called "die for a tie," torpedoed Truman's domestic support, cratered his reputation, and led to the election of Dwight D. Eisenhower from the opposition party. Eisenhower quickly wrapped up the Korean armistice.

During his tenure in office, the former general carefully avoided other proposed interventions except under the most strictly defined conditions.

You could see Bush as Truman, and Obama as Eisenhower, the former overreaching in misplaced good faith, the latter pulling back and strictly metering when and where the U.S. took action. As in the 1950s, the basic strategy made sense. But containment demanded patience and limits, the long game. It also required a U.S. commitment, restricted in numbers but undeniable, as Eisenhower provided in Korea, where America forces remain to this day. In that aspect, Barack Obama faltered, with consequences we have yet to see.

What went wrong?

It has become somewhat fashionable in the senior ranks of the military to point fingers at the civilian officials, elected and appointed, for mucking up the war. Civilian control of the military means the suits propose and the uniforms dispose. The Stan McChrystal *Rolling Stone* imbroglio aside, those on active duty almost always kept their opinions private when it came to dealing with the civilians. Yet the generals and admirals and, especially, the majors and sergeants were not stupid. They saw the war wasn't working. Naturally, they attributed a share of that to two presidents and several secretaries of defense, among others. In their sarcastic exchanges, McChrystal's men didn't say anything that hadn't been heard during the war inside combat outposts, aboard MRAP trucks and C-130 transports, and in various operations centers. By tradition and law, however, such critical comments rarely surfaced. One former Department of Defense appointee referred to the "silence of the generals."

During the war, the top military officers gave their views to the civilian officials in confidence behind closed doors, disdaining the old Washington games of press leaks and background grousing. As for speaking up in public, well, it just didn't happen. A serving senior commander who held a press conference or went before Congress and spoke his mind contrary to Bush or Obama administration views might be right, but he'd also be fired, a truth underlined by Stan McChrystal's terminal brush with that overly zealous journalist. Even a misstep in a press interview, a loose phrase or a mistake, often resulted in a reprimand or, in some cases, removal. Dealing with news people amounted to handling live snakes—not much good came of it. So with the exception of adept Dave Petraeus and a few game if less sure-footed imitators, most generals avoided the news media or fed them pabulum, predigested talking points. Inside the headquarters, a few of those wearing stars sometimes complained that the ladies and gentlemen of the press were not team players, not patriots, and were overly invested in undermining the war effort. Ernie Pyle and Marguerite Higgins were definitely dead. The majority of generals saw it otherwise, noting that despite the fond hopes of some of the more activist writers and editors, the American news media provided a mirror, not a prism. If you didn't like what you saw, it was because the truth hurt. It hurt a lot during this war.

If those on active duty stayed quiet, retired officers, Army generals in particular, did not. In 2006, for example, several prominent officers, including former Iraq war commanders Ric Sanchez, John Batiste, and Paul Eaton, excoriated Secretary of Defense Donald Rumsfeld for his purported unwillingness to heed military counsel. In 2013, Eaton, joined by other retired generals and anonymous serving senior officers, said similar things about the Obama administration. To read all of this, you'd think that the top suits were getting the right steerage, but not deigning to listen.

Yet all too often, the military offered advice hardly worth hearing. The man widely reputed to be the most capable and far-seeing secretary of defense during the war, Bob Gates, shared these rueful words with cadets at West Point: "Any future defense secretary who advises the president to again send a big American land army into Asia or into the Middle East or Africa should 'have his head examined,' as General MacArthur so delicately put it." Wise words, certainly, but this same secretary oversaw not one, but two, troop surges that reinforced stalemated land campaigns in Asia. And who provided the "best military advice" that suggested such things?

Following the sodden, bloody purgatory of the Somme and Passchendaele, the British trench brotherhood who survived the horror show depicted by Wilfred Owen complained loudly about the failures of generals in command of the British Expeditionary Force on the western front. A common refrain referred to "an Army of lions led by donkeys," baldly accusing the wartime leadership of unimaginative ignorance. It was a grossly unfair characterization then and absolutely untrue of the American generals of the present war. That granted, there is a mulish aspect at work.

Equity demands that we start with what went right in a military sense. Many things did, and these certainly redound to the credit of all ranks, from the generals on down. The senior leadership made some very prudent early calls. Moreover, in these areas, they adapted quickly as the war evolved. Recommendations on how to organize, train, and equip the U.S. Armed Forces, especially the hard-pressed ground services, made it all the way up and produced superb outcomes.

Strategic choices set the table for America to carry on a long war. Leaving aside the wisdom of the twin counterinsurgencies, the decision to mobilize the Guard and Reserve guaranteed support for the troops in every county in America. Guardsman George W. Bush in 2001, like Guardsman Harry S. Truman in 1950, grasped fully the strong ties between American communities and their citizen-soldiers. The war declined steadily in popularity, but public support for those fighting it did not.

A second strategic determination, closely aligned to the first, shaped the scale of the war. The United States allowed a relatively modest increase in the size of the regular Army and Marine Corps and made essentially no additions to the Navy and Air Force. When needed, substantial reinforcements came from Guard and Reserve unit call-ups. This accorded well with a long-term containment strategy and lowered the strain on American society.

The third strategic resolution flowed from the first two as well as from the upper uniformed ranks' hard-won experience in Vietnam. The services chose to rotate forces by unit. With limited Guard and Reserve activations and caps on service end strengths, battalions made multiple trips into theater. Because they prepared, deployed, and returned as teams, they did so over and over effectively. In this, the United States validated the lessons learned by British and French regulars during centuries of overseas duties.

Key tactical ideas allowed U.S. units to perform with confidence under fire. As a rule, American weaponry, equipment, training, and small-unit leadership far outstripped anything arrayed against them. The joint aspects of intelligence, logistics, and especially airpower sent every American platoon of soldiers into action confident that they could slay their antagonists with impunity today, tonight, and as long as it took. The computers and other IT made shots count, as long as the Americans could find the enemy. As usual, therein lay the rub.

The SOF elements and all their clandestine relatives eventually developed superb skills in hunting down the enemy leadership. Those skills didn't exist at the outset of the war, and they reached their fullest potential alongside conventional forces serving as beaters to stir up the prey, police up lesser targets, and finish off the cripples. Intelligence fusion, uniting human and technical means, reached its zenith in the Task Force. The line battalions never equaled that impressive standard, but they proved eager learners, and over time, the entire force improved a great deal. Though this achievement was clearly a labor of many, the vision of General Stan McChrystal drove the train. He and his fellow SOF commanders instilled a man-hunting prowess that continues to terrorize the terrorists in all the dark corners of the globe.

Discrete focus on enemy chieftains, carried out by IT-enabled smart munitions, allowed U.S. forces to severely restrict the use of firepower. Outside of the opening invasions and occasional major pushes like Fallujah, American forces did not overdo the use of artillery and close air support. The record of both campaigns does not feature many cases of towns destroyed in order to save them, in that grimly memorable phrase reported by Peter Arnett in Vietnam. Modern weapons do create a lot of damage, so even "precision" strikes are anything but safe for those nearby. Given an opponent fully intertwined with the population, Americans relied on old-fashioned foot patrolling and house raids more often than not. Protecting the people was always part of the calculus.

Protecting our own forces came harder, especially in light of the enemy's reliance on IEDs, without doubt its weapon of choice. America began the war willing to ride out or avoid roadside bombs, mines, and booby traps. Early in the war, the entire wheeled-vehicle fleet proved unable to shrug off a determined pistol shooter, let alone absorb the effects of massive, buried racks of artillery shells. Losses incurred during the first months of fighting demanded a better solution, and fast. Within months after the Iraq campaign began, American industry fielded an entire new set of up-armored Humvees and affiliated cargo trucks. Then the factories trumped that achievement by producing thousands of V-hull MRAPs in the second half of the war. In addition, to clear key routes, the military organized dedicated IED location-and-clearance elements. These brought together specially trained soldiers, unique equipment, and purpose-built demolitions to find and clear IEDs. These concentrated exertions paid off. With each passing year, though the enemy used more and more IEDs, the things became less and less effective in harming Americans and their allies. In future conflicts, America's enemies will surely use IEDs. The U.S. now has the wherewithal to meet that challenge.

We taught these methods to others. Working under fire and on the clock, America helped Iraq and Afghanistan recruit, organize, train, and equip adequate forces, and then use them on operations. Although some allies taught but did not accompany the Iraqis and Afghans, the U.S. did both. "Do as I do" and leading by example are always the most useful means of advisory work, especially given the acknowledged language barriers. Both Iraqis and Afghans fought effectively alongside their American teammates. With the American advisers gone, the real test will begin. Early returns from Iraq are not encouraging. The results in Afghanistan may well be equally uneven. We shall see.

The wholesale training of two nascent foreign militaries during a war says much about the quality and rigor of U.S. military training. In all services, the United States superbly prepared individuals, leaders, and units for war. Time after time, following a firefight, an American soldier or Marine offered that the event turned out to be just like training. Effective training takes time and money, and as the war continued, all the services optimized to fight insurgents. So they all focused on the critical tasks and drilled them hard. This ensured success in thousands of short, sharp engagements and endurance in dozens of longer, bloodier clashes. Training glued it all together. It gave men and women confidence under fire.

Tough training underwrote a personal bond tested in many firefights. The Army's warrior ethos, derived from the Ranger Creed, put it best: "I will never leave a fallen comrade." Men and women fight as they are trained. Throughout the combat that followed 9/11, the U.S. military stressed that units bring everybody out of the crucible, every time. Some may be hurt, and some may die — the enemy shot back — but in this war,

Americans recovered their casualties. This was a very demanding standard, especially given thousands of small units on foot day and night in brutal terrain, extreme weather, and hostile fire. But Americans did it. When the Iraq campaign ended, all uniformed casualties had been recovered. In Afghanistan, the same ethos applied. This blood oath to bring everyone home does a lot to explain why, despite an increasingly unpopular war, young volunteers kept joining up and reenlisting.

The skill and will to close with the enemy made a difference from the start. Placed in contact with their adversaries, Americans almost always prevailed. It was an absolute and very dearly won strength. In it, though, dwelt a risk, the one Robert E. Lee lamented in the aftermath of the disastrous repulse on the third day at Gettysburg: "It's all my fault. I thought my men were invincible."

Above that tactical excellence yawned a howling waste. The civilians provided an adequate containment strategy. The U.S. Armed Forces, led by its admirals and generals, fielded capable, resilient combat units. With few exceptions, small U.S. units proved lethal to their opponents. A gaggle of one-sided firefights, however, do not victory make, especially against guerrilla enemies.

The war required a way to use a tactically superb force to contain and attrit terrorist adversaries. In this, America's generals failed. We found ourselves impaled and bogged down in not one, but two Middle Eastern countries, and this on the best military advice of educated, experienced senior military men and women who had all studied Vietnam in their service schools. Over time, piece by piece, the generals recommended slogging onward, taking on two unlimited irregular conflicts with limited forces. Absent a realistic campaign concept in both countries, wars of attrition developed.

Some saw it as a failure of imagination. Frustrated subordinates and concerned fellow citizens complained that the generals didn't get it. With most senior officers close-mouthed in front of visitors from Congress, inquisitive journalists, and various think-tank gadflies — few with any days in uniform — it became easy to consider the typically taciturn generals narrow-minded, ill-read, or just plain dumb. That kind of impression was off the mark. But in the absence of contrary evidence, it persisted among interested foreign policy and defense intellectuals and others who saw themselves that way, thinkers inclined to grope for the proper clever idea to rescue a floundering war.

Along these lines, many voices, including a large number in uniform, advocated an earlier, formal adoption of a wide range of counterinsurgency practices. These had some value, and most, if not all, eventually got used in both countries. The shiny objects of counterinsurgency theory, neatly captured in FM 3-24, ended up delivering far less than expected or hoped. Counterinsurgency works if the intervening country demonstrates the will to remain forever. That applies in colonies and territorial annexations with the supervising power in full control. And even then, it doesn't always work, as France learned the hard way in Indochina and then Algeria. Once it becomes clear that the external forces won't stay past a certain date, the guerrillas simply back off and wait it out. Had America treated Afghanistan and Iraq from the beginning as the future fifty-first and fifty-second states, FM 3-24 offered a way to pacify them. Saddled with incomplete authority over Afghan and Iraqi internal affairs, inept host governments, and ticking clocks, we could not do it. By the time the manual came out, the techniques had already been tried and found wanting.

A sensible look at American military strengths in 2001 showed a clear alternative to grinding counterinsurgency campaigns. As a joint force and as individual services, the U.S. military recognized the value of short, decisive conventional conflicts waged for limited ends: Panama in 1989, Iraq in 1991, Bosnia in 1995, and Kosovo in 1999. Force composition and training reflected this short-war bias. Employed thusly, American airpower and SOF in Afghanistan in 2001, and airpower and armor in Iraq in 2003, worked as advertised.

Had that ended our efforts, we would have been fighting well within our means. Admiring war colleges would have studied the brilliant opening rounds as models of lightning war. But here success undid us. Rightly impressed by the innovation and speed of the initial attacks in Afghanistan and Iraq and thoroughly convinced of the quality of our volunteer troops, successive generals in command at the four-, three-, and two-star levels signed on for more, a lot more, month by month, then year by year. In so doing, we did not heed Sun Tzu's caution. We did not understand our enemies. Indeed, drawn into nasty local feuds, we took on too many diverse foes, sometimes confusing opponents with supporters and vice versa. Then we compounded that ignorance by misusing our conventionally trained military to comb through hostile villages looking for insurgents. Once it became evident that we would not stay, something we knew in Iraq by 2008 and in Afghanistan by 2011, we continued to press on and lose people in the vain hope that something might somehow improve. Our foes waited us out.

Many generals, including some at the very top, saw this problem as it developed. They shared their views, both inside the military and with civilian leaders. In 2003, Tommy Franks and Jack Keane thought we should back out of Afghanistan and Iraq and expected to do so, but neither pressed the case very hard. That same year, John Abizaid warned of a growing insurgency in Iraq, and by 2004 he joined George Casey to warn that at best, America could buy time for the Iraqis to win their own war. They assumed the Bush administration understood that this would be an effort of decades. They assumed wrongly. Successor commanders in both theaters — Petraeus, Odierno, McChrystal, Allen, Dunford — made equally poor assumptions that, despite every indication to the contrary, the Obama administration would commit to major long-term U.S. troop deployments. Somehow, on this most vital issue of all, during hundreds of hours of meetings, the uniforms and the suits managed to talk past each other. Sergeants and captains, not to mention our fellow citizens, count on generals to sort out such fundamental strategy. We didn't.

It's noteworthy that in the 2006 surge decision, President Bush overruled all of his senior field commanders and the Joint Chiefs of Staff, who urged him to limit the American effort, not escalate. In the 2009 surge decision, President Obama sought recommendations from all the top admirals and generals, then made up his mind and ordered a different solution than the one the military wanted. In these two prominent cases, and many others, senior military voices were definitely heard. But we need to be clear. Limits, curbs, and reductions came up. But simply cutting our losses and pulling out did not. The record to date shows that no senior officers argued for withdrawal. Instead, like Lee at Gettysburg, overly impressed by demonstrated U.S. military capabilities and our superb volunteers, commander after commander, generals up and down the chain, kept right on going. We trusted our invincible men and women to figure it out and rebuild two shattered Muslim countries, and do so under fire from enraged locals.

Once we'd gone down that bad road, our options shrank: Stay the course. Add forces. Pull out. Over time, in both countries, all three approaches were tried. Only the third one, pulling out, worked, and that in the finite sense that it ended U.S. involvement. But it left both friends and foes behind, sowing the seeds for future troubles.

Campaigning with an account ledger, narrowly focused on adding or subtracting forces, aligned well with the way the Joint Chiefs and services did business. They provide resources to forward commanders. Having missed the bus on limiting the campaigns in the early months, once stuck in the morass, all the Pentagon leadership could do was hang in there, send more, or pull the plug. It's noteworthy that the one truly bold new idea—the Iraq surge—originated with retired general Jack Keane, not from among serving senior officers. Yet we shouldn't overstate the value of that effort. The 2007 surge did not save the Iraq campaign. It was too late for that. But it gave America space to withdraw on its own terms and provided Iraq, however ungrateful, a chance to chart a course as a responsible nation-state.

The 2010 surge in Afghanistan achieved less. In a lot of ways, it reflected the near bankruptcy of military planning by that point. As the old joke goes, when all you own is a hammer, everything looks like a nail. Military options presented to President Obama included some, more, or even more (enough, according to the senior officers) reinforcing troops. More hammer. His vice president, not a commanding general or service chief, offered an alternative counterterrorism concept. Having seen the scenario play out in Iraq and uncertain if it would work at all in Afghanistan, Obama surged, but with a crucial caveat: The military had until 2011 to do what it could with the extra forces. The Afghan surge blocked the Taliban and rolled them back a good bit. But in the end, the U.S. chose to get out.

Despite the unmatched courage of those in U.S. uniform—including a good number of generals who led their people under fire—our generals did not stumble due to a lack of intellect. Rather, we faltered due to a distinct lack of humility. Certain we knew best, confident our skilled troops would prevail, we persisted in a failed course for far too long and came up well short, to the detriment of our trusting countrymen.

• • •

The United States of America has already pressed past the Global War on Terrorism. As combat continued in Afghanistan, and the shadow of war went on across a wide range of distant hellholes, the Obama administration formally moved on. On November 17, 2011, in an address to the Australian Parliament regarding American interests in the Pacific Ocean and its littorals, the president promised that "the United States will play a larger and long-term role in shaping this region and its future." The speech signaled a major shift in American strategy, a conscious "pivot to Asia." Although Obama and his key cabinet secretaries avoided the issue, and in some cases even denied it, this latest incarnation of the favored U.S. containment strategy had a new object, which of course was an old object, one Harry Truman, Dwight Eisenhower, Jack Kennedy, or Lyndon Johnson would have recognized. The Chinese government recognized it immediately, protestations from Washington notwithstanding.

Military consequences followed. Confronting nuclear-armed China and its massive army would be done by Air-Sea Battle, pitting American strengths against the competitor's known and assumed weaknesses. That *Land* was missing from that title wasn't an oversight. Having had a go at numerous Chinese Communist divisions in Korea from 1950 to 1953, and with no common land boundary to defend — unlike America's NATO role in West Germany during the Cold War — the United States chose not to use ground units, except perhaps to secure islands or hold key coastal bases. That sure sounded like Marine work, maybe SOF or paratroopers. But it didn't demand legions of tanks and soldiers.

With little need for ground forces in Air-Sea Battle and little interest in committing them elsewhere, the Department of Defense in its January 2012 strategic review delineated the future in no uncertain terms: "However, U.S. forces will no longer be sized to conduct large-scale, prolonged stability operations." Translated, that meant no more counterinsurgencies with thousands of conventional troops. If the operations could be done on the cheap with a few hundred Special Forces, maybe. But mostly, forget it. What Army remained would prepare for the next Desert Storm, although against whom looked fuzzy, for sure. To pay for all of this Air-Sea Battle stuff, cuts in ground forces planned to reduce the regular Army from 570,000 to at least 450,000, probably much lower. The Marines also shed strength, but not as much, as their amphibious nature gave them a major role in the new Pacific strategy. The Air Force and Navy stood pat with some adjustments. It was back to the future, Eisenhower's New Look after Korea or the Nixon Doctrine in the wake of Vietnam. Win the war from the air and sea. How Americans wanted that to be so, then, now, and seemingly ad infinitum. Left unsaid was the ugly truth that if you want to put your boot on an enemy's neck, it has to be attached to a soldier's foot.

With new marching orders, the U.S. Armed Forces began to retool even as the last months played out in Afghanistan. Good as it was — and it was tremendous — all of the thorough preparation for the 9/11 war meant that over time, everything revolved around finding and killing three guys with an RPG launcher. The Air Force deemphasized air-defense suppression and air-to-air tasks in favor of dropping JDAMs and making low passes to spook the guerrillas. The Navy spent less time on antisubmarine and anti-mine missions and more on shooting Tomahawk cruise missiles and generating carrier airstrikes. The Army lost touch with major combined-arms armored tactics and the ability to mass the fires of dozens of tubes of artillery on a single target but grew all too familiar with raiding houses and digging up IEDs. The Marine Corps has done precious few amphibious landings as they struggled to control unruly villages in two landlocked areas of operation. By 2011, as the Iraq conflict ended, American forces might have been hard-pressed to re-create the decisive opening offensive of March 2003. Developing the skills to carry out Air-Sea Battle will not come overnight.

As the United States military returns focus to its core strength, rapid, decisive conventional operations, it must come to grips with the war fought since 2001. Good ideas and bad, lessons learned, relearned, and unlearned, all deserve thorough scrutiny and discussion. In the 1970s and 1980s, following the end in Vietnam, the American military took a very uncompromising look at itself. The wartime generals had passed on, many of them unable or unwilling to see the flaws in the institution they had built. Those who fixed it were the younger leaders in Vietnam, Chuck Horner and Ron Fogleman of the Air Force, Norm Schwarzkopf and Gordon Sullivan of the Army, Leighton "Snuffy" Smith and Jay Johnson of the Navy, and Al Gray and Carl Mundy of the Marine Corps, among many, many, many other NCOs and officers of all ranks. Today, similar work is under way. The colonels have started it, but the younger men and women, ennobled by the war and scarred by the war, will bring it to fruition. They, not today's generals, will figure it out.

As we look back on this long war, we must be frank about our shortcomings. That said, we must also remain alert to the possibilities — and even the hope — the future holds. One of the features of such conflicts involves their indeterminate outcomes. It takes decades to be sure. Today, having risen from devastation in 1953, the Republic of Korea is a vibrant economic miracle, a strong democracy that stands as one of America's staunchest allies. More surprisingly, America also enjoys steadily improving relations, including military ties, with its implacable foe of 1965–75, the Socialist Republic of Vietnam. Perhaps some long-term good may arise from the river valleys of Mesopotamia or the lower slopes of the Hindu Kush. It's still possible, a final irony atop so many others that have defined this hard war.

U.S. and Coalition Military Fatalities in Afghanistan and Iraq, 2001–2014

	AFGHANISTAN		IRAQ	
	U.S.	COALITION	U.S.	COALITION
2001	12	0	0	0
2002	49	21	0	0
2003	48	10	486	94
2004	52	8	849	57
2005	99	32	846	51
2006	98	93	823	50
2007	117	115	904	57
2008	155	140	314	8
2009	317	204	149	1
2010	499	212	60	0
2011	418	148	54	0
2012	310	92	1	0
2013	127	33	0	0
2014	34	11	0	0
Total	2,335	1,119	4,486	318

Sources: Casualties are announced in www.defense.gov/releases/ and tabulated in www.icasualties.org. These numbers reflect losses through June 30, 2014.

Average U.S. and Coalition Troop Strength in Afghanistan and Iraq, 2001–2014

	AFGHANISTAN		IRAQ	
	U.S.	COALITION	U.S.	COALITION
2001	5,200	500	0	0
2002	5,200	4,700	0	0
2003	10,400	5,000	67,700	23,000
2004	15,200	8,500	130,600	24,000
2005	19,100	9,000	143,800	23,000
2006	20,400	21,000	141,100	20,000
2007	23,700	26,703	148,300	12,112
2008	30,100	31,400	157,800	9,907
2009	67,500	38,370	135,600	5,000
2010	98,000	41,730	88,300	0
2011	96,500	40,313	42,800	0
2012	87,500	39,469	260	0
2013	60,000	27,207	170	0
2014	33,500	17,500	750	0

Sources: These average numbers are derived from www.defense.gov/releases/, www.mnf-iraq.com, and www.isaf.nato.int. On May 27, 2014, the United States announced that following the conclusion of combat operations on December 31, 2014, the military plans to reduce troop strength to 9,800 during 2015, 4,900 in 2016, and less than 200, all affiliated with the U.S. embassy in Kabul, by 2017. NATO contributions, called Resolute Support, will start with as many as 3,000 troops in 2015 and decline over the next two years, proportionate to U.S. strength. The mission of remaining Coalition troops will be to train, advise, and assist Afghan security forces, leading to full withdrawal (less the small U.S. embassy element) by 2017. All of this is subject to final agreement by the Afghan government. In June of 2014, the U.S. deployed a small number of security forces and advisers to protect the U.S. embassy and work with embattled Iraqi security forces.

ACKNOWLEDGMENTS

No book writes itself. This one reflects the contributions and wisdom of many. Special thanks go to my very patient agent E. J. McCarthy. I must also salute the great team at Houghton Mifflin Harcourt and Eamon Dolan Books, led by the one and only Eamon himself. Ben Hyman, Larry Cooper, and Tracy Roe all deserve gratitude. My colleagues at North Carolina State University have also been immensely supportive. This book reflects their collective wise counsel. Any shortcomings are mine alone.

Among the multitude who helped, I must particularly thank the late T. R. Fehrenbach, a Korean War veteran and acclaimed author of *This Kind of War: A Study in Unpreparedness.* I met Lieutenant Colonel Fehrenbach once, when he came to Korea to visit his former outfit, the Second Infantry Division. His service, and his writing, have inspired generations of Americans in uniform. If this book in any small way echoes Fehrenbach's masterwork, then it has accomplished its purpose.

Military life has its hardships. Among its pleasures is the opportunity to serve with wonderful men and women. This book recounts a small fraction of their heroism and sacrifice, the part for the sterling whole. It was an honor to be in their presence. In Iraq and Afghanistan, the brothers and sisters of my immediate military families made it all worth it. Special gratitude goes to Drew Bayro, Steve Beisner, Adam Boland, Gabriel Cervantes, Carl Coffman, Rick Fink, DeWayne Growt, Vincent Gunter, Adrienn Halland, Jeffrey Hof, Chris Johnson, Danny Laakmann, Morgan Lamb, Rory Malloy, Ray Myers, Ralph "Nap" Napolitano, Tom Odom, John Opladen, Tim Parks, Vic Scharstein, Alan Simmons, Tiffany Tapia, Carl Tenbrink, Pat Tierney, Mark Ulrich, and Steve Yarborough. Some were wounded. All know war.

NOTES

Author's Note

page

xvi "*Know the enemy*": Sun Tzu, *The Art of War,* ed. and trans. Samuel B. Griffith (London: Oxford University Press, 1963), 84. Sun Tzu goes on to say, "When you are ignorant of the enemy but know yourself, your chances of winning and losing are equal. If ignorant of both your enemy and of yourself, you are certain in every battle to be in peril."

Prologue

xviii "*First we're going*": Dan Balz and Rick Atkinson, "Powell Vows to Isolate Iraqi Army and 'Kill It,'" *Washington Post,* January 24, 1991.

"*Contact, five armored vehicles*": This reconstruction draws on Douglas Macgregor, *Warrior's Rage* (Annapolis, MD: Naval Institute Press, 2009), 139–41; Rick Atkinson, *Crusade* (Boston: Houghton Mifflin, 1993), 442–44; Tom Clancy, *Armored Cav* (New York: Berkley Books, 1994), 257–61; *Frontline,* PBS, transcript #1408T: "The Gulf War, Part B," aired January 10, 1996; Michael R. Gordon and Bernard E. Trainor, *The Generals' War* (Boston: Little, Brown, 1995), 410–12; Tom Clancy and Fred Franks, *Into the Storm* (New York: G. P. Putnam's Sons, 1997), 556–58. Major Macgregor was a squadron S-3 (operations officer) serving with McMaster in the 1991 campaign. His *Warrior's Rage* is the definitive account of a much-studied action. Macgregor's 2009 book benefits from the perspective offered by the passage of time. The U.S. Army carefully documented what the service dubbed the "Battle of 73 Easting," named after a map reference. Some of the first computer-generated imagery in the army was devoted to re-creating the fight as a training scenario. H. R. McMaster, always willing to share his view of events, talked to many official and unofficial interviewers, including reporter Rick Atkinson, author Tom Clancy, and the *Frontline* television team.

The Iraqi element at 73 Easting belonged to the Eighteenth Brigade, Third Tawakalna ala-Allah (meaning "in God we trust") Mechanized Division, Iraqi Republican Guard. The Republican Guard stood outside the regular Iraqi army and reported directly to Saddam Hussein as elite Baathist Party shock troops. They fulfilled much the same function as Adolf Hitler's Waffen Schutzstaffel (armed protection element) in Nazi Germany. As with the Waffen SS in World War II, the Iraqi Republican Guard received more and better equipment and special party indoctrination for all ranks. Divisions like the Tawakalna were considered the most reliable and strongest in the Iraqi order of battle.

xxi *"Stormin' Norman" and "the Bear":* Atkinson, *Crusade,* 68–69; Gordon and Trainor, *The Generals' War,* 42–44. Both accounts reflect subordinates' critical views of Schwarzkopf's personality.

patience and self-discipline: Atkinson, *Crusade,* 73, 120–23.

xxiii *"Despite the final failure":* William C. Westmoreland, *A Soldier Reports* (Garden City, NY: Doubleday, 1976), 515.

"This will not stand": See www.bushlibrary.tamu.edu/research/papers/1990/90080502.html; accessed August 12, 2013.

xxiv *General (and Prince) Khalid bin Sultan:* Richard Jupa and Jim Dingeman, *Gulf Wars* (Cambria, CA: 3W Publications, 1991), 17, 67–68, 70–71; Atkinson, *Crusade,* 72. Khalid began the war as a lieutenant general.

xxv *a very strict code of conduct:* Thomas Taylor, *Lightning in the Storm* (New York: Hippocrene Books, 1994), 64–73. Taylor's account of the 101st Airborne Division's deployment addresses many aspects of life in Saudi Arabia before operations began.

xxvi *The Coalition forces:* U.S. Department of Defense, *Final Report to Congress: Conduct of the Persian Gulf War* (Washington, DC: U.S. Government Printing Office, April 1992), 160, 237–38, 282–86, 725; Jupa and Dingeman, *Gulf Wars,* 65. The Saudis paid more than $16 billion of $53 billion in funds provided by U.S. partners. Kuwait contributed a like amount, and the United Arab Emirates gave $4 billion. Japan offered $10 billion, and the Germans delivered more than $6 billion. Other countries, including the Republic of Korea, sent the rest.

the other side: Anthony H. Cordesman and Abraham R. Wagner, *The Lessons of Modern War,* vol. 2, *The Iran-Iraq War* (Boulder, CO: Westview Press, 1990), 428–30, 466–67, 504–5. NATO observers called the Soviet R-17E rocket Scud because all such surface-to-surface types received catalog names starting with an *S.* The Iraqis actually called their R-17Es Al Abbas and Al Hussein, depending on the modifications made to increase range.

xxviii *War games:* Gregory Fontenot, E. J. Degen, and David Tohn, *On Point: The United States Army in Operation Iraqi Freedom* (Fort Leavenworth, KS: Combat Studies Institute Press, 2004), 4.

predictions of U.S. casualties: Gordon and Trainor, *The Generals' War,* 139.

knew what to expect: Norvell B. De Atkine, "Why Arabs Lose Wars," *Middle East Quarterly* (December 1999): 3.

xxx *"winning wars without battles":* Quoted in ibid.

overwhelming force and numbers: Clancy and Franks, *Into the Storm,* 172–73. The U.S. Army Seventh Corps, commanded by Lieutenant General Fred Franks and including the Second Armored Cavalry Regiment, deployed from Germany, starting in November 1990, to add needed punch to the proposed U.S. offensive.

The Iraqis cracked: Atkinson, *Crusade,* 477; U.S. Department of Defense, *Conduct of the Persian Gulf War,* 248; Gordon and Trainor, *The Generals' War,* 235.

xxxi *"Well, Denham":* *King Kong,* directed by Merian C. Cooper and Ernest B. Schoedsack (New York: RKO Radio Pictures, 1933).

Conservative estimates: U.S. Department of Defense, *Conduct of the Persian Gulf War,* 358–68, 663; Gordon and Trainor, *The Generals' War,* 546.

xxxii *At half past eleven:* Atkinson, *Crusade,* 8–10; Clancy and Franks, *Into the Storm,* 466–69. Franks was present at the Safwan meeting. Atkinson obtained the official transcript of the meeting and, with fellow *Washington Post* reporter Steve Coll, talked to participants. See also H. Norman Schwarzkopf with Peter Petre, *It Doesn't Take a Hero* (New York: Bantam Books, 1992), 473–91, for the general's thinking regarding the event.

xxxiv *Although some plans were made:* Taylor, *Lightning in the Storm,* 264–65, 366. Taylor discusses what the 101st Airborne Division planners called "the Baghdad sequel." Gordon and Trainor, *The Generals' War,* depicts part of the scheme of maneuver in a map on page 476.

President Bush asked: Atkinson, *Crusade,* 488.

In the north: Dennis R. Chapman, *Security Forces of the Kurdistan Regional Government* (Costa Mesa, CA: Mazda Publishers, 2011), 16–19. Colonel Dennis Chapman, a West Point graduate, attorney, and infantry officer, advised a Kurdish brigade in the Iraq War of 2003–11.

xxxv *Shiite southern Iraq:* Gordon and Trainor, *The Generals' War,* 456–57; Bob Woodward, *Plan of Attack* (New York: Simon and Schuster, 2004), 64. Woodward notes that repercussions from the U.S. failure to support the Shiites affected planning for operations in 2003.

xxxvi *a large portion of the Republican Guard:* Gordon and Trainor, *The Generals' War,* 450–52; James G. Burton, "Pushing Them Out the Back Door," *U.S. Naval Institute Proceedings* (July 1993): 37–42.

xxxvii *This construct endured:* Fontenot, Degen, and Tohn, *On Point,* 4–22. The Desert Storm equivalents were termed *major regional conflicts,* then *major theater wars,* but the idea was always the same: two conventional, near-simultaneous fights.

The quality of the volunteers: Clancy and Franks, *Into the Storm,* 88–91. Franks lived the transition from the draft army to the volunteer force and saw how it came together in the 1990–91 campaign.

xxxviii *An air example:* Gordon and Trainor, *The Generals' War,* 199.

xxxix *the favored American way of war:* Russell F. Weigley, *The American Way of War* (Bloomington: Indiana University Press, 1977), xxii–xxiii. Weigley ably explained that although Americans much prefer a Desert Storm–model war of annihilation, far more often, they back into lengthy, indecisive wars of attrition.

xxix *"the Vietnam syndrome":* As quoted in Harry G. Summers Jr., *On Strategy II: A Critical Analysis of the Gulf War* (New York: Dell Books, 1992), 19.

xl *Saddam's subordinates:* Jupa and Dingeman, *Gulf Wars,* 52–54.

xli *a single Scud:* Gordon and Trainor, *The Generals' War,* 251.

Another interested observer: U.S. National Commission on Terrorist Attacks Upon the United States, *The 9/11 Commission Report* (Washington, DC: U.S. Government Printing Office, 2002), 55–57, 61.

1. Harbingers

3 *"A destroyer":* Quoted in Akiva J. Lorenz, "Analyzing the USS *Cole* Bombing," www.maritimeterroism.com, December 27, 2007; accessed August 15, 2013. Bin Laden

delivered the verse at his son's wedding celebration in Kandahar, Afghanistan, in January of 2001.

A glance at a map: U.S. Energy Information Administration, "World Oil Transit Chokepoints," August 22, 2012, at www.eia.govcountrics/analysisbriefs/World_Oil_Transit_Chokepoints/wotc.pdf; accessed on August 17, 2013.

4 *Aden attracted the British:* Piers Brendon, *The Decline and Fall of the British Empire, 1781–1997* (New York: Vintage Books, 2007), 507–13.

5 *In 1990, Yemen unified:* Lawrence Wright, *The Looming Tower* (New York: Alfred A. Knopf, 2006), 322–31.

The USS Cole: Kirk S. Lippold, *Front Burner: Al Qaeda's Attack on the USS* Cole (New York: Public Affairs Books, 2012), 9–10. The ship typically loaded its ninety launching cells with antiaircraft and antisubmarine missiles with the Tomahawk land attack. The destroyer was named for Sergeant Darrell S. Cole, who earned the Medal of Honor at the cost of his life on Iwo Jima in 1945. It was part of the seventy-seven-ship Arleigh Burke class of U.S. Navy guided missile destroyers.

"Even though you Navy officers": William Manchester, *Goodbye. Darkness* (Boston: Little, Brown, 1979), 223. A fine historian, Manchester served as a Marine sergeant on Okinawa, where he was wounded.

An Annapolis graduate: Lippold, *Front Burner,* 3–6.

From the perspective of the U.S. embassy: Ibid., 30–31, quotes Lippold. For details, see U.S. Department of the Navy, U.S. Naval Forces Central Command/Fifth Fleet, "Command Investigation into the Actions of USS *Cole* (DDG 67) in Preparing for and Undertaking a Brief Stop for Fuel at Bandar at Tawahi (Aden Harbor) Aden, Yemen on or about 12 October 2000," November 27, 2000, 15–16.

8 *"fool . . . who tried to hustle the East":* Rudyard Kipling, "The Naulahka," in *The Complete Verse* (London: Kyle Cathie, 1990), 434. See Lippold, *Front Burner,* 38, for the wording of the radio messages.

It all took until 10:31 a.m.: U.S. Naval Forces Central Command/Fifth Fleet, "Command Investigation," 41.

Aboard USS Cole: Lippold, *Front Burner,* 19–20.

a well-rehearsed routine: U.S. Naval Forces Central Command/Fifth Fleet, "Command Investigation," 46–47.

9 *Enterprising Yemenis had offered:* Lippold, *Front Burner,* 44–45. See also the USS *Cole* Commission, *USS* Cole *Commission Report* (Washington, DC: U.S. Government Printing Office, January 9, 2001), 8; it refers to the risks of dealing with unknown local contractors. These sources describe the sequence of events prior to the hostile attack.

10 *A quarter of the* Cole's *sailors:* U.S. Naval Forces Central Command/Fifth Fleet, "Command Investigation," 13–15.

11 *Below decks, Command Master Chief Parlier:* Lippold, *Front Burner,* 51.

The Yemen government: National Commission on Terrorist Attacks, *The 9/11 Commission Report,* 192–93; Wright, *The Looming Tower,* 369.

Bitterly condemned: Bob Woodward, *Bush at War* (New York: Simon and Schuster, 2002), 5, 38. The film *Wag the Dog* opened on January 9, 1998, more than eight months ahead of the missile strikes on Afghanistan and Sudan.

12 *Dispatched in the wake of the attack:* Donald P. Wright and the Contemporary

Operations Study Team, *A Different Kind of War* (Fort Leavenworth, KS: Combat Studies Institute Press, 2010), 28.

The institutional U.S. Navy: Lippold, *Front Burner,* 327–32.

13 *It took much time:* Ibid., 277–78, 290–91, 294.

"The pieces of the bodies": Lisa Beyer, "The Most Wanted Man in the World," *Time* (September 24, 2001): 34.

His designated cameraman: National Commission on Terrorist Attacks, *The 9/11 Commission Report,* 191.

15 *Born in 1957 in Jiddah, Saudi Arabia:* Ibid., 55–57; Wright, *The Looming Tower,* 294.

That man: Wright, *The Looming Tower,* 38–55.

16 *He made arrangements:* National Commission on Terrorist Attacks, *The 9/11 Commission Report,* 57.

al-Qaeda gathered money and recruited operatives: Ibid., 57–58; Stephen Tanner, *Afghanistan: A Military History from Alexander the Great to the War Against the Taliban* (Philadelphia: Da Capo, 2002), 285–86.

Throughout the 1990s: National Commission on Terrorist Attacks, *The 9/11 Commission Report,* 59–60.

17 *He reached out:* Ibid., 61; Wright, *The Looming Tower,* 334–35.

In an even more difficult initiative: Ibid. See also Woodward, *Bush at War,* 298.

The squeeze worked: National Commission on Terrorist Attacks, *The 9/11 Commission Report,* 63; Wright, *The Looming Tower,* 251–52.

18 *Ethnic Pashtuns dominated:* Wright et al., *A Different Kind of War,* 17–21.

Strategy came first: Tanner, *Afghanistan,* 286–87; Wright, *The Looming Tower,* 272.

the Sheikh issued a fatwa: Osama bin Laden, "Declaration of War Against the Americans Occupying the Land of the Two Holy Places," August 23 1996, at www.pbs.org/newshour/updates/military/july-dec96/fatwa_1996.html; accessed August 19, 2013.

21 *these people didn't train:* Wright, *The Looming Tower,* 341–43. In addition to their religious instruction, terrorists gained motivation and even tactical ideas from Western action movies. Arnold Schwarzenegger films were a favorite.

22 *"bunch of grapes":* Ibid., 48–49.

people followed orders: National Commission on Terrorist Attacks, *The 9/11 Commission Report,* 172–73; Wright, *The Looming Tower,* 209.

23 *On August 7, 1998:* National Commission on Terrorist Attacks, *The 9/11 Commission Report,* 115–16.

they must have been collaborators: Wright, *The Looming Tower,* 248.

the U.S. response: Ibid., 116–17. See also Richard J. Newman et al., "America Fights Back," *U.S. News and World Report,* August 23, 1998; Tanner, *Afghanistan,* 286; Wright, *The Looming Tower,* 322–23.

they came close: National Commission on Terrorist Attacks, *The 9/11 Commission Report,* 116–17.

24 *The first attempt:* Wright, *The Looming Tower,* 359; U.S. Naval Forces Central Command/Fifth Fleet, "Command Investigation," 36. One U.S. Navy ship refueled in May and two in August 2000. USS *The Sullivans* is named for five brothers killed in action in the loss of the light cruiser USS *Juneau* off Guadalcanal on November 13, 1942.

moved around to different secure locations: National Commission on Terrorist Attacks, *The 9/11 Commission Report,* 191.
"Does Al Qaeda*":* Ibid., 196.

2. 9/11

25 *"Death is art":* Wright, *The Looming Tower,* 125. Hasan al-Banna was one of the founders of the Muslim Brotherhood in Egypt
four flights: National Air Traffic Controllers Association, *NATCA: A History of Air Traffic Control* (Washington, DC: National Air Traffic Controllers Association, 2013), 15.
"We have some planes": National Commission on Terrorist Attacks, *The 9/11 Commission Report,* 1–34, offers a thorough yet concise reconstruction of the attack.
27 *"Everything in war":* Carl von Clausewitz, *On War,* ed. and trans. Michael Howard and Peter Paret (Princeton, NJ: Princeton University Press, 1976), 119–20.
28 *sudden death or serious injury:* Clausewitz, *On War,* 113–14.
29 *Friction left the team:* National Commission on Terrorist Attacks, *The 9/11 Commission Report,* 10–11, 456, note 73. Mohammed al Kahtani was refused entry by immigration agent Jose Melendez-Perez, a U.S. Army Vietnam veteran. It appeared Kahtani intended to overstay his visa; he'd arrived on a one-way ticket with only a small amount of cash. He was later captured in Afghanistan and moved to the detention facility at Guantánamo, Cuba.
The hijackers acted: Ibid., 11–14. As nobody survived the crash of United 93, all accounts draw on records of telephone calls, cockpit voice-recorder transcripts, and interviews with those who talked with the passengers and crew. The 9/11 Commission offered a great synthesis of the most reliable evidence available.
30 *The uniformed military:* Ibid., 20–21, 23–24, 26–28, 30–31.
31 *land them all:* Ibid., 29; National Air Traffic Controllers Association, *NATCA,* 15.
NORAD struggled: Woodward, *Bush at War,* 17–19, 26–28.
On the ground: National Commission on Terrorist Attacks, *The 9/11 Commission Report,* 311, 314, 552.
32 *"We will make":* Text of Bush's address at http://archives.cnn.com/2001/US/09/11/bush.speech.text; accessed August 21, 2013.
His CIA director: Woodward, *Bush at War,* 26–27, 40.
While Middle Eastern governments: National Commission on Terrorist Attacks, *The 9/11 Commission Report,* 333.
"The Pearl Harbor": Ibid., 37; Tanner, *Afghanistan,* 291.
33 *Bush spoke bluntly:* U.S. Office of the President, "Address to the Joint Session of the 107th Congress, September 20, 2001," in *Selected Speeches of President George W. Bush 2001–2008* (Washington, DC: White House, 2009), 65–68.
34 *It adhered to the diplomatic niceties:* Woodward, *Bush at War,* 98–99, 150, 209.
the scope of the war: Office of the President, "Address to the Joint Session of the 107th Congress," 69.
36 *They would deal with Afghanistan first:* Woodward, *Bush at War,* 43, 48–49.
37 *Some of the first orders:* Ibid., 196; National Commission on Terrorist Attacks, *The 9/11 Commission Report,* 326–27; Wright et al., *A Different Kind of War,* 30.

More important for the long term: R. Barry Cronin, "U.S. Northern Command and Defense Support of Civil Authorities," *Civil Support and the U.S. Army Newsletter* (December 2009): 25–32.

38 *Pressed to do something:* Tommy R. Franks with Malcolm McConnell, *American Soldier* (New York: HarperCollins, 2004), 260; Woodward, *Bush at War,* 79–80; Wright et al., *A Different Kind of War,* 44–45.

The eventual plan had four phases: Wright et al., *A Different Kind of War,* 46–48.

39 *the lineup:* Tanner, *Afghanistan,* 293; Woodward, *Bush at War,* 179.

engineered a broad coalition: Wright et al., *A Different Kind of War,* 32–33; Tanner, *Afghanistan,* 293–94.

40 *more capable allies contributed:* Wright et al., *A Different Kind of War,* 34–35.

arranging international assistance: Ibid., 37–39; Woodward, *Bush at War,* 228, 239, 280.

41 *Powell called him:* Woodward, *Bush at War,* 59.

Phase one led to phase two: Michael DeLong and Noah Lukeman, *Inside CentCom* (Washington, DC: Regnery Publishing, 2004), 34. Lieutenant General Michael DeLong served as the deputy commander of U.S. Central Command at the beginning of both the Afghan and Iraq campaigns. See also Franks, *American Soldier,* 243; Tanner, *Afghanistan,* 295.

Their camera exploded: Wright, *The Looming Tower,* 400–401.

the Northern Alliance: Wright et al., *A Different Kind of War,* 72–73.

42 *the U.S. needed some Pashtuns:* Ibid., 94–95; Tanner, *Afghanistan,* 307–8.

Tommy Franks had seen American airpower: Franks, *American Soldier,* 151, 165.

phase three: Wright et al., *A Different Kind of War,* 46.

43 *The military definition:* U.S. Department of the Army, *Field Manual 1-02/Marine Corps Reference Publication 5-12A: Operational Terms and Graphics* (Washington, DC: U.S. Government Printing Office, September 2004), 1–54. Although the Army and Marines agreed on the definition of *defeat,* no approved joint (Army, Navy, Air Force, Marine Corps) definition was adopted.

44 *He wanted to hand off:* Franks, *American Soldier,* 273, 338; Wright et al., *A Different Kind of War,* 46–47.

a failed state: Wright et al., *A Different Kind of War,* 23, 50.

45 *Infinite Justice:* Woodward, *Bush at War,* 134–35.

3. The Hindu Kush

46 *"I think everybody thought":* Wright et al., *A Different Kind of War,* 37. Ms. Rice served as the national security adviser in the first Bush term (2001–05) and the secretary of state in the second (2005–09).

47 *"This is General Dostum speaking":* Michael Smith, *The Killer Elite* (New York: St. Martin's, 2007), 217–18; Bryan Glyn Williams, "Report from the Field: General Dostum and the Mazar-i-Sharif Campaign," *Small Wars and Insurgencies* (December 2010): 618–20; William M. Knarr Jr., Robert F. Richbourg, and John Frost, *Learning from the First Victories of the 21st Century: Mazar-e Sharif—a Preview* (Alexandria, VA: Institute for Defense Analysis, 2004), part 3, 13, 15, 18, 24. In the 2014 Afghan elections, General Dostum ran for vice president on the Ashraf Ghani ticket.

"We killed the bastards": Quoted in Smith, *The Killer Elite*, 218. Captain Mark Nutsch made this report on October 25, 2001. See also Kalev I. Sepp, "Meeting the G-Chief: ODA 595," *Special Warfare* (September 2002): 10–11, and Williams, "Report from the Field," 619–20.
U.S. CIA and Special Forces: Gary Schroen, *First In* (New York: Random House, 2005), 92, 347. Schroen led the initial CIA effort in northern Afghanistan.
From the outset: Tanner, *Afghanistan*, 295; Smith, *The Killer Elite*, 217; Woodward, *Bush at War*, 210.

48 *For about two weeks:* Smith, *The Killer Elite*, 217; Woodward, *Bush at War*, 251–53, 260–61, 278–79; Tanner, *Afghanistan*, 298.
a few CIA officers: Schroen, *First In*, 194.
Regional SOF: Andrew Feickert, *U.S. Special Operations Forces: Background and Issues for Congress* (Washington, DC: Congressional Research Service), February 6, 2013, 1–5.
Army Special Forces: Tom Clancy with John Gresham, *Special Forces* (New York: Berkley Books, 2001), 148.

49 *On arrival in the mountains:* Williams, "Report from the Field," 616–19.
Six ODAs form an SF company: Clancy, *Special Forces*, 148–51.

50 *gave his teams a mission:* Wright et al., *A Different Kind of War*, 75.
Mulholland's Task Force Dagger: Ibid., 79–81.

51 *the lack of Pashtuns:* Woodward, *Bush at War*, 232–34; Schroen, *First In*, 347.
"give it back": Carlo D'Este, *Patton: A Genius for War* (New York: HarperCollins, 1995), 708. This well-known and characteristic quote was used in the 1970 film *Patton* during a similar exchange regarding the seizure of Palermo, Sicily, in 1943 by Patton's Seventh Army.
Victory developed momentum: Wright et al., *A Different Kind of War*, 102–7.

52 *Brigade 055:* Williams, "Report from the Field," 618; Alex Perry, "Inside the Battle at Qala-i-Jangi," *Time* (December 10, 2001): 51.
brick Fighters' Fortress: Interview with Captain Michael Sciortino at Qala-i-Jangi Fortress, Afghanistan, September 10, 2012. The interview included a guided tour of the fortress and a detailed discussion of the events of November 2001.

54 *Cowed, nervous:* Perry, "Inside the Battle at Qala-i-Jangi," 51; Tanner, *Afghanistan*, 305; Doug Stanton, *Horse Soldiers* (New York: Scribner, 2009), 2–3, 6. Stanton's book offers a superb account of the fight at Qala-i-Jangi as well as a detailed look at the operations of U.S. Special Forces in northern Afghanistan.

55 *They had been active from the outset:* Williams, "Report from the Field," 616; Perry, "Inside the Battle at Qala-i-Jangi," 51.

56 *Now all the prisoners stood up:* Perry, "Inside the Battle at Qala-i-Jangi," 51; Stanton, *Horse Soldiers*, 11. See also Oliver North, *American Heroes in Special Operations*, ed. Chuck Holton (Nashville, TN: Fidelis Books, 2010), 32.
Behind Davis: Kalev I. Sepp, "Uprising at Qala-i-Jangi: The Staff of the 3/5 SF Group," *Special Warfare* (September 2002): 16–17.
"He burst in": Perry, "Inside the Battle at Qala-i-Jangi," 51.

57 *They contacted K2 airfield:* Sepp, "Uprising at Qala-Jangi," 17.
The battle spawned: Mark Bowden, *Black Hawk Down* (New York: Penguin, 2000), 284–85; Ken Nolan, *Black Hawk Down: The Shooting Script* (New York: Newmarket Press, 2002), xv.

At one point in Mogadishu: Bowden, *Black Hawk Down,* 165, 189–90. The urban-warfare training site at Fort Polk, Louisiana, is named in honor of Master Sergeant Gary I. Gordon and Sergeant First Class Randall D. Shughart.

58 *Mitchell tagged others:* Sciortino interview; Sepp, "Uprising at Qala-Jangi," 17; North, *American Heroes in Special Operations,* 32–33.

59 *"You could hear the bullets":* Gretchen N. McIntyre, "A Living Hero," Defense Video and Imagery Distribution System, October 18, 2012, www.dvidshub.net/news/97596/living-hero#UhjlgBsuiSo; accessed August 24, 2013.

A JDAM used: Tom Clancy, *Fighter Wing* (New York: Berkley Books, 1995), 160.

Sciortino started calling: Sciortino interview; Perry, "Inside the Battle at Qala-i-Jangi," 51. Perry quoted the conversations word for word. Sciortino confirmed these exchanges.

60 *Sciortino talked on the radio:* Sciortino interview; Perry, "Inside the Battle at Qala-i-Jangi," 51.

Six more times: Sciortino interview; Perry, "Inside the Battle at Qala-i-Jangi," 51; Sepp, "Uprising at Qala-Jangi," 17.

61 *Morning brought reinforcements:* Wright et al., *A Different Kind of War,* 84.

62 *A massive hot light:* Sciortino interview; Perry, "Inside the Battle at Qala-i-Jangi," 52; David Lightman, "Connecticut Native Calmly Tells Story of a Ferocious Moment," *Hartford Courant,* December 20, 2001. Lightman quotes Sciortino from a 2001 interview.

"We have men down": Sciortino interview; Perry, "Inside the Battle at Qala-i-Jangi," 52.

63 *An AC-130 aerial gunship:* Clancy, *Special Forces,* 168. The AC-130H Spectre mounted 40 mm and 105 mm cannon linked to a targeting computer. The newer AC-130U Spooky (also called the U-Boat) had a 25 mm Gatling gun and a 105 mm cannon, also tied to a precision targeting system.

The explosion lit the night: Sepp, "Uprising at Qala-Jangi," 17.

On the morning of November 27: Perry, "Inside the Battle at Qala-i-Jangi," 52.

took turns clearing: Sepp, "Uprising at Qala-Jangi," 18.

64 *"On the field below":* Ibid.; Perry, "Inside the Battle at Qala-i-Jangi," 52.

Geoje Island: T. R. Fehrenbach, *This Kind of War* (New York: Macmillan, 1963), 571–94. The uprising at Geoje Island (May 7, 1952, to June 12, 1952), spurred by Communist North Korean and Chinese agitators, required U.S. tanks, paratroopers, and infantry battalions to suppress it.

Rozi's men: Wright et al., *A Different Kind of War,* 85. Yaser Esam Hamdi later became the named party in a famous U.S. Supreme Court decision of June 28, 2004, *Hamdi v. Rumsfeld.* The court affirmed the U.S. military's authority to detain enemy combatants, including U.S. citizens. But the court also directed due process for detained U.S. citizens. Hamdi was released and allowed to join his family in Saudi Arabia.

the USCENTCOM commander got a call: Franks, *American Soldier,* 314–15.

66 *merely detainees:* Wright et al., *A Different Kind of War,* 220–23.

The CIA and military SOF don't care for publicity: Woodward, *Bush at War,* 316. In light of the immediate press coverage of Qala-i-Jangi, CIA director George Tenet released the name of Johnny Micheal Spann.

67 *Those needed conventional battalions:* Wright et al., *A Different Kind of War,* 115, 127, 182.
America's chosen Pashtun, Hamid Karzai: Ibid., 109–10; Woodward, *Bush at War,* 314–15; North, *American Heroes in Special Operations,* 29–30.

68 *General Tommy Franks exulted:* Franks, *American Soldier,* xvi.
"Defending the cities": Tanner, *Afghanistan,* 302.

69 *As for al-Qaeda:* Wright et al., *A Different Kind of War,* 119. For a detailed narrative that reflects the military side, see Dalton Fury, *Kill bin Laden* (New York: St. Martin's, 2008). The CIA perspective can be found in Gary Berntsen and Ralph Pezzullo, *Jawbreaker* (New York: Broadway Books, 2006).
An American Raiding Force: Helene Cooper, "Obama Announces Killing of Osama bin Laden," *New York Times,* May 1, 2001. Due to time differences, many U.S. sources refer to the event's occurring on May 1. In the U.S., President Obama announced the results of the raid after 11:00 p.m. on May 1, 2011.

70 *Three major attacks:* These follow-on attacks are well covered in Abdel Bari Atwan, *After Bin Laden: Al-Qa'ida, the Next Generation* (New York: New Press, 2013), 18, 25, 31–32, 228–29.

71 *Iraq came up:* Woodward, *Bush at War,* 287.
Colin Powell cautioned: Ibid., 331–34.
Like President Harry S. Truman: Fehrenbach, *This Kind of War,* 276.

72 *"I will not wait on events":* Woodward, *Bush at War,* 329.

4. Anaconda

73 *"O gods, from the venom":* Quoted by Minister Abdul Rahim Wardak in Daniel P. Bolger, "Notes on Meeting with Afghan Minister of Defense Abdul Rahim Wardak," Kabul, Afghanistan, November 2, 2011.
Colonel Frank Wiercinski: Sean D. Naylor, *Not a Good Day to Die* (New York: Berkley Caliber Books, 2005), 241–42; Wright et al., *A Different Kind of War,* 142–43. Naylor's book offers a definitive account of the first few days of the battle and a decent summary of the preparation and follow-on operations. He was there.

74 *Zia's people:* Kalev I. Sepp, "The Campaign in Transition: From Conventional to Unconventional War," *Special Warfare* (September 2002): 26; Richard Kugler, *Operation Anaconda in Afghanistan: A Case Study of Adaptation in Battle* (Washington, DC: National Defense University, 2007), 10, 14–15.

75 *"I've heard this sound before":* Naylor, *Not a Good Day to Die,* 243; Kugler, *Operation Anaconda in Afghanistan,* 16.
The operation's name: Joseph Miranda, "Operation Anaconda: Battle in Afghanistan, 2002," *Strategy and Tactics* (September/October 2012): 11. The code name also harked back to the 1861 scheme developed at the outset of the American Civil War by the senescent Major General Winfield Scott. Though Scott was old, fat, and feeble by that time, his mind remained brilliant. He charted the future course of the Union forces with his plan to use a naval blockade and multiple converging land offensives to strangle the Confederacy.

76 *"There were no civilians":* Wright et al., *A Different Kind of War,* 143.
Despite the common image: Naylor, *Not a Good Day to Die,* 264–66.

78 *"Who are those guys"*: *Butch Cassidy and the Sundance Kid,* directed by George Roy Hill (Los Angeles: Twentieth Century Fox Film Corporation, 1969). The brilliant William Goldman wrote the screenplay.
The IMU: Naylor, *Not a Good Day to Die,* 156–57, 376.
The other five hundred or so: Kugler, *Operation Anaconda in Afghanistan,* 6–7; Naylor, *Not a Good Day to Die,* 138–39, 376.
79 *five neighbors:* Wright et al., *A Different Kind of War,* 143.
"I owe my life": Naylor, *Not a Good Day to Die,* 190, 244.
80 *"in his element"*: Ibid., 245.
"You guys saved us": Ibid., 245–46; Wright et al., *A Different Kind of War,* 151; Rebecca Grant, "The Echoes of Anaconda," *Air Force* (April 2005): 50.
81 *In that busy day:* Grant, "The Echoes of Anaconda," 50–51.
"Protect them": Naylor, *Not a Good Day to Die,* 180.
82 *"I do recall"*: Wright et al., *A Different Kind of War,* 152.
"clearly saved the day": Naylor, *Not a Good Day to Die,* 256–59.
83 *In Bagram that busy morning:* Marilyn Ware, "Backgrounder: Major General Franklin 'Buster' Hagenbeck," in *Operation Anaconda Overview: Experiencing the Fog of War Military Simulation* (Washington, DC: American Enterprise Institute, 2012), 7–8. Hagenbeck served as a major in the brief Grenada operation in 1983.
approach to command: Ware, "Major General Franklin 'Buster' Hagenbeck," 1–2, 8. A football player at West Point, Hagenbeck assisted with the Seminoles while earning his master's degree in physiology from 1976 to 1978.
84 *Zia Lodin's Afghan militia:* Kugler, *Operation Anaconda in Afghanistan,* 15.
"I didn't want a shoot-down": Naylor, *Not a Good Day to Die,* 265.
"except in one place": Ibid., 266.
"Higher made the decision": Ibid., 256; Ware, "Major General Franklin 'Buster' Hagenbeck," 8; Wright et al., *A Different Kind of War,* 154.
"I've killed enough for today": Naylor, *Not a Good Day to Die,* 284.
Takur Ghar: Kugler, *Operation Anaconda in Afghanistan,* 17; Grant, "The Echoes of Anaconda," 52.
86 *Things dragged on:* Grant, "The Echoes of Anaconda," 52; Wright et al., *A Different Kind of War,* 154–57, 160–72.
People who liked numbers: Tanner, *Afghanistan,* 316–17; Wright et al., *A Different Kind of War,* 173.
Rudyard Kipling warned: Rudyard Kipling, "Arithmetic on the Frontier," *The Complete Verse* (London: Kyle Cathie, 1990), 36.
87 *A few Russian airborne companies:* David Isby, *War in a Distant Country* (London: Arms and Armour Press, 1989), 46–47.
the awful results: Naylor, *Not a Good Day to Die,* 376.
88 *Khushal Khan Khattak:* Ali Ahmad Jalali, "Rebuilding Afghanistan's National Army," *Parameters* (Autumn 2002): 73. The author fought the Soviets and went on to serve as interior minister in the Afghan government from 2003 to 2005.
"Enemy advances": Mao Zedong, *Strategic Problems of China's Revolution* (Beijing: Foreign Languages Press, 1954), 96.
89 *"I thought it was"*: Thomas E. Ricks, *The Generals* (New York: Penguin, 2012), 399.
Now came phase four: Wright et al., *A Different Kind of War,* 46–47.

90 *"I was out of my mind":* Ibid., 173, 184; Robert H. McElroy and Patricia Slayden Hollis, "Afghanistan: Fire Support for Operation Anaconda," *Field Artillery Journal* (September/October 2002): 5–9.

90 *Joe Stilwell:* Barbara Tuchman, *Stilwell and the American Experience in China, 1911–45* (New York: Bantam Books, 1970), 314.

91 *Both commanders:* Wright et al., *A Different Kind of War,* 190–91, 211.

92 *their own little formula:* Andrew J. Krepinevich Jr., *The Army and Vietnam* (Baltimore: Johns Hopkins University Press, 1986), 57.

93 *you had a huge funnel:* Stanley McChrystal, *My Share of the Task* (New York: Penguin, 2013), 91.

94 *"How do we organize":* Smith, *The Killer Elite,* 230.
"butcher and bolt": William Manchester, *Winston Spencer Churchill: The Last Lion,* vol. 2, *Visions of Glory 1874–1932* (New York: Dell, 1983), 256–57.
"Don't you do anything": Wright et al., *A Different Kind of War,* 190.

95 *"We will stay":* Cable News Network, "President Bush Speaks at VMI, Addresses Middle East Conflict," April 17, 2002, at http://transcripts.cnn.com/TRANSCRIPTS/0204/17/se.02.html; accessed September 4, 2013.

96 *"It's been one of initial success":* Ibid.

5. A Weapon of Mass Destruction

97 *"Fliers with premonitions":* Tom Wolfe, "The Truest Sport: Jousting with Sam and Charlie," in *The Purple Decades: A Reader* (New York: Berkley Books, 1982), 237. This article, first published in *Esquire* in 1975, described a 1967 mission launched against North Vietnam from the aircraft carrier USS *Coral Sea.* Along with a superb, rollicking account of flying into the teeth of very active air defenses, Wolfe explores some of the themes that he followed more extensively in his 1979 book *The Right Stuff,* about the early U.S. space program

98 *Russian-made 23 mm:* David C. Isby, *Weapons and Tactics of the Soviet Army,* rev. ed. (London: Jane's Publishing, 1988), 316, 320–32.
"If you watched": Janet Bresenham, "Lt. Col. Jones Glad to Be Back," *Amarillo Globe News,* February 4, 1999.
To the pilots: Associated Press, "Cannon Air Force Base Pilot Shares First Experiences of Dodging Anti-Aircraft Fire," *Lubbock Avalanche-Journal,* December 31, 1998; Clancy, *Fighter Wing,* 89.
"It was pitch-black": Associated Press, "Cannon Air Force Base Pilot."
The single-engine F-16 Falcons: "Gallery of USAF Weapons," *Air Force* (May 2013): 85; Clancy, *Fighter Wing,* 85, 90.

99 *"After the first bomb detonation":* Associated Press, "Cannon Air Force Base Pilot."
that useful envelope: Clancy, *Fighter Wing,* 93.

100 *massive initial series of airstrikes:* Gordon and Trainor, *The Generals' War,* 112–15.
no-shows in 1998: Kevin M. Woods et al., *Iraqi Perspectives Project: A View of Operation Iraqi Freedom from Saddam's Senior Leadership* (Norfolk, VA: U.S. Joint Forces Command, 2005), 40. The translated documents are invaluable.
What about the SAMs: Ibid.; Isby, *Weapons and Tactics of the Soviet Army,* 335–41.
"It's no kidding": Associated Press, "Cannon Air Force Base Pilot."

The Iraqis admitted: Anthony H. Cordesman, *The Military Effectiveness of Desert Fox* (Washington, DC: Center for Strategic and International Studies, December 26, 1998), 16–17, 23.

101 *Slam 11 and Slam 12:* David Noland, "The Bone Is Back," *Air and Space Smithsonian* (May 2008): 20–21; Ashley J. Thum, "B-1 Bomber Makes Historic Combat Debut During Operation Desert Fox," *Air Force Print News Today,* March 7, 2013; Clancy, *Fighter Wing,* 95–97.

"a very healthy pairing": Thum, "B-1 Bomber Makes Historic Combat Debut."

102 *"like we had trained to do":* Noland, "The Bone Is Back," 20; Thum, "B-1 Bomber Makes Historic Combat Debut."

The strike photo: Thum, "B-1 Bomber Makes Historic Combat Debut." Thum's article includes the strike photo. The original label read "Al Kut Barracks West-Northwest, Iraq." The target was in fact located near the city of Al Kut, Wasit Province, in east-central Iraq.

"I was just doing my job": Marco R. della Cava, "Alaskan Is First Female Pilot in Combat," *USA Today,* December 19, 1998. Others highlighted were F/A-18 naval aviator Lieutenant Carol Watts and U.S. Air Force First Lieutenant Cheryl Lamoureux, who fired cruise missiles from her B-52 Stratofortress.

103 *The aggressive Americans:* Patrick Higby, *Promise and Reality: Beyond Visual Range (BVR) Air-to-Air Combat* (Maxwell Air Force Base, AL: Air War College, March 10, 2005), 14.

In the east: Kenneth H. Bacon, Department of Defense news briefing, January 5, 1999, 1–2.

to "outfox" Saddam Hussein: Tom Clancy with Anthony Zinni and Tony Koltz, *Battle Ready* (New York: G. P. Putnam's Sons, 2004), 19–20.

104 *more from the usual direction:* Mark J. Conversino, "Operation Desert Fox: Effectiveness with Unintended Effects," *Air and Space Power Journal: Chronicle Online Journal* (July 5, 2005): 1, at http://www.airpower.maxwell.af.mil/airchronicles/cc/conversino.html; accessed September 7, 2013.

Intrinsic Action exercises: Fontenot, Degen, and Tohn, *On Point,* 29.

105 *air policing:* Manchester, *Visions of Glory 1874–1932,* 700–701. See also John Keegan, *The Iraq War* (New York: Alfred A. Knopf, 2004), 9–11, 14–15.

"Hesitation or delay": Graham Chandler, "The Bombing of Waziristan," *Air and Space Smithsonian* (July 2011): 40.

106 *When the appointed day came:* Ibid. See also Great Britain Ministry of Defence, *Air Policing During the Interwar Years: Research Guide Bibliography* (Oxfordshire, UK: Joint Services Command and Staff College Library, 2011), 2.

109 *The great Linebacker II B-52 offensive:* Norman Friedman, *The Fifty-Year War: Conflict and Strategy in the Cold War* (Annapolis, MD: Naval Institute Press, 2000), 363.

The U.S. saw it as containment: John A. Tirpak, "Legacy of the Air Blockades," *Air Force* (February 2003): 46. Operation *Desert Storm* featured about 101,370 U.S. sorties.

110 *Constant daily skirmishes:* Ibid., 52.

the cost to America: Ibid., 50.

111 *"We did not identify":* Cordesman, *The Military Effectiveness of Desert Fox,* 5.

Butler estimated: Conversino, "Operation Desert Fox," 1.

112 *persistent threat to Kuwait:* W. Eric Herr, *Operation Vigilant Warrior: Conventional Deterrence Theory, Doctrine, and Practice* (Maxwell Air Force Base, AL: School of Advanced Airpower Studies, June 1996), 25, 27; Fontenot, Degen, and Tohn, *On Point,* 100.

Inside Iraq: Chapman, *Security Forces of the Kurdistan Regional Government,* 141; Keegan, *The Iraq War,* 65–66, 99.

backed terrorists of all stripes: Kevin M. Woods with James Lacey, *Iraq Perspectives Project: Saddam and Terrorism, Emerging Insights from Captured Iraqi Documents* (Alexandria, VA: Institute for Defense Analysis, November 2007), 4, 13–15. This extremely useful study reproduces numerous key Iraqi documents. See also McChrystal, *My Share of the Task,* 83–84. McChrystal describes Zarqawi's background.

113 *The law's text:* U.S. Congress, *Iraq Liberation Act of 1998,* Public Law 105-338, 105th Congress, Second Session, House Resolution 4655, 1.

114 *The Kurds up north:* Chapman, *Security Forces of the Kurdistan Regional Government,* 141; Robert Baer, *See No Evil: The True Story of a Ground Soldier in the CIA's War on Terrorism* (New York: Three Rivers Press, 2003), 207.

Ahmed Abdel Hadi Chalabi: Woodward, *Plan of Attack,* 19–20; Dexter Filkins, *The Forever War* (New York: Vintage Books, 2009), 261–63.

115 *"He tried to kill my dad":* Woodward, *Plan of Attack,* 27, 187.

faster options: Fontenot, Degen, and Tohn, *On Point,* 45.

116 *allied support:* Keegan, *The Iraq War,* 107–8; Woodward, *Plan of Attack,* 226. United Nations resolutions that endorse military action use the term "all necessary means." This one, UN 1441, referred only to "serious consequences."

the vote ran true to form: Woodward, *Plan of Attack,* 203–4.

the text of the 2002 authorization: U.S. Congress, *Authorization for the Use of Military Force Against Iraq Resolution of 2002,* Public Law 107-243, 107th Congress, Second Session, House Joint Resolution 114, 1.

117 *an intriguing mash-up:* Franks, *American Soldier,* 389–91.

phase three: Fontenot, Degen, and Tohn, *On Point,* 30, 77–79.

118 *all the way to Baghdad:* Ibid., 30; Franks, *American Soldier,* 437–40.

"There was seriously not anything": Donald P. Wright, Timothy R. Reese, and the Contemporary Operations Study Team, *On Point II: Transition to the New Campaign, the United States Army in Operation Iraqi Freedom, May 2003–January 2005* (Fort Leavenworth, KS: Combat Studies Institute, 2008), 66, 70–72. Fuzzy Webster served as the deputy commander of the Coalition Land Forces Component Command in the 2003 invasion. He later commanded the Third Infantry Division in Baghdad from 2005 to 2006.

"The Americans will use": Woods et al., *Iraqi Perspectives Project,* 30.

Shinseki offered a prescient warning: Eric Schmitt, "Pentagon Contradicts General on Iraq Occupation Force's Size," *New York Times,* February 28, 2003.

6. Apocalypse Then Redux

120 *"We're going to grab him by the nose":* D'Este, *Patton: A Genius for War,* 623. This famous utterance was used word for word in the dramatic opening scene of the 1970 film *Patton.* The *him* refers to the German enemy.

digging all night: Jim Lacey, *Takedown: The 3rd Infantry Division's Twenty-One-Day Assault on Baghdad* (Annapolis, MD: Naval Institute Press, 2007), 32. A former U.S. Army infantry officer, Lacey served as a *Time* magazine embedded reporter with the 101st Airborne Division during the 2003 attack on Baghdad. He also wrote major portions of the assessment compiled by the Institute for Defense Analyses from the extensive debriefings of Iraqi senior officers

121 *Their thermal sights:* Mark K. Schenck, *Unit History, Operation Iraqi Freedom* (Fort Stewart, GA: Headquarters, Second Battalion, Seventh Infantry, 2003), 5–6. Schenck's history, a superb account of the battalion's experiences, was compiled shortly after operations concluded.
hit directly by a box: Fontenot, Degen, and Tohn, *On Point,* 108, 112, 113.

122 *For Iraqi Freedom, 2-7 Infantry:* Schenck, *Unit History, Operation Iraqi Freedom,* 1–5.

123 *Lieutenant Colonel Scott E. Rutter:* Lacey, *Takedown,* 28–29. Scott Rutter knew all too much about terrorists even before 9/11. From 1990 to 1991, Rutter's rifle company (Company C, Second Battalion, Sixteenth Infantry) included Timothy McVeigh, the key felon convicted and executed for the 1995 Oklahoma City federal building bombing.
Rutter expected contact: Fontenot, Degen, and Tohn, *On Point,* 116–17.

126 *a fuel stop:* Clancy, *Armored Cav,* 66, 71–78; Lacey, *Takedown,* 35. Clancy's book offers excellent details on the Bradley. According to Jim Lacey's account, the Third Infantry Division used fifty-six gallons an hour in calculating the fuel consumption rate for an M-1 tank.
his computer display: Fontenot, Degen, and Tohn, *On Point,* 101, 182.

127 *In the rear of the stopped column:* Lacey, *Takedown,* 62–64.
ten more Iraqis: Schenck, *Unit History, Operation Iraqi Freedom,* 10.

128 *They were irregulars:* Woods et al., *Iraqi Perspectives Project,* 48.
Following the two engagements: Lacey, *Takedown,* 64–65.

129 *Use the machine guns:* Schenck, *Unit History, Operation Iraqi Freedom,* 10–11.

130 *discarded Iraqi uniforms:* Ibid., 15.
a line of violent thunderstorms: Fontenot, Degen, and Tohn, *On Point,* 150.
movements became treacherous: Schenck, *Unit History, Operation Iraqi Freedom,* 16.
Two days later: Ibid., 16–17.

131 *Other Third Infantry Division battalions:* David Zucchino, *Thunder Run* (New York: Atlantic Monthly Press, 2004), 80–81.
None of the division's three maneuver brigades was fresh: Fontenot, Degen, and Tohn, *On Point,* 299–300, 332.

132 *The people in 3-69 Armor:* Ibid., 303. The 3-69 Armor movement had one bad stretch. A Humvee tumbled into a creek. Two died: Staff Sergeant Wilbur Davis and *Atlantic Monthly* reporter Mike Kelly.
another convoluted rat's nest: Schenck, *Unit History, Operation Iraqi Freedom,* 21.

134 *Dawn brought trouble:* Caroline Glick, "How U.S. Forces Captured Saddam International Airport," *Jerusalem Post,* April 6, 2003. A former officer in the Israeli Defense Forces, Glick was embedded with 2-7 Infantry during the campaign. Artist Steve Mumford was also embedded with 2-7 Infantry. His sketches and comments are available at http://www.glasschord.com/steve-mumford/correspondent/. Mumford drew a great picture of riflemen in the back of a Bradley fighting vehicle,

among many other unique pieces of art. When he first met Lieutenant Colonel Scott Rutter, the battalion commander was going out on a patrol with his men.

135 *Called the Javelin:* Clancy, *Armored Cav,* 158–60.

watched the missile race: Fontenot, Degen, and Tohn, *On Point,* 306–7.

136 *To the west:* Schenck, *Unit History, Operation Iraqi Freedom,* 23–24.

138 *Paul Smith's heroism:* Ibid., 25; Lacey, *Takedown,* 206–7; Fontenot, Degen, and Tohn, *On Point,* 308; Wright et al., *On Point II,* 18.

Rutter stayed out on the ground: Glick, "How U.S. Forces Captured Saddam International Airport." Lieutenant Colonel Scott Rutter earned the Silver Star for his leadership under fire.

Special Republican Guard: This organization was formed in the early 1990s. It protected Saddam Hussein and fielded four brigades and affiliated tank, artillery, and air-defense battalions. Sometimes known as the Golden Division, the Special Republican Guard was the only military organization permitted to operate inside the city of Baghdad. If the Republican Guard could be compared to the Nazi German Waffen SS, the Special Republican Guard resembled the organization that later fought as the First SS Panzer Division Leibstandarte (Lifeguard) Adolf Hitler.

a very sharp clash: Zucchino, *Thunder Run,* 313; Schenck, *Unit History, Operation Iraqi Freedom,* 28–29. That final major firefight on April 7 proved costly for 2-7 Infantry. The battalion lost Staff Sergeant Lincoln D. Hollinsaid and Private First Class Jason M. Meyer, both engineers. Hollinsaid had replaced Paul Smith on April 4, meaning the engineers lost two platoon sergeants in a few days. Major Rod Coffey, among others, was wounded.

139 *"The Americans came, saw, conquered":* Keegan, *The Iraq War,* 1.

the 507th Maintenance Company: Fontenot, Degen, and Tohn, *On Point,* 154–60, 179–92. On March 23, 2003, in Nasiriyah, eleven men of the 507th Maintenance Company were killed; six were captured, and five were wounded. Five of eighteen trucks were abandoned. Prisoner of war Private First Class Jessica Lynch was later rescued by U.S. SOF. On the night of March 23–24, thirty Apache helicopters attempted to attack the Iraqi Republican Guard Medina Division. The mission failed. Ferocious ground fire damaged twenty-nine helicopters. One went down, with both aviators captured.

"without discernible effect": Keegan, *The Iraq War,* 1–2. For a good example of a conventional map depiction of a most irregular battle, see Fontenot, Degen, and Tohn, *On Point,* 257.

140 *"different from the one we war-gamed against":* Yossef Bodansky, *The Secret History of the Iraq War* (New York: HarperCollins, 2004), 205.

to secure the routes: Fontenot, Degen, and Tohn, *On Point,* 209, 245. For Marine operations in March and April of 2003, see Bing West and Ray L. Smith, *The March Up* (New York: Bantam Books, 2003). John Keegan provided the introduction.

"the tribal nature": Woods et al., *Iraqi Perspectives Project,* 38.

141 *Ansar al-Islam:* Fontenot, Degen, and Tohn, *On Point,* 250; Woods et al., *Iraq Perspectives Project,* 13–15, 18, 42.

Saddam's WMD plans: Fontenot, Degen, and Tohn, *On Point,* 171–74, 358–59. For pictures of some of the chemical rounds located, see U.S. Central Intelligence Agency, *Addendums to the Comprehensive Report of the Special Advisor to the DCI on Iraq's WMD* (Langley, VA: Central Intelligence Agency, March 2005), 18–19.

Iraqi insurgents did use some chemical rounds in improvised explosive devices. The Israelis bombed and destroyed the Tuwaitha nuclear reactor, also called Osirak, in 1981.

142 *"the turning of the tide"*: U.S. President George W. Bush, "Remarks by the President from the USS *Abraham Lincoln*" (Washington, DC: The White House, May 1, 2003), 1.

143 *"You know where a war begins"*: Quoted in Robert K. Massie, *Dreadnought* (New York: Random House, 1991), 76.

7. "Mission Accomplished"

147 *"But I suppose"*: Jean Lartéguy, *The Centurions*, trans. Xan Fielding (New York: E. P. Dutton, 1962), 460. Lartéguy's novel concerns the first few years of the French-Algerian War of 1954–62, culminating with the Battle of Algiers in 1956–57. The character Captain Julien Boisfeuras is the French airborne unit's intelligence officer. The man he addresses, Major Jacques de Glatigny, is another French airborne officer. Both disillusioned veterans of the earlier Indochina War are determined to win in Algeria. Their paratroopers employ harsh measures to gain intelligence. The book has long been a favorite of U.S. military men and women interested in the perils of counterinsurgency.

"The American soldiers": Lacey, *Takedown*, 149.

149 *"Send me a West Point football player"*: Frank Deford, "Code Breakers," *Sports Illustrated* (November 13, 2000): 67.

The battalion commander: Filkins, *The Forever War*, 160. Dexter Filkins embedded with 1-8 Infantry during parts of the deployment. For a view from inside the battalion, see Darron L. Wright, *Iraq Full Circle* (Oxford, UK: Osprey Publishing, 2012), 110. Wright served as the battalion S-3 (operations officer).

150 *Paliwoda was more:* Nathan Sassaman with Joe Layden, *Warrior King* (New York: St. Martin's, 2008), 122–23. See also Wright, *Iraq Full Circle*, 135. For another company commander's perspective, see Todd S. Brown, *Battleground Iraq: Journal of a Company Commander* (Washington, DC: U.S. Army Center of Military History, 2007), 199–201.

the night's raid: Sassaman and Layden, *Warrior King*, 209, 221.

151 *the Balad cannery:* Ibid., 225. For one assessment of the ways of socialist Iraq, see Keegan, *The Iraq War*, 47–48.

heavy-handed measures: Brown, *Battleground Iraq*, 187.

152 *"Incoming"*: Wright, *Iraq Full Circle*, 176–77; Sassaman and Layden, *Warrior King*, 228–30.

153 *"his breaking point"*: Wright, *Iraq Full Circle*, 178.

the rage of Achilles: Homer, *The Iliad*, trans. Robert Fitzgerald (Garden City, NY: Anchor Books, 1974), 528.

The raid went off late: Sassaman and Layden, *Warrior King*, 232–33.

155 *moved out for Samarra*: Brown, *Battleground Iraq*, 203–4, 257–58; Sassaman and Layden, *Warrior King*, 233–34. Brown places this firefight on either January 7 (in the main text) or January 8 (in the appendix). His daily diary, composed on the spot and revised much later, tends to be a bit off. He recorded Eric Paliwoda's January 2 death as occurring on January 4 and Saddam Hussein's capture as hap-

pening on December 12. It occurred on December 13. Sassaman's book places the Samarra firefight on January 3. Wright's book doesn't discuss the firefight and places Paliwoda's death on January 4. All of these combat leaders were busy, tired, and emotionally drained. Sassaman's account has been, throughout, the most reliable, although strongly colored by his opinions.

Around midnight: Filkins, *The Forever War,* 162–65; Sassaman and Layden, *Warrior King,* 240–43, 255–57; Wright, *Iraq Full Circle,* 179–81.

156 *"a swirly":* Filkins, *The Forever War,* 166. Filkins interviewed Ralph Logan and Nate Sassaman after the incident and followed up with the Iraqis as well. He watched what was purported to be a video of Zaydoon's funeral. He also went to the riverbank with Marwan Fadil to discuss the incident.

The rumor mill: Sassaman and Layden, *Warrior King,* 267.

157 *incident at Abu Ghraib:* Wright et al., *On Point II,* 257.

courts-martial: Sassaman and Layden, *Warrior King,* 285–92.

158 *The Sunni Arabs:* Anthony H. Cordesman with Patrick Baetjer, *Iraqi Security Forces: A Strategy for Success* (Westport, CT: Praeger Security International, 2006), xvii–xviv.

ranged up to sixteen thousand: Ibid., 30.

"former regime elements": Wright et al., *On Point II,* 105.

159 *Abu Musab al-Zarqawi:* Ibid., 103; Keegan, *The Iraq War,* 8.

the Shia militants: Wright et al., *On Point II,* 109–10.

160 *Almost from the start:* Ibid., 140–41; Ricardo S. Sanchez with Donald T. Phillips, *Wiser in Battle* (New York: HarperCollins, 2008), 168–71.

161 *Iraqis made themselves available:* Filkins, *The Forever War,* 254–60. See also Wright et al., *On Point II,* 151. When Jay Garner arrived in Baghdad, Chalabi and Allawi were among the first Iraqis he met.

"I had the requisite skills": L. Paul Bremer III with Malcolm McConnell, *My Year in Iraq* (New York: Simon and Schuster, 2006), 7.

He acted decisively: Wright et al., *On Point II,* 593–99, includes both documents in full. See also Bremer, *My Year in Iraq,* 19.

Bremer defended the measure: Bremer, *My Year in Iraq,* 36–37, 42.

162 *Lieutenant General Ricardo S. Sanchez:* Sanchez and Phillips, *Wiser in Battle,* 197.

"dream team": Wright et al., *On Point II,* 145.

"Let me get this right": Ibid.

Sanchez might have been: Sanchez and Phillips, *Wiser in Battle,* 83, 16, 121, 167.

163 *General John Abizaid:* Ibid., 1, 120. For General John Abizaid, see David Cloud and Greg Jaffe, *The Fourth Star* (New York: Three Rivers Press, 2009), 10, 48, 84, 128–29.

many heaped scorn: Bremer, *My Year in Iraq,* 90; Sanchez and Phillips, *Wiser in Battle,* 224, 309, 321, 338–39.

164 *"I believe in my heart of hearts":* Cloud and Jaffe, *The Fourth Star,* 127, 245.

"the invisible horde of people": Russell F. Weigley, *Eisenhower's Lieutenants* (Bloomington: Indiana University Press, 1981), 13.

165 *A few hiccups:* Wright et al., *On Point II,* 166–67, 523; Sassaman and Layden, *Warrior King,* 255. *Stop-loss* meant that soldiers who had completed their enlistments and should have been discharged remained with their units and deployed. More than fifty thousand people were affected over the course of the war. Stop-loss kept

deploying formations at high strength but proved tragic indeed when men and women who might have been safely out of the service were killed or wounded in action. In a typical example, the First Cavalry Division deployed in 2009 with more than 2,393 soldiers (equivalent to the bulk of a brigade combat team) extended through stop-loss provisions. Despite good intentions to cease stop-loss as early as 2006, it wasn't until 2011 that this highly unpopular program ended.
more troops in order to swap out: Wright et al., *On Point II*, 166.

166 *Guard and Reserve:* Ibid., 165–70, 262–63, 511–12.
many foreign contingents: Stephen A. Carney, *Allied Participation in Operation Iraqi Freedom* (Washington, DC: U.S. Army Center of Military History, 2011), 1, 17, 32–123.

167 *"a window of opportunity lost":* Sanchez and Phillips, *Wiser in Battle*, 303.
count of detainees: Wright et al., *On Point II*, 263.
ranged across Iraq: McChrystal, *My Share of the Task*, 107.
The CPA approach: Cordesman and Baetjer, *Iraqi Security Forces*, 55, 61.

168 *an average of eighty detainees a day:* Wright et al., *On Point II*, 263.
continued resistance: Ibid., 252–53.
These abuses: Ibid., 257; Sanchez and Phillips, *Wiser in Battle*, 276–77.

169 *when the story and the graphic photos broke:* Ibid., 369.
manipulation of his environment: Wright et al., *On Point II*, 212.

170 *the favored statistic:* For enemy attack statistics in Iraq, see Michael Eisenstadt and Jeffrey White, *Assessing Iraq's Sunni Arab Insurgency* (Washington, DC: Washington Institute for Near East Policy, December 2005), 26–27. See also Daniel P. Bolger, "Notes on Multi-National Force–Iraq Battlefield Update Assessment," Baghdad, Iraq, January 6, 2006.

8. What Happened in Fallujah

171 *"We are not only fighting":* William T. Sherman, *The Memoirs of General William T. Sherman* (New York: Appleton, 1875; repr., New York: Da Capo, 1984), 227. Sherman wrote these words in a report from Savannah, Georgia, on December 24, 1864.
"Men, we're ordered": For an excellent description of Corporal Ethan Place in Fallujah, see Milo S. Afong, *HOGs in the Shadows* (New York: Berkley Caliber, 2007), 110–12. The overall context appears in the masterly Bing West work *No True Glory: A Frontline Account of the Battle for Fallujah* (New York: Bantam Books, 2005), 204–7. West provides the definitive narrative of the battle from the Marine perspective.

172 *"First we're ordered in":* West, *No True Glory*, 120–21. Mattis said this to Colonel Joseph Dunford, the First Marine Division chief of staff and, in 2013, the four-star U.S. commander in Afghanistan.
For the U.S. war effort: Wright et al., *On Point II*, 418–19.

173 *The second major transition:* Ibid., 173–77.
Their enemies, of course: Ibid., 109; Sanchez and Phillips, *Wiser in Battle*, 246–47; Bodansky, *The Secret History of the Iraq War*, 377–78.

174 *preferred to negotiate:* Richard Holmes, *Dusty Warriors* (London: Harper Perennial, 2006), 173, 257. Holmes, a British Territorial Army senior officer and noted

military historian, spent weeks in Basrah with the First Battalion, Princess of Wales Royal Regiment.

"We can handle that": Sanchez and Phillips, *Wiser in Battle,* 329.

175 *"the kind you can never really foresee"*: Clausewitz, *On War,* 119.

wasn't supposed to happen in Fallujah: West, *No True Glory,* 51.

three-star higher commander: In 2004, First Marine Expeditionary Force ran operations in Anbar as Multi-National Force–West. After the Fallujah operation, the MEF headquarters was reduced to a two-star command, to conform with the U.S. Army, British, Korean, and Polish divisions in country.

176 *"No better friend, no worse enemy"*: West, *No True Glory,* 50. For more on the distinct Marine small-wars tradition and its application in Vietnam, see Krepinevich, *The Army and Vietnam,* 172–77.

In Washington: Sanchez and Phillips, *Wiser in Battle,* 331; West, *No True Glory,* 59–60.

a horrific street fight: Martha Raddatz, *The Long Road Home* (New York: Berkley Books, 2007). This book offers a superb account of the fighting in Sadr City on April 4, 2004. Raddatz also covers the repercussions in the homes of U.S. soldiers killed and wounded in the fight. One of those killed was Specialist Casey Sheehan, whose mother, Cindy, became a celebrated antiwar activist.

a rooftop firefight: Sanchez and Phillips, *Wiser in Battle,* 336–41.

177 *For Marty Dempsey:* Wright et al., *On Point II,* 324.

The enemy knew it: West, *No True Glory,* 91.

178 *The tragic Al Jazeera footage:* Ibid., 123. See also Sanchez and Phillips, *Wiser in Battle,* 352.

179 *"Al Jazeera kicked our butts"*: West, *No True Glory,* 322.

"a strategic disaster": Sanchez and Phillips, *Wiser in Battle,* 369. Despite promises of a potential four-star position after his command of CJTF-7 in Iraq, Sanchez did not advance. Told of the decision by General Peter Pace, the chairman of the Joint Chiefs of Staff, Sanchez replied, "Well, sir, you all have betrayed me." See ibid., 435.

a worse situation: George W. Casey Jr., *Strategic Reflections: Operation Iraqi Freedom July 2004–February 2007* (Washington, DC: National Defense University Press, 2012), 44–46. Interestingly enough, counterinsurgencies that worked saw shorter conflicts (nine years on average) than those that did not (an average of thirteen years).

180 *"Okay, who's my counterinsurgency expert"*: Cloud and Jaffe, *The Fourth Star,* 161.

182 *"Ulysses don't scare worth a damn"*: This well-known anecdote is highlighted in J.F.C. Fuller, *Grant and Lee: A Study in Personality and Generalship* (Bloomington: Indiana University Press, 1957), 60.

a joint mission statement: Casey, *Strategic Reflections,* 26.

184 *Lieutenant General David H. Petraeus:* The July 5, 2004, *Newsweek* cover is reproduced in full in John T. Correll, "The Second Coming of Counterinsurgency," *Air Force* (September 2013): 98. See also Cloud and Jaffe, *The Fourth Star,* 172, which politely describes the relationship between the two as "strained."

a multinational command: Carney, *Allied Participation in Operation Iraqi Freedom,* 17, 20–21.

185 *senior Coalition officers:* Casey, *Strategic Reflections,* 20–21.

Prime Minister Ayad Allawi: Cloud and Jaffe, *The Fourth Star,* 173–74.

186 *In August in Najaf:* Casey, *Strategic Reflections,* 41; West, *No True Glory,* 238.
in Sunni Arab Samarra: Wright et al., *On Point II,* 343.
Fallujah remained: West, *No True Glory,* 227–28, 256.

187 *Phantom Fury:* Ibid., 350; Casey, *Strategic Reflections,* 42. For the British role, see Tim Newman, *Highlander* (London: Constable and Robinson, 2009), 258–59.
The Jolan District: West, *No True Glory,* 273–74. For details of street clearing in Fallujah, see E. J. Catagnus Jr. et al., *Lessons Learned: Infantry Squad Tactics in Military Operations in Urban Terrain During Operation Phantom Fury in Fallujah, Iraq* (Camp Pendleton, CA: Headquarters, Third Battalion, Fifth Marines, March 8, 2005), 4–6. These Marines served in the Scout/Sniper Platoon during the clearance of the Jolan District.

188 *In the Jolan:* Catagnus et al., *Lessons Learned,* 3.
didn't see any civilians: West, *No True Glory,* 274.

189 *"There's four guys":* Gidget Fuentes, "6 Navy Crosses for Darkhorse," www.militarytimes.com, June 11, 2007; accessed September 25, 2013. Promoted to lance corporal, Christopher Adlesperger was killed in action in December of 2004. For his valor in Fallujah, he received the Navy Cross.
In close-quarters infantry combat: West, *No True Glory,* 274–75; McChrystal, *My Share of the Task,* 157.
There was worse: West, *No True Glory,* 275.

190 *The international news media:* Ibid., 315–16; Wright et al., *On Point II,* 357.

191 *the bridge proclaimed:* West, *No True Glory,* photograph facing page 193. See also Richard J. Lowery, "Recording History — Part XV — Lt. Col Patrick Malay," at http://op-for.com/2008/04/recording_history_part_xv_ltco.html#ixzz2fvhbU5nZ; accessed September 25, 2013.

9. The Color Purple

192 *"The ground was always in play":* Michael Herr, *Dispatches* (New York: Avon Books, 1978), 14. A war correspondent in Vietnam, Herr wrote the hypnotic, often disturbing narration for the 1979 Francis Ford Coppola film *Apocalypse Now.*
"broken the back": Tony Perry, "Marine General Gives Upbeat Report on Iraq," *Los Angeles Times,* March 29, 2005.
hosted tours: Daniel P. Bolger, "Notes on Visit to I Marine Expeditionary Force with General (Retired) Jack Keane and ABC News Correspondent Martha Raddatz" (Baghdad, Iraq: Headquarters, Multi-National Corps, March 21, 2005), 1–2. Keane warned the Marines about too much optimism.

193 *rising in Mosul:* Kendall D. Gott, "American Advisor in Action, Mosul, 13 November 2013," in William G. Robertson, ed., *Case Studies from the Long War, Volume I* (Fort Leavenworth, KS: Combat Studies Institute, 2006), 30–31. U.S. Army colonel James H. Coffman Jr. earned the Distinguished Service Cross in action alongside Iraqi forces in Mosul.
"This did not bode well": Casey, *Strategic Reflections,* 50.

195 *Babil Province:* Daniel P. Bolger, *Operations with Corps Troops Supporting TF 2-11th Cavalry* (Baghdad: Headquarters, Multi-National Corps–Iraq, May 12, 2005), 1. FOB Kalsu was named for First Lieutenant Robert Kalsu, killed in action

in Vietnam while serving with the 101st Airborne Division. Kalsu played professional football for the Buffalo Bills, then in the American Football League. He was the only former professional football player killed in Vietnam. In the current war, Corporal Pat Tillman, who played for the NFL's Arizona Cardinals, was killed in action in Afghanistan in 2004 while serving with the Seventy-Fifth Ranger Regiment.

197 *F³EA:* McChrystal, *My Share of the Task,* 137, 153–54.
Fusion of intelligence: Ibid., 101, 115. See also Daniel P. Bolger, "Notes from Corps Commander Conference, Baghdad, Iraq," March 23, 2005.

198 *emphasis on exploiting and analyzing:* McChrystal, *My Share of the Task,* 161.
right on schedule: For a daily log of 2-11 Cavalry's time in Iraq in 2005–06, see Brad Lang, *Brad's Adventure,* at http://www.edlang.us/brad'sadventure.htm; accessed September 27, 2013. Each unit fatality is addressed in some detail. For another fine first-person account of the squadron's year in country, specifically the first enemy indirect fire attacks on FOB Kalsu, see David P. Ervin, *Leaving the Wire: An Infantryman's Iraq* (Portland, OR: BookBaby, 2013), 67–68. Ervin was a sergeant in 2-11 Cavalry and participated in many operations. Of note, the Janabi tribe, led by Abdullah al Janabi, took a leading role in the renewed 2013–14 insurgency against Nouri al-Maliki's Shiite-dominated Baghdad government. Under Janabi leadership and adopting al-Qaeda symbols, the Islamic State of Iraq and al-Sham took control of Fallujah and much of Ramadi.

199 *LANTIRN:* Clancy, *Fighter Wing,* 75–77.

201 *MNSTC-I never fielded:* Wright et al., *On Point II,* 458–59; Cordesman and Baetjer, *Iraqi Security Forces,* 180–86.
Muhammed's hidden scouts: Bolger, *Operations with Corps Troops Supporting TF 2-11th Cavalry,* 1–2. See also Ervin, *Leaving the Wire,* 118–19, regarding the death of Sergeant John M. Smith. Ervin was very close to Smith and carried the latter's M-4 carbine for the rest of the time in Iraq.

204 *single cavalry squadron:* Jenna Fike, *Operational Leadership Experiences: Interview with Major Bryan Trexler* (Fort Leavenworth, KS: Combat Studies Institute, October 31, 2011), 8. The squadron had a total of twenty-three soldiers killed in action and more than seventy wounded.

205 *a full-time enemy:* Ibid., 4; see also Clancy, *Armored Cav,* 208–9.

206 *a persuasive book:* Douglas A. Macgregor, *Breaking the Phalanx* (Washington, DC: Center for Strategic and International Studies, 1997), 4.

207 *ten more deployable BCTs:* John Sloan Brown, *Kevlar Legions: The Transformation of the U.S. Army 1989–2005* (Washington, DC: Center for Military History, 2011), 298–310. This book fully explains the major transition wrought under General Schoomaker. Brigadier General John S. Brown commanded an armor battalion during the 1991 Iraq campaign. His son Captain Todd Brown commanded Company B/1-8 Infantry in Samarra in 2003–04 and wrote *Battleground Iraq.* In later refinements, the tank/mechanized battalions sent their engineer companies to a brigade special troops battalion, which eventually grew into an engineer battalion. Engineers proved their worth again and again in both Iraq and Afghanistan. No matter where they are assigned, there never seem to be enough of these skilled, tough soldiers.
National Guard units: Wright et al., *On Point II,* 605, 616–20.

208 *independent procurement authority:* Clancy, *Special Forces,* 17–19.
every Marine a rifleman: Tom Clancy, *Marine* (New York: Berkley Books, 1996), xiv.

209 *the "Warrior Ethos":* Brown, *Kevlar Legions,* 279, 461.
brought all of them home: On May 30, 2014, the last soldier remaining in Afghanistan, Sergeant Bowe Bergdahl, was returned alive in an exchange for five key Taliban commanders. All soldiers were recovered in Iraq.

210 *handed over MNSTC-I:* Cordesman and Baetjer, *Iraqi Security Forces,* 158.
One high-profile effort: Ricardo A. Herrera, "Brave Rifles at Tall Afar," in William G. Robertson, ed., *Case Studies from the Long War,* 142.

211 *the Third Armored Cavalry Regiment's Tal Afar operation:* Thomas E. Ricks, *The Gamble* (New York: Penguin, 2009), 60–61. Along with his many military accomplishments, H. R. McMaster authored *Dereliction of Duty,* a brilliant, incisive history of the Joint Chiefs of Staff's inability to stand up to President Lyndon B. Johnson in the early years of the Vietnam War.

212 *three hotels in Amman, Jordan:* McChrystal, *My Share of the Task,* 197–98; Casey, *Strategic Reflections,* 70–71.
As for "Sunni in": Casey, *Strategic Reflections,* 69.
Abu Mahal tribe: Daniel P. Bolger, "Notes from Meeting with Commander, MNF-I; Commander, MNSTC-I; and Minister of Defense," Baghdad, Iraq, September 8, 2005.

213 *The fall elections:* Casey, *Strategic Reflections,* 80.
almost fifteen thousand captives: Cheryl Bernard et al., *The Battle Behind the Wire: U.S. Prisoner and Detainee Operations from World War II to Iraq* (Santa Monica, CA: RAND Corporation, 2011), 67.

10. Implosion

214 *"Have you ever":* Marty Kufus, "When Private Hendrix Kissed the Sky," *Command* (November/December 1989): 54–56. See also Herr, *Dispatches,* 38, 181. Jimi Hendrix served in the 101st Airborne Division at Fort Campbell, Kentucky, from 1960 to 1962. He made multiple parachute jumps while on active duty.
al-Askari mosque: Bob Woodward, *State of Denial* (New York: Simon and Schuster, 2006), 444.

215 *American pursuers caught up:* McChrystal, *My Share of the Task,* 229–30; Rebecca Grant, "Iraqi Freedom and the Air Force," *Air Force* (March 2013): 41.
"There are no Sunnis against Shiites": Ricks, *The Gamble,* 32–33.

216 *"The guy that is going to win":* Raffi Khatchadourian, "The Kill Company," *New Yorker* (July 6, 2009): 43–44. Khatchadourian's article offers the best objective summary of the operation.
intel guys predicted: Stjepan G. Mestrovic, *The "Good Soldier" on Trial: A Sociological Study of Misconduct by the U.S. Military Pertaining to Operation Iron Triangle, Iraq* (New York: Algora Publishing, 2009), 61–65. Mestrovic has reproduced many key primary documents pertaining to the events of May 9, 2006.

217 *"I will eat you":* Ibid., 20–21. For one version of Colonel Steele's talk, see http://www.youtube/watch?v=fxy-Rsd2wsm; accessed October 4, 2013.

218 *"This is for real":* Khatchadourian, "The Kill Company," 51.

"We don't know": Ibid., 51–52.

219 *They killed all three:* Mestrovic, *The "Good Soldier" on Trial*, 62. See also Khatchadourian, "The Kill Company," 50–51.

The men came by road: Mestrovic, *The "Good Soldier" on Trial*, 40. Mestrovic includes the entire statement of Specialist Jason A. Stachowski, a sniper involved in the incident.

220 *The scuffle with the detainee:* Ibid., 6–9. Mestrovic reproduces the sworn statement of Sergeant David Chavez, which includes a sketch map and a casualty count. Horne later denied issuing orders to shoot the Iraqis at the gas station. In an after-action review, he also stated that he regretted using the term *collateral damage* and thanked Sergeant Beal for calling the cease-fire. Horne received a written reprimand, which was not included in his permanent personnel file. He was promoted to captain, but had left the U.S. Army by 2009.

221 *One was Abu Abdullah:* Khatchadourian, "The Kill Company," 50.

Things did not work out: Mestrovic, *The "Good Soldier" on Trial*, 159. The verbatim testimony of two Iraqi soldiers is included.

223 *revealed nothing of interest:* Khatchadourian, "The Kill Company," 53.

In the official verbiage: Mestrovic, *The "Good Soldier" on Trial*, 64.

"then put a blindfold on them": Khatchadourian, "The Kill Company," 49, 54; Mestrovic, *The "Good Soldier" on Trial*, 198, 242, 254. Girouard was accused of murder, but convicted on the lesser charge of negligent homicide (as well as obstruction of justice and violation of a general order) by court-martial; he was sentenced to ten years in military confinement. In 2011, the negligent homicide charge was overturned by the U.S. Court of Appeals for the Armed Forces. Claggett and Hunsaker pled guilty to murder before a military court-martial; each was sentenced to eighteen years in military confinement. Graber pled guilty to negligent homicide and received a nine-month sentence.

224 *very grim rumors:* Mestrovic, *The "Good Soldier" on Trial*, 56–57.

Lieutenant General Peter W. Chiarelli: Michael R. Gordon and Bernard E. Trainor, *The Endgame* (New York: Pantheon, 2012), 180; Cloud and Jaffe, *The Fourth Star*, 144–45. See also Peter W. Chiarelli and Patrick R. Michaelis, "Winning the Peace: The Requirement for Full Spectrum Operations," *Military Review* (July/August 2005): 4–17.

225 *"some reason to hope"*: Cloud and Jaffe, *The Fourth Star*, 224–25.

events had dropped by 85 percent: Khatchadourian, "The Kill Company," 44.

support for the "honorable resistance": Eisenstadt and White, *Assessing Iraq's Sunni Arab Insurgency*, 9.

226 *incident in Sicily in 1943:* D'Este, *Patton: A Genius for War*, 509.

227 *Hadithah:* Ricks, *The Gamble*, 3–8, 35; Gordon and Trainor, *The Endgame*, 202. For the Hadithah incident, see U.S. Commander, Multi-National Corps–Iraq, *Investigation into the Events of 19 November 2005 at Hadithah, Iraq* (Baghdad: Headquarters, Multi-National Corps–Iraq, June 15, 2006). For the Yusufiyah incident, see Jim Frederick, *Black Hearts* (New York: Harmony Books, 2010). The victims in Yusufiyah were from the Janabi tribe.

The early line: Khatchadourian, "The Kill Company," 57.

228 *"allowed to go unchecked"*: Ibid.; see also Cloud and Jaffe, *The Fourth Star*, 237.

"a fundamental difference of opinion": Khatchadourian, "The Kill Company," 42.

229 *"sectarian violence"*: Casey, *Strategic Reflections,* 110.
MNF-I estimates: Woodward, *State of Denial,* 477; Eisenstadt and White, *Assessing Iraq's Sunni Arab Insurgency,* 26.
"finding dead people": Ricks, *The Gamble,* 55.

230 *Maliki's last name:* Ibid., 461.
preferred to chase "Baathists": Casey, *Strategic Reflections,* 113.
explosively formed penetrator: Ricks, *The Gamble,* 48.
Such separate armies: De Atkine, "Why Arabs Lose Wars," 10–11.

231 *Shia-dominated:* Cloud and Jaffe, *The Fourth Star,* 209–11; Gordon and Trainor, *The Endgame,* 186; Casey, *Strategic Reflections,* 79; Ricks, *The Gamble,* 46–47.
Together Forward I and II: Ricks, *The Gamble,* 46–47; Gordon and Trainor, *The Endgame,* 224–25.

232 *an especially egregious incident:* Casey, *Strategic Reflections,* 122–23. For more on the capture and return of the deceased Staff Sergeant Ahmed Kousay Al-Taie, see Hannah Allam, "U.S. Military Receives Remains of Last U.S. Soldier Missing in Iraq," *Miami Herald,* February 26, 2012. Ahmed Chalabi was involved in arranging the return of the missing soldier's remains.
"I waited too long": Casey, *Strategic Reflections,* 115.
The numbers: Ian Livingston and Michael E. O'Hanlon, *Iraq Index* (Washington, DC: Brookings Institution, January 31, 2011), 4.

233 *British commanders on the western front:* Norman F. Dixon, *On the Psychology of Military Incompetence* (London: Jonathan Cape, 1976), 80–85. Dixon uses the British performance in the Great War as the premier example of what can go wrong when commanders do not adapt.
Iraq Study Group: Bob Woodward, *The War Within* (New York: Simon and Schuster, 2008), 197–98, 205.

234 *summarized in two words:* Ibid., 42–45, 262–63.
massive B-52 strikes on Hanoi and Haiphong: Phillip B. Davidson, *Vietnam at War: The History, 1946–1975* (Novato, CA: Presidio, 1988), 654–58.
a surge in Iraq: Woodward, *The War Within,* 252.

235 *no more U.S. forces:* Cloud and Jaffe, *The Fourth Star,* 244–45.
general used his considerable influence: Ricks, *The Gamble,* 98–101. Ricks includes a partial transcript of the meeting. See also Woodward, *The War Within,* 279–82.

236 *told Bush to pick:* Cloud and Jaffe, *The Fourth Star,* 128, 239, 258; Woodward, *The War Within,* 145–46, 293–94.
victory in Iraq: Bush, *Selected Speeches of President George W. Bush,* 449, 453.
"like Christ come to cleanse the temple": Charles B. MacDonald, *A Time for Trumpets* (New York: Bantam Books, 1984), 424.

11. Malik Daoud

237 *"He talks about the big picture"*: Anton Myrer, *Once an Eagle* (New York: Berkley Medallion Books, 1976), 710. Myrer was a Marine veteran of the 1941–45 war in the Pacific. His novel focuses on two U.S. Army officers and their families, friends, and experiences from just before World War I until the Vietnam era. Myrer's protagonist, Sam Damon, is promoted from the ranks. He is brave, caring, and self-

less, a model commander. His rival, West Pointer Courtney Massengale, is a brilliant, manipulative careerist, a high-level staff officer always maneuvering to his own advantage. Damon ends his service as a lieutenant general on the Army staff. Massengale, of course, reaches four stars and theater command. Military officers like to identify with Sam Damon. Subordinates, however, see a lot more Massengales in the mix.

"the King David thing": Cloud and Jaffe, *The Fourth Star,* 130.

"I will risk the dictatorship": Bruce Catton, *Never Call Retreat* (New York: Washington Square Press, 1965), 64. The risk of dictatorship proved minimal. Major General Joe Hooker suffered a decisive defeat in the May 1863 Chancellorsville campaign. Hooker was reassigned to lesser duties.

Field Manual 3-24, Counterinsurgency: U.S. Department of the Army, *Counterinsurgency* (Washington, DC: U.S. Government Printing Office, December 2006), 1-26 to 1-28. The manual was simultaneously adopted by the U.S. Marine Corps as Marine Corps Warfighting Publication (MCWP) 3-33.5. For the Marines, Lieutenant General James Mattis and Lieutenant General James Amos led the effort. Both senior Marines rose to four-star rank, Mattis as commander of U.S. Joint Forces Command and then U.S. Central Command, and Amos as assistant commandant and then commandant of the Marine Corps.

238 *an introduction by Sarah Sewall:* Fred Kaplan, *The Insurgents: David Petraeus and the Plot to Change the American Way of War* (New York: Simon and Schuster, 2013), 154.

240 *twenty thousand troops:* Ricks, *The Gamble,* 118–20.

"the Baghdad belts": Farouk Ahmed, *Backgrounder #28: Multi-National Division-Center's Operations During the 2007–2008 Troop Surge* (Washington, DC: Institute for the Study of War, April 2008), 1. The actual al-Qaeda in Iraq sketch depicting the Baghdad belt region is reproduced in this document.

"turning their own people against them": Robert M. Utley, *Frontier Regulars: The United States Army and the Indian, 1866–1890* (New York: Macmillan, 1973), 55.

giving up on Anbar Province: Ricks, *The Gamble,* 331.

241 *the provincial capital of Ramadi:* Gordon and Trainor, *The Endgame,* 240.

three successive U.S. Army brigades: Ibid., 244; Michael Fitzsimmons, *Governance, Identity, and Counterinsurgency: Evidence from Ramadi and Tal Afar* (Carlisle Barracks, PA: Strategic Studies Institute, March 2013), 24, 34.

242 *"don't do a Fallujah"*: Steven Clay, *Interview with Colonel Sean MacFarland* (Fort Leavenworth, KS: Contemporary Operations Study Team, January 17, 2008), 29.

Starting on June 18: Gordon and Trainor, *The Endgame,* 246–47.

easy to underestimate: William Doyle, *A Soldier's Dream: Captain Travis Patriquin and the Awakening of Iraq* (New York: New American Library, 2011), 87. Doyle's book is a superb study of Patriquin's personality and his impact on the U.S. effort in and around Ramadi in the summer and fall of 2006.

243 *more than met the eye:* Ibid., 56–57, 80–82; Clay, *Interview with Colonel Sean MacFarland,* 38.

244 *Sheikh Sattar Abu Risha:* Ricks, *The Gamble,* 66–67; Doyle, *A Soldier's Dream,* 125–26.

245 *"a new military force"*: Doyle, *A Soldier's Dream*, 125–26.
first assembly for July 4: Bing West, *The Strongest Tribe* (New York: Random House, 2009), 173–74.
the new IPs: Clay, *Interview with Colonel Sean MacFarland*, 26; Doyle, *A Soldier's Dream*, 131. For a perspective by a highly experienced foreign service officer present for the early stages of the Anbar Awakening, see Carter Malkasian, "A Thin Blue Line in the Sand," *Democracy Journal* (Summer 2007): 48–49.

246 *This time, the tribesmen:* Clay, *Interview with Colonel Sean MacFarland*, 26.
assassins nabbed and killed: Gordon and Trainor, *The Endgame*, 250. In addition to the police recruiting, several Anbar tribes began abducting and killing al-Qaeda members, terrorizing the terrorists in true Bedouin tradition.

247 *The Marines had dealt for years:* Timothy S. McWilliams and Kurtis P. Wheeler, eds., *Al Anbar Awakening, Volume I: American Perspectives* (Quantico, VA: Marine Corps University Press, 2009), 143–44. Major General Richard Zilmer highlighted the outreach to the Anbar sheikhs exiled in Jordan.
Sahwa al Anbar: Doyle, *A Soldier's Dream*, 150–51; Malkasian, "A Thin Blue Line in the Sand," 55, 57.
"keep Sattar alive": Gordon and Trainor, *The Endgame*, 252.
an eighteen-slide briefing: Travis Patriquin, "How to Win the War in Al Anbar" (Camp Ramadi, Iraq: Headquarters, First Brigade Combat Team, First Armored Division, 2006); Doyle, *A Soldier's Dream*, 185–86. The slides can be viewed at www.abcnews.go.com/images/US/how_to_win_in_anbar_v4.pdf; accessed October 10, 2013.

248 *"We're under attack"*: Doyle, *A Soldier's Dream*, 217.
the Abu Soda tribe: Ricks, *The Gamble*, 69.
immediate attention: Clay, *Interview with Colonel Sean MacFarland*, 38–39.

249 *"wave them over [their] heads"*: Doyle, *A Soldier's Dream*, 225–26; Gordon and Trainor, *The Endgame*, 258.

250 *Four other trucks:* Doyle, *A Soldier's Dream*, 226–27; Clay, *Interview with Colonel Sean MacFarland*, 38–39.
"a very surreal scene": Doyle, *A Soldier's Dream*, 231.

251 *"the strongest tribe"*: West, *The Strongest Tribe*, 211.
"I'm so happy": Doyle, *A Soldier's Dream*, 266, 273; Cloud and Jaffe, *The Fourth Star*, 263.
did not live to meet Petraeus: Doyle, *A Soldier's Dream*, 272, 284; Gordon and Trainor, *The Endgame*, 259; West, *The Strongest Tribe*, 239. Along with Captain Travis Patriquin, the IED blast in Ramadi on December 6, 2006, killed Major Megan McClung, U.S. Marine Corps, and Specialist Vincent J. Pomante III, U.S. Army. At the time, Major McClung was the highest-ranking female Annapolis graduate to be killed in action since women had begun attending the U.S. Naval Academy in 1976.
Sons of Iraq: Ricks, *The Gamble*, 203–4.

252 *the Biden formula:* Woodward, *State of Denial*, 481.
more strength for the Army and the Marine Corps: Woodward, *The War Within*, 315.

253 *in and around Baghdad:* Ricks, *The Gamble*, 165–71.
outside the Iraqi capital: Gordon and Trainor, *The Endgame*, 370–72, 389, 390.

254 *Petraeus was the man to do it:* For a look at Petraeus before his tenure in command of Multi-National Force–Iraq, see Rick Atkinson, *In the Company of Soldiers* (New York: Henry Holt, 2004), 36–38.

there was another thread in there: Cloud and Jaffe, *The Fourth Star,* 35, 39–40, 96, 101–2; Tom Clancy, *Airborne* (New York: Berkley Books, 1997), 256.

255 *cemented his achievement:* Woodward, *The War Within,* 384–86.

257 *The embarrassed British:* Mark Urban, *Task Force Black* (New York: St. Martin's, 2011), 268–71.

a very nicely phrased instrument: U.S. Department of State, "Strategic Framework Agreement for a Relationship of Friendship and Cooperation Between the United States of America and the Republic of Iraq" (Washington, DC: Department of State, November 17, 2008); U.S. Department of State, "Agreement Between the United States and the Republic of Iraq on the Withdrawal of U.S. Forces from Iraq and the Organization of Their Activities During Their Temporary Presence in Iraq" (Washington, DC: Department of State, November 17, 2008).

12. Requiem on the Tigris

258 *"In Chinatown": Chinatown,* directed by Roman Polanski (Paramount Pictures, 1974). The superb screenplay was written by Robert Towne.

259 *the dark warrens of Fadhil:* Marisa Cochrane, *Backgrounder Baghdad Neighborhood Project: Rusafa* (Washington, DC: Institute for the Study of War, 2007), 5–8. Cochrane includes a tabular summary of key operations and casualties. See also Sinan Salaheddin, "Roadside Bombing Kills Five U.S. Soldiers in Baghdad Area," Associated Press, March 26, 2007, and Sean Ryan, "'Devil in Baggy Pants' Soldier Earns Silver Star with Made-for-Hollywood Heroics," Defense Video and Imagery Distribution System, August 20, 2007.

went with the Sahwa model: Craig A. Collier, "Now That We're Leaving Iraq, What Did We Learn?" *Military Review* (September/October 2010): 88, 91–92.

the Don of Fadhil: Geoff Ziezulewicz, "Empowered by the U.S., Imprisoned by the Iraqis," *Stars and Stripes,* September 24, 2009.

260 *Mashadani knew the foe well:* Frank A. Rodriguez, "COIN Operations and the Rise and Fall of an Iraqi Warlord," *Infantry* (August/December 2009): 37–38.

Rumors surfaced: Ibid., 38; Ziezulewicz, "Empowered by the U.S."; Sudarsan Raghavan and Anthony Shadid, "In Iraq, 2 Key U.S. Allies Face Off," *Washington Post,* March 30, 2009.

required an Iraqi warrant: U.S. Department of the Army, *Command Report Multi-National Division–Baghdad 2009–2010,* ed. Bianka J. Adams (Fort Hood, TX: Headquarters, First Cavalry Division, 2010), 46.

261 *transferred the militias to Iraqi command:* Daniel P. Bolger, "Notes on Fadhil Operations," March 28–29, 2009.

"a hornet's nest": Ziezulewicz, "Empowered by the U.S."

262 *Operation Shwey Shwey:* Rodriguez, "COIN Operations and the Rise and Fall of an Iraqi Warlord," 39. See also Daniel P. Bolger, "Notes on Joint Security Site Bab-al-Sheikh and Fadhil Operations," March 9, 2009.

accomplished something else: Bolger, "Notes on Joint Security Site Bab-al-Sheikh."

The work of Jennifer Thaxton of the Human Terrain Team proved especially useful. Her numerous street and home interviews of Iraqi civilians, especially women, painted a very detailed portrait of Fadhil under Mashadani's oppressive rule.

263 *"What if Naqeeb Raad makes a mistake"*: Rodriguez, "COIN Operations and the Rise and Fall of an Iraqi Warlord," 38.

apprehended him without a shot fired: Alissa J. Rubin, "Guns Go Silent After 24-Hour Face-Off in Baghdad, but Tensions Remain High," *New York Times,* March 30, 2009.

264 *"Too much fire"*: Rodriguez, "COIN Operations and the Rise and Fall of an Iraqi Warlord," 40.

265 *"the volume of fire"*: Ibid., 41.

Airborne cavalrymen: Bolger, "Notes on Fadhil Operations," March 28–29, 2009.

266 *a hard answer:* Rodriguez, "COIN Operations and the Rise and Fall of an Iraqi Warlord," 41.

267 *Just before sunup:* Bolger, "Notes on Fadhil Operations," March 28–29, 2009. See also Geoff Ziezulewicz, "U.S. Troops, Iraqi Army Work to Secure Baghdad District After Militia Leader's Arrest," *Stars and Stripes,* April 5, 2009.

article 4: Department of State, "Agreement Between the United States and the Republic of Iraq."

268 *In all of 2009 in Baghdad:* Department of the Army, *Command Report Multi-National Division–Baghdad 2009–2010,* 4; Daniel P. Bolger, "Summary of MND-B [Multi-National Division–Baghdad] Operations," January 13, 2010.

deployments dropped to nine months: Lance M. Bacon, "Army Announces Switch to 9-Month Deployments," *Army Times,* August 5, 2011.

the Sunni Awakening: Department of the Army, *Command Report Multi-National Division–Baghdad 2009–2010,* 112.

269 *Abu Ayyub al Masri:* Gordon and Trainor, *The Endgame,* 623.

a much reduced rate: Department of the Army, *Command Report Multi-National Division–Baghdad 2009–2010,* 27–29.

On the Shiite side: Ibid., 29–30.

270 *"our combat mission in Iraq will end"*: See http://www.whitehouse.gov/the_press_office/Remarks-of-President-Barack-Obama-Responsibly-Ending-The-War-in-Iraq; accessed October 12, 2013.

271 *Hill got nowhere:* Gordon and Trainor, *The Endgame,* 586–87.

272 *the results he wanted:* Ibid., 590. See also United Kingdom of Great Britain, Ministry of Defence, "Royal Navy Training Mission in Iraq Ends," May 18, 2011, at http://www.royalnavy.mod.uk/News-and-Events/Latest-News/2011/May/18/110518-Royal-Navy-Training-Mission-in-Iraq-Ends; accessed October 14, 2013.

273 *a reverse cycle:* Department of the Army, *Command Report Multi-National Division–Baghdad 2009–2010,* 19–21.

the June 30 deadline: Alissa J. Rubin, "Iraq Marks Withdrawal of U.S. Troops from Cities," *New York Times,* June 30, 2009.

274 *advocated backing Maliki:* Gordon and Trainor, *The Endgame,* 631. By mid-2014, Sunni insurgents led by radical al-Qaeda affiliates had overrun Mosul, Tikrit, and much of Anbar Province.

The last soldier killed: J. Freedom du Lac, "In Iraq, the Last to Fall: David Hickman, the 4,474th U.S. Service Member Killed," *Washington Post,* December 17,

2011. General Lloyd J. Austin III commanded American forces in Iraq during the 2010–2011 withdrawal. A veteran commander with time in Afghanistan and earlier service in Iraq, including the 2003 invasion, Austin capably directed the drawdown and departure of U.S. troops. He did so while the U.S. main effort shifted to Afghanistan.

275 *Army planners:* Gordon and Trainor, *The Endgame*, 670.

Office of Security Cooperation–Iraq: U.S. Department of Defense, Inspector General Special Plans and Operations, *Report No. DODIG-2012–063, Assessment of the DoD Establishment of the Office of Security Cooperation–Iraq* (Washington, DC: Department of Defense Inspector General, March 16, 2012), 4. In December of 2013, all contract trainers (mostly former U.S. military personnel) were formally transferred to Iraqi-run contracts. The Iraqis used U.S. security assistance funds to pay these contractors.

276 *"what is our mission"*: U.S. Department of the Army, Center for Army Lessons Learned, *OSC-I: Office of Security Cooperation–Iraq Interviews* (Fort Leavenworth, KS: Combined Arms Center, May 2012), 17. When resurgent Sunni rebels threatened Baghdad in 2014, the Maliki government agreed to an exchange of notes granting legal immunity to several hundred U.S. advisers and security forces deployed to protect the American embassy and assist the Iraqi troops.

it had not turned out well: John R. Ballard, David W. Lamm, and John K. Wood, *From Kabul to Baghdad and Back: The U.S. War in Afghanistan and Iraq* (Annapolis, MD: U.S. Naval Institute, 2012), 224–25.

13. Undone

279 *"When you're wounded"*: Rudyard Kipling, "The Young British Soldier," in *The Complete Verse* (London: Kyle Cathie, 1990), 331–33.

The Royal Gurkha Rifles: See http://www.rgrra.com/index.php/battle-honours; accessed October 14, 2013.

280 *The Kajaki Dam project:* Rajiv Chandrasekaran, *Little America: The War Within the War for Afghanistan* (New York: Alfred A. Knopf, 2012), 16–17. See also Cynthia Clapp-Wincek, *A.I.D. Special Evaluation Special Study No. 18: The Helmand Valley Project in Afghanistan* (Washington, DC: U.S. Agency for International Development, December 1983), 3.

281 *Royal Air Force Chinooks:* These aircrews stayed very busy in Helmand Province in 2006. See http://www.raf.mod.uk/news/royalairforcech47chinookbravonovember.cfm; accessed October 15, 2013.

The district compound: U.S. Department of the Navy, U.S. Marine Corps Intelligence Activity, *Afghan Insurgent Tactics, Techniques, and Procedures (TTP) Field Guide Volume II: Southern Afghanistan* (Quantico, VA: U.S. Marine Corps Intelligence Activity, November 2009), 36.

282 *to patrol Now Zad:* James Fergusson, "Journey Inside the Taliban: Briton's Dangerous Secret Meeting with the Warlords Who Will Never Surrender," *Mail Online*, July 18, 2008; http://www.dailymail.co.uk/news/article-1034462/Journey-inside-Taliban-Britons-dangerous-secret-meeting-warlords-surrender.html; accessed October 15, 2013. Fergusson interviewed Major Dan Rex and several of his soldiers.

283 *"they were all around us"*: Ibid.
The Taliban initiated: U.S. Marine Corps Intelligence Activity, *Afghan Insurgent Tactics, Techniques, and Procedures (TTP) Field Guide Volume II: Southern Afghanistan,* 38.
"they were testing us out": Fergusson, "Journey Inside the Taliban."

284 *The British were not bad soldiers:* Ibid. Fergusson interviewed a Taliban commander who'd fought at Now Zad in the summer of 2006. The Pashtun gave his name as Abdullah.
the enemy kept probing: U.S. Marine Corps Intelligence Activity, *Afghan Insurgent Tactics, Techniques, and Procedures (TTP) Field Guide Volume II: Southern Afghanistan,* 38.
the enemy changed tactics again: Ibid., 39–40.
"straight down the chimney": Fergusson, "Journey Inside the Taliban."
a major assault: U.S. Marine Corps Intelligence Activity, *Afghan Insurgent Tactics, Techniques, and Procedures (TTP) Field Guide Volume II: Southern Afghanistan,* 39.
Gurkha Corporal Kailash Kebang: Fergusson, "Journey Inside the Taliban."

286 *"more orthodox"*: James Fergusson, *Taliban: The Unknown Enemy* (Boston: Da Capo, 2012), 273.
long emphasized night fighting: Holmes, *Dusty Warriors,* 85, 87, describes the current British army use of night-vision devices in Basrah, Iraq. The Gurkhas in Afghanistan used the same equipment.
Just before midnight: U.S. Marine Corps Intelligence Activity, *Afghan Insurgent Tactics, Techniques, and Procedures (TTP) Field Guide Volume II: Southern Afghanistan,* 39.

287 *The Gurkhas would have none of it:* Fergusson, *Taliban,* 181.

288 *twenty-eight distinct shooting incidents in twenty-two days:* Headquarters, Combined Forces Command–Afghanistan, "Coalition Confirms No Civilian Non-Combatants Killed in Nowzad Operations," July 15, 2006; Fergusson, "Journey Inside the Taliban"; U.S. Marine Corps Intelligence Activity, *Afghan Insurgent Tactics, Techniques, and Procedures (TTP) Field Guide Volume II: Southern Afghanistan,* 37.
"We do not have enough troops": Seth G. Jones, *In the Graveyard of Empires: America's War in Afghanistan* (New York: W. W. Norton, 2009), 220, 255.
slowly forming Afghan security forces: Ian S. Livingston and Michael O'Hanlon, *Afghanistan Index* (Washington, DC: Brookings Institution, July 12, 2012), 6.
By comparison, Afghanistan endured: Ibid., 4–5; Livingston and O'Hanlon, *Iraq Index,* 4, 7, 13.

289 *The U.S. troops joked:* Jones, *In the Graveyard of Empires*, xxiv.
"a decisive year": Rory Stewart and Gerald Knaus, *Can Intervention Work?* (New York: W. W. Norton, 2011), 50. Rory Stewart is a member of Parliament in Great Britain. Gerald Knaus is the founding chairman of the European Stability Initiative.
the first provincial reconstruction teams: Wright et al., *A Different Kind of War,* 227–29.

290 *Casey in Iraq adopted it:* Casey, *Strategic Reflections,* 195.
The resultant debacle: Dixon, *On the Psychology of Military Incompetence,* 71–79.

291 *"poorly informed, organized, and executed"*: McChrystal, *My Share of the Task*, 77; Jones, *In the Graveyard of Empires*, 183–85, 193, 195–96.

"less than the sum of their parts": Bing West, *The Wrong War: Grit, Strategy, and the Way Out of Afghanistan* (New York: Random House, 2011), 132.

"a more classic counterinsurgency strategy": Stewart and Knaus, *Can Intervention Work?*, 50–51.

based on five pillars: Wright et al., *A Different Kind of War*, 245–47. See also Ballard, Lamm, and Wood, *From Kabul to Baghdad and Back*, 117–18.

"This is ridiculous": Ballard, Lamm, and Wood, *From Kabul to Baghdad and Back*, 129–30.

292 *clear direction:* Jones, *In the Graveyard of Empires*, 244–45.

Combined Security Transition Command–Afghanistan: Wright et al., *A Different Kind of War*, 302.

"establishing bases rather than chasing militants": Stewart and Knaus, *Can Intervention Work?*, 51.

General Dan K. McNeill returned: West, *The Wrong War*, 174.

293 *five regional commands:* Wright et al., *A Different Kind of War*, 289–90, 294.

"They had made a mess of things": Stewart and Knaus, *Can Intervention Work?*, 51–52.

more troops: Amy Belasco, *Troop Levels in the Afghan and Iraq Wars, FY2001–FY2012: Cost and Other Potential Issues* (Washington, DC: Congressional Research Service, July 2, 2009), 12.

294 *villagers coined a slogan:* McChrystal, *My Share of the Task*, 317; West, *The Wrong War*, 61, 174; Jones, *In the Graveyard of Empires*, 248–53.

By ISAF estimates: West, *The Wrong War*, 174; for civilian casualty estimates, see Livingston and O'Hanlon, *Afghanistan Index*, 15.

295 *"the Quiet Commander"*: Michael Hastings, *The Operators* (New York: Blue Rider Press, 2012), 7–8, 10. The late Michael Hastings became infamous when his *Rolling Stone* article resulted in the dismissal of General Stanley McChrystal in 2010. Hastings was embedded in Afghanistan for segments of the McKiernan, McChrystal, and Petraeus tenures.

a grim picture: Livingston and O'Hanlon, *Afghanistan Index*, 4–6, 10.

requested four more BCTs: Ballard, Lamm, and Wood, *From Kabul to Baghdad and Back*, 198–201.

"the war that had to be won": Bob Woodward, *Obama's Wars* (New York: Simon and Schuster, 2010), 113; West, *The Wrong War*, xxvi.

296 *Logistics limited:* Wright et al., *A Different Kind of War*, 322–24.

"We're not losing": Woodward, *Obama's Wars*, 70–71.

"I don't know": Ibid., 71.

297 *new strategic visions:* Ibid., 77.

298 *"We will defeat you"*: President Barack H. Obama, "Remarks by the President on a New Strategy for Afghanistan and Pakistan," March 27, 2009, at http://www.whitehouse.gov/the-press-office/remarks-president-a-new-strategy-afghanistan-and-pakistan; accessed October 16, 2013.

299 *"the leadership is wrong"*: Woodward, *Obama's Wars*, 82–83.

"I know I have a better answer": Ibid., 118–19; Hastings, *The Operators*, 7–8, 10, 11.

14. The Good War

301 *"a grim fondness"*: Fehrenbach, *This Kind of War,* 607.
thank Rudyard Kipling as well as director John Huston: The Man Who Would Be King, directed by John Huston (Los Angeles, CA: Allied Artists Pictures, 1975). Accomplished master John Huston directed from a screenplay written by himself and Gladys Hill. The film closely followed Kipling's novella.

302 *"a continuation of political intercourse"*: Clausewitz, *On War,* 87.
They went in: Corporal Jeff Sutton, "Guest Column: Operation Mountain Fire and the Battle for Barg-e Matal," *Williams (AZ) News,* June 29, 2010.

303 *That was an AK-47:* Public-affairs officer, First Battalion, Thirty-Second Infantry, "Battle of the Barge: 1st Battalion, 32nd Infantry Regiment Wages Three-Month Battle to Secure Remote Village," *Mountaineer Online,* November 19, 2009, http://www.drum.army.mil/mountaineer/Article.aspx?ID=3544; accessed October 17, 2013.
a mixed force: Ibid.

304 *his primary mission:* West, *The Wrong War,* 52–53.
a new tactical directive: Stanley A. McChrystal, *Tactical Directive* (Kabul, Afghanistan: Headquarters, International Security Assistance Force, July 6, 2009), 1.

305 *"courageous restraint"*: McChrystal, *My Share of the Task,* 312, 369.
made of sterner stuff: Ibid., 340–41. McChrystal recounts a meeting with local elders after a major civilian-casualty incident. the spokesman told the general, "We need these operations to tell the opposition that 'we mean business.'" The emotion behind the argument surprised McChrystal. He wondered if the elders were "not broadly representative" of the local people.
"very limited and prescribed conditions": McChrystal, *Tactical Directive,* 2.
"Any entry into an Afghan house": Ibid.

306 *discouraged detentions:* Livingston and O'Hanlon, *Iraq Index,* 11; Livingston and O'Hanlon, *Afghanistan Index,* 19. See also West, *The Wrong War,* 61.
"any mosque or any religious or historical site": McChrystal, *Tactical Directive,* 2.
"a matter of self-defense": Ibid.

307 *in the German sector:* McChrystal, *My Share of the Task,* 339–42.

308 *"cultural shift"*: McChrystal, *Tactical Directive,* 2.
"you must do so": British Broadcasting Corporation, "'Shift Needed' in Afghan Combat," June 25, 2009, http://news.bbc.co.uk/2/hi/south_asia/8118013.stm; accessed October 18, 2013. The speaker was McChrystal's deputy Lieutenant General Sir James Benjamin Dutton, a Royal Marine.
Civilian casualties dropped: Livingston and O'Hanlon, *Afghanistan Index,* 11, 15.

309 *"avoiding a single Blue-on-Blue"*: John "Sandy" Woodward with Patrick Robinson, *One Hundred Days: The Memoirs of the Falklands Battle Group Commander* (Annapolis, MD: U.S. Naval Institute Press, 1992), 317.
had been on the ground: McChrystal, *My Share of the Task,* 331.
In Barg-e Matal: Public-affairs officer, "Battle of the Barge," 1.

310 *Something was up:* Sutton, "Guest Column," 1.
"They were everywhere": Ibid. See also West, *The Wrong War,* 76.

"drenched in blood": Sutton, "Guest Column," 1.
Jeff Sutton summarized it best: Ibid.; Public-affairs officer, "Battle of the Barge," 1; West, *The Wrong War,* 77.

311 *two 40 mm grenades:* Sutton, "Guest Column," 1.
tried to breast the cornfield: Public-affairs officer, "Battle of the Barge," 1.
two A-10s arrived: Ibid.; for a description of the A-10 and its massive 30 mm cannon, see Clancy, *Airborne,* 147.
The airpower ended the enemy attack: West, *The Wrong War,* 77; Jake Tapper, *The Outpost* (Boston: Little, Brown, 2012), 448–49.
To attack a Taliban van: West, *The Wrong War,* 81, 88–89.

313 *"echelons of staff":* Ibid., 89.
face-to-face encounters: Public-affairs officer, "Battle of the Barge," 1.
hard-working mortar men: Ibid.

314 *Cleaning the cornfield:* Ibid. Among the nine helicopters damaged were several medical evacuation flights. See also Miles Amoore, "What's the Point? The Taliban Will Be Back in a Week," *Sunday London Times,* October 4, 2009.
decided to deal with the Taliban camp: Ibid. See also West, *The Wrong War,* 82.
A night mission: West, *The Wrong War,* 89.

315 *an annual summer ritual:* Ahmad Waheed and Bill Roggio, "Taliban Seize District in Northeastern Afghanistan," July 25, 2010, at http://www.longwarjournal.org/archives/2010/07/taliban_seize_distri.php; accessed October 19, 2013.
nobody could be sure of that tally: Public-affairs officer, "Battle of the Barge," 1.
What of the election: West, *The Wrong War,* 90, 283: Tapper, *The Outpost,* 471. Tapper reports 128 voters in Barg-e Matal.

316 *The constitution allowed:* Islamic Republic of Afghanistan, *The Constitution* (Kabul, Afghanistan: The Constitutional *Loya Jirga* [Great Assembly], January 26, 2004), 16–17. Articles 60–70 concern the presidency.
to compel a runoff: Ibid., 16. Article 61 describes the runoff requirement. For Abdullah Abdullah, see Woodward, *Obama's Wars,* 164.
Ashraf Ghani Ahmadzai: Chandrasekaran, *Little America,* 91–93; Hastings, *The Operators,* 94. Hastings is particularly critical of Ghani. In the April 2014 election, Ghani reinvented himself and fared much better. He ceased speaking English in public, donned Afghan garb, soft-pedaled his U.S. ties, embraced Uzbek warlord Abdul Rashid Dostum as his running mate, and came in a respectable second in the first round of voting. In the June runoff against Abdullah Abdullah, Ghani and his supporters engaged in massive ballot-box stuffing reminiscent of the worst days in old Cook County, Illinois. Maybe he did learn something from the Americans after all.
The early counts confirmed: Woodward, *Obama's Wars,* 164. Abdullah Abdullah won the preliminary round of the 2014 Afghan presidential election. In the June runoff, he fell into a bitter dispute over hundreds of thousands of falsified ballots favoring challenger Ashraf Ghani.

317 *Lieutenant General David "Rod" Rodriguez:* McChrystal, *My Share of the Task,* 299. Rodriguez had been about to assume duty as the military assistant to Secretary of Defense Robert Gates when he was chosen to run ISAF Joint Command.
Lieutenant General William B. Caldwell IV: Livingston and O'Hanlon, *Afghani-*

stan Index, 6. For Caldwell's biography, see Public Affairs Office, *Lieutenant General William B. Caldwell IV* (Camp Eggers, Afghanistan: Headquarters, NTM-A/CSTC-A, November 2009), 1.

318 *The new ambassador:* Woodward, *Obama's Wars,* 217–18; Hastings, *The Operators,* 34–35. Hastings talked to McKiernan on the subject.

319 *Richard C. Holbrooke:* Woodward, *Obama's Wars,* 72; Chandrasekaran, *Little America,* 54.

three questions: McChrystal, *My Share of the Task,* 317.

320 *focused on strategy:* Ibid., 330.

McChrystal's method: Ibid., 331.

The math suggested: Department of the Army, FM 3-24, *Counterinsurgency,* on page 1-13, see paragraph 1-67. The manual reflects 1:40 as preferred, 1:50 as acceptable, and notes that "no predetermined fixed ratio of friendly troops to enemy combatants ensures success in COIN [counterinsurgency]."

321 *"likely to result in mission failure":* McChrystal, *My Share of the Task,* 332; Woodward, *Obama's Wars,* 178, 273.

leaked McChrystal's assessment: Woodward, *Obama's Wars,* 178–83.

322 *"This is not helping":* Ibid., 193–94; McChrystal, *My Share of the Task,* 349–50.

"never felt entirely the same": McChrystal, *My Share of the Task,* 350.

being asked to repeat the Bush strategy: Woodward, *Obama's Wars,* 168–69, 252.

323 *"not an adequate strategic partner":* Chandrasekaran, *Little America,* 134.

President Obama finally made the call: Woodward, *Obama's Wars,* 385–86. Woodward reproduces the key document verbatim.

324 *Numbers followed:* Ibid., 263, 276, 387–88.

15. Taliban Heartland

325 *"Your Marines seem to have exceeded the ops plan":* Chandrasekaran, *Little America,* 208. Foreign service officer Marc Chretien was the senior U.S. Department of State representative in Helmand Province during the Marjah operation. The British officer was not named in the account.

Getting 10,672 Marines into north Helmand: Ibid., 58–63.

326 *Brigadier General Lawrence D. Nicholson:* U.S. Department of the Navy, First Marine Division Public-Affairs Officer, *Major General Lawrence D. Nicholson* (Camp Pendleton, CA: Headquarters, First Marine Division, 2013), 1.

identified Marjah as a problem: Theo Farrell and Antonio Giustozzi, *The Taliban at War: Inside the Helmand Insurgency, 2004–2012* (London: King's College, October 31, 2012), 22–23, 30, 33–34. This excellent study includes multiple excerpts from interviews with Taliban fighters in Helmand Province.

327 *"too big of a fight":* Chandrasekaran, *Little America,* 67–68.

"an operation with rapid, observable impact": McChrystal, *My Share of the Task,* 363.

"government in a box": Ibid., 363–65.

tightened the noose around Marjah: Jeffrey Dressler, *Operation Moshtarak: Taking and Holding Marjah* (Washington, DC: Institute for the Study of War, March 2, 2010), 1–2.

328 *detailed planning and mission execution:* Harry D. Tunnell IV, *Memorandum for the Honorable John McHugh, Secretary of the Army, Subject: Open Door Policy—Report from a Tactical Commander* (Fort Lewis, WA: Headquarters, Fifth Brigade Combat Team, Second Infantry Division, August 20, 2010), 2, 7. Tunnell clashed repeatedly with Carter, so his highly critical report must be considered in this light. He notes that "RC (S) was incapable of establishing a forward field command post to support operations in Marjah." For more on Carter's approach, see West, *The Wrong War,* 206. West refers to Carter's tactical orders having "descended into self-caricature." For the British perspective, see Marco Giannangeli, "Axe-Man Is Army's New Boss," *UK Sunday Express,* October 14, 2012. The journalist noted that Carter "has his critics." Following another period of service in Afghanistan as a lieutenant general, Carter rose to become a full general and chief of the general staff, the senior officer of the British army.

the Marines had pioneered vertical envelopment: Clancy, *Marine,* 139–40.

329 *The helicopters went straight:* Dressler, *Operation Moshtarak,* 3–4.

"this was real": James W. Clark, *Their D-Day: Marjah Veterans Look Back a Year to Where They Were, Lessons Learned* (Camp Lejeune, NC: Headquarters, Second Marine Expeditionary Force, February 16, 2011), 1.

330 *"I didn't call air":* West, *The Wrong War,* 196–97.

331 *You had to walk carefully:* Chandrasekaran, *Little America,* 134–35. Chandrasekaran walked in with Company C, 1/6 Marines. His description is superb.

A series of AK shots: Ibid., 134–35. For the battle drill "React to Contact," see U.S. Department of the Army, FM 7-8, *Infantry Rifle Platoon and Squad* (Washington, DC: U.S. Government Printing Office, April 22, 1992), 4–18.

332 *ran up a big Afghan flag:* Dressler, *Operation Moshtarak,* 4.

approached Jawad Wardak: Michael M. Phillips, "U.S. and Afghan Troops Expand Control in Marjah," *Wall Street Journal,* February 14, 2010. See also Chandrasekaran, *Little America,* 135.

Something didn't look right: Chandrasekaran, *Little America,* 136.

333 *the day's contacts:* Dressler, *Operation Moshtarak,* 4.

"hit and run": "Afghanistan NATO Operation 'Meets First Objectives,'" *BBC News,* February 14, 2010, http://news.bbc.co.uk/2/hi/south_asia/8514397.stm; accessed October 22, 2013.

immediately suspended all use: Thomas Harding, "Operation Moshtarak: Missiles That Killed Civilians Hit Correct Target," *London Daily Telegraph,* February 16, 2010; C. J. Chivers and Rod Nordland, "Errant Rocket Strike Kills Civilians in Afghanistan," *New York Times,* February 14, 2010. For more on the missile itself, see Clancy, *Armored Cav,* 100–109. For the ISAF response, see McChrystal, *My Share of the Task,* 369. For the RC-South restrictions, see West, *The Wrong War,* 205.

334 *accompanied the new governor:* Chandrasekaran, *Little America,* 143.

"We are all Taliban here": West, *The Wrong War,* 216.

A burst of AK fire: Ibid., 216–17; Alfred de Montesquiou, "General: Maybe Weeks to Reclaim Marjah," *CBS News,* February 14, 2010.

returned fire in volume: Chandrasekaran, *Little America,* 136–37. Chandrasekaran was present for the firefight.

335 *Taking Marjah and holding it:* West, *The Wrong War,* 217.

It was Karzai's show: McChrystal, *My Share of the Task,* 373–74; Chandrasekaran, *Little America,* 143–44. McChrystal saw the Karzai meeting in Marjah as a glass half full; Chandrasekaran emphasized the negative aspects. Chandrasekaran (48–50) explained the serious allegations against Abdul Rahman Jan and Sher Mohammed Akhundzada.

The Marjah operation buoyed McChrystal's spirits: McChrystal, *My Share of the Task,* 375.

336 *"to sharply limit the use":* Ibid.

what the Pashtun villagers really thought: Alexander Jackson, *Operation Moshtarak: Lessons Learned* (London: International Council on Security and Development, May 2010), 2, 11, 18, 23.

Companies and platoons lost people: West, *The Wrong War,* 104–6, 117–20. See also McChrystal, *My Share of the Task,* 332.

337 *McChrystal accepted the invitation:* Hastings, *The Operators,* 232–33, reproduces the e-mail in full, as well as McChrystal's response. For McChrystal's recollections, see *My Share of the Task,* 378–79.

the men worked their way: McChrystal, *My Share of the Task,* 378–79.

338 *Corporal Michael Ingram:* Hastings, *The Operators,* 259, 264. Hastings was present at the meeting. See also McChrystal, *My Share of the Task,* 380.

339 *press assistant Duncan Boothby:* Ernesto Londoño, "Civilian Press Aide Resigns Amid Flap over McChrystal's 'Rolling Stone' Profile," *Washington Post,* June 22, 2010.

saw a lot of good in Michael Hastings: Hastings, *The Operators,* 16.

340 *Hastings followed the ISAF commander:* McChrystal, *My Share of the Task,* 384, 386; Michael Hastings, "The Runaway General," *Rolling Stone,* June 22, 2010, at http://www.rollingstone.com/politics/news/the-runaway-general-20100622; accessed October 23, 2013.

Hastings decided otherwise: Hastings, *The Operators,* 316–17.

341 *Hastings's article "The Runaway General":* The article won the George Polk Award, and the *Huffington Post* political commentary website named him a Game Changer for 2010. A subsequent investigation by the U.S. Army inspector general later cleared McChrystal of charges of insubordination and raised questions about the uncertain ground rules of Hastings's embedded reporting.

"How in the world": McChrystal, *My Share of the Task,* 386.

342 *"there was only one way to play it":* Tom Wolfe, *The Right Stuff* (New York: Bantam Books, 1979), 132.

McChrystal offered his resignation: Woodward, *Obama's Wars,* 371–72.

"assuming this difficult post": Ibid., 373–74.

16. Malik Daoud Again

344 *"brilliant, inventive, tireless":* Myrer, *Once an Eagle,* 878.

The Normandy hedgerows: Clay Blair, *Ridgway's Paratroopers* (Garden City, NY: Dial Press, 1985), 260, 349. The Gethsemane quote and a superb account of the Normandy *bocage* fighting can be found in Rick Atkinson, *The Guns at Last Light: The War in Western Europe, 1944–45* (New York: Henry Holt, 2013), 111–13.

346 *the Arghandab River Valley:* Carl Forsberg, *Afghanistan Report 7, Counterinsur-*

gency in Kandahar: Evaluating the 2010 Hamkari Campaign (Washington, DC: Institute for the Study of War, 2010), 12.

Colonel Arthur A. Kandarian: Eric Snyder, "2 BCT's Colors Have Honored Role in Change: Kandarian Takes Charge of Unit," *Clarksville (TN) Leaf-Chronicle,* March 14, 2009.

"a fortification like a wall": Caesar is quoted in Atkinson, *The Guns at Last Light,* 111. Atkinson delineates the many studies that described the difficult Norman terrain.

operating in the Arghandab for years: Chandrasekaran, *Little America,* 161.

347 *epitomized the experience of most surge units:* Headquarters, Second Brigade Combat Team, 101st Airborne Division, *Book of Valor: Combined Task Force Strike Operation Enduring Freedom 10–11* (Fort Campbell, KY: Headquarters, Second BCT, 101st Airborne Division, June 2011), 234–36.

348 *MRAP–all-terrain vehicle (M-ATV):* Scott R. Gourley, "M-ATV," *Army* (November 2013): 62–63.

minefield-breaching equipment: John K. Chung, "Alpha Company, Second Brigade Special Troops Battalion," in Headquarters, Second Brigade Combat Team, 101st Airborne Division, *CTF Strike OEF 10–11 Company-Troop-Battery Commanders After Action Review* (Fort Campbell, KY: Headquarters, Second BCT, 101st Airborne Division, June 2011), 32–33.

349 *Joint IED Defeat Organization (JIEDDO):* Ibid. Lieutenant General Michael L. Oates directed JIEDDO in 2010 and did much to meet the needs of the Second BCT, 101st Airborne Division and other deploying Afghan surge units. Oates knew Colonel Kandarian well from the 101st Airborne Division's 2005–06 Iraq deployment, in which the general served as a one-star deputy. Oates also commanded the Tenth Mountain Division in central and southern Iraq from 2008 to 2009.

contracted off-leash IED-detector dogs: Headquarters, Second Brigade Combat Team, 101st Airborne Division, *Book of Valor,* 131.

350 *to fight as infantry:* Brown, *Kevlar Legions,* 302–7; John Smith, "Archer Battery, 3/2 SCR," Headquarters, Second Brigade Combat Team, 101st Airborne Division, *CTF Strike OEF 10–11 Company-Troop-Battery Commanders After Action Review,* 49.

The other six battalions: Headquarters, Second Brigade Combat Team, 101st Airborne Division, *Book of Valor,* 4.

Ahmed Wali Karzai: Chandrasekaran, *Little America,* 162–64.

352 *"interlocking fires":* Headquarters, Second Brigade Combat Team, 101st Airborne Division, *CTF Strike OEF 10–11 Company-Troop-Battery Commanders After Action Review,* 101. Captain Andrew Shaffer of the First Battalion, 320th Field Artillery quoted Master Sergeant Robert Pittman and explained the tactic. The Asymmetric Warfare Group represented yet another initiative by U.S. Army chief of staff General Peter Schoomaker to ensure that the latest tactics, techniques, and procedures migrated from the special operators to the conventional battalions.

Two days of hard walking: Ibid., 59–60.

353 *a medevac helicopter:* Headquarters, Second Brigade Combat Team, 101st Airborne Division, *CTF Strike OEF 10–11 Company-Troop-Battery Commanders After Action Review,* 107, 159.

requested immediate heliborne extraction: Headquarters, Second Brigade Combat Team, 101st Airborne Division, *Book of Valor,* 62, 263. Page 263 includes the Bronze Star with V (for Valor) device citation for Sergeant Major John White that addresses the July 11 fight, among other events. White took over as acting brigade command sergeant major when Command Sergeant Major Alonzo Smith was wounded in action on June 13, 2010. Smith returned to duty weeks later.

354 *a more positive course:* Ibid., 250. The source includes the Silver Star citation for Lieutenant Colonel David S. Flynn that describes the events of July 11 as well as subsequent operations.

under Taliban control: David S. Flynn, "Extreme Partnership in Afghanistan," *Military Review* (March/April 2012): 28. For enemy-strength estimates, see Forsberg, *Counterinsurgency in Kandahar,* 15.

355 *"more tired division commanders":* George S. Patton Jr., *War As I Knew It* (New York: Pyramid Books, 1970), 305.

"empowered junior non-commissioned officers": Headquarters, Second Brigade Combat Team, 101st Airborne Division, *Book of Valor,* 250. The quotations come from Lieutenant Colonel David Flynn's Silver Star citation.

"beaten on his own ground": Headquarters, Second Brigade Combat Team, 101st Airborne Division, *CTF Strike OEF 10–11 Company-Troop-Battery Commanders After Action Review,* 159.

a rougher time: Headquarters, Second Brigade Combat Team, 101st Airborne Division, *Book of Valor,* 62–63.

356 *returned two weeks later:* Flynn, "Extreme Partnership in Afghanistan," 30, 33.

oppressed the Alkozai: Forsberg, *Counterinsurgency in Kandahar,* 17.

Before dawn: Megan McCloskey, "'Probably Wasn't the Smartest Thing I've Done,'" *Stars and Stripes,* June 14, 2011.

358 *Apache gun runs:* Chung, "Alpha Company, Second Brigade Special Troops Battalion," 32–33.

insurgents to the south: Headquarters, Second Brigade Combat Team, 101st Airborne Division, *Book of Valor,* 75–76, 81–83. The pages include extracts from the valor citations for Private First Class Jose Rosario, Specialist Cameron Fontenot, and Captain Patrick McGuigan.

359 *another was called:* Ibid., 78.

360 *the Taliban needed to counterattack:* Forsberg, *Counterinsurgency in Kandahar,* 19.

COP Stout took fire: Headquarters, Second Brigade Combat Team, 101st Airborne Division, *Book of Valor,* 84.

Sergeant First Class Kyle Lyon volunteered: Ibid., 83. See also McCloskey, "'Probably Wasn't the Smartest Thing I've Done,'" 1. For details on the AGS-17, see Isby, *Weapons and Tactics of the Soviet Army,* 425–26.

August 1 dawned hot: Headquarters, Second Brigade Combat Team, 101st Airborne Division, *Book of Valor,* 86–87. The pages provide the narrative for Staff Sergeant Christopher Young's Bronze Star with V device.

361 *approach of Lyon's platoon:* McCloskey, "'Probably Wasn't the Smartest Thing I've Done,'" 1.

362 *Staff Sergeant Benjamin Tivao:* Headquarters, Second Brigade Combat Team, 101st Airborne Division, *Book of Valor,* 79–80.

363 *seventeen combined American-Afghan outposts:* Flynn, "Extreme Partnership in Afghanistan," 30.
"no space for the Taliban": Paula Broadwell with Vernon Loeb, *All In: The Education of General David Petraeus* (New York: Penguin Press, 2012), 96. An Army Reserve officer and West Point graduate, Broadwell, with the approval and active support of General David Petraeus, spent time in the Arghandab with 1-320 Field Artillery while researching her book. See also Headquarters, Second Brigade Combat Team, 101st Airborne Division, *Book of Valor,* 175–76, and Daniel P. Bolger, "Notes from Meeting with U.S. and Afghan Commanders, 2nd BCT, 101st Airborne Division," COP Nolen, Arghandab District, Afghanistan, November 16, 2010.
364 *"the greatest achievement":* Headquarters, Second Brigade Combat Team, 101st Airborne Division, *Book of Valor,* 312–13. The numbers and Major General Carter's quote come from the narrative for the Presidential Unit Citation awarded to the Second Brigade Combat Team, 101st Airborne Division.
McChrystal had gone too far: Kaplan, *The Insurgents,* 343–44.
365 *fourteen questions:* Headquarters, Second Brigade Combat Team, 101st Airborne Division, *CTF Strike OEF 10–11 Company-Troop-Battery Command Post SOP* (Fort Campbell, KY: Headquarters, Second BCT, 101st Airborne Division, June 2011), 51.
366 *referred to the other country constantly:* Chandrasekaran, *Little America,* 221.
"Anaconda Strategy vs. Insurgents in Afghanistan": Commander, International Security Assistance Force, *Anaconda Strategy vs. Insurgents in Afghanistan* (Kabul, Afghanistan: Headquarters, International Security assistance Force, 2010), 1. For comparison, see Commander, Multi-National Force–Iraq, *Anaconda Strategy vs. AQI* (Baghdad, Iraq: Headquarters, Multi-National Force–Iraq, 2008), 1.
367 *"It's laughable":* Chandrasekaran, *Little America,* 222.
"It can't work": Woodward, *Obama's Wars,* 332.
General Ashfaq Parvez Kayani: Ibid., 366–67. Kayani had previously headed the Inter-Services Intelligence organization. The ISI is reputed to maintain continuing ties with the Afghan Taliban and sponsored them in the 1990s.
368 *killed Osama bin Laden:* Chandrasekaran, *Little America,* 324.
"I have three main enemies": Kaplan, *The Insurgents,* 346.
Task Force Shafafiyat: Ibid., 348.
369 *the otherwise superb work:* Livingston and O'Hanlon, *Afghanistan Index,* 6–8. For a good in-progress assessment of the early days of NTM-A, see Terrence E. Kelly et al., *RAND Objective Assessment of the Afghan National Security Forces* (Santa Monica, CA: RAND Corporation, April 2010), 21–22.
ran afoul of Michael Hastings: Michael Hastings, "Another Runaway General: Army Deploys Psy-Ops on U.S. Senators," *Rolling Stone,* February 23, 2011, at http://www.rollingstone.com/politics/news/another-runaway-general-army-deploys-psy-ops-on-u-s-senators-20110223; accessed November 7, 2013.
370 *the final ISAF structure:* Kelly, *RAND Objective Assessment of the Afghan National Security Forces,* 82; Chandrasekaran, *Little America,* 280.
Afghan local police (ALP): Kaplan, *The Insurgents,* 341.
371 *"Seek out and eliminate":* David H. Petraeus, *COMISAF's Counterinsurgency Guidance* (Kabul, Afghanistan: Headquarters, International Security Assistance Force, August 1, 2010), 1–4.
Petraeus told a journalist: West, *The Wrong War,* 226.

not in evidence: David Ignatius, "Can Petraeus Handle the CIA's Skepticism on Afghanistan?" *Washington Post,* September 1, 2011. In 2011, in the assessment of CIA analysts, the war was a stalemate, a view strongly contested by ISAF.

372 *"Hold for years":* Woodward, *Obama's Wars,* 349.

"a responsible end": President Barack Obama, "Remarks by the President on the Way Forward in Afghanistan," June 22, 2011, at http://www.whitehouse.gov/the-press-office/2011/06/22/remarks-president-way-forward-afghanistan; accessed November 7, 2013.

He departed Kabul: Kaplan, *The Insurgents,* 349–50.

Something distracted: Hastings, *The Operators,* 296. Hastings describes how General Petraeus passed out during congressional testimony on June 15, 2010. The general attributed the incident to dehydration and jet lag. See also Kaplan, *The Insurgents,* 342, regarding an episode of "early-stage prostate cancer" that afflicted Petraeus during his time at U.S. Central Command.

17. Attrition

373 *"studying all those statistics":* David Halberstam, *The Best and the Brightest* (Greenwich, CT: Fawcett Crest Books, 1972), 304.

les cafards*:* Douglas Porch, *The French Foreign Legion* (New York: HarperCollins, 1991), 426–27.

Similar things happened: Utley, *Frontier Regulars,* 24, 84–91. Desertions ran as high as 32.6 percent of the force in 1871; see Evan S. Connell, *Son of the Morning Star: Custer and the Little Bighorn* (New York: North Point Press, 1984), 149–55. Connell includes numerous anecdotes regarding the hardships and stresses affiliated with army life on the Great Plains.

374 *bored czarist soldiers:* For a fictional reference to these pursuits, see Georges Surdez, "Russian Roulette," *Collier's Illustrated Weekly* (January 30, 1937): 16, 57.

le cafard *crept through:* For suicide rates, see Timothy Williams, "Suicides Outpacing War Deaths for Troops," *New York Times,* June 8, 2012. For the January 2012 Marine urination incident, see Rowan Scarborough, "Marine Corps Drops Urination Desecration Case," *Washington Times,* September 7, 2013.

375 *COP Palace:* U.S. Department of the Army, Mr. Haytham Faraj, "Defense Motion to Dismiss the Charges for a Violation of the Right to Counsel at the Article 32 Hearing," *United States v. Sergeant Jeffrey T. Hurst* (Fort Bragg, NC: Second Judicial Circuit, June 19, 2012), 1–2. This document includes a wealth of operational detail about Company C in Panjwai in the summer and early autumn of 2011.

most of Panjwai: Lindsey Kibler, "There's No Place Like Home," *Task Force Arctic Wolves to Root Out the Taliban from the Horn of Panjwai* (Sperwan Ghar, Afghanistan: Headquarters, First Brigade Combat Team, Twenty-Fifth Infantry Division, October 27, 2011), at http://www.dvidshub.net/news/79082/theres-no-place-like-home-third-story-three-part-series-actions-task#.UoO90vmsj0s; accessed November 13, 2013. The story includes some key details about both Doab and Mushan villages. See also Sam Friedman, "Fort Wainwright Soldiers Move into Taliban Stronghold," *Fairbanks Daily News-Miner,* October 10, 2011.

376 *Private Danny Chen:* Jennifer Gonnerman, "Pvt. Danny Chen, 1992–2011," *New*

York, January 6, 2012, at http://nymag.com/news/features/danny-chen-2012-1/index2.html; accessed November 13, 2013. For more details on the views of the soldiers who bullied Private Chen, see Faraj, "Defense Motion to Dismiss the Charges for a Violation of the Right to Counsel at the Article 32 Hearing," 4.

377 *The subsequent investigation:* Drew Brooks, "Pvt. Danny Chen's Platoon Leader to Be Removed from Army," *Fayetteville (NC) Observer,* December 18, 2012.

The new captain took charge: Daniel P. Bolger, "Notes on Forward Operating Base Zangabad and Panjwai Operations," November 28, 2011.

West Panjwai, the Horn proper, continued to fester: Ibid.

378 *Staff Sergeant Robert Bales slipped out:* Jack Healy, "Soldier Sentenced to Life Without Parole for Killing 16 Afghans," *New York Times,* August 23, 2013.

379 *The evidence all pointed:* Bill Ardolino, "The Taliban Are Worried About the Uprising Happening Here: An Interview with the Panjwai District Governor," *Long War Journal,* March 30, 2013; at http://www.longwarjournal.org/archives/2013/03/property_ownership_i.php. Governor Hajji Faizal Mohammed was one of the first Afghan officials on the scene of the massacre on the morning of March 11, 2012.

even Bales wasn't sure: Ibid.

battalion after battalion of *resolute Canadians:* Matthew Fisher, "Panjwai Handover to U.S. Marks 'Page in Canadian History,'" *Postmedia News,* July 5, 2011, at http://www.canada.com/news/Panjwaii+handover+marks+page+Canadian+history/5050844/story.html; accessed November 13, 2013. For an excellent account of the fierce 2006 fighting in Panjwai, see Rusty Bradley and Kevin Maurer, *The Lions of Kandahar* (New York: Bantam Books, 2011). Major Rusty Bradley fought in Panjwai west of Kandahar in 2006.

Lieutenant Colonel Steve Miller: Sheryl Nix, "1-25th Stryker Brigade Soldiers Bore Through History, Continue Legacy," U.S. Army Alaska (Fort Wainwright, AK: Fort Wainwright Public Affairs Office, April 5, 2011), at http://www.usarak.army.mil/main/Stories_Archives/Apr5-9/100405_FS6.asp; accessed November 13, 2013.

380 *into Mushan village:* U.S. Marine Corps Intelligence Activity, *Afghan Insurgent Tactics, Techniques, and Procedures (TTP) Field Guide Volume II: Southern Afghanistan,* 88–89.

381 *V-hull variants:* Seth Robson, "Fleet of 'V-Hull' Strykers Growing in Afghanistan," *Stars and Stripes,* December 9, 2011.

The morning didn't start well: Bolger, "Notes on Forward Operating Base Zangabad, Mushan Village, and Panjwai Operations," March 26–27, 2012, 1–2.

Sergeant Abdul Hamid: Bolger, "Notes on Forward Operating Base Zangabad, Mushan Village, and Panjwai Operations," March 26–27, 2012, 2.

382 *how people stayed alive:* Carl Forsberg, *Afghanistan Report 3, the Taliban's Campaign for Kandahar* (Washington, DC: Institute for the Study of War, 2009), 33.

385 *First Lieutenant Patrick Higginbottom:* Ibid. Lieutenant Higginbottom recovered from his wound. He was in good company. In 1918 during the Meuse-Argonne offensive, Lieutenant Colonel George S. Patton Jr. suffered a similar wound. See D'Este, *Patton: A Genius for War,* 259.

386 *two U.S. Marine Corps F/A-18D Hornet fighter jets:* Clancy, *Marine,* 131, 143, 231–32, offers some useful descriptions of Marine Corps F/A-18 aircraft.

the intelligence people later determined: Bolger, "Notes on Forward Operating Base Zangabad, Mushan Village, and Panjwai Operations," March 26–27, 2012, 1–2.

387 *"He is always alert":* Clancy, *Marine,* 251.

a deputy commander in Anbar Province: Ricks, *The Gamble,* 220.

388 *"None in the Marine Corps":* Lewis Sorley, *Thunderbolt: General Creighton Abrams and the Army of His Times* (New York: Simon and Schuster, 1992), 209.

389 *"I am going to manage you by slides":* Cloud and Jaffe, *The Fourth Star,* 260.

things that defied numbers: Livingston and O'Hanlon, *Afghanistan Index,* 20–21.

created numeric charts: Ibid., 8.

391 *stayed about the same:* Anthony H. Cordesman, *Afghanistan: The Uncertain Course of War and Transition* (Washington, DC: Center for Strategic and International Studies, January 18, 2013), 18, 20, 21. Cordesman says, "There is no clear pattern of military success after 2011, and there are as many metrics that show a constant or increasing level of violence as there are that show any progress."

"a carefully rigged portrayal": Ibid., 18, 20, 21. Cordesman shows how ISAF selectively claimed percentage decreases in hostile attacks from April 2008 to October 2012. He states, "ISAF and U.S. have tended to focus on the 'positive' trends in the fight and ANSF [Afghan National Security Forces] in ways that are all too close to the 'follies' in Vietnam." Cordesman alludes here to the regular U.S. military press briefings in Saigon referred to as the "Five O'Clock Follies" by reporters.

392 *Joint Afghan-NATO Inteqal Board:* North Atlantic Treaty Organization, Supreme Allied Commander, Europe, *Inteqal: Transition to Afghan Lead* (Brussels, Belgium: Headquarters, Supreme Allied Commander, Europe, June 18, 2013), 1, at http://www.nato.int/cps/en/natolive/topics_87183.htm; accessed November 14, 2013. Ghani parlayed his role on the Inteqal Board, with travels all over Afghanistan, into a campaign for the Afghan presidency in 2014.

Ahmed Wali Karzai was no more: Chandrasekaran, *Little America,* 320.

"in together, out together": Ballard, Lamm, and Wood, *From Kabul to Baghdad and Back,* 266.

393 *a tax of about ten thousand:* Daniel P. Bolger, "Notes on ISAF Commanders Strategic Consultations Regarding Coalition Cohesion," May 30, 2012.

Karzai went ballistic: Ballard, Lamm, and Wood, *From Kabul to Baghdad and Back,* 267. See also Daniel P. Bolger, "Notes on ISAF Commanders Strategic Consultations Regarding Civilian Casualties," May 13, 2012. The role of Afghan soldiers in the incident did not get much, if any, publicity, either in or out of country.

394 *agreed to significant limits:* Ballard, Lamm, and Wood, *From Kabul to Baghdad and Back,* 267–68.

"Enduring Strategic Agreement": U.S. Department of State, *Enduring Strategic Agreement Between the United States of America and the Islamic Republic of Afghanistan* (Washington, DC: U.S. Department of State, May 2, 2012).

what Iraq had gotten: Woodward, *Obama's Wars,* 376–77. As early as a July 10, 2010, interview with Bob Woodward, President Obama explicitly described his long-term plan for Afghanistan in terms of what had happened in Iraq: "Our commitment to your [Afghan] long-term security and stability will extend for a very long time, and in the same way that our commitment to Iraq will extend beyond our combat role there."

a public admission: Kaplan, *The Insurgents,* 365.

395 *the results came back:* Rajiv Chandrasekaran, "General Allen Cleared in Misconduct Investigation," *Washington Post,* January 22, 2013.

18. Green on Blue

396 *"We are content with discord":* Wright et al., *A Different Kind of War,* 9.
you could rent it: Smith, *The Killer Elite,* 214, offers one version of the comment. For examples of using money to gain tribal support early in the war, see Woodward, *Bush at War,* 155, 214, 230.

397 *ill-fated 1839 to 1842 expedition:* Tanner, *Afghanistan,* 136, 139–43.
"another starts up": Ibid., 159.

398 *rebels overran and looted:* Dixon, *On the Psychology of Military Incompetence,* 72–75.
"the most incompetent soldier": Ibid., 73.
humiliating Ghilzai terms: Wright et al., *A Different Kind of War,* 14.

399 *"attendant train of horrors":* Tanner, *Afghanistan,* 179.
Elphinstone surrendered personally: Dixon, *On the Psychology of Military Incompetence,* 77–78.
the pathetic remnants: Ibid., 78.
The Russians, too: Timothy Gusinov, "Green on Red: The Soviet Experience," *COIN Common Sense* (Kabul, Afghanistan: COMISAF Advisory and Assistance Team, March 2013), 7–9. Gusinov served as a Soviet adviser during the 1979–89 Afghan campaign. A retired Russian army major, Gusinov emigrated to the United States and now works with U.S. military forces. His article includes excellent documentation on the Herat rising of March 1979. For an example of the usual, though vague, version of the Herat rebellion, see Tanner, *Afghanistan,* 232. Tanner reported "one hundred Soviet advisors and their families" killed and Soviet heads paraded around Herat on the ends of poles. For a much more balanced Western account, see Isby, *War in a Distant Country,* 21.

400 *admitted to two events:* Gusinov, "Green on Red," 9.
better than expected: Lester W. Grau, *Breaking Contact Without Leaving Chaos: The Soviet Withdrawal from Afghanistan* (Fort Leavenworth, KS: Foreign Military Studies Office, 2011), 17, 19.

401 *When the Soviet Union collapsed:* Mark N. Katz, "Lessons of the Soviet Withdrawal from Afghanistan" (Washington, DC: Middle East Policy Council, 2013). See also Tanner, *Afghanistan,* 272–73.
Colonel Keith A. Detwiler: U.S. Headquarters, Fifth Army, *Colonel Keith A. Detwiler, Chief of Staff, United States Army North (Fifth Army)* (Fort Sam Houston, TX: Headquarters, Fifth Army, 2013), 1.

402 *trends became obvious:* The numbers come from R. Hossain, *Afghanistan: Green-on-Blue Attacks in Context* (Washington, DC: Institute for the Study of War, November 13, 2012), 1. Slightly different numbers can be found in U.S. Department of Defense, *Report on Progress Toward Security and Stability: Report to Congress* (Washington, DC: Department of Defense, November 2013), 23–24.

403 *Afghans also assaulted one another:* Department of Defense, *Report on Progress Toward Security and Stability,* 24.
U.S. footprint out in RC-West: Daniel P. Bolger, "Notes on "Camp Stone and Camp

Zafar Operations," July 9–10, 2012. The Albanians are particularly active new NATO members. Their officers voiced the hope that being placed in Italian-run RC-West was a nod to cross-Adriatic ties rather than a conscious commemoration of Benito Mussolini's ill-fated 1940 Fascist invasion. The reasoning behind NATO force assignments can be obtuse.

404 *the NTM-A commander supporting RC-West:* Monika Comeaux, *Regional Support Command West Welcomes New Commander* (Camp Eggers, Kabul, Afghanistan: Headquarters, NTM-A?CSTC-A, July 16, 2012), 1.

$3,991,635 of U.S. taxpayer funds: New Jan Group Corporation and Supply Service, "ANCOP [Afghan National Civil Order Police] Camp Shouz," at http://www.newjangroup.com/index.php?option=com_content&view=article&id=56&Itemid=76; accessed November 17, 2013. The New Jan Group is an Afghan-owned construction firm based in Kabul. For a brief look at the poverty and lack of education in the surrounding area, see Riza Caporros, "Supply Drop to Shouz Valley School Promotes Counterinsurgency," *International Security Assistance Force Mirror* (September 2009): 19.

405 *the Fourth Alabama:* Tom Gordon, "Heroes Past, Present, and Future," *Birmingham Weld,* August 14, 2012. Like most Guard units, the 1-167 Infantry drew on a rich history and kept the lineage and battle honors of the Fourth Alabama Regiment, one of the premier Confederate fighting outfits during the Civil War. Their successor 167th Infantry Regiment in the Great War of 1917–18 fought as part of the famous Forty-Second "Rainbow" Division, for a time under the command of young Brigadier General Douglas MacArthur. In World War II, the regiment liberated the Philippines under the command of that same General MacArthur, by then wearing five stars. To commemorate this long fighting tradition, the modern soldiers of the 1-167 Infantry proudly wore a Fourth Alabama patch.

"Vigilance is how we do our job": Daniel P. Bolger, "Notes on Camp Stone, Camp Zafar, and Herat Area Operations," November 12–13, 2012.

408 *Afghan civilian traffic:* For an overview of the Afghan economy, see NATO Training Mission–Afghanistan, "Economy of Afghanistan," *Weekly Spotlight,* October 17, 2012, 6.

409 *One of those rare attacks:* Daniel P. Bolger, "Notes on Events of June 25, 2012," *NTM-A Commander's Journal.*

principal training base: Stewart Nusbaumer, "The New Afghan Air Force," *Air and Space Smithsonian* (January 2011): 52–53. In their day, the Soviets stationed the Fifth Guards Motor Rifle Division at Shindand. A huge, rusting pile of Soviet-era tanks and aircraft can be found inside the fence line.

410 *guardian angels:* Marc V. Schanz, "Nine Americans Die After Attack in Kabul," *Air Force Magazine* (June 2011): 18.

Two ISAF Joint Command advisers had died: Daniel P. Bolger, "Notes on Events of February 25, 2012," *NTM-A Commander's Journal.*

412 *the green-on-blue ambush on July 22:* Maria al-Habib, "After Afghan Shooting, Delay Raises Questions," *Wall Street Journal,* August 1, 2012.

As the tumult subsided: Bolger, "Notes on Camp Stone, Camp Zafar, and Herat Area Operations," November 12–13, 2012. For his valorous actions that day, Specialist John Yates received the Army Commendation Medal with V device.

413 *That day, ISAF recorded:* Bill Roggio and Lisa Lundquist, "Green-on-Blue Attacks in Afghanistan: The Data," *Long War Journal,* October 26, 2013, at www.longwarjournal.org/archives/2012/08/green-on-blue_attack.php#timeline; accessed November 18, 2013.

major drop in green-on-blue events: Lisa Lundquist, "Senior ISAF Member Killed in Green-on-Blue Attack in Southern Afghanistan," October 5, 2013, at www.longwarjournal.org/archives/2013/10/3rd_green-on-blue_at.php; accessed November 18, 2013. Lundquist explained: "The relative downturn is likely due to the introduction of security measures in 2012 such as the use of 'guardian angel' soldiers to overwatch US troops, and to reduced partnering between Afghan and Coalition forces as the drawdown continues." For the 2013 statistics, see Department of Defense, *Report on Progress Toward Security and Stability,* 24.

"my personal apology": See Alissa J. Rubin, "11 Afghans Killed in Military Actions Near Pakistan Border," *New York Times,* February 13, 2013, and Rod Nordland, "Two Afghan Boys Accidentally Killed by NATO Helicopter," *New York Times,* March 2, 2013.

414 *General Joe Dunford:* Headquarters, International Security Assistance Force, *General Joseph F. Dunford, Jr.* (Kabul, Afghanistan: Headquarters, International Security Assistance Force, February 10, 2013), 1, at http://www.isaf.nato.int/about-isaf/leadership/general-joseph-f.-dunford-jr.html; accessed November 18, 2013. In mid-2014, General Dunford was chosen to be the next commandant of the U.S. Marine Corps, the first American theater commander in Afghanistan in more than a decade to move to another key four-star billet. His predecessors all retired after their time in country. Two (McKiernan and McChrystal) were relieved of duty. General John F. Campbell, vice chief of staff of the U.S. Army and veteran of both Iraq and Afghanistan, was named to take command and oversee the announced U.S. withdrawal.

return of General Rod Rodriguez: General David Rodriguez became the commander of U.S. Africa Command. John Vandiver, "Rodriguez, Experienced in Afghanistan, Becomes New AFRICOM Boss," *Stars and Stripes,* April 5, 2013.

"The tide has turned": Ballard, Lamm, and Wood, *From Kabul to Baghdad and Back,* 270, quotes President Obama's speech at Bagram Airfield. In addition, the president limited U.S. military support to Afghanistan after December 31, 2014, to a narrow noncombat role supporting Afghan security forces.

415 *"a range of options":* U.S. Office of the Press Secretary, Office of the President, *Background Briefing by Senior Administration Officials on Afghanistan — Via Conference Call* (Washington, DC: White House, June 18, 2013), at www.whitehouse.gov/the-press-office/2013/06/18/background-briefing-senior-administration-officials-afghanistan-conferen; accessed November 18, 2013. On May 27, 2014, the United States announced a withdrawal and transition plan subject to agreement by the Afghans to respect U.S. troop immunities from prosecution. Americans prepared to shift from combat operations to a training, advisory, and assistance mission on December 31, 2014. In 2015, U.S. strength will drop to 9,800, then 4,900 in 2016. By 2017, fewer than 200 uniformed Americans will remain, all affiliated with the U.S. embassy in Kabul. NATO announced a complementary effort called Resolute

Support that will start with up to 3,000 troops in 2015. Over time, NATO intends to draw down proportionate to the U.S. contingent.

Epilogue

416 *"Bitter as hell"*: Joseph W. Stilwell, *The World War II Diaries of General Joseph W. Stilwell*, vol. 2: *1942* (Stanford, CA: Hoover Institution, 2013), 68, at http://media.hoover.org/sites/default/files/documents/1942Stilwell20120515.pdf; accessed November 19, 2013. The quote comes from the entry for July 29, 1942.

417 *"a strong message"*: Lang, *Brad's Adventure*, at http://www.edlang.us/brad'sadventure.htm; accessed November 19, 2013. Company C, 1-155 Infantry was attached from the 155th Armored Brigade Combat Team, Mississippi Army National Guard.

418 *"It took me four days"*: Daniel P. Bolger, "Notes on Operations with 2-11 Cavalry in and Around FOB Kalsu," May 26–27, 2005, 1. See also Ervin, *Leaving the Wire*, 123–25. Sergeant David P. Ervin was onsite for the recovery of the vehicle and the four soldiers. Intelligence experts later determined the IED employed ammonium nitrate to enhance the explosive power and added flame accelerants as well. The four soldiers killed in action in the enemy attack were Specialist Bryan D. Barron, Sergeant Audrey D. Lunsford, Staff Sergeant Saburant Parker, and Sergeant Daniel Ryan Varnado.

419 *Great War veteran Wilfred Owen:* Wilfred Owen, "Dulce et Decorum Est," in *The Collected Poems of Wilfred Owen* (New York: New Directions Books, 1964), 55. The Latin phrase, from the poet Horace, translates to "it is sweet and fitting" and refers to death in battle for one's country. A British infantry officer, Owen was killed in action one week before the 1918 Armistice.

What do we know: For Iraq's estimated civilian losses, see Ballard, Lamm, and Wood, *From Baghdad to Kabul and Back*, 224–25. For Afghan losses, see Susan G. Chesser, *Afghanistan Casualties: Military Forces and Civilians* (Washington, DC: Congressional Research Service, February 29, 2012), 3. The latest numbers can be found in Department of Defense, *Report on Progress Toward Security and Stability*, 22–23.

cost the U.S. a lot of money: Ballard, Lamm, and Wood, *From Baghdad to Kabul and Back*, 224–25, 283–84.

420 *"secret even in success"*: Office of the President, "Address to the Joint Session of the 107th Congress, September 20, 2001," in *Selected Speeches of President George W. Bush, 2001–2008*, 69.

423 *"silence of the generals"*: Rosa Brooks, "Obama vs. the Generals," *Politico*, November 2013, at http://www.politico.com/magazine/; accessed November 19, 2013.

gave their views: For an example, see Rod Nordland, "General Fired over Karzai Remarks," *New York Times*, November 5, 2011. Told that Afghan president Hamid Karzai would side with Pakistan in a dispute with the United States, Major General Peter Fuller of NATO Training Mission–Afghanistan told an interviewer: "Why don't you just poke me in the eye with a needle? You've got to be kidding me. I'm sorry, we just gave you $11.6 billion and now you're telling me, 'I don't really care'?" Fuller was dismissed from his position.

424 *retired officers:* Woodward, *State of Denial,* 454; Brooks, "Obama vs. the Generals," 1–3.

rueful words: Brad Knickerbocker, "Gates Warning: Avoid Land War in Asia, Middle East, or Africa," *Christian Science Monitor,* February 26, 2011.

"lions led by donkeys": Dixon, *On the Psychology of Military Incompetence,* 80 81.

426 *towns destroyed in order to save them:* This famous remark can be found in many sources, among them Neil Sheehan, *A Bright Shining Lie: John Paul Vann and America in Vietnam* (New York: Vintage Books, 1989), 719. The U.S. Army major was referring to the devastation inflicted on Ben Tre during the 1968 Tet Offensive.

enemy's reliance on IEDs: Anthony H. Cordesman, *Coalition, ANSF* [Afghan National Security Forces], *and Civilian Casualties in the Afghan Conflict: From 2001 Through August 2012* (Washington, DC: Center for Strategic and International Studies, September 4, 2012), 17–18. see also Livingston and O'Hanlon, *Afghanistan Index,* 12.

427 *"I will never leave a fallen comrade":* Brown, *Kevlar Legions,* 279, 461.

428 *"It's all my fault":* Douglas Southall Freeman, *Lee's Lieutenants,* vol. 3, *Gettysburg to Appomattox* (New York: Charles Scribner's Sons, 1944), 166.

432 *"a larger and long-term role":* Mark E. Manyin et al., *Pivot to the Pacific? The Obama Administration's "Rebalancing" Toward Asia* (Washington, DC: Congressional Research Service, March 28, 2012), 1.

Air-Sea Battle: U.S. Department of Defense, *Air-Sea Battle: Service Collaboration to Address Anti-Access and Area Denial Challenges* (Washington, DC: Air-Sea Battle Office, May 2013), 8.

"no longer be sized": Manyin et al., *Pivot to the Pacific,* 12. For the report itself, including the cited sentence, see U.S. Department of Defense, *Sustaining U.S. Global Leadership: Priorities for 21st Century Defense* (Washington, DC: U.S. Department of Defense, January 3, 2012), 6.

433 *come to grips with the war:* For examples of the debate about the war now well under way, see Gian Gentile, *Wrong Turn: America's Deadly Embrace of Counterinsurgency* (New York: New Press, 2013) and Peter Mansoor, *Surge* (New Haven, CT: Yale University Press, 2013). Both of these U.S. Army officers commanded in Baghdad during the Iraq campaign. Gentile questions the validity of counterinsurgency methods. Mansoor stresses their value during the 2007–08 surge. In this, they echo a similar debate after Vietnam between two other officers, Harry Summers, who emphasized conventional aspects in his book *On Strategy,* and Andy Krepinevich, who addressed counterinsurgency considerations in *The Army and Vietnam.* The Summers and Krepinevich books were widely read in the military.

INDEX